AF342300

Property of SWIRS

HOW TO DISPOSE OF TOXIC SUBSTANCES
AND INDUSTRIAL WASTES

HOW TO DISPOSE OF TOXIC SUBSTANCES AND INDUSTRIAL WASTES

Property of SWIRS

Philip W. Powers

NOYES DATA CORPORATION
Park Ridge, New Jersey   London, England
1976

Copyright © 1976 by Philip W. Powers
No part of this book may be reproduced in any form
without permission in writing from the Publisher.
Library of Congress Catalog Card Number: 76-2317
ISBN: 0-8155-0615-5
Printed in the United States

Published in the United States of America by
Noyes Data Corporation
Noyes Building, Park Ridge, New Jersey 07656

FOREWORD

This book discusses all recognized and allowed ultimate disposal methods in detail and contains a long list of specific recommendations for specific substances plus alternative disposal or recovery methods.

Ultimate waste disposal implies the final disposition of nondegradable, persistent, harmful, and cumulative wastes that may be solid, liquid, or gaseous. Workable solutions to ultimate disposal problems as described in this book include conversion to harmless end products, subsurface storage in ponds or landfills, and disposal in the ocean. The five general categories of hazardous wastes are:

> (1) Toxic Chemical
> (2) Radioactive
> (3) Flammable
> (4) Explosive
> (5) Toxic Biological

Many toxic or hazardous wastes contain valuable materials. Whenever this is the case, recovery and reuse is one of the most desirable methods of hazardous waste avoidance.

The book is based on government-sponsored reports (often elaborated by highly qualified personnel from industrial companies), U.S. Patents, and on pertinent articles in authoritative journals. In the United States, we are fortunate in receiving direct help from the numerous surveys, together with active research and development programs that are being supported by the Federal Government to help industry control its wastes and harmful emissions.

In this book are condensed vital data that are scattered and often difficult to assemble. Important techniques are interpreted and explained by actual case histories. This condensed information will enable you to establish a sound background for action towards disposal of toxic and hazardous materials with safety.

Advanced composition and production methods developed by Noyes Data are employed to bring our new durably bound books to you in a minimum of time. Special techniques are used to close the gap between "manuscript" and "completed book." Industrial technology is progressing so rapidly that time-honored, conventional typesetting, binding and shipping methods are no longer suitable. We have bypassed the delays in the conventional book publishing cycle and provide the user with an effective and convenient means of reviewing up-to-date information in depth.

Foreword

The Table of Contents is organized in such a way as to serve as a subject index and provides easy access to the information contained in this book. Bibliographic reference lists follow each chapter in order to provide the reader with easy access to further information on this timely topic. A valuable appendix gives the names and addresses of disposal contractors.

CONTENTS AND SUBJECT INDEX

IRON AND STEEL INDUSTRY WASTES 257

METAL FINISHING INDUSTRY WASTES 263

NUCLEAR INDUSTRY WASTES. 273

ORDNANCE INDUSTRY WASTES 301

ORGANIC CHEMICAL INDUSTRY WASTES 316

INTRODUCTION

The final disposition of residual or refractory wastewaters and sludges resulting from various processes is generally termed ultimate disposal. Solutions to ultimate disposal problems include subsurface storage, conversion of wastes to innocuous end products, storage in ponds or land spreading, and ocean disposal.

Hazardous wastes have been defined as any "wastes or combinations of wastes which pose a substantial present or potential hazard to human health or living organisms because they are lethal, nondegradable, persistent in nature, can be biologically magnified, or otherwise cause, or tend to cause, detrimental cumulative effects"(5). The five general categories of hazardous wastes are: [1] toxic chemical, [2] radioactive, [3] flammable, [4] explosive, and [5] biological.

There is overlap, of course. For example, flammable and explosive wastes may be toxic as well; however, in this case, the primary waste characteristics of concern are flammability and explosiveness, rather than toxicity. The same logic applies to many radioactive and some biological wastes as well. Most of the nonradioactive hazardous waste generated in this country (about 10 million tons annually) falls into the toxic chemical category. Most toxic wastes can be subcategorized as: (a) inorganic toxic metals, salts, acids or bases, and (b) synthetic organics (5).

Many toxic or hazardous chemical wastes contain valuable materials. Whenever this is the case, recovery and reuse is one of the most desirable methods of hazardous waste avoidance. Silver, selenium, thallium, mercury, copper, lead, chromium, zinc and other heavy metals, once they are removed from water or air streams, are often recovered rather than land disposed. It is uneconomical to reclaim these metals only when they are very minor constituents of the waste or when the quality of the product is adversely affected as in chrome pigments manufacture. There are also private contractors that specialize in hazardous waste reclamation. For details of processes for recovery and recycling of industrial wastes, the reader is referred to a companion volume (1).

OCCURRENCE OF WASTES

The probability that a hazardous waste stream is produced at an industrial site can be estimated based on a set of material production-consumption characteristics. Table 1, on the following page, indicates these characteristics and the evaluation of such characteristics for different segments of the industrial community.

TABLE 1: HAZARDOUS MATERIAL USE, PRODUCTION AND WASTES BY INDUSTRY

Industry	Properties of Products Used					Properties of Products Made					Toxic Materials Used					Toxic Materials Produced					Disposal Methods				Toxic Wastes Produced				
	C	E	F	R	P	C	E	F	R	P	S	H	G	O	I	S	H	G	O	I	DW	SEA	CON	AEC	S	H	G	O	I
Mining	x	x		x										x	x											x	x		x
Food			x							x				x												x		x	
Textiles	x		x	x							x	x		x								x					x		x
Paper and allied products	x		x																			x							x
Alkalis and chlorine						x						x						x		x		x			x	x	x	x	x
Cyclic intermediates	x	x	x			x	x	x			x	x	x	x	x	x		x	x		x	x	x		x	x	x	x	x
Organic chemicals	x	x	x			x	x	x			x	x	x	x	x	x		x	x		x	x	x			x	x	x	x
Inorganic chemicals	x		x			x	x	x			x	x		x	x		x			x	x	x	x		x		x	x	
Plastic materials	x	x	x				x	x			x		x	x	x			x									x	x	x
Drugs	x		x	x	x				x		x	x		x	x		x		x	x	x	x	x					x	
Soaps and cleaners	x		x			x					x			x	x				x						x			x	x
Paints and allied products			x					x			x				x	x			x	x			x	x	x			x	x
Agricultural chemicals	x	x	x				x	x			x	x		x	x			x	x	x		x						x	x
Explosives	x						x	x			x			x	x				x	x	x	x	x		x	x	x	x	x
Petroleum and coal products	x	x	x				x	x			x	x	x	x	x	x		x	x	x	x	x	x			x		x	x
Leather tanning	x		x								x	x		x	x					x						x			x
Asbestos products														x	x					x			x		x	x	x	x	x
Blast furnaces and steel	x													x	x					x						x			x
Nonferrous metals	x													x	x								x						x
Metal services	x		x								x	x		x	x								x		x	x			x
DOD	x	x	x	x			x	x	x		x	x		x	x			x			x	x		x	x	x	x	x	x
AEC	x			x					x					x	x						x		x		x	x	x	x	
NASA	x	x	x				x	x			x		x	x	x			x			x		x		x	x		x	
Consumers	x	x	x	x	x						x	x		x	x								x		x	x		x	x

Hazard Code

C = Corrosive
E = Explosive
F = Flammable
R = Radioactive
P = Pathogenic

Toxic Code

S = Toxic solvents
H = Toxic metals
G = Toxic gases
O = Toxic organics
I = Toxic inorganics

Disposal Code

DW = Deep well
SEA = Deep sea dump
CON = Contract disposal
AEC = Radioactive disposal

Source: PB 221 465

Some industries produce waste streams that are typical of the industry, i.e., textile, pulp and paper, leather tanning, and plating wastes. The characteristics of these waste streams are similar. Other industries have highly variable waste streams. The chemical industry produces thousands of different materials by a wide variety of processes and may produce over a thousand different products at a single facility.

The quantities of wastes produced are dependent on the materials processed. The extractive industries produce large quantities of spoil from which the valuable ores have been extracted; for example, an average of seven barrels of brine is produced for each barrel of oil. Some industrial operations produce little waste; for example, the electrolysis of brine to produce chlorine and caustic, but such waste (mercury) may be significant although small in quantity.

The list of hazardous wastes that has been prepared identifies those substances which are significant wastes, even in relatively small amounts, because of their toxicity or other hazardous properties. It is not, of course, an exhaustive list of hazardous material, but does represent the type materials that create conditions which are potentially hazardous. This list provides a means to evaluate the potential hazards involved in a specific waste stream. If these compounds are present, it is presumptive evidence that a hazard exists which requires control.

The list can be used in several ways: to identify the producers of particularly hazardous materials, to identify the industrial consumers of such materials, and to identify the extent to which these materials enter the municipal waste streams.

Those waste streams which are now transported to offsite disposal plants or dumps include: ocean disposal of barge delivered liquid and solid wastes; deep well injection streams; contract disposal wastes; munition disposal operations; radioactive wastes; and industrial chemical containers.

The economics of waste disposal will determine, ultimately, the amounts and types of wastes that will be moved to distant disposal sites. It is unlikely that industry will move wastes if they can be treated satisfactorily and more economically at the point of origin.

The data in Table 2 illustrates the type and quantities of materials now moved to offsite disposal locations. Should deep well disposal and deep sea disposal be restricted, a substantial improvement in onsite capabilities for tertiary treatment will be required. Most wastes now deep welled will require tertiary treatment to remove toxic contaminants.

TABLE 2: QUANTITATIVE DATA ON OFFSITE DISPOSAL TONNAGES

Marine Disposal Operations - 1968

Waste Type		Amount (tpy)
Dredge spoils		52,200,000
Industrial wastes, total		4,690,500
Waste acid	2,720,000	
Refinery wastes	562,000	
Pesticide wastes	328,000	
Paper mill wastes	142,500	
Other	938,000	
Sewage		4,477,000
Construction and demolition		574,000
Explosive		15,200
Radioactive		4
Garbage		26,000
Miscellaneous		200
		61,982,904

(continued)

TABLE 2: (continued)

Deep Well Injection Streams - 1968

Waste Type	Amount (bpd)
Oil field brines	10,000,000
Chemical wastes	>1,000,000
Inorganic salt solutions	
Mineral and organic acids	
Basic solutions	approximately 175 deep wells
Chlorinated and oxygenated hydrocarbons	
Municipal sewage	

Contract Disposal Wastes

	Amount (tpy)
	30,000 to 40,000

Munitions Disposal - 1970

Waste Type	Amount (tpy)
Explosive munitions	80,000 to 120,000
Chemical munitions	8,000

Radioactive Wastes - 1963

Waste Type	Amount
High level wastes stored, gallons	70,000,000
Solid waste buried at Atomic Energy Commission sites, ft^3	7,000,000
Low activity waste processed, gallons	400,000,000

Industrial Chemical Containers - 1971

Waste Type	Amount
Plastic containers*	193,000,000

*75%, <33 ounces; 5%, 33 to 127 ounces; 20%, >127 ounces.

Source: PB 221 465

In the disposal of hazardous wastes, advantage is often taken of existing mines, quarries, abandoned government property and other facilities which have fortuitous geological and environmental isolation. For example, abandoned missile silos in Idaho are utilized by a private contractor for disposal of hazardous wastes. Clay pits in Ohio and California provide impervious liquid waste disposal sites.

STORAGE, COLLECTION AND TRANSPORTATION OF WASTES

Nonprocess Wastes

The systems used by industrial chemical plants for the storage, collection and transportation of plant nonprocess waste are the same as those used by other industrial plants and commercial establishments. Containers for waste storage are located throughout the plant near generating areas. The containers consist of a variety of receptacles such as boxes and barrels, or those designed for solid waste storage, such as dumpster boxes or concrete bins. The majority of plants salvage metal waste and provide separate containers for metal and trash. Normally the wastes are collected from the storage facilities by truck.

Systems vary for nonprocess waste collection and transportation to final disposal. A plant may use trucks to pick up solid waste from each container and haul it directly to disposal, or the smaller containers may be emptied into larger ones and then hauled to disposal. In some cases one or two large containers, sometimes large stationary compactors, are used to receive all the wastes of the plant, which are periodically hauled to disposal. Many plants have instituted a completely containerized system where large specially designed containers are located throughout the plant. They are picked up periodically by trucks equipped to

haul the entire container to disposal, after which the container is returned to its location.
The system chosen by a plant depends in part on the quantities of trash to be handled
and on costs (3).

Process Wastes

Normally process wastes are handled separately from nonprocess wastes. Storage facilities
for process wastes consist of a variety of bins, barrels, fiber drums, tanks, and luggers or
dumpsters, as well as ponds, tank trucks and railroad cars; they are sometimes piled casually on the ground.

Heavy sludges that cannot be pumped are generally stored as described above. Sludges
stored in bins, or stored casually, must be removed and deposited in trucks for transportation to disposal. If they are contained in fiber drums, the drum is usually disposed of
along with the sludge. Luggers and dumpsters are metal containers which can be lifted
and their contents dumped into a truck, or the container carried to the disposal area and
dumped. Sludges stored in railroad cars are generally off-quality product which is stored
longer than usual to allow time for possible marketing opportunities. In many cases, the
railroads will haul sludges stored in their cars to land which they own. Most filter residues
would be handled in the same manner as sludges.

Tars and aqueous sludges that can be pumped are stored in barrels, tanks, ponds, tank
trucks and railroad cars. Barrels are either emptied at the disposal site or in landfills;
sometimes they are buried along with the waste. Tanks are often used both at the point
of waste generation and at the disposal site. They are used with tar incinerators not only
to provide storage but also to provide an opportunity to mix various wastes to form a
desirable blend for burning. Some tanks are heated to liquify tars that would solidify at
normal temperatures. Similarly, tanks are often equipped with agitating means to prevent
settling of suspended solids and to assist in preventing tars from solidifying. When the
waste is stored in ponds, it is either pumped by pipeline directly from the pond to disposal or into tank trucks. As with sludge, railroad tank cars sometimes are used for off-
quality tar wastes.

Transportation equipment used for hauling process solid waste from storage to disposal
consists of open dump style trucks, tank trucks, railroad cars, barges and pipelines. As
previously mentioned, some tank trucks must be capable of heating and agitating the
waste to prevent solidification. Barges are used to transport some waste to ocean disposal.
In most cases where pipelines are used, the waste is not stored, since the pipeline carries
the waste directly from the generating process to disposal in a continuous operation (3).

ALTERNATIVES FOR ULTIMATE DISPOSAL

Nonprocess Wastes

Disposal methods for nonprocess wastes as used by the chemical industry are essentially
the same as those used for municipal and commercial refuse. The methods are all variations of land disposal or incineration. No plants were found that employed composting,
and the few governmentally operated composting plants in the country receive an insignificant amount of nonprocess waste from industrial chemical plants.

Land disposal methods fall into the categories of burning dumps, open dumps, landfills
and sanitary landfills. Dumps exist where the solid wastes are deposited on the ground
and left in the open for considerable periods of time. In some cases the dumps are deliberately fired to reduce waste volume. This practice is increasingly running afoul of state
regulations. In a landfill, the solid wastes are periodically covered with fill material. A
sanitary landfill is a particular type of landfill which must meet certain criteria, i.e., [1] It
must be operated without creating a nuisance or a hazard to public health or safety;
[2] The solid wastes must be confined to the smallest practical area, reduced to the

smallest practical volume, and covered with a prescribed layer of earth at the conclusion of each day's operation; and [3] Groundwater contamination from leaching must be controlled. This is the most desirable form of land disposal.

Incineration of nonprocess waste is also accomplished by a number of methods, including open burning, burning in tepees, and incineration in single and multiple chamber incinerators with or without stack air pollution control equipment.

In the past, many chemical plants disposed of nonprocess waste by open burning or in tepee burners, which are large tepee-shaped structures serving to protect the burning material from blowing about, but having little beneficial effect on reducing air pollution. Air pollution regulations have virtually eliminated these two waste disposal forms at chemical plants, and they have been replaced by incineration, landfilling, or hauling waste to nearby private or municipal facilities.

Most landfills operated by industrial chemical plants do not meet all the criteria to be considered sanitary landfills. They are often not covered each day, but this is usually not as critical as with municipal refuse since the small amount of garbage in the plant trash is not as likely to attract rodents. In addition, most are not properly located or sealed to prevent possible groundwater pollution. These same conditions have been reported for private and municipal landfills. In addition, plant landfill operators may often experience difficulty in providing proper compaction of the waste, thereby leaving voids which may cause uneven settlement. The difficulty stems from the large quantities of discarded cardboard boxes, crates and fiber drums which are unusually hard to compact.

Plant nonprocess waste is generally compatible with municipal refuse, and the decision of whether to use municipal or private facilities or build plant facilities usually is based on local regulations, comparative costs, and the adequacy of municipal or private facilities.

The choice of storage, collection, and transportation systems for nonprocess waste involves such factors as the length of haul, the number of sources of waste, and the quantity of waste. In addition, choice of equipment depends on comparative costs, flexibility, conformance to health and safety guidelines, ease of maintenance, and compatibility with other plant operations and facilities.

FIGURE 1: SCHEMATIC DIAGRAM OF CHEMICAL PLANT NONPROCESS SOLID WASTE DISPOSAL ALTERNATIVES

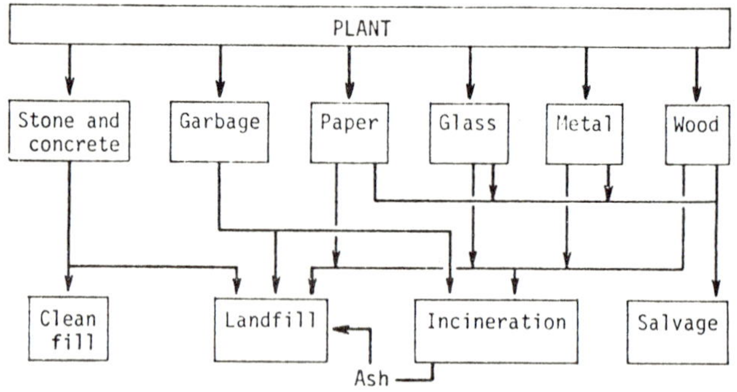

Source: PB 226 420

Disposal alternatives are relatively few and are illustrated in Figure 1. Stone and concrete can be used as clean fill or dumped in a landfill. The remaining nonprocess waste can be disposed by landfill or incineration. If combustible and noncombustible waste is separated prior to disposal, the glass and metal waste can be sent directly to a landfill, while garbage, paper and wood waste are generally sent to an incinerator. Most chemical plants salvage certain valuable metal wastes for sale. Other wastes such as paper, glass and wood are capable of salvage but in most cases it is uneconomical and not practiced. The objectives of national solid waste management plans include recycling this material. Salvage operators are available in some areas to take both process and nonprocess plant wastes for recovery and recycling.

Process Wastes

Disposal of process solid wastes is generally accomplished by incineration, land disposal, settling in lagoons or ponds, and ocean dumping.

In the management of solid wastes, a plant can elect to buy collection, transportation and disposal services from either private or municipal sources, rather than handle everything itself. Some plants, mainly small plants, are provided with trash collection and/or disposal by their local municipality, financed through general taxation. Most often, however, the plant must provide transportation of trash to municipal facilities and pay by the ton for disposal.

Some chemical process wastes, including sludges and barreled tars, are disposed of at public facilities. Many chemical plants which dispose of trash at public facilities strive to provide private disposal for their process wastes to avoid problems. Due to growing concern in recent years, regulations on the types of waste that may be deposited at public facilities exclude chemical waste. These regulations have increased the burden on chemical plants to provide their own disposal facilities.

Along with municipal waste services, there are also many private companies which provide either waste hauling, disposal facilities, or both, for process and nonprocess waste. A plant may elect to have certain of its wastes handled by such a company. Sometimes these private companies are also the main agency for disposal of refuse for the surrounding community.

Recovery or utilization of wastes by the plant involves possible use in other processes within the plant, sale to companies dealing in waste salvage, or sale directly to consumers. The process of finding a market for a waste is expensive and time consuming. It is an extra task for the marketing staff, who may not have experience in the necessary marketing area, and storage costs for the waste are usually high. In addition, the waste may compete with a customer's product, and too, there is the chance of divulging process secrets by inference.

There are only a few independent chemical waste salvage companies operating in the country, but demand for their services is growing. This type of company may be in the best position to find and develop markets for chemical wastes. Their marketing staff is specifically oriented toward this area, and they are better able to provide consistent and sufficient quantities to their customers.

These firms buy wastes and either process them into saleable materials or sell them directly. The wastes must generally be obtained from a number of plants so that their investment in processing equipment and marketing is not dependent on the operations of one company. Marketing of wastes is hampered by high storage costs and high freight rates for both shipping and receiving which increase the final price of the waste product to the consumer. In addition, some wastes come in quantities that are too small to be attractive, or are cyclic, prohibiting a continuous market, or their chemical content is variable, making it difficult to meet required chemical specifications.

The potential consumers of salvaged waste products are the most important economic factor in development of markets. Their substitution of salvaged wastes for production chemicals is based almost entirely on economics. The waste must be available at an attractive price, meet chemical specifications, contain no deleterious impurities, and be available in significant quantities.

A chemical plant's selection of which method to use for storage, collection, transportation and disposal involves many factors. The plant must choose a management system which is economical, efficient and conforms to health and safety standards.

The type and characteristics of plant solid waste are the most important factors in the development of a management system. Nonprocess and process wastes exhibit markedly different characteristics and are handled almost entirely by separate systems, although certain operations, such as disposal, may be combined.

Solid waste management systems for process wastes require specialized equipment to handle the variety of waste types. Equipment for storage, collection and transportation of process wastes must be designed for the chemical and physical characteristics of the waste to handle it efficiently. The same factors that influence the choice of nonprocess waste equipment as listed above influence the choice of process waste equipment. It is of great importance that equipment conform to the intended disposal method. For example, ocean disposal will require barges; ponds used for storage must be lined, and will have to be pumped out into tank trucks or pipelines; and incinerators must be provided with storage adequate for the particular wastes to enable proper feeding and mixing.

If the plants elect to use private or municipal disposal facilities, they are freed from the task of developing their own. However, certain process wastes may cause problems with the public or private disposal facility, resulting in adverse public reaction. There are a number of private disposal facilities throughout the country, however, that do employ proper disposal equipment to handle a variety of plant wastes, and where they exist, industries are inclined to use these private facilities rather than build their own.

FIGURE 2: SCHEMATIC DIAGRAM OF CHEMICAL PLANT PROCESS SOLID WASTE DISPOSAL ALTERNATIVES

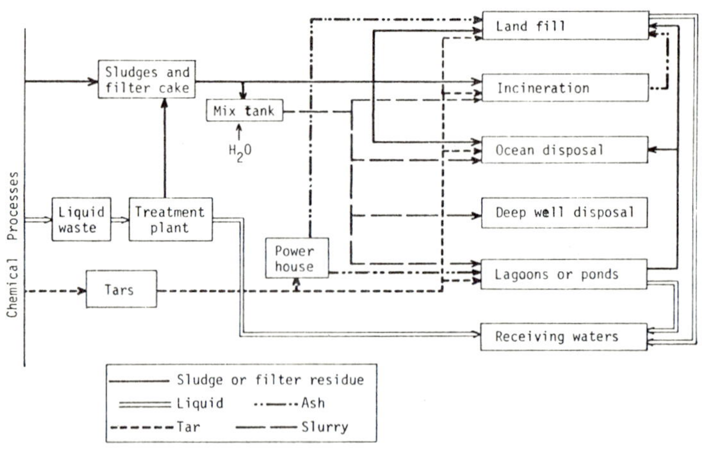

Source: PB 226 420

The disposal methods for process wastes are more varied than those for nonprocess waste, since they depend on the type and characteristics of the waste to be disposed. Disposal alternatives based on the common waste types are shown in Figure 2. The method used for disposal of sludge or filter cake depends on the consistency of the sludge. If it is a heavy sludge with low water content, three methods are used: [1] It can be landfilled; [2] It can be incinerated either in bulk or in fiber drums; [3] It can be dumped in the ocean either in bulk or in barrels. If the sludge has a high liquid content or is slurried with water to allow pumping, it can be fired through a nozzle and burned in an incinerator, pumped aboard barges and disposed at sea, pumped into lagoons or ponds, or in few cases, pumped into underground wells. Sludges generated by process wastewater treatment may be handled in a manner similar to other sludges.

Seven methods predominate among the alternatives for ultimate solid waste disposal. Considering each briefly:

[1] Incineration: This involves oxidative conversion of combustible solid material to harmless gases suitable for atmospheric release. Undesired gaseous products, such as HCl, SO_2, NO_x, must be removed prior to release. Heats of combustion may sometimes be recovered, and occasionally additional combustible material must be added to insure adequate combustion. Residual solids, i.e., ash, are landfilled. To the extent that residual solids remain, incineration is a waste reduction step rather than an ultimate disposal step.

[2] Dispersal into contiguous water bodies: Historically, this practice has been widely employed, but EPA action has now curtailed it in many industries. Usually, the waste solid is temporarily stored in a lagoon, wherein it is treated to minimize environmental impact. For example, acids and bases are neutralized and suspended solids are permitted to settle. In the case of river discharge, attempts are made to release the waste during periods of high flow.

[3] Ocean dumping: This practice involves transporting the material out to sea, then releasing it. Usually the material is conveyed by barge; it may be released directly or within containers. Costs depend upon the distance of transport, and whether the material can be loaded in-plant or must be intermediately transported.

[4] Lagooning: Lagoons may be employed for either temporary or permanent storage. As previously mentioned, lagoons are frequently employed for interim storage and water treatment prior to release to contiguous water bodies. Permanent storage occurs when the solid material tends not to settle, e.g., phosphate slimes, or would have undesirable properties if water cover were not maintained, e.g., red mud (blows) from alumina and phossy water (burns) from phosphorus. Phossy water is water which contains particles of phosphorus metal. The metal bursts into flames when exposed to oxygen; thus water cover is required.

[5] Sanitary landfill: This pertains to the burial of nonhazardous solid wastes under controls sufficient to preclude degradation of the surrounding environment. Controls include: proper site selection, exclusion of hazardous materials, periodic provision (usually daily) of soil cover, etc.

[6] Chemical landfill (termed Class 1 landfill in California—no connection with groundwater): This is the extension of sanitary landfill to enable acceptance of hazardous materials. Included are: the provision of careful material identification, adequate safety standards (firefighting, etc.,), surface water channeling, leachate collection and treatment, materials recycling, etc.

[7] Subsurface injection: In this case the solid material is slurried and then pumped into underground cavities. The implications of many aspects of this practice are not yet clearly understood. Subsurface injection of hazardous waste should only occur when the leachate can be fully contained.

The environmental problems associated with present disposal methods can be described in terms of the seven techniques which now predominate. Considering the environmental implications of each:

[1] Incineration: Unless adequate controls are exercised, incineration can lead to atmospheric release of undesired materials. The increasing stringency of air pollution regulations is leading to more expensive, environmentally acceptable practice in most instances. For example, hydrogen chloride generated during incineration of chlorinated hydrocarbons is now being removed through absorption in water. The hydrochloric acid so formed is then either neutralized or is concentrated for plant use.

[2] Dispersal into contiguous water bodies: If water salinity is unduly increased, water biota are endangered and ultimately agricultural land is harmed. Interim water pollution regulations are acting to curtail discharge of dissolved solids. Most firms are preparing to lagoon material previously discharged.

[3] Ocean dumping: There have been a number of instances of fish kill and beach spoilages due to ocean dumping. The Marine Protection, Research and Sanctuaries Act of 1972 establishes a permit system for control of future ocean dumping. In addition, the Secretary of Commerce has been instructed to conduct and encourage a broad effort aimed at minimizing or ending all dumping of materials within five years of the effective date of the act (October 23, 1973).

[4] Lagooning: Potential environmental problems that may result from lagooning relate to subsurface transport of the leachate and surface water runoff. For example, during the rainy season lagoons may overflow, and occasionally the containment dikes are breached. Such an instance is reported in EPA's Report to Congress on Hazardous Waste Disposal:

> "Phosphate Slime Spill. On December 7, 1971, at a chemical plant site in Fort Meade, Florida, a portion of a dike forming a waste pond ruptured releasing an estimated two billion gallons (7.58 billion liters) of slime composed of phosphatic clays and insoluble halides into Whidden Creek. Flow patterns of the creek led to subsequent contamination of Peace River and the estuarine area of Charlotte Harbor. The water of Charlotte Harbor took on a thick milky white appearance. Along the river, signs of life were diminished, dead fish were sighted and normal surface fish activity was absent. No living organisms were found in Whidden Creek downstream of the spill or in Peace River at a point eight miles downstream of Whidden Creek. Clam and crab gills were coated with the milky substance and in general all benthic aquatic life was affected in some way."

[5] Sanitary landfill: Sanitary landfill protects against surface runoff but the concern remains that the leachate may cause environmental damage. For this reason, hazardous wastes cannot be accommodated in such a facility. The potential duration of danger from leachate is illustrated by the following experience.

> "As a result of arsenic burial 30 years ago on agricultural land in Perham, Minnesota, several people who recently consumed water contaminated by the deposit were hospitalized. The water came from a well that was drilled near this 30 year old deposit of arsenic material. Attempts to correct this contamination problem are now being studied."

[6] Chemical landfill: If properly managed, this method of disposal poses no environmental hazard. Care must be taken to contain contaminated surface runoff and leachate.

[7] Subsurface disposal: The dangers with this technique are that large quantities of material may be emplaced before difficulties arise. EPA opposes such disposal practice for hazardous wastes,

> "unless all other alternatives have been found to be less satisfactory in terms of environmental protection, and unless extensive hydraulic and geologic studies are made to insure that groundwater contamination will be minimized."

The overall effect of environmental considerations for the seven disposal techniques is to identify incineration, sanitary landfill and chemical landfill as the preferred modes of disposal. Those alternatives which lead to dispersal in water bodies will be increasingly curtailed. Lagooning will be continued for temporary storage and in those cases in which it is not possible to convert the waste to a form suitable for ultimate emplacement or destruction. Subsurface disposal may ultimately become established as a viable alternative for nonhazardous wastes as more experience is gained in this technique.

ECONOMICS OF ULTIMATE DISPOSAL

The unit costs of treatment via each of the seven techniques were ascertained from the literature. Actual costs vary widely depending upon waste quantity, geographic location, company size and expertise, etc. The values shown in Table 3 were employed in estimating the disposal costs for 33 selected chemicals. (Subsurface injection is not included since this is not the predominant disposal method for any of the wastes considered.) It was assumed that slurries and sludges are temporarily stored in lagoons pending final disposal, and that acids and bases are neutralized at that point. Hazardous wastes were relegated to chemical landfill.

TABLE 3: ULTIMATE DISPOSAL COSTS

	Method	Cost ($/T)	Qualifications
[1]	Incineration	8	HC still bottoms, slight inorganic content
		15	High water, ash, sulfur or chlorine content
		95	Hazardous waste
[2]	Dispersal to contiguous water	-	Storage costs attributed to lagooning
[3]	Ocean dumping	1-3	Excluding containment and transport
[4]	Lagooning	1-3	Neutralization
		10	Precipitation
[5]	Sanitary landfill	3-5	Excluding inplant collection and storage
[6]	Chemical landfill	40	
[7]	Disposal fee	19-82	Quote from Rollins

Source: PB 233 464

It is useful to place the disposal figures for the 33 chemicals in perspective. The total waste quantity from the 33 chemicals is calculated to be 55.9 million tons. This compares with an estimate of 140 million tons for all industrial solid wastes. The correspondence seems reasonable, in view of the high contribution from phosphorus and phosphoric acid solid wastes.

Figure 3 compares the costs for environmentally adequate treatment and disposal with dumping costs for land disposal and sea disposal. It is to be noted that relatively unrefined dumping techniques represent possible huge savings, especially when large tonnages are involved.

Land usage requirements for disposal of solid wastes from the 33 chemicals are estimated in Table 4. This table accounts for the disposition of over 99% of the solid wastes. The remainder are incinerated or sold as byproduct. The estimate for chemical landfill assumes most tars are handled in this manner. Since tars are incinerated whenever possible, the land requirements for chemical landfill may be overestimated. However, Union Carbide's Institute, West Virginia plant, alone allocates 24,000 yd^3 per year to chemical landfill.

Lagooning accounts for 90% of the land use. This is traceable to its use in control of ore residuals, such as red mud, and for storage of large volume inorganic byproduct salts prior to their dispersal to water bodies. As noted in the table, 2,080 acres are available for reuse and 5,855 acres are added to permanent inventory. In comparison, 5×10^6 acres of land are presently occupied by the extractive industries, and roughly 5% of this figure, i.e., 250,000 acres, is consumed each year. Industrial chemical annual land usage is, therefore, less than 3% that of the extractive industries. A significant factor is that, like the extractive industries, most industrial chemical facilities producing large quantities of solid waste are located in rural areas, e.g., phosphorus and phosphoric acid plants. When this is not the case, as for carbide-process acetylene manufacture, solid waste disposal land requirements do pose a problem.

FIGURE 3: COST COMPARISON OF LAND AND OCEAN DISPOSAL vs ENVIRONMENTALLY ADEQUATE TREATMENT

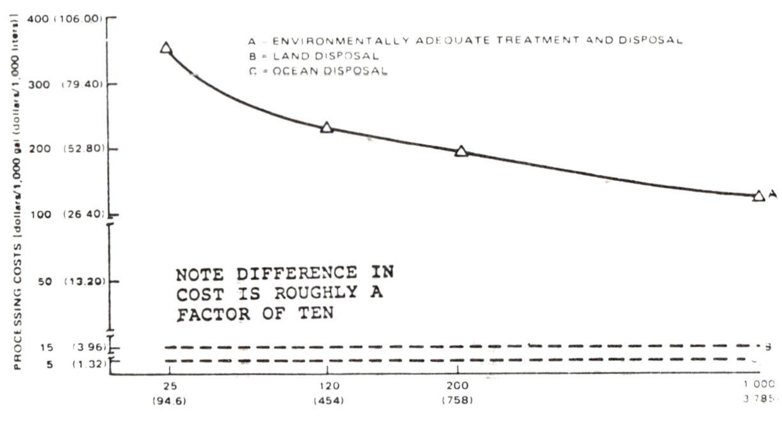

Note: 1,000 gallons = approximately 4.2 tons

Source: Environmental Publication SW-115

TABLE 4: LAND USE FOR SOLID WASTE DISPOSAL

Storage or Disposal Mode	Waste Quantity $(10^3$ S.T.)	Waste Volume* $(10^6$ yd^3)	Land Area** (acres)
Temporary lagooning	12,064	161	2,080
Permanent lagooning	29,460	393	5,075
Sanitary landfill	14,930	60	769
Chemical landfill	217	1	11
Total	56,671	615	7,935***

*Based on 504 lb/yd^3 and 30% solids content in lagoon.
**Based on 48 ft depth for lagoons and landfills.
***This represents land used during the year. The 2,080 acres employed for temporary lagooning can be reused. The remaining 5,855 acres are added to permanent inventory.

Source: PB 233 464

Land usage for solid wastes from the 33 chemicals following curtailment of water discharge is shown in Table 5. At present, material temporarily stored in lagoons is subsequently discharged to the river, ocean dumped, or sent to sanitary landfill. About 760 x 10^3 S.T. are estimated to fall into the latter category, with the remaining 11,300 S.T. discharged to water bodies. It was assumed that when discharge to water bodies is curtailed, half the material now temporarily stored in lagoons will be retained in permanent lagoons and the remainder will be consolidated in landfill. Comparing Tables 4 and 5 one sees that land use actually decreases. This is due to the consolidating effect of solids transfer from lagoons to sanitary landfill. Land consumption increases by 22%, however, since the land becomes the final repository of material previously released to the water bodies. Nevertheless, land consumption remains a small percentage of that for the extractive industries.

TABLE 5: LAND USE FOR SOLID WASTE DISPOSAL IN ABSENCE OF WATER DISCHARGE

Storage or Disposal Mode	Waste Quantity $(10^3$ S.T.)	Waste Volume* $(10^6$ yd^3)	Land Area** (acres)
Permanent lagooning	35,110	468	6,053
Sanitary landfill	20,580	82	1,062
Chemical landfill	217	1	11
Other	24	------	---------
Total	55,931***	551	7,126

*Based on 504 lb/yd^3 and 30% solids content in lagoon.
**Based on 48 ft depth for lagoons and landfills.
***Another 584 x 10^3 S.T. are sold as by-product Na_2SO_4, $CaCl_2$ and $CaSO_4$.

Source: PB 233 464

REFERENCES

(1) M. Sittig, *Resource Recovery and Recycling Handbook of Industrial Wastes,* Park Ridge, N.J., Noyes Data Corp. (1975).

(2) Booz-Allen Applied Research, Inc., *A Study of Hazardous Waste Materials, Hazardous Effects and Disposal Methods,* Vol. I, Report PB 221 465, Springfield, Va., Nat Tech Information Service (July 1973).

(3) J.K. Holcombe and P.W. Kalika, *Solid Waste Management in the Industrial Chemical Industry,* Report PB 226 420, Springfield, Va., Nat Tech Information Service (1973).

(4) J.C. Saxton and M. Kramer, *Industrial Chemicals Solid Waste Generation,* Report PB 233 464, Springfield, Va., Nat Tech Information Service (June 1974).

(5) Office of Solid Waste Management Programs, *Report to Congress: Disposal of Hazardous Wastes,* Environmental Publication SW-115, Wash., D.C., U.S. Government Printing Office (1974).

PRETREATMENT FOR ULTIMATE DISPOSAL

CEMENTATION

As pointed out in a recent article (1), environmental objections to the disposal of raw industrial sludges at U.S. landfill sites is growing. One way of improving this situation is to immobilize sludges so that they can be disposed of more conventionally. Here the term cementation has been applied to the fixation or immobilization of toxic materials so they can be more easily handled and are less easily leached.

As pointed out by Weismantel (1), Chemfix, Inc. (Pittsburgh, Pa.) and Crossford Pollution Services, Ltd. (Sale, England) have each developed processes that are now being used commercially to harden residues into inert rock-like materials said to be safe for use in landfill.

The Chemfix process uses soluble silicates to fix the wastes. The Crossford process converts the waste into a monomeric mixture which polymerizes at the landfill site within three days to a rock-hard solid. TRW, Inc. of Redondo Beach, Calif. is reported by Weismantel (1) to also be working on a polymeric sludge solidification scheme.

A process developed by J.R. Connor (2) is one in which an aqueous solution of an alkali metal silicate is mixed with waste material and a silicate setting agent to cause the silicate and setting agent to react with each other. The proportions of the silicate and setting agent are such that the reaction converts the mixture into a consolidated chemically and physically stable solid product that is substantially insoluble in water and in which pollutants are entrapped in the solidified silicate so that the waste material is rendered nonpolluting and fit for ultimate disposal.

A process developed by D.J. Opacic, et al (3) is based on a mobile unit for treating liquid waste to render it nonpolluting and fit for ultimate disposal. Such a unit includes a mixing hopper having an inlet for receiving liquid waste, and an outlet for connection to a conduit extending to a disposal area. The unit also includes a bin for a powdered setting agent for a liquid alkali metal silicate. The bin has an outlet at its bottom and a porous floor spaced above its bottom. Compressed air is delivered to the space beneath the floor to form a fluidized bed of setting agent above the floor.

The setting agent is delivered to the mixing hopper in which it is mixed with the liquid waste. After the mixture leaves the hopper, liquid alkali metal silicate from a tank in the mobile unit is delivered to the mixture. The treated waste will set after it has reached the disposal area.

International Utilities/Conversion Systems, Inc. of Whitemarsh, Pa. has reportedly (4) developed a process for adding fly ash and hydrated lime to industrial by-product sludges to fix sludges. The process is called the Poz-o-tec process.

A process developed by J.V. Dunlea, Jr. (5) is one in which bulk rubbish is shredded, saturated with a binding agent, compacted on a continuously extruded or piecemeal basis, reduced in size, rinsed and disposed of either as landfill or dumped at sea. Alternative steps include treatment of the rubbish to kill bacteria, drying prior to compaction and cutting or chipping the compacted rubbish into small pieces. Figure 4 is a block flow diagram of this process.

FIGURE 4: SCHEME FOR COMPACTING BULK RUBBISH PRIOR TO DISPOSAL

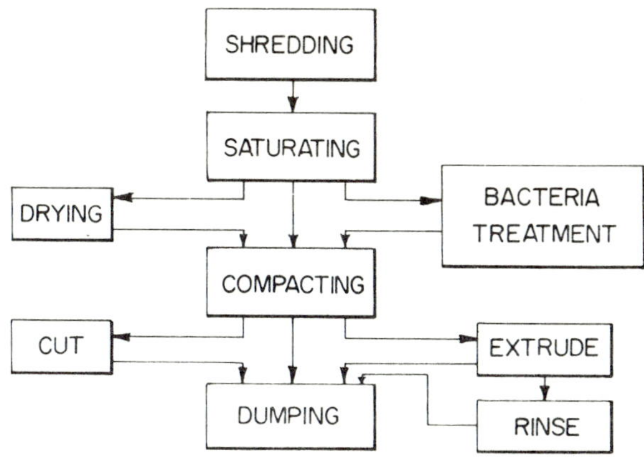

Source: U.S. Patent 3,721,183

A process developed by K. Tezuka (6) is designed to convert industrial wastes such as sludges, deposits and cyanides, which cause environmental pollution, to innoxious solid blocks. This is done by incorporating a coagulant into such industrial waste, kneading the mixture, press-shaping the kneaded mixture, sealing the outer periphery of the shaped article by means of concrete coating or the like, and aging and solidifying the sealed block. The resulting innoxious solid blocks may be utilized for reclamation, construction and the like. Further, when such sealed solid blocks are discarded in the sea, no environmental pollution is brought about.

Figure 5 is a block flow diagram of this process and Figure 6 is a perspective view of a suitable form of apparatus for the conduct of this process. An industrial waste discharged in a large quantity from a chemical factory or the like, such as sludges, deposits and the like is thrown into a hopper **2** of a compression casing **1** which has an opening on the upper surface thereof and has a rectangular parallelepiped form. On one end of a longer side of the casing **1** there are provided oil pressure presses **3** and **3**. A pressure block **5** is mounted on a ram **4** of each oil pressure press **3** so that it can slide in the compression casing **1** in the longitudinal direction.

On the other end of a longer side of the compression casing **1** there is provided a silo **6** maintained at a predetermined height. Four partitioned rooms **A, B, C** and **D** are formed in the silo **6** to contain therein coagulants such as cement. The lower portion of the silo **6** is connected to the compression casing **1**. Handles **7** and **7** are mounted at this connecting area to feed the contents of the silo **6** into the casing **1** as occasion demands.

FIGURE 5: BLOCK FLOW DIAGRAM FOR PRODUCTION OF SOLID BLOCKS FOR OCEAN DISPOSAL OF INDUSTRIAL WASTES

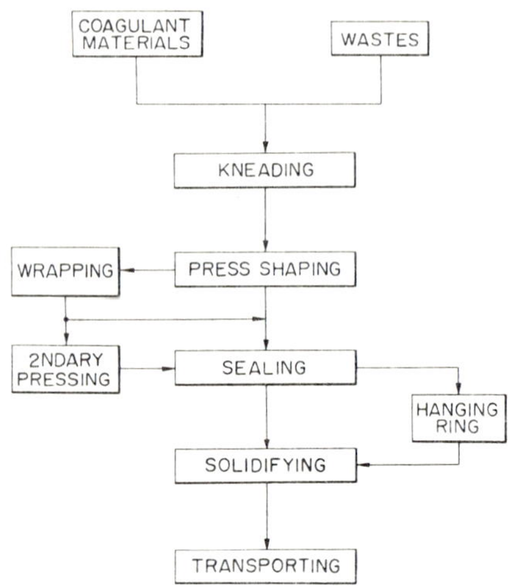

FIGURE 6: PERSPECTIVE VIEW OF APPARATUS FOR CONVERTING INDUSTRIAL WASTES TO SOLID BLOCKS FOR OCEAN DISPOSAL

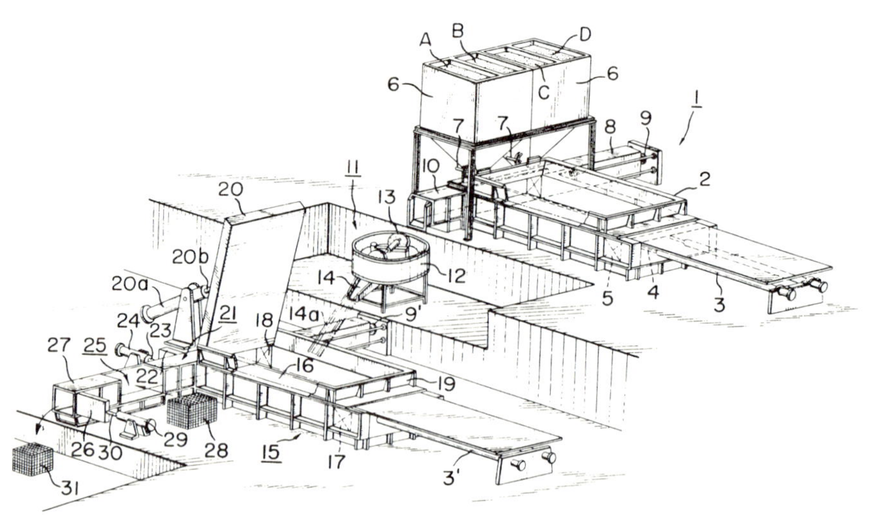

Source: U.S. Patent 3,848,392

On the side of the compression casing **1** where the silo **6** is disposed, an extrusion block **8** is mounted in a manner such that it can move forwardly and backwardly in the casing **1** through an oil pressure press **9** in the direction vertical to the moving direction of the pressure block **5**. On the side of the casing **1** opposite to the side where the extrusion block is intruded into the casing **1**, there is provided an extrusion opening **10**.

The starting waste thrown into the compression casing **1** from the hopper **2** is pressed by the pressure block **5** actuated by the oil pressure presses **3** and **3** and is forwarded through the interior of the compression casing **1** toward the position below the silo **6**. In this manner the starting waste is compressed. After the pressure block has been retreated to such an extent that the extrusion block **8** can be actuated, the coagulant contained in the silo **6** is fed in a predetermined amount into the compression casing **1** by opening handles **7** and **7** appropriately. Then, the extrusion block **8** is actuated to mix the industrial waste with the cement coagulant, and mixture is extruded and allowed to fall on the mixer **11** through the opening **10**.

The mixer is disposed below the extrusion opening **10** and comprises a cylindrical receiver **12** having an opening and an agitator **13** mounted rotatably in the interior of the cylindrical receiver. The agitator is connected to a suitable drive source, for instance, an electric motor, and a chute **14** is mounted below the receiver.

The drive source is actuated to rotate the agitator **3**, and the mixture which has fallen on the receiver is kneaded by the agitator **13** which is thus rotated. As a result, there is formed a homogeneous mixture of the cement coagulant and industrial waste. The kneaded mixture is then forwarded to a shaping press **15** through the chute **14**. It is preferred that a dismountable auxiliary chute **14a** is provided to connect the chute **14** to a shaping casing **16** of the shaping press **15**.

The kind of the coagulant to be used in this process is not limited to cement, but gypsum, lime and other synthetic coagulants can be used. It is preferred that the mixing ratio of such coagulant and the industrial waste is within a range of from about 30:7 to about 4:6.

The shaping press comprises the shaping casing, a pressure block **17** movable forwardly and backwardly in the lengthwise direction in the casing and an extrusion block **18** movable forwardly and backwardly in the widthwise direction in the casing. Each of these blocks includes drive means such as oil pressure presses **3'** and **9'** and ram.

A hopper **19** and an optionally openable and closeable lid **20** are mounted on the shaping casing, and a driving means including an oil pressure press **20a** and a ram **20b** is mounted on the lid to open and close the lid. On the side of the shaping casing opposite to the side of the extrusion block, there is formed an extrusion opening **21**. A slide door **22** acting as a pressure-resistant wall of the extrusion block is connected to an oil pressure press **24** via a ram **23** to optionally open and close the extrusion opening **21**. A wrapping device is attached to this extrusion opening **21**.

More specifically, this extrusion opening **21** extends along a certain predetermined length and has an opening **25**. At the end of the extended portion of the extrusion opening **21**, there is provided a wrapping chamber **27** having a slide door **26**. The opening **25** is formed by cutting away the side walls and ceiling of the extended portion of the extrusion opening **21** along a length sufficient to receive therein a preformed wire net case **28** having an inlet opening defined by projecting slides. An oil pressure press **29** is mounted on the slide door **26** through a ram **30** so as to optionally open and close the outer side end of the wrapping chamber **27**, while the inner side end of the wrapping chamber is always opened.

It is not always necessary that the wrapping device should be provided in the state attached to the shaping press, but the wrapping device may be provided separately. Accordingly, press-shaped blocks need not always be wrapped. In case the wrapping device is provided separately, there may be employed those previously proposed in U.S. Patents, 3,451,185,

3,451,190 and 3,514,921. These wrapping device are not disclosed as independent devices but as attachments to apparatuses for disposal of wastes and refuses.

The kneaded mixture fed in the shaping casing of the shaping press is press-shaped. More specifically, the pressure block is moved in the shaping casing in the lengthwise direction and the extrusion block is actuated to press the kneaded mixture toward the side of the slide door **22**. When the oil pressure press is actuated to draw the ram **23**, the slide door **22** opens the extrusion opening **21**. Accordingly, when the extrusion block is further advanced, a press-shaped article having a substantially cubic form is discharged from the extrusion opening **21**.

In this case, if a wire net case **28** is fitted in the opening **25**, with the opening operation of the slide door **22** the press-shaped article is inserted into the wire net case by an action of the extrusion block, and when the extrusion block is further advanced, the wire net case containing the press-shaped article inserted therein is forwarded to the wrapping chamber and received by the slide door **26**. When the extrusion block is retreated, the projecting side of the inlet opening of the wire net case is appropriately bent inwardly and extrusion block **18** is advanced again to perform compression, the wrapping operation is completed.

After completion of the wrapping operation, the oil pressure press is driven to draw out the slide door **26** from the wrapping chamber to form an opening on the outer side of the wrapping chamber, and the wrapped, press-shaped block **31** is withdrawn from this opening by further advance of the extrusion block. In case the wire net case is not employed, only a press-shaped unwrapped block is withdrawn.

The wrapping material to be used is not limited to a wire net case but synthetic resin sheets, metallic sheets such as zinc-plated steel sheets, vinyl resin-coated wire nets and vinyl resin-coated metallic sheets can be used. In case a concrete coating is applied on a press-shaped block of an industrial waste, a wire net urges the adhesion of the concrete coating to the press-shaped block. Since a synthetic resin sheet has a water-impermeability and a high resistance, if such synthetic resin sheet is used as a wrapping material, these preferred properties are given to the resulting wrapped block.

A metallic sheet can be used for coating the entire surface of a block to be wrapped. In case such metallic sheet or wire net is coated with a vinyl resin or the like and is subjected to the rust-preventive treatment, the durability is further enhanced. After such wrapping step, the press-shaped block of the industrial waste is forwarded to the next sealing step.

As shown in Figure 7, a concrete case **32** having an opening on the upper surface thereof is used for inserting the press-shaped block therein. This case **32** is preformed in a separate factory or the like. The press-shaped block **31** formed by means of the shaping press is packed in the case **32** and the upper opening of the case **32** is sealed with raw concrete.

FIGURE 7: SECTION OF SOLID BLOCK FOR OCEAN DISPOSAL CONTAINING INDUSTRIAL WASTES

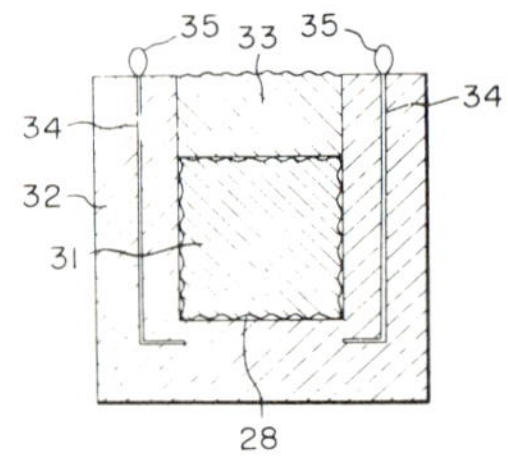

Source: U.S. Patent 3,848,392

The configuration and inner capacity of the case **32** are decided appropriately depending on the size of the press-shaped block and the like. It is preferred that reinforcing steel bars **34** are embedded in the case **32** at the time of preparation thereof and hanging rings **35** are attached to expose ends of these reinforcing bars **34**. These rings **34** are effective as hanging means for transportation of the concrete case.

CENTRIFUGATION

Basically, centrifuges separate solids from the liquid through sedimentation and centrifugal force. The most effective centrifuges for dewatering sewage sludges are horizontal, cylindrical-conical, solid bowl machines. In a typical unit, sludge is fed through a stationary feed tube along the center line of the bowl to the hub of the screw conveyor. The screw conveyor is mounted inside the rotating conical bowl and rotates at a slightly lower speed than the bowl. Sludge leaves the end of the feed tube, is accelerated, passed through the ports of the conveyor shaft, and is distributed to the periphery of the bowl. The solids are settled through the liquid and are moved along the bowl wall by the blades of the screw conveyor.

The solids move out of the liquid bowl and onto a drainage deck on which they are continuously conveyed by a screw to the end of the machine, at which point they are discharged. The liquid effluent is discharged through effluent ports after traveling the length of the pool under centrifugal force. The depth of the liquid or pool volume can be varied by adjustment of weir plates located at opposite ends of the bowl.

Centrifugation has been used for sludge thickening and sludge dewatering, and has been used with and without chemical conditioning or the addition of polyelectrolytes. The centrifuge is most effective in recovering primary solids, and least effective when recovering waste activated sludge. Design considerations and process variables involved are:

> [1] Feed rate
> [2] Sludge solids characteristics
> [3] Feed consistency
> [4] Temperature
> [5] Chemical additives
> [6] Bowl design
> [7] Bowl speed
> [8] Pool volume
> [9] Conveyor speed

Unlike filtration, it is not easy to predict centrifugal performance by the use of laboratory techniques. Pilot studies must be made and appropriate scaling factors determined for process design. When scaling pilot plant data to larger units, it is necessary to consider the residence time of the liquid in the bowl, the solids thickness and the residence time of the solids on the deck, and particularly the effective centrifugal force on the liquids and solids (Albertson and Guidi, 1969).

It may be possible to determine the feasibility of centrifugation from simple laboratory tests. The sludge to be dewatered may be spun in a laboratory hand centrifuge for thirty seconds at 2,000 rpm, and after centrifugation, the tube filtered to determine the stability of the deposited cake (7).

FILTRATION

Filter pressing presently is not widely practiced for sludge handling in the United States, probably because of the high labor and maintenance costs involved in the batch operation. However, the applicability of filter pressing sludges is becoming more promising. The press does not dewater solids by squeezing, but rather acts as a pressure filter similar to a rotary vacuum filter with the exception that higher pressures are used. Like the vacuum filter, the filter press usually requires chemical conditioning, bench scale filtration studies, and careful selection of the filtering media. Economics of the process dictate sludge concentrating by some other means prior to pressing. The Sludge All System developed by Beloit-Passavant Corporation is a completely automated process and the customary shutdown for cake removal is not necessary. The system includes thickening, pressing, and incineration; and the accumulated filter cake from the press contains about 50% solids.

Thickened sludge is ground prior to filtration and often a precoat is added and mixed with the feed sludge before pressing. The sludge enters the press through a center feed and is forced into the filter modules. The filtrate is continuously pressed through the fabric and discharged for recycle. Design considerations and process variables involved are:

        [1]   Sludge volumes
        [2]   Raw sludge solids concentration
        [3]   Nature of the sludge
        [4]   Chemical conditioning
        [5]   Ultimate disposal of filter cakes
        [6]   Filter pressing pressure
        [7]   Filter mesh aperture size
        [8]   Filter pressing time

The basic design criteria for sizing a filter press unit can be developed by use of a pilot plant unit. Pilot plant data should also provide thickening and chemical conditioning requirements. Means for ultimate disposal of the cake solids often will dictate the required solids concentration in the final sludge cake. A final cake solids of 40% may be expected in the sludge cake provided feed solids are maintained near 10%.

FREEZING

A process which has been developed (8) is one in which wastes to be conveyed to an incinerator in containers are pretreated either by deep freezing the wastes or by volatilizing the wastes under exclusion of air, or by mixing the wastes with substances of addition or with leaning substances, e.g., mud, ash, sand, wood shavings, whereupon the thus obtained wastes, either reduced in size or in volatilized condition are fed into the fire chamber of the incinerator. A very similar process has been developed by E. Schuster (9).

SAND DRYING BEDS

Dewatering of digested sludge is commonly accomplished by the use of open sand drying beds. Sewage sludge should be properly digested before application to atmospheric drying as digestion enhances sludge drainability and raw sludge may cause undesirable odors. Most beds are open and completely exposed, while others are glass-covered to reduce the effects of inclement weather on the drying process.

Drying beds usually consist of 4 to 9 inches of sand over 8 to 18 inches of graded gravel or stone. Six to twelve inches of digested sludge are applied to the bed and allowed to drain through the sand beds where the filtrate is collected by underground laterals, conveyed to sumps, and subsequently recycled through the plant. Experiments have shown that tile-drained sludge beds dry 25% faster than beds with impervious bottoms. However, other experiments and studies indicate that nonunderdrained beds can be built and operated at one-fifth the cost of a tile-drained bed.

The sludge drying process incorporates drainage of the sludge moisture through the bed and the simultaneous removal of water by evaporation. The amount of moisture lost by evaporation during the initial stages is insignificant but becomes very important one to two days after the sludge is applied to the bed.

The sludge is allowed to dry until it reaches a liftable condition at which time the dried sludge is removed from the beds for final disposal. This condition normally occurs when the moisture content is 70 to 80%. The moisture content of the dried sludge is on the order of 60%. Design considerations and process variables involved are:

 [1] Climatic and atmospheric conditions
 [2] Depth of sludge application
 [3] Presence or absence of chemical conditioning
 [4] Sludge moisture content
 [5] Source and type of sludge
 [6] Extent of sludge digestion
 [7] Sludge age
 [8] Sludge composition
 [9] Solids concentration when applied
[10] Sludge bed construction

Only recently have attempts been made to formulate rational design standards for sludge drying beds. In the past, bed requirements have been based only on empirical relationships or experience factors. Investigators have attempted to derive design criteria from laboratory experiments and pilot plant operations. However, correlation between experimental data and actual plant data has been relatively poor. Selected variables of the sludge drying processes have been successfully studied in the laboratory as well as on a pilot plant scale as described by B.A. Carnes in a master's thesis, University of Texas (1966).

SUBMERGED COMBUSTION

Submerged combustion offers possibilities for the concentration of hazardous wastes prior to ultimate disposal as noted by Booz-Allen Applied Research, Inc. (10). Submerged combustion is the burning of a gaseous fuel in a specially designed burner with the burner chamber submerged in the wastewater. This device has been used successfully in totally or partially evaporating waste streams, concentrating any dissolved solids, either which have reuse value or which are easier to dispose of than large volumes of the liquid waste.

A submerged combustion unit reduced 75% of the volume of a nylon waste stream, the remainder of which was mixed with other process streams and treated biologically. A polymeric waste stream containing suspended synthetic rubber particles, organic solvents, inorganic salts, and synthetic detergents was not amenable to biological treatment and, consequently, treated by submerged combustion. This waste stream was evaporated to about 10% of its original waste volume with the resulting slurry emptied to a drying bed. Volatile organic compounds in the polymeric waste, such as alcohols and amines, were oxidized or burned so that no odors were detected in the surrounding area.

THICKENING

Sludges from primary or secondary processes usually require thickening prior to dewatering by air drying, vacuum filtration, or centrifugation. If a sludge can be thickened, the process offers the advantages of improving digester operation, reducing sludge volumes for direct land or sea disposal, and enhancing the efficiency of process sludge dewatering systems.

Biological sludges usually can be thickened to a solids concentration of 4 to 6%, depending on the type of thickener and the nature of the sludge. As the ability of many sludges to thicken can be defined only in general terms, bench or pilot thickening studies should be initiated prior to finalizing the thickener design. Air flotation, mentioned previously, is also used for sludge thickening in certain applications. Design considerations and process variables involved are given on the following page.

[1] Sludge loading
[2] Diameter and configuration of thickening unit
[3] Depth of thickener
[4] Temperature
[5] Raking mechanism
[6] Detention time
[7] Sludge depth in thickener
[8] General sludge characteristics

The most common approach in evaluating thickening requirements for sludges is to use batch settling cylinders. The sludge in question is put into the batch settling cylinder and the settling and thickening characteristics are evaluated under quiescent conditions, serving as the basis for scale-up and design (7). A process developed by R.C. MacKensie (11) involves the dewatering of a sludge derived from the treatment of an industrial wastewater stream. The sludge is mixed with a latex comprising rubbery polymer and, on addition of a coagulating agent, a coagulum is formed which is separable from the aqueous phase.

VACUUM FILTRATION

Vacuum filtration is used to dewater wastewater sludges in which water is removed under an applied vacuum through a porous media which retains the solids. In the operation of vacuum filters, a rotary drum passes through a slurry tank in which solids are retained on the drum under an applied vacuum.

When the drum emerges from the slurry tank, the deposited cake is further dried by the transfer of liquid to the air which is drawn through the cake by the applied vacuum. The filter cake is then removed to a collection hopper for hauling or incineration. Design considerations and process variables involved are:

[1] Sludge feed concentration
[2] Sludge conditioning (chemical additives)
[3] Degree of thickening preceding filtration
[4] Sludge viscosity
[5] Filtrate viscosity
[6] Operating vacuum
[7] Chemical and physical composition including partial size, shape, etc.
[8] Type and porosity of filter media

The most rapid means of estimating the vacuum filter yield for a given sludge is by use of the leaf test apparatus. The test variables such as an applied vacuum, sludge concentration, sludge conditioning, etc., are established and the sludge yield for each condition is estimated. Continuous pilot scale tests are recommended if design information for large and costly vacuum filtration installation is required.

REFERENCES

 (1) G.E. Weismantel, *Chemical Engineering*, 82, No. 22, pp. 76, 78, 80 (October 13, 1975).
 (2) J.R. Conner; U.S. Patent 3,837,872; September 24, 1974; assigned to Chemfix, Inc.
 (3) D.J. Opacic, A.L. Lengyel, E.A. Zawadzki and F.H. Jackson; U.S. Patent 3,893,656; July 8, 1975; assigned to Chemfix, Inc.
 (4) *Environmental Science and Technology*, 6, No. 10, 874-875 (October 1972).
 (5) J.V. Dunlea, Jr.; U.S. Patent 3,721,183; March 20, 1973.
 (6) K. Tezuka; U.S. Patent 3,848,392; November 19, 1974; assigned to Tezuka Kosan KK
 (7) M. Sittig, *Pollution Control in the Organic Chemical Industry*, Park Ridge, N.J., Noyes Data Corp. (1974).
 (8) E. Schuster; U.S. Patent 3,902,435; September 2, 1975; assigned to L&C Steinmuller GmbH.
 (9) E. Schuster; U.S. Patent 3,918,372; November 11, 1975; assigned to L&C Steinmuller GmbH.
(10) Booz-Allen Applied Research, Inc., *A Study of Hazardous Waste Materials, Hazardous Effects and Disposal Methods,* Vol III, Report PB 221 467, Springfield, Va., Nat Tech Inf Ser (July 1973).
(11) R.C. MacKenzie; U.S. Patent 3,729,412; April 24, 1973; assigned to Polymer Corp., Ltd.

SOME ULTIMATE DISPOSAL METHODS

There are some ultimate disposal methods which are worthy of brief mention here even though they are not techniques of major importance. The major techniques include:

> —deep well disposal
> —incineration
> —land application, land burial
>   and landfills
> —ocean disposal

BIOLOGICAL DEGRADATION

Biological degradation, while a common technique for reducing the concentration of undesirable constituents in wastewaters is not commonly considered an ultimate disposal method, even though it may completely dispose of ease-to-degrade pollutants.

CHEMICAL DEGRADATION

Chemical treatment is again a common treatment procedure to reduce pollutant concentrations but is not commonly thought of in terms of ultimate disposal. An exception is chlorination or chlorinolysis which can be an ultimate disposal technique for refractory materials.

Chlorination

Destructive chlorination or chlorinolysis can effectively be applied to the disposal of organic compounds (1). In the chlorinolysis process, the undisposables, including DDT and other pesticides and herbicides and any nondistillable chlorinated residues left over from organic chemical manufacturing, are converted with high yields into a principal product, carbon tetrachloride.

According to R.S. Ottinger et al (2) chemicals which may be disposed of by chlorination are the cyanides in general. They are treated by oxidation by the hypochlorite ion (chlorination under alkaline conditions) for both dilute and concentrated wastes. Concentrated wastes should be diluted before chlorination. Hydrocyanic acid can be handled by chemical conversion to ammonia and carbon dioxide using chlorine or hypochlorite in a basic media. Controlled incineration is also adequate to totally destroy cyanide. Metallic cyanides can be handled by chlorination under alkaline conditions (after the waste is diluted). Additional treatment may be to remove the metallic ion.

Composting

Another treatment that also is a way of utilizing solid industrial waste is composting (3). Composting is essentially the biodegradation of refuse or a portion of it to produce a humus-like material. Although plastics are not compostible, because they are not susceptible to rapid biodegradation, they have not been reported to cause significant problems if the refuse is shredded. Fine plastic particles merely behave as any other inert material like grit, small stones, or slate; therefore, they would not be expected to have any detrimental effects on the root development and growth processes of plants.

Composting has not been an effective way of disposing of municipal solid waste in this country. Despite its very early use as a soil conditioner and fertilizer in Europe, composting has made little progress in the U.S. This lack of progress has been due to the low cost and wide use of chemical fertilizers and the high operating and distribution cost of compost. Thus, the number of successful composting operations in the United States has been limited and municipalities have been reluctant to enter into this type of commercial operation.

Composting is more commonly applied to municipal waste disposal than to the disposal of toxic or hazardous industrial wastes, in any event. For a review of large-scale composting techniques as applied to mixed wastes, which can include industrial wastes, the reader is referred to a volume by M.J. Satriana (4).

A process developed by K.F. Petersen (5) for the fermentation of solid organic waste materials for making compost utilizes an apparatus comprising an elongated horizontally disposed rotatable container. There are provided apertures or nozzles, which are open to the interior of the container, the apertures or nozzles being distributed in the longitudinal direction of the shell and being in connection with an outside source of aeration medium under positive pressure.

A device developed by D.B. Waldenville (6) is an apparatus for controlled compost processing of waste materials to provide rapid and complete reduction to compost material. The apparatus consists of a ground-supported enclosure which includes slide-covered access openings; the enclosure further including a longitudinally arrayed, rotatable agitation assembly and material wetting assembly, each of which is periodically actuatable to provide the requisite mixing and wetting functions.

Further, the apparatus is particularly adaptable to a novel shredder-grinder apparatus which is attachable through access opening to the composting apparatus, and which serves to receive waste material through a hopper input for introduction to an input cylinder which includes a plurality of axially arrayed, rotationally driven cutting elements operating in coaction with alternating screening discs which provide final forced entry of chopped waste material into the interior of the composting apparatus enclosure.

Electrolytic Processes

According to R.S. Ottinger et al (2), inorganic wastes which may be subjected to ultimate disposal by electrolytic processes include lead azide. This process converts the lead azide to metallic lead and nitrogen.

A process developed by J. Gordy et al (7) involves the electrolytic destruction of waste material such as sewage, organic process waste materials, and the like. The process more specifically involves oxidation by electrolysis in the anode compartment of an electrolytic cell, in the presence of hydrochloric acid and cuprous chloride. This results in the formation of oxidation products. The resulting liquid may be reduced by electrolysis in the cathode compartment of the electrolytic cell, and this will permit recovery of the copper catalyst.

By the use of this process the formation of organic sludge is largely avoided. Organic compounds are converted to carbon dioxide, ammonia, nitrates and phosphates. The process is adapted to small or large scale operations as required and will find use in the form of self-contained units for home use and for use in vehicles. Cities and metropolitan areas will be able to install large highly efficient treatment plants and to operate them effectively and economically, without the production of large volumes of organic sludge.

Many industries that have serious problems with organic waste disposal and reclamation of water, and will be able to overcome these problems by the use of the process, which has obvious benefits for operators of pulp mills, fish processing plants, fruit and vegetable canneries, cereal and grain processing plants, breweries, wineries, distilleries, detergent manufacturers, and similar industrial plants.

Lagooning

Lagoons or ponds are both natural water bodies or man-made water bodies constructed either by digging out a depression in the earth or by erecting dikes. Lagoons normally refer to basins where the overflow easily passes into receiving waters, and ponds are usually those basins with no overflow. The liquid portion becomes permanently entrapped in the pond and is reduced only through evaporation. Lagoons are better suited for wastes with low solids content and ponds for those with high solids content. Lagoons are used for clarification of both chemical plant process waters and wastewaters, and along with ponds are used for slurries and solids deliberately slurried to enable transport by pipeline.

As the basins fill, they are either cleaned and the solids removed and discarded, or the solids are left to dry and new basins are constructed. If the liquid portion of the waste can contaminate underground water, special precautions must be used to seal the lagoon or pond bottom to prevent leakage.

Lagoons sometimes cover many acres and receive thousand of tons of solid wastes annually. Where a high degree of solids clarification is desired, lagoons may be arranged in series with each successive lagoon providing treatment until the desired water quality is reached. Under some conditions aerobic or possibly anaerobic decomposition of wastes may occur in the lagoons, thereby providing further treatment (8).

Lagoons may be divided into 3 classes: (a) thickening, storage, and digestion lagoons; (b) drying lagoons; and (c) permanent lagoons. Lagooning is the most popular sludge disposal technique at industrial wastewater treatment plants. Lagooning may be considered as a stage process in the handling of sludge or as a final sludge disposal process (9).

The first type of lagoon is used when conventional digestion units are overloaded or sometimes as substitutes for conventional processes. Drying lagoons are used as substitutes for sand drying beds and a permanent lagoon where sludge is never removed is one of the cheapest methods of sludge disposal.

Permanent lagoons may be considered as a landfilling method of sludge disposal. A supernatant decanting system enhances the operation of permanent lagoons by providing additional storage. Sludge stored in lagoons may be dewatered from about 95% moisture to 55 or 60% moisture in 2 to 3 year periods. The standard operating procedure for permanent lagoons is to discharge digested sludge to the lagoons at regular intervals, allowing a drying and cleaning period.

Waste stabilization ponds can be used effectively to treat excess activated sludge. Lagoons are often used for sludge dewatering prior to ultimate disposal and in effect the lagoon acts as a gravity thickener. When a lagoon is used for dewatering and not permanent storage, recommended filling depths are 2.5 to 4 feet. However, ponds 5 to 6 feet in depth have been used to stabilize waste activated sludge. Design considerations and process variables involved in the stabilization of waste activated sludge are listed on the following page.

[1] available land area
[2] climatic atmospheric conditions
[3] subsoil permeability
[4] lagoon depth
[5] lagoon surface loading
[6] sludge characteristics
[7] moisture content of sludge

Sludge lagoons have been designed on the basis of surface or volumetric loadings or based upon ultimate disposal requirements. Model ponds have proven useful in evaluating several variables of pond design. These model ponds were developed for studies of waste stabilization pond treatment of wastewaters, but can also be used to study sludge stabilization. Development of design criteria is dependent on lagoon type, e.g., whether the process is to be designed for thickening, drying, digestion or storage.

Approximately 90% of the hazardous waste in the United States is in liquid or sludge form. It is likely that a similar percentage of close to 90% nondry solid hazardous waste exists for the inorganic chemicals industry as well (10). Whenever feasible, pond or lagoon disposal of hazardous wastes is one of the simplest and most economical of approaches. It suffers from a number of restrictions, however:

[1] The pond needs to provide protection from both surface and groundwater contamination. With the exception of some pits in naturally impervious soil or in some very dry climates this means a lined pond. Liners include clay, plastic sheeting, asphalt, concrete, epoxy and other impervious materials, all of which are relatively expensive.

[2] Except for dry climates such as found in the western U.S., the ponds without discharge will eventually overflow from rainfall accumulation.

[3] Ponds are prone to be flushed out whenever massive rainfall and floods inundate the pond area. Here again, it is difficult and expensive to provide flood protection, particularly in wet climates.

In view of these restrictions, disposal ponds and lagoons are usually feasible only in the western United States where evaporation disposal of a large volume per year of liquid waste is carried out. Ponds, in other regions, must be considered as storage ponds which will eventually need drainage cleaning or treatment and disposal action. Pond storage and disposal probably accounts for 20 to 30% of the total land-destined hazardous wastes. Pond storage of wastes occur, for example, in the manufacture of borax, sodium silicofluoride, dichromates, boric acid, and chlorine (10).

Long-Term Storage

In those few instances where a hazardous waste cannot be treated or disposed of adequately, the best alternative is engineered storage until adequate methods are developed. However, only a very small percentage of the total quantity of hazardous wastes generated in this country should require permanent storage. An engineered storage facility must provide for safe storage of hazardous wastes for long periods of time, and retrievability of the wastes at any time during this storage. Solidification of wastes prior to storage may be desirable to eliminate leakage. The storage facility should be routinely monitored and deteriorating drums or other containers replaced as required. Ultimately, the goal is to reclaim these wastes or transform them so they are acceptable in a permanent disposal facility.

Long-term storage may involve techniques such as those described in a chapter found under the heading of "Land Burial." It may also involve the use of aboveground bins or silos.

According to R.S. Ottinger et al (2), inorganic chemicals which may best be subjected to the following ultimate disposal by long-term storage are:

> Antimony Pentachloride: When dissolved in water and neutralized, the slightly soluble oxide is formed. Removal of the oxide is followed by sulfide precipitation to ensure the removal of the metal ion from solution. The antimony oxides can be sent to a refiner or placed in long-term storage.
>
> Antimony Potassium Tartrate: Dissolve wastes in water, acidify and precipitate the sulfide using hydrogen sulfide as the reactant. The antimony sulfide precipitate should be returned to suppliers or manufacturers for reprocessing or be placed into long-term storage.
>
> Antimony Trichloride: Same as Antimony Pentachloride above.
>
> Arsenic: Elemental arsenic wastes should be placed in long-term storage or returned to suppliers or manufacturers for reprocessing.
>
> Arsenic Pentaselenide: Wastes should be placed in long-term storage or returned to suppliers or manufacturers for reprocessing.
>
> Arsenic Trichloride: Hydrolyze to arsenic trioxide utilizing scrubbers for hydrogen chloride abatement. The trioxide may then be placed in long-term storage.
>
> Arsenic Trioxide: Long-term storage in large siftproof and weatherproof silos.
>
> Cacodylic Acid: Long-term storage in concrete vaults or weatherproof bins; landfill in a California Class 1 site for small amounts.

The following chemicals are best suited to ultimate disposal by long-term storage in large, weatherproof, and siftproof storage bins or silos; landfill in a California Class 1 site.

Calcium Arsenate	Manganese Arsenate
Calcium Arsenite	Sodium Arsenate
Copper Acetoarsenite	Sodium Arsenite
Copper Arsenates	Sodium Cacodylate
Lead Arsenate	Zinc Arsenate
Lead Arsenite	Zinc Arsenite
Magnesium Arsenite	

Pyrolysis

Pyrolysis is generally defined as the thermal decomposition of a compound. With respect to waste carbonaceous materials, pyrolysis represents a means of converting the unwanted waste into a usable commodity with economic value. That is, most municipal and industrial wastes which are basically organic in nature can be converted to coke or activated charcoal and gaseous mixtures which may approach natural gas in heating values through the utilization of pyrolysis.

Pyrolysis has only recently been applied to the conversion of organic wastes. The process has traditionally been used to convert low economic value homogeneous materials, such as wood chips and heavy hydrocarbon still bottoms, to compounds with higher overall economic value, such as fuel gas, pitch, creosote, acetic acid, crude methanol and charcoal (as in the case with wood chips), and coke, fuel gas and gas oil (as in the case with still bottoms).

In view of increasing energy-use and raw material availability constraints, pyrolysis is getting a new look as an ultimate disposal technique.

The heart of the pyrolytic waste conversion process is the pyrolytic converter (Figure 8). The unit consists of a sealed, airtight retort cylinder inside a heavy insulating jacket. The gas-fired retort revolves slowly on a slight decline from infeed to outfeed. Wastes are injected through a seal area that intermittently opens (the flapper valve seal is designed to minimize entry of oxygen).

FIGURE 8:   PYROLYTIC WASTE CONVERSION

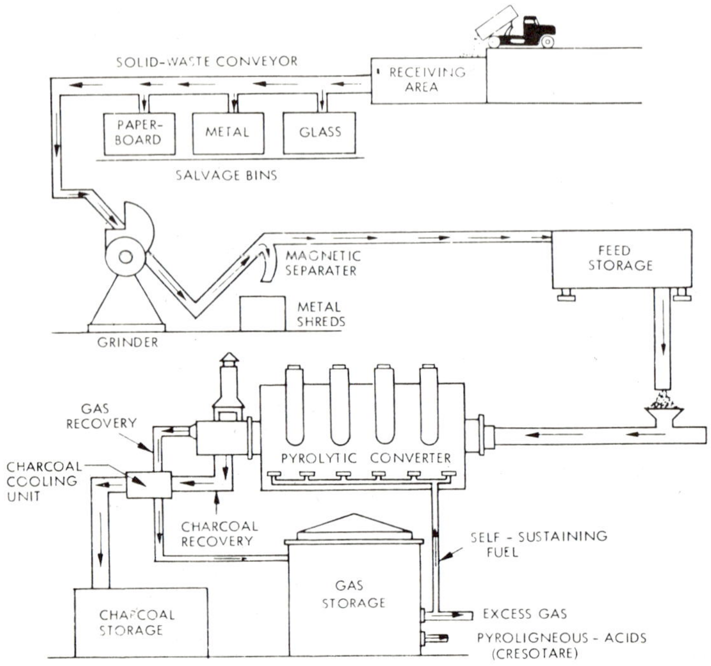

Source: PB 224 582

Inside the retort, ground-up wastes are subjected to temperatures of about 1200° ± 300°F, depending upon the nature of the wastes, in an essentially oxygen-free atmosphere. Without oxygen, the wastes cannot burn and are broken down (pyrolyzed) into steam, carbon oxides, volatile vapors and charcoal.

Steam, carbon dioxide and carbon monoxide are the first gases to emerge. They are trapped and then: [1] vented after flaring; [2] used to dry incoming feed; or [3] the steam can be condensed and the effluent gas burned for fuel value. As the wastes are heated further, volatile gases are distilled. Typically, these gases will include hydrogen, methane, ethylene and ethane. About 25 ft^3 of combustible gases are recovered from every pound of industrial and municipal refuse. Energy value per cubic foot is usually between 400 and 500 Btu.

The crude, combustible gas is drawn off and used in part (30 to 40%) to fuel the converter while the remainder may be utilized to fuel other equipment or to make steam for such uses as heating and power generation. In fact, a 100-ton per day municipal waste conversion facility will produce enough excess gas to create 400,000 kw of electricity per day. At a quarter-cent per kw, that amounts to $1,000 of electrical energy per day. Thus, recovery of combustible gas usually offsets the total operating costs of a waste-conversion facility.

The surplus gas cannot be piped over any great distance and retain its original heat value. Nor can it be stored. For, allowed to cool, pyroligneous acids begin to condense, reducing heat value per cubic foot to around 350 Btu and reducing volume by as much as 80%. However, the liquid condensate is of economic value. It constitutes a form of cresotar which is used for, among other things, dust control on unpaved roads. Moreover, the liquid represents a potentially valuable source of organic compounds. Chemical analyses have identified such constituents as methanol, ethanol, isobutanol, n-pentanol, t-pentanol, 1,3-propanediol, 1-hexanol, acetic acid and various other alcohols, ketones and tars.

The last by-product is charcoal. Generally 30 to 35% of input by weight is recovered in this form, and each pound of charcoal has an approximate heat value of 12,000 Btu. Used as fuel, it can create about half the electricity expected from the recovered gas.

Other major pieces of equipment required for a municipal or industrial waste-conversion facility are a hogger/grinder, a magnetic separator, conveyors, and storage facilities. A hogger/grinder is necessary because converters generally operate best when solid waste materials are reduced to no more than a few inches in size. Since pyrolysis units do not usually vent any products to the atmosphere, they require no air pollution control equipment.

Pyrolytic conversion processes are generally custom-engineered according to input volumes and types of waste being treated. For this reason, there is not a great deal of specific design information available. Converters have been made with capacities ranging from ¼ to 12 tph. Units can be installed in batteries to handle more than 12 tph, e.g., one municipal pyrolytic process utilizes four 4 tph converters.

Intake-to-discharge cycles vary with heat intensity, the converters length, and its rotational speed. The average for industrial and municipal refuse is 12 to 15 min. Such hard to pyrolyze materials as coal (which is converted into coke) and rubber may take as long as 30 min. Operational temperatures will vary with waste type and desired products. They are usually in the $1200° \pm 300°F$ range with the lower operating temperature generally resulting in greater residue (coke), tar and light oil yields and lower gas yields.

The primary factors determining the capital cost of pyrolysis systems include waste flow rate, waste composition, secondary effluent treatment for product recovery, and materials of construction. Operating costs are determined by labor rates, maintenance requirements, and conversion product values. An exhaustive economic analysis of a commercial size (5,000 tpd) municipal refuse pyrolysis plant was made in 1970 by Cities Service Oil Company. The study indicated that a total investment including working capital of 32.6 million dollars was required and that the associated operating cost would be $2.14 per ton of refuse pyrolyzed (Table 6). There is no economic data available with respect to industrial waste pyrolysis processes.

TABLE 6: INVESTMENT AND OPERATING COSTS FOR A 5,000 TPD MUNICIPAL PYROLYSIS PLANT

Capital Investment

Investment	$ Million
Inside battery limits	16.0
Offsites	6.4
Total	22.4
Additional Costs	
Contingency (25%)	5.6
Construction interest, first year (8% x 33%)	0.7
Construction interest, second year (8% x 67%)	1.5
Startup extraordinary (assumed 8% x 100%)	2.4
Total additional costs	10.2
Total investment (includes working capital)	32.6

(continued)

TABLE 6: (continued)

Operating Costs

Fixed Costs	$ Million (per year)	
Labor	0.49	
Supervision	0.15	
Overhead	0.64	
Maintenance, onsite	1.00	
Maintenance, offsite	0.24	
Taxes and Insurance	0.49	
Total fixed costs	3.01	$1.65/ton
Variable Costs		
Water and power	0.20	
Fuel	0.70	
Total Variable Costs	0.90	$0.49/ton
Total Operating Costs	3.91	$2.14/ton

Source: PB 224 582

Modifications to the pyrolysis process generally involve treatment of converter effluents (Figure 9). The pyrolysis oils may be sent through a hydrotreating unit and converted to industrial fuel oil. The pyrolysis effluent gas may be cooled and the resultant condensate separated into its components, namely acetic acid, methanol, furfural, acetone, butyric acid, propionic acid, methyl ethyl ketone, light fuel oil, and other water soluble volatile organics, through the use of conventional separation techniques. The cooled wet gas may be dried and utilized as fuel gas. The char-like pyrolysis residue can be further treated and converted into activated carbon.

FIGURE 9: SECONDARY TREATMENT ALTERNATIVES IN PYROLYSIS PROCESSING

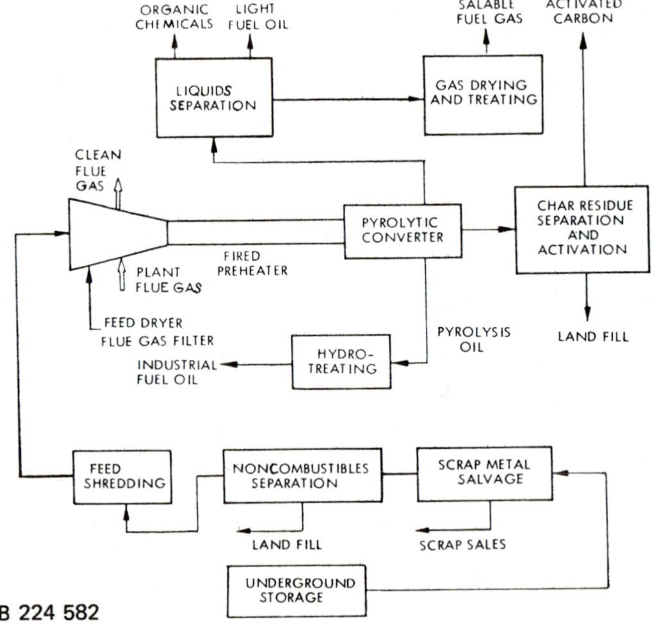

Source: PB 224 582

The utilization of pyrolytic processes to convert waste into useful materials is a relatively new concept. For this reason, there are few processes in operation today. Those processes which are in operation convert municipal refuse, industrial refuse, paint sludges, tires, plastics and other organic polymers into materials with economic value.

Although the pyrolytic converter is a versatile piece of equipment that can be operated under varying conditions with various feed materials, its auxiliary equipment tends to be specific for various feeds and desired end products. For that reason, the overall pyrolytic process tends to lack versatility.

A process developed by W.T. Hage (12) involves the disposal of solid material, utilizing a pyrolytic decomposition chamber for heating the material to its decomposition temperature in the absence of air and utilizing the vaporized material therefrom with air and a combustible gas, such as propane, in a combustion chamber for the further reduction of the vaporized material. The heat generated in the combustion chamber is utilized through a heat exchanger to transfer heat to the primary heating chamber.

A process developed by F.G. McMullen (13) involves combusting solid waste material on grates in combination with comminuted waste material being combusted in suspension within a combustion zone. Conduits are positioned external to the combustion zone around which exhaust gases from the combustion zone pass. Comminuted waste material is driven through the conduits and by pyrolytic action emit gases which are collected and passed to a recovery system. Additionally, predried sludge and sludge vapors derived from sanitary sewage are incorporated into the combustion zone for burning. The combined waste materials processing system may be incorporated within a steam generating system in order to increase the overall efficiency of the boiler.

A process developed by P.R. Kelly (14) is one in which shredded waste material is stacked in a pyrolyzing chamber or column through which hot gases, initially containing no oxygen pass upwardly, driving off volatile material. A condenser reclaims the condensable materials, and the noncondensable gases are recirculated through the combustor and pyrolyzing chamber carbonizing the pyrolyzed waste in the lower reaches of the chamber.

After pyrolyzing is completed in the lower portions of the waste material, excess air is admitted into the recirculating hot gas stream. Contact of the oxygen in the excess air with the carbonized material causes substantially immediate combustion. Ash from the combustion zone is then dumped into a discharge chute, or, optionally, removed by traveling scraper bars.

This process is stated to be particularly applicable to the disposal of agricultural wastes including animal feed lot wastes.

A process developed by E.A. Grannen and L. Robinson (15) is one in which recovery of valuable organic compounds from organic industrial and household wastes is effected by rapid and relatively low temperature pyrolysis of the organic material in a microwave discharge. The organic wastes are separated from inorganic materials and shredded to a fine particle size. This material is mixed with a gas stream at a low pressure, typically about 10 mm of mercury, and the wastes are then conveyed through a pyrolysis zone where they are subjected to microwave discharge in the low gigahertz frequency range.

Molecular decomposition occurs principally at carbon-oxygen and other polar bonds and the many reactive fragments may react with the gas stream which preferably has reactive species of gas such as hydrogen or hydrocarbons. Vaporized products are rapidly swept from the pyrolysis zone by the gas stream and passed to separation stages where various organic products are isolated and residual fluids may be employed as fuels. A portion of the gases may be recycled to the pyrolysis zone.

A device developed by J. von Klenck et al (16) is one in which industrial or municipal waste is pyrolytically treated by causing it to drift and float, up and through a bath of molten metal or glass. The reduction products are taken from the surface level and above the bath for subsequent use. A feeder pipe has a flared end submerged in the bath for introducing the waste in a low level, while receiving waste through a lateral port. The waste while dropping in the pipe is dried and heated. The molten metal or glass is heated to maintain the molten state. Heavy residue is collected in and extracted from the bottom of the bath containing vessel.

Salt Deposit Disposal

A category of ultimate disposal which is similar to, yet quite different from, deep-well disposal is salt deposit disposal.

A process developed by D.A. Shock (17) provides a method by which predominantly organic waste materials of a highly toxic or noxious nature, such as mustard gases, nerve gases, harmful biological agents, phenols and the like, may be permanently disposed of in a safe manner. Basically, and in its broadest aspect, the method comprises the mixing of a toxic organic waste material with an oil base, fluid material which is compatible with the waste material, and which forms a pumpable slurry therewith.

The oil base material contains appropriate gelling agents to impart thixotropic properties to the slurry, or at least to assure that the mixture will set up to a solid state upon standing. The slurry may thus be transported to, and allowed to solidify in, a location which is inaccessible to human and animal life, and which is isolated from contact with useful mineral deposits. The physical and chemical properties of the slurry are such that underground cavities may be utilized for the permanent disposition of the material if desired.

A process developed by Z.W. Rogers et al (18) is one in which refuse is collected, transported and then deposited in an impermeable cylindrical solution cavity formed within a subterranean rock salt formation. The air containing noxious odors and excess heat is extracted from the solution cavity purified and then released into the atmosphere.

Stream Discharge

This form of waste disposal is becoming less and less popular with regulatory authorities and the trend is toward the mandatory primary and secondary treatment of all wastewaters. However, certain plants are allowed to discharge their wastewaters to receiving bodies of water without treatment providing: (a) sufficient receiving water is available as a diluent; (b) there are no toxic or refractory compounds in the waste stream; and (c) the assimilative and recovery capacity of the receiving water based on the pollutional input is adequate.

Rapid wastewater diluent mixing is required, and this often is accomplished by the use of jets or diffusers. The quality and use of the receiving water often dictates direct discharge practices. For example, regulatory authorities may permit the discharge of spent caustic wastes directly into salt lakes, brackish streams, and estuaries.

According to R.S. Ottinger et al (2), inorganic chemicals which may be disposed of (after indicated pretreatment in some cases) by stream discharge are as follows:

> Aluminum Sulfate: Pretreatment involves hydrolysis followed by neutralization with NaOH. The insoluble aluminum hydroxide formed is removed by filtration and can be heated to decomposition to yield alumina which has valuable industrial applications. The neutral solution of sodium sulfate can be discharged into sewers and waterways as long as its concentration is below the recommended provisional limit of 250 mg/l.
>
> Ammonium Chloride: Pretreatment involves addition of sodium hydroxide to liberate ammonia and form the soluble sodium salt. The liberated ammonia can be recovered and sold. After dilution to the permitted provisional limit, the sodium salt can be discharged into a stream or sewer.

Ammonium Nitrate: Pretreatment involves addition of sodium hydroxide
to liberate ammonia and form the soluble sodium salt. The liberated
ammonia can be recovered and sold. After dilution to the permitted
provisional limit, the sodium salt can be discharged into a stream or
sewer.

Borax, Dehydrated: The material is diluted to the recommended provisional limit (0.10 mg/l) in water. The pH is adjusted to between
6.5 and 9.1 and then the material can be discharged into sewers or
natural streams.

Boron Chloride: Pretreatment involves addition of soda ash-slaked lime
solution to form the corresponding sodium and calcium salt solution. This solution can be safely discharged after dilution.

Calcium Chloride: Pretreatment involves precipitation with soda ash to
yield the insoluble calcium carbonate. The remaining brine solution,
when its sodium chloride concentration is below 250 mg/l, may be
discharged into sewers and waterways.

Calcium Hydride: In a pretreatment step, the waste material is mixed
with dry sand before adding to water. The hydrogen gas liberated
is burned off with a pilot flame. The remaining residue is a hydroxide and should be neutralized by an acid before being disposed of.

Calcium Hydroxide: Pretreatment involves neutralization with hydrochloric acid to yield calcium chloride. The calcium chloride formed
can be treated by the method described earlier for this compound.

Calcium Hypochlorite: Dissolve the material in water and add to a large
volume of concentrated reducing agent solution, then acidify the
mixture with H_2SO_4. When reduction is complete, soda ash is
added to make the solution alkaline. The alkaline liquid is decanted
from any sludge produced, neutralized, and diluted before discharge
to a sewer or stream. The sludge is landfilled.

Chlorosulfonic Acid: Pretreatment involves chemical decomposition using
sodium bicarbonate, and ammonium hydroxide as reactants followed
with dilution with water, neutralization and discharge into the
sewer system.

Hydrazoic Acid: Pretreatment involves chemical decomposition utilizing
nitrous acid followed by neutralizaton and dilution with water and
discharge into the sewer system.

Hydrochloric Acid: Soda ash-slaked lime is added to form the neutral
solution of chloride of sodium and calcium. This solution can be
discharged after dilution with water.

Hydrocyanic Acid: Chemical conversion to ammonia and carbon dioxide
using chlorine or hypochlorite in a basic media. Controlled incineration is also adequate to totally destroy cyanide.

Hydrogen Peroxide: Dilution with water to release the oxygen. After
decomposition the waste stream may be discharged safely.

Hypochlorite, Sodium: Dissolve the material in water and add to a large
volume of concentrated reducing agent solution, then acidify the
mixture with H_2SO_4. When reduction is complete, soda ash is added
to make the solution alkaline. The alkaline liquid is decanted from
any sludge produced, neutralized, and diluted before discharge to a
sewer or stream. The sludge is landfilled.

Magnesium Chlorate: Dissolve the material in water and add to a large volume of concentrated reducing agent solution, then acidify the mixture
with H_2SO_4. When reduction is complete, soda ash is added to make
the solution alkaline. The alkaline liquid is decanted from any sludge
produced, neutralized, and diluted before discharge to a sewer or
stream. The sludge is landfilled.

Nitric Acid: Soda ash– slaked lime is added to form the neutral solution
of nitrate of sodium and calcium. This solution can be discharged
after dilution with water.

Phosphorus Oxychloride: Decompose with water forming phosphoric and
hydrochloric acids. Neutralize acids and dilute if necessary for discharge into the sewer system.

Phosphorus Pentachloride: Same as Phosphorus Oxychloride

Phosphorus Pentasulfide: Decompose with water forming phosphoric acid, sulfuric acid and hydrogen sulfide. Provisions must be made for scrubbing hydrogen sulfide emissions. The acids may then be neutralized and diluted if necessary, and discharged into the sewer system.

Phosphorus Trichloride: Same as Phosphorus Oxychloride.

Potassium Bifluoride: Pretreatment involves reaction with excess of lime, followed by lagooning, and either recovery or landfill disposal of the separated calcium fluoride. The supernatant liquid from this process is diluted and discharged to the sewer.

Potassium Binoxalate: Pretreatment involves ignition–to convert it to a carbonate. Since carbonates are nontoxic, the material may be sent to a landfill or simply sewered.

Potassium Fluoride: Pretreatment involves reaction with an excess of lime, followed by lagooning, and either recovery or landfill disposal of the separated calcium fluoride. The supernatant liquid from this process is diluted and discharged to the sewer.

Potassium Hydroxide: Dissolve in water followed by neutralization with an acid and sewering.

Potassium Oxalate: Pretreatment involves ignition-to convert it to a carbonate. Since carbonates are nontoxic, the material may be sent to a landfill or simply sewered.

Potassium Peroxide: Neutralize liquid waste if necessary and dilute for discharge into the sewer system.

Potassium Phosphate: The material is diluted to the recommended provisional limit in water (0.05 mg/l). The pH is adjusted to between 6.5 and 9.1 and then the material can be discharged into sewers or natural streams.

Potassium Sulfate: Potassium sulfate is relatively harmless and can be diluted to a concentration below 250 mg/l and released to sewers and waterways.

Potassium Sulfide: Pretreatment involves precipitation with ferric chloride solution. The insoluble FeS formed is removed by filtration. The remaining potassium chloride solution can be diluted to a concentration below 250 mg/l and discharged to sewers and waterways.

Silicon Tetrachloride: Pretreatment involves addition of soda ash-slaked lime solution to form the corresponding sodium and calcium salt solution. This solution can be safely discharged after dilution.

Sodium Acid Sulfate: Dilution with large volumes of water followed by reaction with soda ash, calcium hypochlorite and HCl followed by discharge into the sewer system.

Sodium Amide: Hydrolyzes rapidly to form sodium hydroxide and ammonia, both of which can be neutralized by hydrochloric or sulfuric acid. The neutral solution can be safely discharged if the salt content is below the limits set to maintain water quality (0.10 mg/l).

Sodium Bifluoride: See Potassium Bifluoride

Sodium Carbonate: The material is diluted to the recommended provisional limit in water (0.02 mg/l). The pH is adjusted to between 6.5 and 9.1 and then the material can be discharged into sewers or natural streams.

Sodium Fluoride: See Potassium Fluoride.

Sodium Hydride: The waste material is mixed with dry sand before adding to water. The hydrogen gas liberated is burned off with a pilot flame. The remaining residue is a hydroxide and should be neutralized by an acid before being disposed of.

Sodium Nitrate: The material is diluted to the recommended provisional limit in water.(The pH is adjusted to between 6.5 and 9.1 and then the material can be discharged into sewers or natural streams.

Sodium Nitrite: Dilution with large volumes of water followed by reaction with soda ash, calcium hypochlorite and HCl followed by discharge into the sewer system.

Sodium Orthophosphates: The material is diluted to the recommended provisional limit in water (0.05 mg/l as H_3PO_4. The pH is adjusted to between 6.5 and 9.1 and then the material can be discharged into sewers or natural streams.

Sodium Perchlorate: Dissolve the material in water and add to a large
volume of concentrated reducing agent solution, then acidify the
mixture with H_2SO_4. When reduction is complete, soda ash is
added to make the solution alkaline. The alkaline liquid is
decanted from any sludge produced, neutralized, and diluted before discharge to a sewer or stream. The sludge is landfilled.
Sodium Peroxide: Neutralize liquid waste if necessary and dilute for
discharge into the sewer system.
Sodium Silicates: Acidification with HCl followed by neutralization,
dilution with water and release into the sewer system.
Sodium Sulfite: Dilution with large volumes of water followed by reaction
with soda ash, calcium hypochlorite and HCl followed by discharge
into the sewer system.
Sodium Thiocyanate: Dissolve in a large quantity of water, buffer with
a slight excess of soda ash, neutralize with an acid, and sewer.
Sulfuryl Fluoride: Addition of soda ash–slaked lime solution to form
the corresponding sodium and calcium salt solution. This solution
can be safely discharged after dilution.
Thiocyanates: Dissolve in a large quantity of water, buffer with a slight
excess of soda ash, neutralize with an acid, and sewer.

According to R.S. Ottinger et al (2), organic chemicals which can be disposed of (after
indicated pretreatment in some cases) by stream discharge are as follows:

Benzoyl Peroxide: Pretreatment involves decomposition with sodium
hydroxide. The final solution of sodium benzoate, which is very
biodegradable, may be flushed into the drain. Disposal of large
quantities of solution may require pH adjustment before release
into the sewer; controlled incineration after mixing with a noncombustible material.

Wet Air Oxidation

According to a recent article (19), wet air oxidation provides some of the best practical
available technology to destroy waste substances, recover valuable inorganic materials from
the waste stream, and condition sludge solids for easy disposal. Since the oxidation is exothermic, substantial energy recovery is also feasible.

Wet air oxidation (WAO) is based on the discovery that any organic material in aqueous
solution or suspension can be oxidized to any desired extent by air, under pressure, at temperatures from 350° to 700°F. The degree of oxidation (from 0 to 100%) depends on the
temperature and the amount of air supplied. The WAO of some compounds may be effectively catalyzed.

In addition to pulp and paper and municipal waste treatment, WAO is being increasingly
applied to other pollution and recovery problems. WAO is being used to detoxify acrylonitrile wastes by destroying cyanide. Recent process developments have made it possible to
produce a water-white effluent containing less than 2% of the influent COD. Ammonium
sulfate of marketable quality can be recovered from the effluent. Sufficient heat is liberated during the oxidation not only to make the process thermally self-sustaining, but also
to concentrate the ammonium sulfate by evaporation. To date, four acrylonitrile waste
plants have been built.

WAO will be used by the U.S. Navy to safely destroy, without air pollution off-specifica-
tion and outdated propellants, explosives, and munitions. Such materials have been disposed
of in the past by open-pit burning. In WAO, however, a slurry of ground explosives is oxidized with air; the residue is a small volume of inert ash and salts. The Naval Ordnance
Station at Indian Head, Md. is the site of the first such installation.

REFERENCES

(1) *Environmental Science and Technology,* 8, No. 1, 18-19 (January 1974).

(2) R.S. Ottinger, J.L. Blumenthal, D.F. Dal Porto, G.I. Gruber, M.J. Santy and C.C. Shih;
      *Recommended Methods of Reduction, Neutralization, Recovery or Disposal of Hazardous
      Waste,* Vol. I, Report PB 224 580 (August 1973).

(3) M. Sittig, *Pollution Control in the Plastics and Rubber Industry,* Park Ridge, N.J., Noyes Data
      Corp. (1975).

(4) M.J. Satriana, *Large Scale Composting,* Park Ridge, N.J., Noyes Data Corp. (1974).

(5) K.F. Petersen; U.S. Patent 3,055,744; September 25, 1962; assigned to Dano Ingeniorforretning og
      Maskinfabrik Ingenior Kai Petersen's Fond.

(6) D.B. Waldenville; U.S. Patent 3,845,939; November 5, 1974; assigned to Waste-Treat, Inc.

(7) J. Gordy and D.R. Harris; U.S. Patent 3,703,453; November 21, 1972; assigned to Stellar
      Industries, Ltd.

(8) J.K. Holcombe and P.W. Kalika; *Solid Waste Management in the Industrial Chemical Industry,*
      Report PB 226 420; Springfield, Va., Nat Tech Information Service (1973).

(9) M. Sittig, *Pollution Control in the Organic Chemical Industry,* Park Ridge, N.J., Noyes Data Corp.
      (1974).

(10) R.G. Shaver, L.C. Parker, E.F. Rissman, K.M. Slimak and R.C. Smith, *Assessment of Industrial
      Hazardous Waste Practices—Inorganic Chemicals Industry,* Report PB 244 832, Springfield,
      Va., Nat Tech Information Service (March 1975).

(11) R.S. Ottinger, J.L. Blumenthal, D.F. Dal Porto, G.I. Gruber, M.J. Santy and C.C. Shih,
      *Recommended Methods of Reduction, Neutralization, Recovery or Disposal of Hazardous
      Waste,* Vol III, Report PB 224 582, Springfield, Va., Nat Tech Information Service
      (August 1973).

(12) W.T. Hage; U.S. Patent 3,768,424; October 30, 1973; assigned to Mechtron International Corp.

(13) F.G. McMullen; U.S. Patent 3,769,921; November 6, 1973.

(14) P.R. Kelly; U.S. Patent 3,771,468; November 13, 1973.

(15) E.A. Grannen and L. Robinson; U.S. Patent 3,843,457; October. 22, 1974; assigned to Occidental
      Petroleum Corp.

(16) J. von Klenck, E. Michel and K.D. Gerstenäcker; U.S. Patent 3,890,908; June 24, 1975; assigned
      to Mannesmann AG.

(17) D.A. Shock; U.S. Patent 3,196,619; July 27, 1965; assigned to Continental Oil Co.

(18) Z.W. Rogers and W.L. Kirk; U.S. Patent 3,665,716; May 30, 1972.

(19) *Environmental Science and Technology,* 9, No. 4, 300-301 (April 1975).

DEEP WELL DISPOSAL

Subsurface disposal of liquid wastes is not a new concept as the oil and gas producers have been using this method for disposal of oil field brines for half a century. Only recently have the process industries realized the applicability of deep well injection for disposing of concentrated and relatively untreatable waste streams (1). The depth at which these wastes are discharged vary from a few hundred feet to over 2,000 feet and wellhead injection pressures for some wells have approached 4,000 psi. A suitable disposal formation is the very heart of a good disposal system and if water can be disposed of with a vacuum on the injection wellhead, operating expenses would be much less than when injection pressures are required. Deep well disposal requires the injection of liquid waste into a porous subsurface stratum which contains noncommercial brines. The wastes are merely stored below ground in strata which are sealed by impervious strata, thus isolated from usable underground water supplies or mineral resources.

As pointed out in the literature subsurface storage of wastes is a form of potential environmental pollution. Further, the unknowns concerning subsurface storage far outnumber the knowns according to 1972 review on this topic (2).

A year later, it was pointed out (3) that injection wells for industrial waste continued to increase slowly in number, and still a year later (4) in spite of the uncertainties, the number of such wells continued to increase, albeit with reluctance.

DEFINITION

Deep well disposal is a system of disposing of raw or treated, filtered hazardous wastes by pumping the wastes into deep wells where they are contained in the pores of the permeable subsurface rock, separated from other groundwater supplies by impermeable layers of rock or clay. A generalized flow sheet is shown as Figure 10.

Because adequate surface disposal of wastes is usually quite expensive, the disposal of wastes in deep wells has been selected in many cases as being a practical and economical alternative for limiting pollution hazards.

Subsurface injection has been extensively and successfully used in the disposal of oil field brines; there are now somewhere between 10,000 and 40,000 brine injection wells in the United states. The same principles can also be utilized in the design and installation of industrial waste disposal well systems.

FIGURE 10: DEEP WELL DISPOSAL

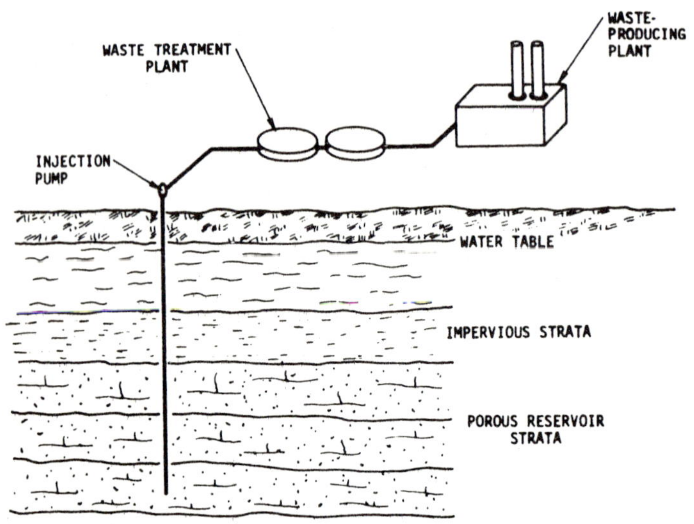

Source: PB 224 582

EXTENT OF DEEP WELL DISPOSAL

Partly because of more stringent pollution legislation, the number of industrial waste injection wells completed in the United States has increased considerably over the recent past. From 1950 to 1963, 36 wells were completed; in the following 3 years, 39 more wells were installed; and by 1968, there were some 100 disposal wells in operation.

Injection wells can be used by virtually any type of industry which is located in the proper geologic environment and which has a waste product amenable to this method. A number of industries presently using the deep well injection method are shown below. The largest users of deep well disposal systems are the chemical and pharmaceutical industries.

	Percent
Chemical and pharmaceutical plants	50
Refineries	22
Steel and metal plants	7
Other, including paper mill, coke plants, etc.	21

A "census" of industrial waste injection wells has been published by D.L. Warner (6).

DESIGN OF DEEP WELL DISPOSAL SYSTEMS

The disposal system consists of a well and surface equipment, such as pumps and pretreatment units. Design considerations and process variables involved are: [1] subsurface geology, (a) permeability, (b) porosity and (c) fracture gradient; [2] subsurface hydrology; [3] fluid compatability; [4] pretreatment considerations; [5] flow and [6] characteristics of waste, including temperature.

Initially, a feasibility study is required which includes acquisition of basic information concerning subsurface geology and hydrology. Based on these data, the operating wellhead pressure may be estimated and for steady-state single-phase flow, the injection rate may be predicted. Pilot wells are often constructed prior to the design and construction of the waste injection well system.

Preliminary estimates of injection rates and hole pressure may be verified by these pilot studies. These studies are particularly important for the design of wastewater pumps and appurtenances. The test hold will also serve to identify subsurface geology and allow for sampling of the formation fluids. The duplication of formation hydrologic and geologic conditions in the laboratory is highly impractical and in most cases impossible.

Fluid compatibility can be evaluated in the laboratory by mixing the waste and formation waters and observing any changes in the physical appearance. Preliminary investigations should also include bacteriological analyses of the formation water and wastewater as micro-organisms may also cause plugging of the well formation.

The following discussion provides an overview of the various factors considered in the selection, design, construction, and operation of a deep well waste disposal facility. Much of the technical and procedural information presented (5) was obtained from Mr. John Heckard of Dames and Moore, Consulting Engineers and from a report on deep well disposal by the National Industrial Pollution Control Council. The discussion of the various factors is followed by an assessment of the application of deep well disposal to hazardous wastes.

Deep well injection is actually a storage system, since the waste materials injected into the subsurface formations remain there indefinitely. The question of major importance is, therefore, "Under what conditions can deep subsurface strata be utilized for the storage of liquid wastes?"

To serve as an adequate liquid storage reservoir, an injection stratum must have sufficiently high porosity and permeability. Although under certain conditions all types of rocks are capable of storing injected fluids, porous sedimentary rocks (such as sandstone, limestone, and dolomite) are most likely to have the proper geologic characteristics required for waste injection.

An injection horizon must be separated from fresh water aquifers or any other usable natural resources by impervious confining strata such as clay or shale. The selection of a site, therefore, must provide for the protection of developed and undeveloped mineral resources, including groundwater.

The design of an injection well is based on the depth of the well, the anticipated injection pressures, and the anticipated future maintenance requirements. In addition, state regulatory agencies often maintain specific requirements concerning the construction of waste injection wells; and in all cases, the final design of the well must be approved by the appropriate state agency.

The construction details of a typical injection well are shown in Figure 11. In most cases, two or more well casings are used in the injection well. The customary procedure is to drill a large diameter hole through all fresh water aquifers. Casing is then inserted in the well, and the annular space is filled with cement.

If the formation to be used for injection is known in advance, drilling proceeds to the top of that formation, where a second string of smaller pipe is cemented in the hole from top to bottom. Then a smaller hole—usually about eight inches in diameter—is drilled through the injection formation. Injection tubing is placed in the casing and sealed at the top and bottom with packers. The well is then ready for testing.

If several rock zones need to be tested, the appropriate tests are performed in each zone as the hole is being drilled. Once the hole reaches the planned total depth, casing is installed and the annular space is grouted with cement. The well is then completed by perforating the casing and the cement at the appropriate zones.

FIGURE 11: TYPICAL INJECTION WELL

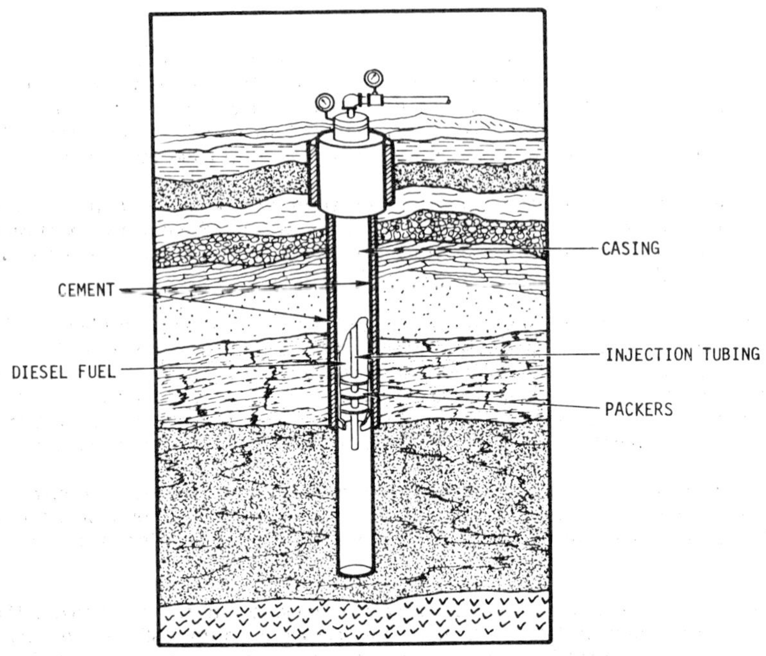

Source: PB 224 582

The annular space between the injection tubing and the casing may be filled with an inert
fluid such as diesel fuel. A pressure-recording gauge is installed to measure changes in pressure
in the annular space. Should either the tubing or the casing develop a leak, a pressure change
would be recorded. In some instances, the fluid in the annular space is maintained under
a pressure higher than that in the injection tubing. Then, if a break in the tubing occurs,
the effluent will not leak into the annular space. Wells presently in use range from about
300 feet to more than 12,000 feet in depth; the depths of a sample of 75 wells now in
use are as follows:

Depth of Well, feet	Percent of Total Wells
0 - 1,000	7
1,000 - 2,000	29
2,000 - 4,000	22
4,000 - 6,000	31
6,000 - 12,000	9
>12,000	2

A scheme developed by S.L. Sutton (7) involves disposing of contaminated liquid effluent
in the region of a body of water flowing over the surface of a salt water sand layer resting
over a plurality of other such layers stacked vertically one below the other, the bottom
layer resting on a solid bed. Figure 12 shows a suitable form of apparatus.

There is shown a body of water 10 having a bottom consisting of a plurality of layers 12
of salt water sand stacked vertically one below the other, the bottom layer resting on a
solid bed 14. A vertical concrete cylinder 16 having a central or axial bore extends between
the top surface of the uppermost layer 12 and the upper surface of bed 14.

FIGURE 12: SECTIONAL VIEW OF WASTEWATER DISPOSAL WELL INSTALLATION

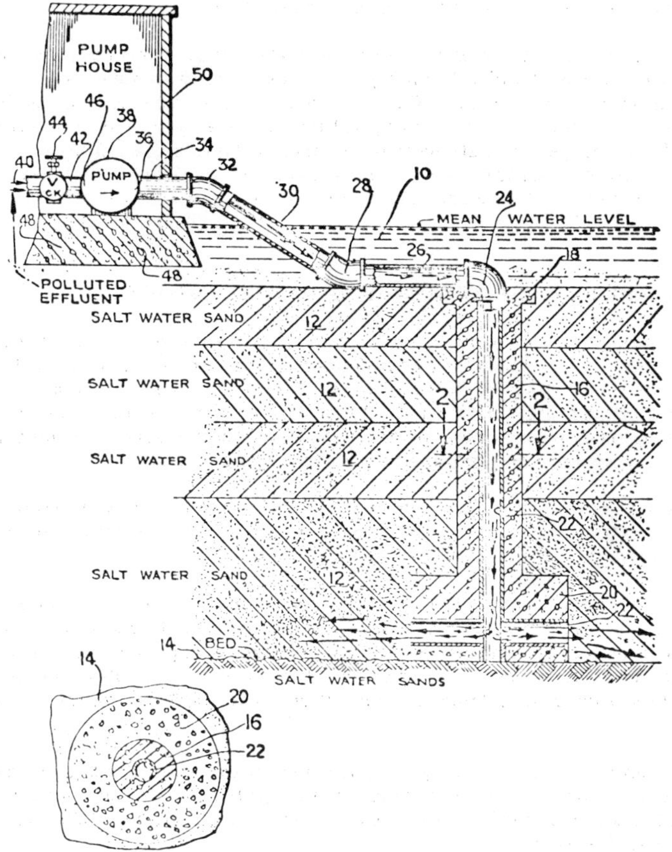

Source: U.S. Patent 3,375,666

The top of cylinder **16** has an enlarged horizontal annular section **18** which is relatively shallow. The bottom of cylinder **16** has an even more enlarged annular section **20** which is relatively thick. The center of section **20** is provided with a horizontal bore which extends all the way through and communicates with the vertical bore. Both bores are lined with hollow joined pipe sections **22** forming a continuous conduit. The top of the vertical section is connected by a right angle elbow **24** to one end of a hollow pipe section **26** horizontally lying on top of the uppermost layer **12**.

The other end of section **26** is connected by a less than 90° elbow **28** to the bottom end of a hollow pipe section **30** which extends inclinedly upward out of the body of water and is connected at its upper end by a less than 90° elbow **32** to one end of a horizontal pipe section **34**. The other end of section **34** is connected to the outlet port **36** of a high-pressure pump **38**. Polluted or contaminated or wastewater or effluent **40** is fed via horizontal pipe section **42** (containing a check valve **44**) into the inlet port **46** of the pump. The pump rests on a concrete slab **48** resting on the top layer **14** and extending above the water line and in a pump house **50** having slab **48** as a floor.

In use, the effluent is forced under pressure from the pump downward through the vertical pipe section **22** and outward through the horizontal pipe section **22** into the bottom salt water layer where the effluent mixes with the salt water and sand is dispersed in a safe and harmless manner without contaminating water **10**.

The pipe sections can be cemented above top water sands, using Lane Bryant or similar guns to shoot holes in the pipe **1** forcing the disposal into each water sand as necessary. If desired, a plurality of cylinders **16** can be used with a suitable interconnecting pipe system whereby the effluent is first discharged through one cylinder until the surrounding sands are fully charged with effluent; the system is then connected to an adjacent cylinder and so on in sequence. After the last cylinder has been so used the sand surrounding the first cylinder will have been purified by natural circulation and can be reused as required.

PRETREATMENT OF EFFLUENTS FOR DEEP WELL DISPOSAL

Often it is necessary to treat the liquid waste to avoid detrimental reactions during the injection process. The required treatment depends on the amount and size of the solids suspended in the waste, the pore sizes of the formation to be injected, the chemical compatibility of the effluent and the formation fluids and the corrosiveness of the effluent. The removal of suspended solids may not be necessary if the injection zones are composed of limestone or dolomite, since these rocks have rather large pores.

Surface storage facilities are usually included in the design of deep well disposal systems. Commonly, cement-lined sumps or steel tanks are used. An oil layer is frequently used to prevent contact of the effluent with the air.

However, if there is oil in the effluent, it is generally removed before injection because it tends to plug the injection zone. Oil may be removed by first passing the waste through a settling tank equipped with internal baffles and then through a clarifier or a sedimentation tank designed to remove the suspended solids. The sedimentation process can be accelerated by using a flocculation or coagulation agent such as aluminum sulfate or ferric sulfate.

Coagulation and sedimentation may not adequately prepare the effluent for injection. Where sand and sandstone injection zones are susceptible to plugging, filtration is included as a part of waste treatment. The filters may consists of a series of metal screens coated with diatomaceous earth.

If the waste contains microoganisms, some chemical treatment may also be required. Generally, five types of microorganisms can interfere with the subsurface injection system: slime formers, algae, iron bacteria, sulfate-reducing bacteria, and fungi.

Often the cost of the effluent treatment facilities exceeds the cost of drilling, testing, and constructing the injection well; but nevertheless this treatment is less expensive than the treatment which would be required to render the effluent acceptable for dishcarge into streams.

The characteristics of the wastewater and the nature of the receiving formation dictate the exact pretreatment requirements. Conditions that require surface treatment and those facilities commonly employed are summarized in Table 7. The pH adjustment of wastewaters is often of primary importance when considering the performance of the injection well. Neutralization is often required, but adjustments to the acid or base range are sometimes advantageous. For example the acidification of a petrochemical wastewater prior to injection in a limestone formation resulted in higher injection rates at lower wellhead pressures. In another case, rigid pH control was required, as a wastewater containing acetic acid and its chlorinated derivatives formed polymeric tars at pH values greater than five, while precipitates were formed at a pH of less than four.

TABLE 7: POLLUTANTS REQUIRING SURFACE TREATMENT PRIOR TO DEEP WELL INJECTION

Wastewater Characteristics	Surface Treatment
1. Suspended solids	settling, centrifugation, or filtration *
2. Dissolved gases	degasification by purging, air stripping, etc.
3. Colloidal matter, turbidity	coagulation and precipitation, followed by sedimentation and/or filtration
4. Ions which precipitate on contact with formation waters	possible pH adjustment, coagulation and precipitation followed by sedimentation and/or filtration
5. Oils and oil-like polymers	skimming devices, oil separators
6. Corrosive character	neutralization or installation of corrosion-resistant piping and appurtenances
7. Biological activity and growth	chlorination, filtration

* Filter aids may be required.

Source: Jones, H.R. *Environmental Control in the Organic and Petrochemical Industries*

A technique developed by J.F. Tate (9) involves disposing of certain process effluent waste streams by injecting them into subterranean formations and entails inhibiting the formation of solid precipitates which plug the subterranean formation. The method of inhibiting the formation of precipitates involves lowering the pH of the mixed streams and optionally removing any organic phase created thereby before injection.

DEEP WELL OPERATING FACTORS

Three critical factors which control the operation of an injection well are: the compatibility of the effluent with the formation and the formation fluids, the injection pressure and the injection rate.

Effluent Compatibility

The physical and chemical properties of the effluent are extremely important. The pores of the injection horizon can be plugged by suspended solids or dissolved gas contained in the effluent. Plugging can also be caused by chemical reactions between the effluent and the aquifer materials, or between the effluent and the native water in the injection zone. Plugging of the pores results in a decrease in porosity of the storage formations which, in turn, causes a reduction in well capacity.

Precautions to minimize the possibility of chemical reactions between the effluent and the aquifer materials were discussed in the preceding section. The various chemical reactions between injected and native water have been studied in some detail by various researchers. Although the exact influence of such reactions on aquifer permeability is uncertain, they will often cause undesirable results.

Sometimes when chemical reactions between injected and intersticial fluids are anticipated, the injection of a neutral fluid such as treated water has been successful in forming a buffer zone between the injected wastes and the intersticial fluids. Mathematical calculations substantiated by laboratory experiments have shown that longitudinal effluent dispersion will increase with the square root of the time or distance of flow.

The size of the neutral buffer zone necessary to prevent reaction can be related to the undiluted width of the buffer zone by taking into account the total pore space in the buffer zone. Generally, preventing a chemical reaction from occurring within 100 feet of the well bore is sufficient. For more critical conditions it may be necessary to consider the dispersion coefficient and the viscosities of the fluids. It is important, however, that the two injected fluids do not bypass each other in the formation.

Injection Pressure

The injection pressure of a well consists of the sum of the injection zone pressure and the friction head losses due to the flow of the fluid through the well and into the injection zone. It is possible to minimize such head losses by proper design of the injection well. For example, they can be held at a minimum if the formation porosity is not reduced by sedimentation or flocculation within the injection formation. Furthermore, artificial stimulation of the aquifer can sometimes increase the porosity of the formation in the vicinity of the well.

Generally, injection pressures vary with the depth of the well, but most injection pressures are less than 200 psi. The range of injection pressures for wells presently in use is shown below.

Injection Pressure, psi	Percent of Total Wells
Partial vacuum	14
0 - 150	29
150 - 300	27
300 - 600	9
600 - 1,500	20
>1,500	1

High injection pressures are usually undesirable not only because they restrict the rate of the effluent injection but also because they require considerably more expensive equipment. The well itself must then be designed to withstand higher pressures.

Injection Rate

Most of the problems that arise during the operation of an injection well are related to the rate of injection. Usually, an optimum injection rate can be established for each well; to exceed this rate might result in operational problems. Data regarding the injection rates of wells presently in use are as follows:

Injection Rate, gpm	Percent of Total Wells
0 - 50	27
50 - 100	17
100 - 200	25
200 - 400	26
400 - 800	4
>800	1

Sometimes it is necessary to increase the porosity and permeability of the injection formation to produce an increase in the injection rate. This can be achieved by acidizing or fracturing the injection zone. Acidizing increases the effective permeability of limestone and dolomite formations in the vicinity of the well bore by dissolving certain minerals such as calcium carbonate. Fracturing increases the permeability by breaking up the rock or by enlarging preexisting fractures by hydraulic or detonation methods. It is important, however, that fractures created by this process do not extend vertically through confining layers, since groundwater contamination might then occur.

DEEP WELL ECONOMICS

In 1963, the cost of complete deep well disposal installations ranged from $30,000 to $1,400,000. In the least expensive system, no surface equipment was required for treating the waste; and the well was only 1,800 feet deep. The most expensive system included a treatment plant with a clarifier, dual filters, and four positive displacement pumps for injection into a 12,000-foot-deep well.

As of 1973, the cost of deep well disposal ranged from 50 cents to $2.00 per thousand gallons injected. This cost depends upon many variables including the depth of the well, the type of well completion, injection pressure, and treatment equipment required.

Cost analysis for a typical well—about 3,000 feet deep—indicates that drilling, completing, and testing would cost less than $150,000. The cost of necessary treatment facilities would be additional expenses and dependent on the particular requirements of the wastes and the site (5).

LEGAL REQUIREMENTS

Before starting a deep well disposal project, it is essential to discuss the plan with the appropriate regulatory agency and to obtain the necessary approvals. In about 34 states the construction of waste injection wells is subject to certain requirements. Only 3 states, Missouri, Ohio and Texas, have laws specifically governing wells for the disposal of industrial wastes.

Most other states do not rule out the use of injection wells, but they often lack favorable geologic conditions and, therefore, do not have suitable sites for injection wells. Some states do not allow the use of deep well injection. If the geology is suitable and if the plans for well construction are adequate, authorization to proceed is generally given, but if a reasonable doubt exists in the minds of regulatory officials, subsequent hearing may be required.

In some cases, conditional permits are issued. For example, a permit may limit the maximum injection pressure which may be used, or it may stipulate that one or more formation pressure-monitoring wells must be included in the project.

REGIONAL CONSIDERATIONS

The specific location of a waste injection well must be evaluated by a detailed geologic subsurface investigation. However, regional geologic conditions can be used to evaluate whether certain areas are generally suitable for injection wells.

The regional favorability map (Figure 13) indicates that certain areas of the continental United States such as the Rocky Mountains are generally unsuitable for waste injection wells because igenous or metamorphic rocks lie near the ground surface (gray areas in the figure). Such rocks do not have sufficiently high porosity to warrant their use as a disposal formation. Areas underlain by extensive layers of volcanic rock (triangles on map) generally are not suitable fur waste disposal wells. Even though these rocks have porous zones, they generally contain fresh water. The waste disposal potential of the Basis and Range Provinces (see angled lines on map) is largely unknown owing to complex geologic conditions.

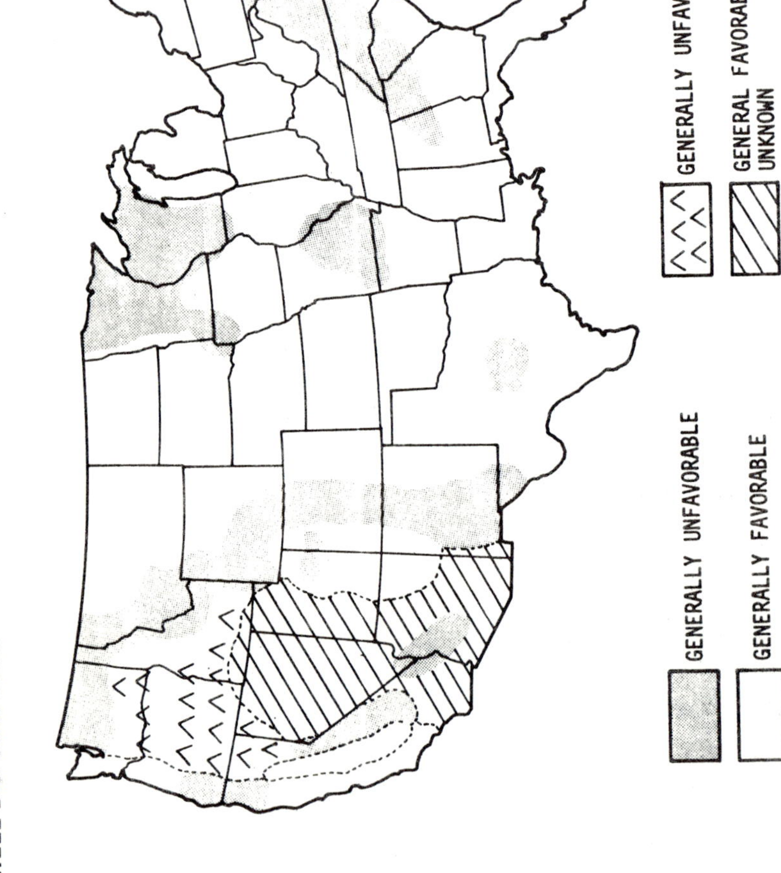

FIGURE 13: DEEP WELL DISPOSAL SITES

Source: PB 224 582

While the central valley of California is geologically well suited for the installation of disposal wells, several factors discourage their use. Thick sequences of sandstone in the region provide suitable injection horizons; but discontinuities in pervious strata, earthquake hazards, and presence of extensive oil and gas accumulations are negative factors. Although the geology of the West Coast is complex, coastal areas north of Los Angeles may contain satisfactory potential sites for injection wells.

The Atlantic and Gulf coastal plains are underlain by thick sequences of sedimentary rock which, except in oil and gas-producing areas, are generally suitable for deep well injection. The midcontinent and much of the Midwest are underlain by rather thick sequences of sedimentary rocks. Most of the injection wells in use today are located in these areas.

GEOLOGIC INVESTIGATIONS AND FIELD TESTS

The final appraisal of a disposal well site is usually determined by a two-phased geologic investigation. The first phase includes an evaluation of potential sites on the basis of available data. The second phase consists of a more detailed evaluation of subsurface conditions based on information obtained from drilling a pilot hole or the injection well.

Information sought during the first phase of the investigation and prior to the installation of an injection well includes the extent, thickness, depth, porosity, permeability, temperature, water quality and hydrostatic pressure of potential injection zones. The presence of impermeable confining beds, lateral changes in rock properties, the existence of faults or joints, and the occurrence of any mineral resource in the area must also be evaluated. Existing wells in the area which may penetrate the potential injection zones must be located since, if not properly plugged, liquid wastes could escape through these wells.

The second phase of the investigation is conducted during the drilling and testing of the injection well. Often the actual injection zone is not selected until the well has been drilled and a number of potential zones have been tested for porosity, permeability, temperature, and hydrostatic pressure, and the chemical quality of water in the potential injection zones has been evaluated. Pumping tests are used to measure the permeability, and water samples are obtained for chemical analysis.

Other important rock properties are measured by geophysical logging, drill-stem testing tolls, or by laboratory tests on the core samples. The results of these geologic investigations are used not only in evaluating the feasibility of subsurface waste disposal but also to provide basic data for designing the injection well and the optimum rate of injection.

Wastes may have different requirements relating to their extent of horizontal and vertical travel with time. For example, a chemically stable dilute waste may require only injection into, and dispersion in, a body of rapidly circulating groundwater that is recharged continually. However, biochemically unstable effluent may require a residence time within the injection zone to permit further reaction. As an example, a very concentrated waste may require a long residence time without dispersion. A system of zone classification has been proposed as follows:

Zone of Rapid Circulation: The zone of rapid circulation extends from the land surface downward some tens or hundreds of feet. Injection into this zone is normally precluded.

Zone of Delayed Circulation: This zone is generally composed of fresh water which circulates continually and freely, but is retarded sufficiently so that the residence time is of the order of several decades or even centuries. Certain innocuous wastewaters have been injected into this zone successfully with suitable monitoring.

Subzone of Lethargic Flow: In this subzone the native liquid is commonly saline and has very low movement measured in hundreds or even thousands of years. This subzone of lethargic flow is a primary zone for potential storage of the more concentrated wastes.

Stagnant Subzones: These subzones are, with few exceptions, several thousand feet below land surface, and the fluid is hydrodynamically trapped. This zone would seem ideal for injection of very toxic waste. However, the capability to accept and retain injected fluids needs to be assessed with extreme caution.

Dry Subzones: A common type of dry subzone would be a salt bed or dome in which free water is virtually nonexistent, and which may be impermeable in a finite sense. Waste injected in such a zone would be wholly isolated from natural hydrodynamic circulation. However, since movement could occur through hydrofractures, performance of a dry sub-zone under injection should be assessed cautiously.

It is recommended that research be done to coordinate the limits of the various zones mentioned and to associate such zones with the various categories of wastes. In this manner, further information can be gathered.

A process developed by F.H. Poetteman (10) is one in which contamination of underground aquifers by waste materials injected through disposal wells is limited by injecting a mobility buffer ahead of the waste material. The mobility buffer can be water containing a mobility-reducing agent, e.g., water-soluble polymer, a water-external emulsion, or any other fluid containing a mobility-reducing agent. A more favorable mobility ratio can be achieved by injecting a series of consecutively mobility-graded fluids into the aquifer. Similar ground is covered by Poetteman (11).

SUGGESTIONS FOR FUTURE USE

In the past, there has not been enough attention given to the monitoring of deep well disposal systems. It is desirable to monitor injection wells to determine the extent of travel of injected waste permitting the detection of well casing or cement failures, the escape of waste through fractured or faulted cap rocks, or through other abandoned or operating wells and the loss of permeability in the injection horizon during injection.

Monitoring is also required to determine the pressure needed to maintain a constant injection rate, since this increases with time. An increase in pressure probably indicates decreased permeability. A sudden increase in the intake rate of the injection well might indicate the opening of horizontal or vertical fractures in the injection horizon and possibly in the confining beds, or the failure of such well facilities as the casing, cement, or packers. Such monitoring activities need to be documented and be made requirements of state and federal laws relating to deep well disposal.

Related to monitoring requirements is the necessity for developing adequate planned methods and procedures to be followed to rapidly institute corrective actions in the event of a system failure. It is recommended that research be done to establish a list of the proper monitoring methods and implementation methods associated with deep well disposal and to develop procedures for instituting corrective actions in the event of a system failure.

In addition, complete operating records are required to denote quantities and types of waste injected into a particular stratum. Requirements for such records need to be part of state and federal legislation.

The use of deep well disposal techniques should be limited at the present state-of-the-art to those waste stream constituents which have low toxicity in themselves and which also do not have breakdown or expected reaction products demonstrating high toxicity. This recommendation is based primarily on the apparent lack of control over wastes following injection. Without proper and adequate monitoring techniques the migration of hazardous materials from the storage area may not be detected until there is an effect on the non-storage area (groundwater contamination, etc.) when it might be too late. Furthermore, given that an unexpected migration is detected there are currently no tested procedures which will reverse the migration or allow total recovery of the materials, or seal the peripheries to insure halting the migration.

In summary, deep well disposal methods can be utilized subject to detailed geological investigations and selection, rational selection of wastes to be so disposed and proper monitoring of the sites so that disposal can be stopped at the proper time without fear of migration.

SPECIFIC INDUSTRIAL APPLICATION AREAS

The reader is referred to particular sections of this volume which follow for application of the deep well disposal technique, specifically to Acrylonitrile Process Wastewaters (under Organic Chemical Industry Wastes) and Sodium Azide Solutions (under Inorganic Chemical Industry Wastes).

Some specific examples of disposal wells used for the subsurface injection of wastewater are listed in Table 8 from a report by Booz-Allen Applied Research, Inc. (12).

TABLE 8: PETROCHEMICAL WASTE DISPOSAL BY DEEP WELL INJECTION— TYPICAL INSTALLATIONS

Type Waste	Flow (gpm)	Depth (ft)	Injection Pressure (psia)	Formation	Required Pretreatment
Acrylonitrile and Detergent Manuf. Wastes: COD = 17,500 mg/1; pH = 5.4, SO_4 = 10,000 mg/1	650	7,203	Up to 2,000	Sat. Brine, Miocene Sands	Neutralization; Settling and Equalization in Pond; Coagulation pH Adjustment and Clarification; Gravity Sand Filters
10-15% NaC1; Diss. Metal Salts; Trace Organics; pH 7.5-8.5	500-600	4 wells:	200	Unconsolidated Brine Sands	Oil Separation; Settling; Pressure Leaf Filtration; Diatomite Filtration
Refinery and Petrochem. Cooling Water Blow-down Boiler Blow-down, Process Waters	500	1,200	75	Sandstone	
Petrochem. Waste Organic Nitrogen Nitrites COD = 20,000 ppm pH = 12 Uranium 238	400	6,700	400	Sands	Neutralization, Precipitation - Sedimentation, Filtration
Phenolic Waste: COD= 12,000 ppm; 850 ppm Phenol; 150 ppm Oil; pH 10.8	300	6,330	1,000	Sat. Brine, Miocene Sands	Neutralization with H_2SO_4; Clarifier, Pressure Sand Filter
Aromatics Phenols 1,000-2,000 ppm COD 10,000 ppm pH 10.7	300	6,100	1,000	Miocene Sands	
0.3% Acetic Acid Chlorinated Derivatives	204	3,700	2,000	Miocene Brine Sand	Cool to 150°F. Adjust pH to 4.0-5.0, Settling Coal Filter; Cartridge Filter for Solids 10
Terephthalic Acid Manuf. Cooling and Boiler Blow-down, Process Wastes Containing Organic Acids, H. C., inorganics	150	5,600		Sands	Settling, Filtration

(continued)

TABLE 8: continued

Type Waste	Flow (gpm)	Depth (ft)	Injection Pressure (psia)	Formation	Required Pretreatment
Nylon, Ammonia, Olefins, Polyolefins Refinery, Butadiene, Sytrene, Synthetic Rubber 1	Small-96	5,802	800-1,100	Limestone	Conventional Waste Treatment, 0.3% by Volume of Acid, Added Before Injection
Cuprous Ammonium Acetate from Butadiene Pond; Caustic Waste from Ethylene Prod. Caustic and Phenols from Refinery	85	4,000	1,500-2,000		Equalization, Settling
Refinery Cooling & Boiler Blowdown, Process Wastes, Brines	50	5,000	600	Sandstone	Settling and Storage
Ammonia Prod.	45	1,000	225	Sandstone	API Separator
Hydrochloric Acid	40	1,200	14.7	Sandstone	None
Detergent Product 32% HCl Benzene Chlorinated HC	35	3,400		Miocene Sands	Dilution with Equal Volume Fresh Water
Spent Alkylation Acid- 90% H_2SO_4; 7% Oil; 3% H_2O	1	5,100			
Filtrates and Distillates from Chloromycetin Manuf: $BOD_{20} = 45,000$ ppm; pH 3.5, Diss. Solids = 50,000 ppm		1,400		Saturated Brine, Sand	
Saturated NaCl, Conc. Ca-Mg, Liquors, Phenols Chloro-Phenols, Bis-Phenols, Methocel, Weak Caustic Washes		3,000		Limestone	Suspended Solids Removed

Source: PB 221 467

A process developed by D.L. Miller (13) involves initially treating wastes, prior to their discharge into underground strata, with water derived from such strata to cause chemical reaction between such water and such wastes. This is followed by the separation of the solids formed by such chemical reaction by any means known in the art, such as by settling followed by decantation of the supernatant liquid, or by filtration, or by centrifuging, or otherwise. The liquid thus clarified is then discharged into the same stratum from which the water was obtained, or into any other water-yielding stratum in which the water has substantially the same chemical composition as the water in which the industrial waste has been treated.

While this leaves for disposal the precipitates formed as a result of the process, it is found that the quantities thereof represent a very small percentage of the initial waste to be disposed of. These precipitates may be disposed of by any suitable means, such as by discharge onto a dump, or by burial, or by partial or total burning in the case of partially or completely combustible precipitates, etc.

The amounts of such precipitates are held to a virtual minimum in the practice of this process, for in the case of the use of reagents from other sources, it is necessary to over-treat and obtain large quantities of precipitates, in order to insure against possible plugging of the strata.

Table 9 shows some of the process streams from the Sharples plant and the reagents used by their pretreatment prior to deep well disposal. In the table, reagent A was groundwater obtained from the Sylvania stratum at Riverview, Michigan, by means of a well which is approximately 300 feet deep, and which leads down into this stratum. This groundwater has a pH of approximately 7.4, and contains approximately 6.8 g/l of dissolved solids.

The alkaline portion of these solids is equivalent to approximately 1.2 g of sodium hydroxide, as found by titration in the presence of phenolphthalein indicator. This water also contains traces of hydrogen sulfide. Reagent B was a solution which was compounded to closely approximate the chemical composition of groundwater occurring at a depth of approximately 3,400 feet in the so-called Prairie du Chien stratum at Riverview, Michigan. This groundwater has a total soluble solids content of approximately 175 g/l.

TABLE 9: TREATMENT OF INDUSTRIAL WASTES

Type of Waste	Reagent	Type of Precipitate
Discharge liquids from amyl chloride vent scrubbers containing traces of amyl chloride, hydrogen chloride, dichloropentanes, and miscellaneous chemicals.	A B	White solid. White solid.
Caustic water obtained from washing amyl chloride and containing traces of amyl chloride, caustic, and miscellaneous organic chemicals	A B	Dark precipitate, primarily magnesium hydroxide. Dark precipitate, primarily magnesium hydroxide.
Tank vent scrubber liquid containing small quantities of hydrogen chloride and organic chemicals.	A B	None. None.
Amyl alcohol caustic washings containing primarily amyl alcohol, caustic, and sodium chloride.	A B	Heavy flocculent precipitate, containing primarily magnesium hydroxide. White precipitate containing primarily magnesium hydroxide.
Still discharge containing traces of amyl chloride, amyl alcohol, and other organic chemicals.	A B	Brown to white precipitate, composition unknown. Brown to white precipitate, composition unknown.
Amyl acetate still discharge containing traces of amyl acetate, amyl alcohol, acetic acid, and sulfuric acid.	A B	Small amount of white precipitate, containing primarily magnesium hydroxide and sulfide. None.
Amylphenol vacuum jet discharge containing traces of phenol, amylphenol, amylene, and other organic materials.	A B	Slight turbidity. Slight turbidity.
Amylphenol acid waters.	A B	Heavy white precipitate containing primarily magnesium hydroxide. Light white precipitate.
Still discharge containing traces of amyl mercaptan, amyl chloride, sodium hydrosulfide, hydrogen sulfide, ethanol, and sodium chloride.	A B	Heavy black precipitate. Heavy black precipitate.
Still discharge containing traces of amyl chloride, amylnaphthalene, caustic, and other chemicals.	A B	Light brown precipitate. Light brown precipitate.
Ethylamines still discharge containing traces of ethylamines and tarry substances.	A B	Light amber precipitate. Light amber precipitate.
Vacuum jet discharge containing alkylaminoethanols and related organic chemicals.	A B	None. None.
Still discharge containing traces of alkylphenol and related materials.	A B	None. Light precipitate, composition unknown.
Mercaptan decanter liquid containing traces of isobutylene, tertiary dodecyl mercaptan, hydrogen sulfide, boron trifluoride, acids and related materials.	A B	White to gray precipitate containing primarily calcium sulfide and magnesium hydroxide. Heavy white precipitate.

Source: U.S. Patent 2,707,171

Its approximate equivalent in magnesium oxide is 6 g/l, in calcium oxide 30 g/l, in chlorine 108 g/l, in sulfur dioxide 0.27 g/l, in sulfur trioxide 0.12 g/l, and in bromine 0.11 g/l.

The amount of reagent used in each instance was approximately 20% by volume of the particular waste treated. The procedure was to mix the water and the waste at room temperature, and to let the mixture stand for a considerable period of time to permit settling of the precipitate formed as a result of the chemical reaction produced. This was followed by separation of the supernatant liquid from the precipitate by decantation. For test purposes, to determine whether a sufficient quantity of reagent had been used, the separated supernatant liquid was mixed with approximately 20% by volume of reagent water, and in each instance no further precipitation took place.

Figure 14 is a diagrammatic illustration in cross section of certain strata occurring to a depth of approximately 600 feet at Riverview, Michigan, as well as a diagrammatic illustration of the apparatus required for this process. In the drawings at **10** is illustrated a well which extends down through surface drift illustrated at **11**, a limestone stratum illustrated at **12**, into a sandstone stratum illustrated at **13**. At **14** is shown a pump for pumping water from well **10**. This water flows through line **15** into mixer **16**, wherein the water is mixed with waste flowing into mixer **16** through line **17**, from waste receiving tank **18**.

If desired, proportioning pumps (not shown) may be employed for proportioning the flow of water and waste to the mixer **16**. From mixer **16**, the mixture of water and waste flows through line **19** into separator **21**. In separator **21** the solid phase is separated from the liquid phase, the solids being discharged through line **22** for disposal as desired. The liquid effluent from separator **21** flows through line **23**, and is delivered to well **24** which as illustrated extends down into stratum **13**. Any other arrangement of apparatus may be employed for the intended purpose, that particularly described being merely for purposes of illustration.

**FIGURE 14: SCHEME FOR DEEP WELL DISPOSAL OF SYNTHETIC ORGANIC
CHEMICAL PLANT WASTES**

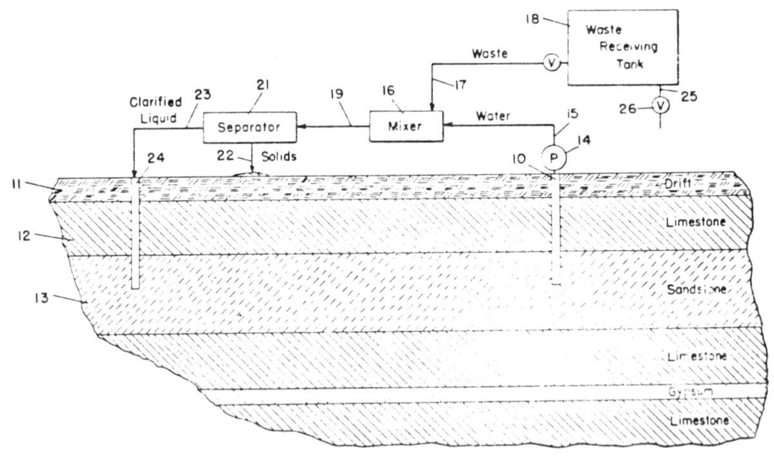

Source: U.S. Patent 2,707,171

The residence of the water and waste in mixer **16** in the apparatus illustrated is, of course, such as to permit substantially complete chemical reaction between the water and the waste. Mixer **16** may take any suitable form, and may be a continuous mixer, or a tank having stirring paddles, or otherwise. Mixer **16** and separator **21** may be combined into a single device, such as a tank having staggered baffles arranged in a manner so as to permit the settling of solids to the bottom of the tank and their removal therefrom.

Any suitable commercial clarifier may be employed for both mixing and separating. Thus, the reaction may be conducted continuously, semicontinuously, or batchwise, such as in a holding or mixing tank, or in any other suitable manner. As pointed out, the liquid effluent from the separating stage may be returned to the underground stratum under any desired pressure, generated by means of a pump, or otherwise, the hydraulic head developed in a well, such as illustrated in **24**, for example, frequently in itself being suitable for the purpose.

The reaction between the water and the waste may be conducted at any desired temperature. Ordinary atmospheric temperature is suitable for the purpose. However, if desired, the reaction may be conducted at an elevated temperature by the application of heat, or at a reduced temperature, or at the temperature resulting from the mixture of the water and the waste, the water usually being at a somewhat lower temperature than atmospheric, and the waste frequently being at a somewhat higher temperature, particularly if the latter comes from a process which is operated at elevated temperature.

From a strictly scientific standpoint, the reaction might be carried out at the temperature of the water in the stratum from which it is obtained, for this will be the temperature of the clarified waste mixture when it is returned to the same stratum. However, it is found that the temperature at which the reaction is conducted is not particularly critical with the ordinary industrial wastes. However, the temperature of the reaction is preferably not so high as to substantially convert the chemicals naturally occurring in the water into some other form, such as the conversion of calcium and/or magnesium bicarbonates into their carbonates, the latter in themselves being insoluble.

As pointed out above, any desired proportion of water to waste may be employed, the substantially minimum proportion of water to waste being that necessary to bring out complete reaction. Thus substantial excesses of water may be employed if desired for any reason. As further pointed out, any water-yielding stratum may be employed. As illustrated in the drawing, a sandstone stratum is used, the particular stratum illustrated being the Sylvania stratum as it occurs at Riverview, Michigan.

Limestone strata, if sufficiently porous to be water-yielding, can be employed as well as other water-yielding strata, such as shale when water-yielding. A water-yielding stratum is, of course, a stratum which is sufficiently porous to yield water for pumping purposes. While strata capable of yielding water in relatively large quantities are preferred, it will be understood that any stratum capable of yielding water in pumpable quantities comes within the broad scope of the process.

It will be understood that industrial wastes vary widely in composition, depending upon the waste of a particular process. It will also be understood that the chemical composition of various waters derived from various strata, and from various geographical areas, will vary widely. In fact, where a single stratum extends over many square miles, there may be some difference in the chemical composition of the water derived therefrom at widely separated points.

In view of this, it is preferred to return the clarified waste mixture to the stratum from which the water is derived at a point within a radius not greater than one mile, and more preferably within a radius of one-fourth mile. In fact, the clarified waste mixture might be returned to the same well from which the reaction water is obtained, particularly when the treatment and disposal of wastes is not carried on continuously, but with substantial time intervals elapsing between treatments.

Wastes may be treated individually or in admixture. The latter is sometimes preferred when the clarified waste mixtures are discarded into a stratum through a single well. This is particularly true when different wastes are capable of chemical reaction between themselves to form insoluble precipitates. This avoids chemical reaction of a character capable of possible plugging within a stratum. Thus different wastes may be collected in waste receiving tank **18**, wherein any precipitate formed may be permitted to settle. Such precipitate may be removed by any suitable means, for example through line **25** controlled by valve **26**.

Obviously, any other means for separating precipitate from the mixed wastes may be employed. Thus a pond containing the treating water can be employed to receive the different wastes, and the supernatant liquid may be withdrawn, such as by means of a weir. If the treating water is pumped continuously into the pond, flow over the weir may be continuous.

The precipitate collects on the bottom of the pond. When the pond becomes so full of precipitate as to seriously interfere with the operation of the process, a new pond can be placed in use, and the old pond covered with earth, or the old pond may be cleaned out and placed back in use. In such an arrangement, the pond takes the place of receiving tank **18**, mixer **16**, and separator **21**, as will be obvious.

With respect to the ability of underground strata to absorb clarified waste mixtures, it has been geologically estimated that 200,000 gpd of clarified waste mixture can be pumped into the Sylvania stratum at Riverview, Michigan, and that 50 years hence, evidences of clarified waste will not be detected beyond the radius of one mile from the point at which such clarified waste mixtures are introduced into this stratum.

A report by Donaldson (14) gives details of 14 examples of deep well disposal systems for industrial wastes. Deep well disposal of industrial wastes has been discussed in detail with reference to pretreatment of wastes and well design and cost considerations by B.G. Liptak (15).

REFERENCES

(1) M. Sittig, *Pollution Control in the Organic Chemical Industry,* Park Ridge, N.J., Noyes Data Corp. (1974).

(2) *Environmental Science & Technology,* 6, No. 2, 120-122 (Feb. 1972).

(3) B. Greek, *Environmental Science & Technology,* 7, No. 13, 1106-1108 (Dec. 1973).

(4) *Environmental Science & Technology,* 9, No. 1, 24 (Jan. 1975).

(5) R.S. Ottinger, J.L. Blumenthal, D.F. Dal Porto, G.I. Gruber, M.J. Santy and C.C. Shih; *Recommended Methods of Reduction, Neutralization, Recovery or Disposal of Hazardous Wastes,* Vol III, Report PB 224 582, Springfield, Va., Nat Tech Information Service (Aug. 1973).

(6) D.L. Warner, *Survey of Industrial Waste Injection Wells* (3 volumes), Reports AD-756, 641-3, Springfield, Va., Nat Tech Information Service (June 1972).

(7) S.L. Sutton: U.S. Patent 3,375,666 (Apr. 2, 1968).

(8) H.R. Jones, *Environmental Control in the Organic and Petrochemical Industries,* Park Ridge, N.J., Noyes Data Corp. (1971).

(9) J.F. Tate; U.S. Patent 3,817,859; June 18, 1974; assigned to Texaco, Inc.

(10) F.H. Poetteman; U.S. Patent 3,606,925; September 21, 1971; assigned to Marathon Oil Co.

(11) F.H. Poetteman; U.S. Patent 3,722,593; Mar. 27, 1973; assigned to Marathon Oil Co.

(12) Booz-Allen Applied Research, Inc., *A Study of Hazardous Waste Materials, Hazardous Effects and Disposal Methods,* Vol. III, Report PB 221 467, Springfield, Va., Nat Tech Information Service (July 1973).

(13) D.L. Miller; U.S. Patent 2,707,171; Apr. 26, 1955; assigned to Sharples Chemicals, Inc.

(14) E.C. Donaldson, *Subsurface Disposal of Industrial Wastes in the United States,* Information Circular 8212, Washington, D.C., U.S. Bureau of Mines (1964).

(15) B.G. Liptak; *Environmental Engineers Handbook,* Radnor, Pa., Chilton Book Co. (1974).

INCINERATION

DEFINITION OF TERMS

Incineration is a controlled process that uses combustion to convert a waste to a less bulky, less toxic, or less noxious material. The principal products of incineration from a volume standpoint are carbon dioxide, water and ash while the products of primary concern, due to their environmental effects are compounds containing sulfur, nitrogen and halogens. When the combustion products from an incineration process contain undesirable compounds, a secondary treatment such as afterburning, scrubbing or filtration is required to lower concentrations to acceptable levels prior to atmospheric release. The solid and liquid effluents from the secondary treatment processes will occasionally require treatment prior to ultimate disposal.

FIGURE 15: TYPES OF INCINERATORS AND THEIR APPLICATIONS

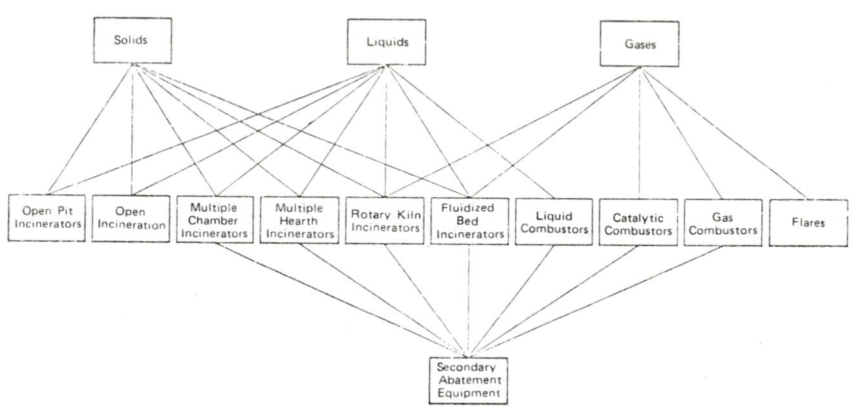

Source: PB 224 582

The later sections of this volume will discuss in detail the operating principles, process applicability, process design, and process economics of each of the various types of incineration systems. Incinerators are generally classified by the form of waste which they burn. There are ten basic types of incinerator units according to R.S. Ottinger et al (1): open pit incinerators, open burning, multiple chamber incinerators, multiple hearth incinerators, rotary kiln incinerators, fluidized bed incinerators, liquid combustors, catalytic combustors, gas combustors and stack flares. The type of waste for which each of these incineration units is best suited is detailed diagrammatically (Figure 15). The types of incinerator systems which are amenable to secondary abatement equipment application are also shown.

To the types of incinerators listed above might be added afterburners (a type of incinerator which can be used to follow other incinerators to insure total disposal) and molten salt incinerators (a newer type of device). Some of the various types of incinerators used in the incineration of process wastes have been reviewed by J.I. Frankel (2).

CHEMICALS WHICH CAN BE DISPOSED OF BY INCINERATION

According to R.S. Ottinger et al (3), waste organic chemical stream constituents which may be subjected to ultimate disposal in concentrated form by controlled incineration are:

Acetaldehyde

Acetic Acid

Acetic Anhydride

Acetone

Acetone Cyanohydrin: oxides of nitrogen are removed from the effluent gas by scrubbers and/or thermal devices.

Acetonitrile: oxides of nitrogen are removed from the effluent gas by scrubbers and/or thermal devices.

Acetyl Chloride

Acetylene

Acridine: oxides of nitrogen are removed from the effluent gas by scrubber, catalytic or thermal device.

Acrolein: 1500°F, 0.5 sec minimum for primary combustion; 2000°F, 1.0 sec for secondary combustion, combustion products CO_2 and H_2O.

Acrylic Acid

Acrylonitrile: NO_x removed from effluent gas by scrubbers and/or thermal devices.

Adipic Acid

Allyl Alcohol

Allyl Chloride: 1800°F, 2 seconds minimum.

Aminoethylethanolamine: incinerator is equipped with a scrubber or thermal unit to reduce NO_x emissions.

Amyl Acetate

Amyl Alcohol

Aniline: oxides of nitrogen are removed from the effluent gas by scrubber, catalytic or thermal device.

Anthracene

Benzene

Benzene Sulfonic Acid: incineration followed by scrubbing to remove the SO_2 gas.

Benzoic Acid

Benzyl Chloride: 1500°F, 0.5 second minimum for primary combustion; 2200°F, 1.0 second for secondary combustion;

elemental chlorine formation may be alleviated through injection of steam or methane into the combustion process.

Butadiene

Butane

Butanols

1-Butene

Butyl Acrylate

n-Butylamine: incinerator is equipped with a scrubber or thermal unit to reduce NO_x emissions.

Butylenes

Butyl Phenol

Butyraldehyde

Camphor

Carbolic Acid (Phenol)

Carbon Disulfide: a sulfur dioxide scrubber is necessary when combusting significant quantities of carbon disulfide.

Carbon Monoxide

Carbon Tetrachloride: preferably after mixing with another combustible fuel; care must be exercised to assure complete combustion to prevent the formation of phosgene; an acid scrubber is necessary to remove the halo acids produced.

Chloral Hydrate: same as carbon tetrachloride.

Chlorobenzene: same as carbon tetrachloride.

Chloroform: same as carbon tetrachloride.

Creosote

Cresol

Crotonaldehyde

Cumene

Cyanoacetic Acid: oxides of nitrogen are removed from the effluent gas by scrubbers and/or thermal devices.

Cyclohexane

Cyclohexanol

Cyclohexanone

Cyclohexylamine: incinerator is equipped with a scrubber or thermal unit to reduce NO_x emissions.

Decyl Alcohol

Di-n-Butyl Phthalate

Dichlorobenzene: incineration, preferably after mixing with another combustible fuel. Care must be exercised to assure complete combustion to prevent the formation of phosgene. An acid scrubber is necessary to remove the halo acids produced.

Dichlorodifluoromethane (Freon): same as dichlorobenzene

Dichloroethyl Ether: same as dichlorobenzene

Dichloromethane: (methylene chloride) same as dichlorobenzene

1,2-Dichloropropane: same as dichlorobenzene

Dichlorotetrafluoroethane: same as dichlorobenzene.

Dicyclopentadiene

Diethanolamine: incinerator is equipped with a scrubber or thermal unit to reduce NO_x emissions.

Diethylamine: same as diethanolamine.

Diethylene Glycol

Diethyl Ether: concentrated waste containing no peroxides: discharge liquid at a controlled rate near a pilot flame. Concentrated waste containing peroxides: perforation of a container of the waste from a safe distance followed by open burning.

Diethyl Phthalate

Diethylstilbestrol

Diisobutylene

Diisobutyl Ketone

Diisopropanolamine: incinerator is equipped with a scrubber or thermal unit to reduce NO_x emissions.

Dimethylamine: same as diisopropanolamine.

Dimethyl Sulfate: incineration ($1800°F$, 1.5 seconds minimum) of dilute, neutralized dimethyl sulfate waste is recommended. The incinerator must be equipped with efficient scrubbing devices for oxides of sulfur.

2,4-Dinitroaniline: controlled incineration whereby oxides of nitrogen are removed from the effluent gas by scrubber, catalytic or thermal device.

Dinitrobenzol: incineration ($1800°F$, 2.0 seconds minimum) followed by removal of the oxides of nitrogen that are formed using scrubbers and/or catalytic or thermal devices. The dilute wastes should be concentrated before incineration.

Dinitrocresol: incineration ($1100°F$ minimum) with adequate scrubbing and ash disposal facilities.

Dinitrophenol: incinerated ($1800°F$, 2.0 seconds minimum) with adequate scrubbing equipment for the removal of NO_x.

Dinitrotoluene: pretreatment involves contact of the dinitrotoluene contaminated waste with $NaHCO_3$ and solid combustibles followed by incineration in an alkaline scrubber equipped incinerator unit.

Dioxane: concentrated waste containing no peroxides; discharge liquid at a controlled rate near a pilot flame. Concentrated waste containing peroxides: perforation of a container of the waste from a safe distance followed by open burning.

Dipropylene Glycol

Dodecylbenzene

Epichlorohydrin: incineration, preferably after mixing with another combustible fuel. Care must be exercised to assure complete combustion to prevent the formation of phosgene. An acid scrubber is necessary to remove the halo acids produced.

Ethane

Ethanol

Ethanolamine: controlled incineration; incinerator is equipped with a scrubber or thermal unit to reduce NO_x emissions.

Ethyl Acetate

Ethyl Acrylate

Ethylamine: controlled incineration; incinerator is equipped with a scrubber or thermal unit to reduce NO_x emissions.

Ethylbenzene

Ethyl Chloride: incineration, preferably after mixing with another combustible fuel. Care must be exercised to assure complete combustion to prevent the formation of phosgene. An acid scrubber is necessary to remove the halo acids produced.

Ethylene

Ethylene Cyanohydrin: controlled incineration (oxides of nitrogen are removed from the effluent gas by scrubbers and/or thermal devices).

Ethylene Diamine: same as ethylene cyanohydrin.

Ethylene Dibromide: controlled incineration with adequate scrubbing and ash disposal facilities.

Ethylene Dichloride: incineration, preferably after mixing with another combustible fuel. Care must be exercised to assure complete combustion to prevent the formation of phosgene. An acid scrubber is necessary to remove the halo acids produced.

Ethylene Glycol

Ethylene Glycol Monoethyl Ether: concentrated waste containing no peroxides; discharge liquid at a controlled rate near a pilot flame. Concentrated waste containing peroxides: perforation of a container of the waste from a safe distance followed by open burning.

Ethyl Mercaptan: incineration ($2000°F$) followed by scrubbing with a caustic solution.

Fatty Acids

Formaldehyde

Formic Acid

Furfural

Glycerin

n-Heptane

Hexamethylene Diamine: incinerator is
equipped with a scrubber or thermal
unit to reduce NO_x emissions.

Hexane

Hydroquinone: incineration ($1800°F$, 2.0
sec. minimum) then scrub to remove
harmful combustion products.

Isobutyl Acetate

Isopentane

Isophorone

Isoprene

Isopropanol

Isopropyl Acetate

Isopropyl Amine: controlled incineration
(incinerator is equipped with a scrubber
or thermal unit to reduce NO_x emissions).

Isopropyl Ether: concentrated waste containing no peroxides; discharge liquid
at a controlled rate near a pilot flame.
Concentrated waste containing peroxides:
perforation of a container of the waste
from a safe distance followed by open
burning.

Maleic Anhydride: controlled incineration:
care must be taken that complete oxidation to nontoxic products occurs.

Mercury Compounds:(Organic): incineration
followed by recovery/removal of mercury
from the gas stream.

Mesityl Oxide

Methanol

Methyl Acetate

Methyl Acrylate

Methyl Amine: controlled incineration (incinerator is equipped with a scrubber or
thermal unit to reduce NO_x emissions).

Methyl Amyl Alcohol

n-Methylaniline: controlled incineration
whereby oxides of nitrogen are removed
from the effluent gas by scrubber, catalytic or thermal device.

Methyl Bromide: controlled incineration
with adequate scrubbing and ash disposal
facilities.

Methyl Chloride: same as methyl bromide

Methyl Chloroformate: incineration, preferably after mixing with another combustible
fuel. Care must be exercised to assure
complete combustion to prevent the formation of phosgene. An acid scrubber
is necessary to remove the halo acids
produced.

Methyl Ethyl Ketone

Methyl Formate

Methyl Isobutyl Ketone

Methyl Mercaptan: incineration followed
by effective scrubbing of the effluent
gas.

Methyl Methacrylate Monomer

Morpholine: controlled incineration (incinerator is equipped with a scrubber or
thermal unit to reduce NO_x emissions).

Naphtha

Naphthalene

β-Naphthylamine: controlled incineration
whereby oxides of nitrogen are removed
from the effluent gas by scrubber, catalyst or thermal device.

Nitroaniline: incineration ($1800°F$, 2.0 seconds minimum) with scrubbing for NO_x
abatement.

Nitrobenzene: same as nitroaniline

Nitrocellulose: incinerator is equipped with
scrubber for NO_x abatement.

Nitrochlorobenzene: incineration ($1500°F$,
0.5 second for primary combustion;
$2200°F$, 1.0 second for secondary combustion). The formation of elemental chlorine can be prevented through injection
of steam or methane into the combustion
process. NO_x may be abated through the
use of thermal or catalytic devices.

Nitroethane: incineration, large quantities of
material may require NO_x removal by
catalytic or scrubbing processes.

Nitromethane: same as nitroethane

Nitrophenol: controlled incineration: care
must be taken to maintain complete combustion at all times. Incineration of large
quantities may require scrubbers to control
the emission of NO_x.

Nitropropane: same as nitroethane

4-Nitrotoluene: same as nitrophenol

Nonyl Phenol

Octyl Alcohol

Oleic Acid

Oxalic Acid: pretreatment involves chemical
reaction with limestone or calcium oxide
forming calcium oxalate. This may then
be incinerated utilizing particulate collection equipment to collect calcium oxide
for recycling.

Paraformaldehyde

Pentachlorophenol: incineration ($600°$ to
$900°C$) coupled with adequate scrubbing
and ash disposal facilities.

n-Pentane

Perchloroethylene: incineration, preferably
after mixing with another combustible
fuel. Care must be exercised to assure
complete combustion to prevent the formation of phosgene. An acid scrubber
is necessary to remove the halo acids produced.

Phenylhydrazine Hydrochloride: controlled
incineration whereby oxides of nitrogen
are removed from the effluent gas by
scrubber, catalytic or thermal device.

Phthalic Anhydride

Polychlorinated Biphenyls (PCBs): incineration ($3000°F$) with scrubbing to remove
any chlorine containing products.

Polypropylene Glycol Methyl Ether: concentrated waste containing no peroxides: discharges liquid at a controlled rate near a
pilot flame. Concentrated waste containing
peroxides: perforation of a container of the

waste from a safe distance followed by open burning.

Polyvinyl Chloride: incineration, preferably after mixing with another combustible fuel. Care must be exercised to assure complete combustion to prevent the formation of phosgene. An acid scrubber is necessary to remove the halo acids produced.

Propane

Propionaldehyde

Propionic Acid

Propyl Acetate

Propyl Alcohol

Propyl Amine: controlled incineration (incinerator is equipped with a scrubber or thermal unit to reduce NO_x emissions).

Propylene

Propylene Oxide: concentrated waste containing no peroxides: discharge liquid at a controlled rate near a pilot flame. Concentrated waste containing peroxides: perforation of a container of the waste from a safe distance followed by open burning.

Pyridine: controlled incineration whereby oxides of nitrogen are removed from the effluent gas by scrubber, catalytic or thermal devices.

Quinone: controlled incineration (1800°F, 2.0 seconds minimum).

Salicylic Acid

Sorbitol

Styrene

Tetrachloroethane: incineration, preferably after mixing with another combustible fuel. Care must be exercised to assure complete combustion to prevent the formation of phosgene. An acid scrubber is necessary to remove the halo acids produced.

Tetraethyl Lead: controlled incineration with scrubbing for collection of lead oxides which may be recycled or landfilled.

Tetrahydrofuran: concentrated waste containing peroxides: perforation of a container of the waste from a safe distance followed by open burning.

Tetrapropylene

Toluene

Toluene Diisocyanate: controlled incineration (oxides of nitrogen are removed from the effluent gas by scrubbers and/or thermal devices).

Toluidine: same as toluene diisocyanate.

Trichlorobenzene: incineration, preferably after mixing with another combustible fuel. Care must be exercised to assure complete combustion to prevent the formation of phosgene. An acid scrubber is necessary to remove the halo acids produced.

Trichloroethane: same as trichlorobenzene.

Trichloroethylene: same as trichlorobenzene.

Trichlorofluoromethane: same as trichlorobenzene.

Triethanolamine: controlled incineration (incinerator is equipped with a scrubber or thermal unit to reduce NO_x emissions).

Triethylamine: same as triethanolamine.

Triethylene Glycol

Triethylene Tetramine: same as triethanolamine.

Turpentine

Urea: same as triethanolamine.

Vinyl Acetate

Vinyl Chloride: incineration, preferably after mixing with another combustible fuel. Care must be exercised to assure complete combustion to prevent the formation of phosgene. An acid scrubber is necessary to remove the halo acids produced.

Xylene

Also according to R.S. Ottinger et al (3), inorganic chemicals which may be disposed of (after indicated pretreatment in some cases) by controlled incineration are:

Boron Hydrides: with aqueous scrubbing of exhaust gases to remove B_2O_3 particulates.

Fluorine: pretreatment involves reaction with a charcoal bed. The product of the reaction is carbon tetrafluoride which is usually vented. Residual fluorine can be combusted by means of a fluorine-hydrocarbon air burner followed by a caustic scrubber and stack.

Hydrazine: controlled incineration with facilities for effluent scrubbing to abate any ammonia formed in the combustion process.

Hydrazine/Hydrazine Azide: the blends should be diluted with water and sprayed into an incinerator equipped with a scrubber.

Mercuric Chloride: incineration followed by recovery/removal of mercury from the gas stream.

Mercuric Nitrate: same as mercuric chloride.

Mercuric Sulfate: same as mercuric chloride.

Phosphorus (white or yellow): controlled incineration followed by alkaline scrubbing and particulate removal equipment.

Sodium Azide: disposal may be accomplished by reaction with sulfuric acid solution and sodium nitrate in a hard rubber vessel. Nitrogen dioxide is generated by this reaction and the gas is run through a scrubber before it is released to the atmosphere. Controlled incineration is also acceptable (after mixing with other combustible wastes) with adequate scrubbing and ash disposal facilities.

Sodium Formate: pretreatment involves conversion to formic acid followed by controlled incineration.

Sodium Oxalate: pretreatment involves conversion to oxalic acid followed by controlled incineration.

Sodium-Potassium Alloy: controlled incineration with subsequent effluent scrubbing.

Further, according to R.S. Ottinger et al (3), waste pesticide streams which may be subjected to ultimate disposal by incineration are:

Aldrin: (1500°F, 0.5 seconds minimum for primary combustion; 3200°F, 1.0 second for secondary combustion) with adequate scrubbing and ash disposal facilities.

Chlordane: same as aldrin.

DDD: incineration (1500°F, 0.5 second minimum for primary combustion; 2200°F, 1.0 second for secondary combustion) with adequate scrubbing and ash disposal facilities.

DDT: same as DDD.

Demeton: same as DDD.

2,4-D: same as DDD.

Dieldrin: same as aldrin.

Guthion: same as DDD.

Heptachlor: same as aldrin.

Hexachlorophene: incineration, preferably after mixing with another combustible fuel. Care must be exercised to assure complete combustion to prevent the formation of phosgene. An acid scrubber is necessary to remove the halo acids produced.

Methyl Parathion: same as DDD.

Parathion: same as DDD.

Finally, according to R.S. Ottinger et al (3), ordnance waste streams which may be subjected to ultimate disposal by incineration are:

Ammonium Picrate: incineration followed by adequate particulate abatement and wet scrubbing equipment.

1,2,4-Butanetriol Trinitrate: the current method of absorption in sawdust, wood pulp or fullers earth followed by open pit burning is feasible but unsatisfactory because of the NO_x evolved. Methods currently under investigation for minimum environmental impact include bacterial degradation and controlled incineration with afterburners and scrubbers for abatement of NO_x.

Chlorates with Red Phosphorus: incineration followed by effluent scrubbers to abate NO_x, P_4O_{10}, HCl, SO_2 and metal oxides.

Chloropicrin: incineration (1500°F, 0.5 second minimum for primary combustion; 2200°F, 1.0 second for secondary combustion) after mixing with other fuel. The formation of elemental chlorine may be prevented by injection of steam or using methane as a fuel in the process.

Copper Chlorotetrazole: controlled combustion employing a rotary kiln incinerator equipped with appropriate scrubbing devices. The explosive is fed to the incinerator as a slurry in water. The scrubber effluent would require treatment for recovery of particulate metal compounds formed as combustion products.

Diazodinitrophenol: incinerator is equipped with suitable afterburner or alkaline scrubbing systems for the abatement of the NO_x liberated.

Dipentaerythritol Hexanitrate: controlled incineration in rotary kiln incinerators equipped with afterburner or flue gas scrubbers.

GB (Nonpersistent Nerve Gas): incineration followed by adequate gas scrubbing equipment; chemical reaction with sodium hydroxide.

Gelatinized Nitrocellulose (PNC): controlled incineration in rotary kiln incinerators equipped with afterburners or flue gas scrubbers.

Glycerolmonoacetate Trinitrate (GLTN): current method of absorption in sawdust, wood pulp or fullers earth followed by open pit burning is feasible but unsatisfactory because of the NO_x evolved. Methods currently under investigation for minimum environmental impact include bacterial degradation and controlled incineration with afterburners and scrubbers for abatement of NO_x.

Glycol Dinitrate (DDN): controlled incineration in the scrubber equipped Deactivation Furnace incinerator (The Chemical Agent Munition Disposal System).

Gold Fulminate: controlled combustion employing a rotary kiln incinerator equipped
with appropriate scrubbing devices. The explosive is fed to the incinerator as a
slurry in water. The scrubber effluent would require treatment for recovery of
particulate metal compounds formed as combustion products.

Lead 2,4-Dinitroresorcinate (LDNR): controlled combustion—the lead dinitroresorcinate
is fed to the incinerator as slurry in water. The scrubber effluent requires treatment
for recovery of the particulate lead oxide formed as a product of combustion; U.S.
Army Materiel Command's Deactivation Furnace.

Lead Styphnate: controlled incineration—the lead styphnate is fed to the incinerator as
a slurry in water. The scrubber effluent would then require treatment for recovery
of the particulate lead oxide formed as a combustion product.

Mannitol Hexanitrate: incineration followed by an afterburner to abate NO_x, and cyclones
and scrubbing towers for removal of metallic dusts and fumes.

Mercuric Fulminate: incineration (Army Materiel Command's Deactivation Furnace) followed
by caustic or soda ash gas scrubbing. The mercury is removed from the scrubbing
solution.

Nitrogen Mustards: incineration—combustion products are carbon dioxide, water, HCl and
nitrogen oxides. The nitrogen oxides require scrubbing or reduction to nitrogen and
oxygen before the combustion gases are released to the atmosphere.

Nitroglycerin: incineration—exit gases should be scrubbed in a packed tower with a solution
of caustic soda or soda ash. (U.S. Army Materiel Command Deactivation Furnace)

Pentaerythritol Tetranitrate (PETN): The PETN is dissolved in acetone and incinerated.
The incinerator should be equipped with an afterburner and a caustic soda solution
scrubber.

Picric Acid: controlled incineration in a rotary kiln incinerator equipped with particulate
abatement and wet scrubber devices.

Silver Styphnate: controlled combustion employing a rotary kiln incinerator equipped
with appropriate scrubbing devices. The explosive is fed to the incinerator as
a slurry in water. The scrubber effluent would require treatment for recovery of
particulate metal compounds formed as combustion products.

Silver Tetrazene: same as silver styphnate.

Smokeless Powder: controlled incineration—incinerator is equipped with scrubber for NO_x
abatement.

Sulfur Mustards: sulfur mustard may be dissolved in gasoline and incinerated using the
U.S. Army Materiel Command's Deactivation Furnace (Chemical Agent Munition
Disposal System). The combustion products are removed by alkaline scrubbing.

TNT: TNT is dissolved in acetone and incinerated. The incinerator should be equipped
with an afterburner and a caustic soda solution scrubber.

Tear Gas (CN) (Chloroacetophenone): tear gas-containing waste is dissolved in an organic
solvent and sprayed into an incinerator equipped with an afterburner and alkaline
scrubber; reaction with sodium sulfide in an alcohol-water solution. Hydrogen sulfide is liberated and collected by an alkaline scrubber.

Tear Gas, Irritant: hydrolysis in 95% ethanol and 5% water followed by incineration and
then by a caustic scrubber.

Tetranitromethane: open burning at remote burning sites. This procedure is not entirely
satisfactory since it makes no provision for the control of the toxic effluents,
NO_x and HCN. Suggested procedures are to employ modified closed pit burning,
using blowers for air supply and passing the effluent combustion gases through
wet scrubbers.

VX (persistent nerve gas): incineration followed by adequate gas scrubbing equipment.

TYPES OF WASTES TO BE HANDLED

Gaseous Waste Incineration

The type and form of waste will dictate the type of combustion unit required. A number
of control methods have been successfully developed for applications where the pollutants
are in the form of fume or gas. If the waste gas contains organic materials which are combustible, then incineration should be considered as a final method of disposal. Direct flame,
thermal, or catalytic oxidation of such wastes can produce an effluent of carbon dioxide,
nitrogen, and water vapor which can be vented safely to the atmosphere. Economic considerations are paramount in the selection of incineration systems because of the high fuel
costs when concentrations of organic constituents are low.

Direct flame incineration is used normally with materials which are at or near their lower combustibility limit. In a well-designed commercial combustor or burner, gases having heating values as low as 100 Btu/ft^3 can be burned without auxiliary fuel. Gases having even lower heating values, which are preheated to 600° or 700°F, often will sustain combustion without help from auxiliary fuel. Hydrogen cyanide, which is an extremely toxic gas, may be burned in air; carbon monoxide, which is also a deadly gas and a by-product of many partial combustion reactions, can be burned in this manner. Solvent vapors mixed in high quantities with air may produce a combustible mixture which can be burned in a conventional forced draft combustion system.

When the amount of combustible material in the mixture is below the lower flammable limit, it may be necessary to add small quantities of natural gas or other auxiliary fuel to sustain combustion in the burner. But in either case, whether the material burns with or without the assistance of auxiliary fuel, combustion occurs at high temperatures (about 2500°F), good mixing is achieved with the oxygen in the air, and the resultant products of combustion generally are carbon dioxide, nitrogen, and water vapor. Here the contaminant, whether it is a solvent vapor or pure gas, is serving as a part of the fuel. It is contributing a significant portion of the total heat released to the system and can be burned with a minimum of auxiliary fuel and therefore a minimum of operating cost. Direct flame combustion should be employed only where the amount of auxiliary fuel needed to sustain combustion is low and where the contaminant supplies at least 50% of the fuel value of the mixture.

Equipment for direct flame incineration may be a conventional industrial burner or combustor and combustion chamber (either forced or induced draft), or it may be a flare type burner as found in many petroleum refineries and petrochemical plants. Most waste gas incineration problems involve mixtures of organic material and air in which the amount of organic material is very small. This means that if it were injected directly through a burner, along with auxiliary fuel such as natural gas, the amount of natural gas required to achieve complete combustion would be quite high. Most conventional industrial burners require temperatures of 2200°F or greater to sustain combustion, whereas thermal incineration can be carried out at much lower temperatures, sometimes as low as 900°F, but generally between 1000° and 1500°F.

Weak mixtures of organic material and air will usually have very low heating values, on the order of 1 to 20 Btu/ft^3. Some of the most common applications may be found in drying ovens which drive off a solvent or plasticizer in low concentrations in air, or form lithographing ovens or other process drying operations. Here it is more economical to heat a combustion chamber, using a conventional fuel in an industrial burner, and inject the contaminated air into this chamber just downstream from the burner flame, or even into the burner flame. Usually the waste gas is essentially air and therefore contains enough oxygen to complete combustion of the organic contaminant. But in some cases, where sufficient oxygen is not present in the fume, it can be added by means of a fan or blower, either by premixing with the fume or by injecting into the secondary combustion chamber along with the fume.

Incineration systems for thermal oxidation of gaseous wastes are of many different types and forms. Some utilize line burners when the fume contains sufficient oxygen for its own combustion. Here the waste gas passes over and through the flame of the line burner in a refractory lined duct. Other systems utilize an external burner, either natural, forced draft, or aspirating type. In this system the flame passes into the duct from the burner mounted in the duct wall, causing turbulence in the chamber, and the contaminated air passes through and around the flame and is heated to the reaction temperature. Such units may be vertical or horizontal and may be induced or forced draft, depending upon the physical arrangement most desirable for the system.

Catalytic incineration is applied to gaseous wastes containing low concentrations of combustible materials and air. Usually noble metals such as platinum and palladium are the catalytic agents. A catalyst is defined as a material which promotes a chemical reaction without taking a part in it. The catalyst does not change nor is it used up.

These catalysts must be supported in the hot waste gas stream in a manner that will expose the greatest surface area to the waste gas so that the combustion reaction can occur on the surface, producing nontoxic effluent gases of carbon dioxide, nitrogen, and water vapor. Since most waste gases from ordinary industrial processes are at low temperatures up to 300°F, a preheat burner is required to bring these gases up to the reaction temperature.

The advantage of the catalyst is that the reaction temperature in catalytic systems is lower than it is in thermal systems because the catalyst promotes the reaction at a lower temperature. Most catalytic reactions can be carried out at preheat temperatures between 600° and 1000°F. This of course results in a fuel saving when compared with thermal systems but involves a much higher initial investment because of the catalyst cost. Catalytic incinerators usually operate at or below 25% of the LEL (lower explosive limit) of the material in the waste gas and below the normal oxidation temperature of the contaminant.

Care should be taken, however, when analyzing the waste for catalytic combustion, that the waste gas contains a low enough concentration of the contaminant to prohibit burnout of the catalyst. Most catalysts are suitable for maximum operating temperatures of 1500° to 1600°F. A high concentration of contaminant in the waste gas, even with minimal preheat, may release enough heat on the surface of the catalyst to cause catalyst burnout. Therefore, catalytic systems are most applicable to low concentrations of contaminants where the temperature rise across the catalyst will be on the order of several hundred degrees.

Catalytic systems have been used widely in the oxidation of paint solvents, odors arising from chemical manufacture, food preparation, wire enameling ovens, lithographing ovens, and similar applications. Catalyst systems are susceptible to poisoning agents, activity suppressants and fouling agents. These compounds appear as contaminants in the waste gas stream and are specific for different types of catalysts. Catalyst manufacturers can usually state which compounds are detrimental to the operation of specific catalysts. It is therefore necessary to know what contaminants are present in a waste gas stream prior to the selection of an efficient catalytic combustion system.

Liquid Waste Incineration

Incineration is one possibility for the destruction of liquid wastes. Liquid wastes may be classified into two types from a combustion standpoint: [1] combustible liquids, and [2] partially combustible liquids. Noncombustible liquids cannot be treated or disposed of by incineration. The first category would contain all materials having sufficient calorific value to support combustion in a conventional combustor or burner. The second category would include materials that would not support combustion without the addition of auxiliary fuel and would have a high percentage of noncombustible constituents such as water. A partially combustible waste may also contain material dissolved in the liquid phase which, if inorganic in nature, will form an inorganic oxide upon combustion and require secondary collection prior to atmospheric release.

Assuming that either of these types of wastes is primarily organic in nature, even though the quantity of the organic material may be small, incineration of such materials becomes essentially a straightforward combustion problem in which air must be mixed with the combustible at some temperature above its ignition temperature. When starting with a waste in liquid form, it is necessary to supply sufficient heat for vaporization in addition to raising it to its ignition temperature.

Since liquids vaporize and react more rapidly when finely divided in the form of a spray, atomizing nozzles are usually employed to inject waste liquids into incineration equipment whenever the viscosity of the waste permits atomization. There are many wastes which might be classified liquid which are hardly liquid in nature. Slurries, sludges, and other materials of high viscosity can be handled only in special types of incineration systems.

In order that a liquid waste may be considered combustible, there are several rules of thumb which should be used. The waste should be pumpable at ambient temperatures or capable of being pumped after heating to some reasonable temperature level. The liquid must be capable of being atomized under these conditions. If it cannot be pumped or atomized, it cannot be burned as a liquid but must be handled as a sludge or solid. Liquid waste incineration generally involves liquids having viscosities up to approximately 1,000 SSU, although lower viscosities are desirable.

In order to be considered a combustible liquid waste, the material must sustain or support combustion in air without the assistance of an auxiliary fuel. This means that the waste will generally have a calorific value of 8,000 to 10,000 Btu/lb or higher. Below this calorific value, the material would not exhibit properties which would enable it to maintain a stable flame in a commercial combustor or burner. Materials which fall into this category (>8,000 Btu/lb) are light solvents (such as toluene, benzene, acetone, ethyl alcohol) and heavy organic tars and still bottoms similar to residual fuel oil. The wastes may be combinations of both, which would give a mixture having an intermediate viscosity and heating value. These wastes come from cleaning operations in chemical plants and refineries or are the residues from distillation processes and are usually not recovered for economic reasons.

The equipment used to incinerate combustible liquid waste can vary according to the manufacturer, but its basic form will be that of a combustor or burner designed to handle a liquid waste through a steam, air, or mechanical atomizing nozzle. High heat release combustors require minimal secondary incineration chambers, but usually incineration is carried out in combustion chambers having volumes which provide for a heat release of 25,000 Btu per hour per cubic foot of combustion volume. Residence times within an incinerator burning liquid waste will vary from 0.5 to 1 second.

The combustion chamber is usually cylindrical in shape and may be used in a vertical or horizontal arrangement. The vertical chamber has the advantage that the incinerator acts as its own stack, but obviously it is not well adapted to a tall stack arrangement. Horizontal incinerators can be more easily connected to tall chimneys or stacks. Some specially designed rotary kilns have been applied to the disposal of liquid chemical warfare agents. Highly explosive wastes (wet machining wastes from munitions manufacturing) are currently disposed of by open burning.

Many combustible liquid wastes can be utilized as fuel for a boiler, air preheater, or other heat recovery device which can turn the waste heat energy from the incineration system into profit. Heat recovery devices, however, are advisable only when the amount of heat recovered and the cost of the recovery equipment can be economically justified. If the waste liquid should contain noncombustibles such as inorganic salts, or materials which would be converted into corrosive compounds in the combustion reaction, such as chlorides or fluorides, then heat recovery is usually incompatible and should not be considered.

Liquid wastes having a heating value of below 8,000 Btu/lb can be considered in the partially combustible category. It must again be emphasized that this is a rule of thumb and that some materials as high as 10,000 or 11,000 Btu/lb will not sustain combustion by themselves. It is also important with this type of waste that the material handling method be compatible with the equipment selected. Viscosities should be reduced to the point where the material is pumpable and atomizable at either ambient or slightly elevated temperatures.

Waste material in this classification is often aqueous in nature, consisting of organic compounds miscible with water. Such waste may also contain sulfur compounds, phosphorus compounds, or combinations of organic and noncombustible inorganics. These materials may have enough organic content to exhibit visible combustion in a high temperature furnace, or they may be so low in combustible material that no visible combustion is apparent.

Incineration of wastes which are not pure liquids but which might be considered sludges or slurries is also an important waste disposal problem. Because sewage is handled in sludge form, many of the processes and equipment previously applied to handling sewage sludge have found application in the industrial disposal field. The combustion principles are the same, but the manner of achieving the combustion is different. Some of the types of incinerators which are applicable to this type of disposal problem are rotary kilns, multiple hearth furnaces and fluidized bed incinerators.

Some of the various liquid waste materials which are currently incinerated are shown in Table 10.

TABLE 10: DESTRUCTION OF LIQUID WASTES BY INCINERATION

Waste	Remarks
Liquid wastes from manufacture of ammonia, urea, nylon intermediates, ethylene glycol, methylamines, methacrylates.	Wastes concentrated to 50% organic content—no auxiliary fuel required. Steam atomized burners.
Organic tars, catalyst complexes	Metals must be collected when catalysts are burned.
Liquid wastes containing hydrocarbons, high-boiling degradation products, tars from nylon intermediate manufacture.	–
Liquid from acrylonitrile manufacture containing acetonitrile and cyanides + slop oils + phenolic resin wastes (inorganic)	Nitrile wastes have high fuel value. Provision for auxiliary fuel gas was made, 1600°F.
Two soot streams from acetylene manufacture; still residues from acrylonitrile and vinyl chloride processing stripping steam with acrylonitrile.	Natural gas used as auxiliary fuel. Flow and organic content of waste fluctuates greatly, 1500°F.
Heavy sludge acid, sulfonated tars from benzene plant. General refuse, scrap plastic.	Solid wastes fed first, then liquid.
Styrene still residues.	Mixed with fuel oil and used in heating furnaces.
Organic acids, salts, anhydrides, hydrocarbons and chlorinated hydrocarbons from manufacture of chlorinated organics.	Natural gas fuel used in a vortex burner. Wastes neutralized with ammonia prior to incineration to prevent corrosion.
Sludges containing oil, solids from separators, clarifiers, tank bottoms.	Fluidized-bed furnace, 800° to 900°F.
Biological sludges	–
Vent gases—H_2S, mercaptans	–
Spent caustic—50% phenols	Fuel oil used as auxiliary fuel.
High- and low-boiling organics from nylon manufacture.	–

Source: PB 224 582

Solids Incineration

Solids incineration is not a total disposal method, because most solid materials contain non-combustibles and have residual ash. Municipal and industrial incinerators probably account for 30 to 50% of the total trash disposal within the United States at the present time. The object of any incinerator is to provide complete combustion of the material fed into it. The complications are, however, the wide variety of materials which must be burned. There is everything from wet garbage with a heating value of approximately 2,000 Btu/lb to such plastics as polystyrene, which have a heating value of approximately 19,000 Btu/lb. Controlling the proper amount of air to give good combustion of both materials is difficult, and with many incinerator designs it is impossible.

Many types of conventional incinerators designed for combustion of refuse have, however, been adapted for disposal of chemical process solid wastes (4). The incinerators have been used for solid chemicals, sludges, filter residues, off-quality product, and tars. The waste must contain a sufficient percentage of combustible material and be physically compatible with the incineration equipment. For example, certain wastes such as sludges may be dried from 80 to 15% moisture before incineration.

Some incinerators have been designed to burn both process and nonprocess wastes together. Figure 16 shows a typical incinerator equipped to incinerate refuse, solid chemicals, heavy sludges, and tars. Plant refuse is dumped into a storage pit and deposited by a bucket crane into a receiving hopper. The waste is conveyed to a rotary kiln where primary combustion occurs. Other units may have manual charging and either a grate furnace or a rotary hearth furnace instead of the rotary kiln. At the end of the kiln, the ash is dropped to a hopper and the combustion gases pass through a secondary combustion zone, possible spray cooling and air pollution control equipment, and then to the stack.

FIGURE 16: SCHEMATIC DIAGRAM OF COMBINED PROCESS AND NONPROCESS WASTE INCINERATOR

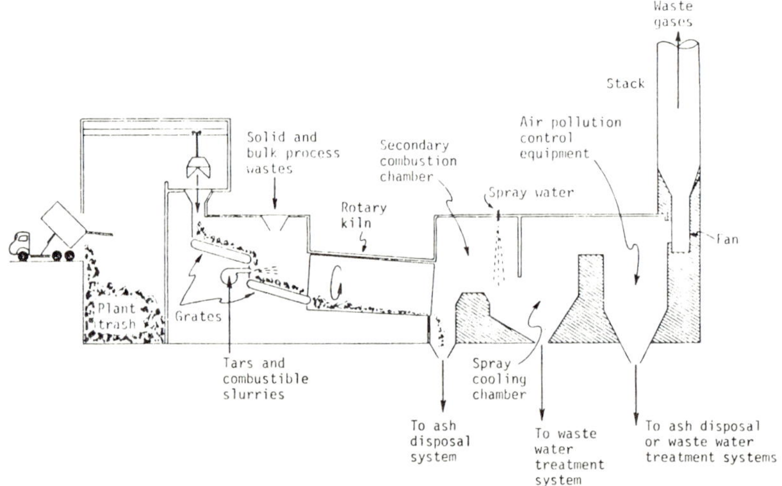

Source: PB 226 420

Tars or other combustible liquids are pumped to a burner and fired in the primary combustion chamber of the furnace. Low ash content tars are most desirable for this type of incineration, since the tars are burned in suspension and any ash produced could cause air pollution problems. Solid chemicals and sludges are also charged to the primary combustion chamber. In many cases they are contained in fiber drums, and the whole package is charged as a unit. Shredding of the waste prior to incineration is normally not done.

Temperatures in these units will range from 1400° to 2000°F. It is desirable to maintain temperatures at the higher end of this range to insure complete breakdown of organic materials. Generally, incinerators of this type cannot handle materials which will produce corrosive combustion products (e.g., HCl from the destruction of chlorine-containing organics), unless special materials of construction are used and precautions are taken to scrub such reactive materials from the exhaust gases.

Various incineration processes have been applied to the disposal of residues resulting from wastewater treatment. These processes may be classified as multiple hearth, fluidized bed, flash combustion and atomized suspension. The wet oxidation process and incineration are considered combustion processes. All these methods depend to some extent on the organic material in the waste as a source of fuel. However, regardless of the heat value of the sludge, some auxiliary fuel system is necessary during start-up of the incineration process. The incinerator may be classified as an open flame water evaporator since the actual incineration does not take place until the sludge has lost most of its moisture through evaporation. The temperature in the burning zone must be at least 1350° to 1400°F for complete combustion of the sludge solids and for complete oxidation of the exiting gases to innocuous compounds.

In the multiple hearth furnace, preconditioned dewatered sludge is conveyed to the upper hearth of the furnace, and mechanical rakes move the sludge from one hearth to the next lower hearth, where the sludge is sequentially dried, burned, and the ashes cooled. In the fluidized bed incinerators, air is utilized to maintain hot sand in an expanded or fluidized state. The violent boiling of the expanded sand allows intimate contact of the sludge and other fuel with oxygen without the aid of mechanical mixing. Dewatered sludge is fed directly into the fluidized bed to which preheated air is introduced.

The atomized suspension process is essentially a high temperature, low pressure thermal process for the oxidation of fine particles of sludge to innocuous ash. Sludge is concentrated, ground, and sprayed as an atomized suspension into a stainless steel cylinder. The walls of the cylinder are maintained at a temperature between 1000° to 2000°F and heat is transferred to the falling droplets of sludge by radiation.

In the cyclone flash dryer and incinerator process, dewatered sludge is mixed with previously heat dried sludge to produce a fluffy material. In this form, it is carried through the drying stage by the high velocity of hot gases and the dried sludge is burned in the furnace. Design considerations and process variables involved are:

  [1] sludge flow
  [2] moisture content of sludge
  [3] volatile solids content of the sludge
  [4] heat value of the sludge
  [5] type of operation, continuous or intermittent
  [6] source and type of auxiliary fuel
  [7] method of final disposal of solid residue
  [8] sludge variability
  [9] sludge loading
[10] possibility of alternate methods

Design criteria are established through incinerating the sludge in a bench or pilot scale burner. The auxiliary fuel requirements, nature of the ash, and composition of the off-gases are the important parameters to be ascertained (5).

Certain basic principles for complete combustion of solid waste with a subsequent low particulate emission from the combustion zone are as follows:

  [1] Excess air: Air quantities should usually be kept on the order of 50 to 150% above the stoichiometric requirements.
  [2] Minimum use of underfire air: This maintains low velocities and therefore reduces the particulate emission from the incinerator because it keeps small particles out of the gas stream.
  [3] Proper use of overfire air: This provides ample oxygen and turbulence in the combustion space above the fuel bed. The overfire air injected into the system may be as high as 50% of the total required.
  [4] Temperature: Temperature in the furnace space should be between 1400 and 1800°F to reduce the rate of smoke formation and odor.

Temperatures below 1400°F will produce smoke and allow odor to escape from the incinerator. Above 1800°F there may be sintering or fusing of the ash with the furnace refractories. Excess air is used to control the furnace temperature.

[5] Sufficient combustion volume: The incinerator should have enough combustion volume to provide sufficient residence time for the burnout of all flying particulate matter. The average heat release per cubic foot of furnace volume should not exceed 25,000 Btu per cubic foot per hour.

[6] Secondary chamber: A secondary chamber zone should be provided in every incinerator and, in fact, is required in most municipal and state codes now being adopted.

[7] Residence time: The residence time in the incinerator should be between 1 and 2 seconds.

[8] Reasonable loading rates: Low loading rates per square foot of grate surface should be adhered to, even in forced draft incinerators. They should be no more than 60 pounds of waste per square foot per hour.

Early solid waste incinerators were charged by a crane from a storage pit onto stationary grates. Some furnaces were hand stoked. Ash handling was arranged through undergrate hoppers in which the ash was water quenched and dumped into trucks and hauled away. Today, because of the awareness of air pollution problems, more modern designs have been developed. A variety of modern stokers (each best suited for a particular size and form of solid waste), which provide uniform and regular agitation of the feed, are used to feed the waste material onto the surface of the grate. The grate may be a fixed grate or a traveling grate system. Where the traveling grate is used, it is also considered part of the stoking apparatus.

The types of incinerators which are applicable to solid wastes are open pit incinerators (when no secondary pollution problem exists) and closed incinerators such as retort and inline multiple chamber units, rotary kilns and multiple hearth incinerators. The incinerator design does not have to be limited to a single combustible or partially combustible waste. Often it is both economical and feasible to utilize a combustible waste, either liquid or gas, as the heat source for the incineration of a partially combustible waste which may be either liquid or gas. Multiple or dual fuel burners for combustible wastes can be utilized in a single incineration chamber. Combination systems can often reduce the operating cost in terms of auxiliary fuel and should be carefully evaluated in the overall waste treatment program of any process plant. Heat recovery in these systems is applicable on an individual basis.

FACTORS IN INCINERATOR SYSTEM SELECTION

In order to determine the proper type of incinerator system (i.e., storage facility, incinerator, and effluent purification equipment) for use in a particular waste disposal situation, certain basic information about the waste material must be known (Table 11).

TABLE 11: BASIC DATA CONSIDERATIONS WHEN CHOOSING A WASTE DISPOSAL INCINERATOR SYSTEM

Type(s) of Waste	Liquid, Solid, Gas, or Mixtures
Ultimate analysis	Carbon, hydrogen, oxygen and nitrogen, water, sulfur, and ash on an "as-received" basis.
Metals	Calcium, sodium, copper, vanadium, etc.
Halogens	Bromides, chlorides, fluorides.
Heating value	Btu/lb on an "as-received" basis.
Solids	Size, form and quantity to be received.

(continued)

TABLE 11: (continued)

Type(s) of Waste	Liquid, Solid, Gas, or Mixtures
Liquids	Viscosity as a function of temperature, specific gravity and impurities.
Gases	Density and impurities.
Special characteristics	Toxicity and corrosiveness, other unusual features.
Disposal rates	Peak, average, minimum (present and future).

Source: PB 224 582

Waste Toxicity

The toxicity of the waste material and its combustion products dictates the safety procedures, safety equipment and monitoring equipment required for personnel safety. For instance, an extremely toxic material might require remote operation of all processing units and the wearing of protective clothing, masks and self-contained breathing apparatuses by all operators in the vicinity of the incineration unit.

Equipment to monitor combustion temperature and/or effluent compound concentration might be utilized in series with feed regulating equipment to ensure total combustion of the toxic material. The overall ability of the disposal system (storage facilities, transport equipment, incinerator, secondary abatement equipment and stack) to maintain low toxic contaminant concentrations at ground level could be determined through periodic or constant monitoring in and around the disposal facility.

Disposal Rate

Regardless of the type of waste to be incinerated, the disposal rate determines the size and number of incinerators (as well as storage equipment) needed to combust a given waste stream.

Economic Factors

It should be noted that when applied to treatment processes handling hazardous waste constitutents, a high degree of efficiency is of primary concern as opposed to economic considerations.

Corrosiveness, Operating Temperature and Material Selection

The waste stream's corrosiveness determines the materials of construction required in both the incineration unit and materials handling equipment. Incineration systems may be fabricated from a wide variety of construction materials. The selection of construction materials depends upon several factors: [1] corrosion, [2] strength, and [3] temperature. Most incinerators are constructed of carbon steel material and are lined with appropriate alumina refractory to withstand the temperatures of the incineration process.

Some catalytic units, however, as well as some thermal incinerators are fabricated without refractory, using only high-temperature stainless steel. The advantage of this approach is the elimination of the refractory, which will eventually need refurbishing or replacement. More costly materials such as Inconel, Incoloy, or Hastelloy are normally utilized only when the waste is corrosive to other materials.

Refractories used in incineration systems are generally of the alumina type. Standard or superduty firebrick backed up by insulating brick is suitable in most cases. Castable refractories are also widely used. In short, the refractories, which are used for most incineration applications are equivalent to those which would be used for high-temperature furnaces.

Secondary Abatement Requirements

Many wastes which lend themselves to incineration cannot be incinerated without some secondary form of treatment, due to the production of compounds which might be toxic in nature and therefore cannot be released to the atmosphere. Normally these wastes can be divided into these categories: [1] waste which contains inorganic salts; [2] waste which contains halogen compounds; [3] waste which contains sulfur compounds; and [4] waste which contains nitrogen compounds.

Those which contain inorganic salts will produce the oxide of the metal ion of that salt upon combustion. The commonest inorganic metal ion is sodium, although potassium, or for that matter any other metal ion, may be found in the waste. The oxide which is formed in the combustion reaction usually will be in a finely divided form and will require subsequent removal by either mechanical methods or wet scrubbing. This type of product usually requires a high-energy scrubber of the venturi type.

The halogen ions most commonly found in wastes are generally chlorine and fluorine, which are often part of halogenated hydrocarbons. Complete combustion of the organic portion of the waste may result in the production of chlorine or fluorine in the products of combustion. These are relatively insoluble in water and therefore cannot be removed by wet scrubbing as long as they are in this form. This type of waste should first be analyzed to determine the amount of hydrogen present, since the hydrogen will react with the halogen forming the halogen acid.

If there is not sufficient hydrogen in the waste material to accomplish complete conversion of the halogen to the halogen acid, the conversion must be accomplished by injection of additional hydrogen in the form of natural gas or other combustible. Once the halogen has been converted to the acid gas, it may be satisfactorily removed in a wet scrubber. Here the low-energy or packed tower type of scrubber is satisfactory. For example, if trichloroethylene is a major component in a waste effluent, it can be incinerated in accordance with the following reaction:

$$CHClCCl_2 + 2O_2 \longrightarrow 2CO_2 + HCl + Cl_2$$

While the hydrogen chloride which is formed in this reaction can be removed by scrubbing with water, the chlorine, which is relatively insoluble, will pass through the water and into the atmosphere. By the addition of natural gas or another hydrocarbon fuel, all the chlorine can be converted to hydrogen chloride as follows:

$$CHClCCl_2 + \tfrac{7}{2}O_2 + CH_4 \longrightarrow 3CO_2 + 3HCl + H_2O$$

The hydrogen chloride formed in this reaction can be removed in a wet scrubber. In this particular case, natural gas or other auxiliary fuel would be required for combustion because of the low calorific value of the trichloroethylene.

Sulfur compounds are often found in wastes either as part of a sulfonated organic molecule or in the form of sulfates or sulfides. Complete combustion of these wastes with air results in SO_2 formation. Complete removal of SO_2 can be handled by caustic scrubbing or a number of more complicated processes designed ultimately to recover sulfur.

One pollutant, NO, is common to all incineration processes which utilize air as the oxidant source (as opposed to pure oxygen or some other oxidant which contains no nitrogen). The NO formation is the result of the oxygen and nitrogen from the air reacting at the elevated flame temperature present in incinerators. The thermodynamic equilibrium concentrations of NO present in combustion effluent streams as a function of flame temperature and excess air have been determined. Generally industrial incineration operations do not abate NO emissions. There are currently abatement techniques available but they involve catalytic or thermal reduction (requiring a reductant such as CH_4 or H_2) of NO to N_2 followed by thermal oxidation of excess reductant. These processes are capable of reducing NO_x concentrations in the stack effluent to the 50 ppm level. It is the reductant cost plus the catalyst cost (when a catalyst is used) which makes application of this type of abatement process economically unattractive in most situations. Of course, wastes which contain nitrogen compounds produce nitrogen oxide in excess of those resulting from combustion alone.

Steam Plume Generation

Another problem common to most incineration processes is that of steam plume generation (1). The appearance of a steam plume has a very important psychological implication which must be considered. An incinerator process effluent may contain no harmful pollutants, however, the presence of combustion product water plus any moisture picked up in scrubbing operations will cause the stack effluent to show a steam discharge plume upon becoming saturated with moisture. Although the steam will have no deleterious effects on the surrounding area, its appearance may cause concern with the public that the air cleaning equipment is inadequate or malfunctioning.

Steam plumes are the result of rapid cooling of moisture containing gases to below their saturation temperature. Gases from wet scrubbers will quickly show a steam plume at the stack discharge. During the cold winter months, the gases are subjected to a colder atmosphere and are thus air-condensed sooner and have shorter plume trails compared with equivalent gases cooled by the summer atmosphere.

As a guide, saturated gases which discharge from the stack below 105°F will have a negligible appearance and will not create a questionable steam plume. At 105°F the volume of moisture content is less than 7%, whereas at higher saturation temperatures, the percentage of moisture by volume is as follows: 130°F, 15.0%; 160°F, 32.5%; 180°F, 51.0%.

In addition to appearance, steam plumes have other side effects which include:

   [1]   SO$_2$ (or other corrosive gases) may be absorbed by the newly formed droplets, forming sulfurous acid and then fall on homes and industrial sites.
   [2]   In some cases odoriferous constituents may be entrapped by falling droplets to increase odor at ground elevation.

There are three basic methods for steam plume minimization.

[1] Indirect Cooling of Hot Gases: Cooling is effected within a continuous S shaped duct with surface exposure for radiation and cooling by atmospheric air surrounding the ducts. Often the ducts may be arranged essentially vertically with U bends and should be furnished with cleanout doors and hoppers to permit intermittent dust removal. The addition of fans, handling large quantities of atmospheric air, blowing or inducing air across banks of ducts which transport the uncleansed gas, has recently been introduced.

[2] Sensible or Direct Cooling of Saturated Gases: This technique cools already 100% saturated gases with cool water. Sufficient coolness is required to dehumidify or condense water vapor down to the desirable lower saturation. By using a standard spray tower cooling water at 70 to 85°F may be introduced at 25 psig. Droplets will fall counterflow to the gas passage and carry away latent heat of the water vapor and sensible heat of the dry gas. As a general rule, the cool water leaving a properly designed tower will approach within 10° or 15°F of the entrance gas temperature. Therefore, pounds of 80°F water needed equals 3.51 lb water/lb dry gas or 0.42 gal/lb dry gas. Use of a tower filled with drip-point grid tiles offers a method to obtain benefit of maximum heat transfer with an approach of approximately 5°F or less (between the gas and liquid discharging).

[3] Cooling by Mixing with Atmospheric Air: In some special cases sensible cooling may be obtained by the addition of atmospheric air having a low dew point temperature. However, this method becomes impractical where already large saturated volumes of gas having high saturation temperatures are involved.

Waste Heat Recovery

Few industrial chemical plants recover waste heat from incineration of solid wastes (4). Although in most cases a plant could use the heat energy for production of steam, the design, operation, and maintenance of a waste heat boiler is fraught with problems. Two boiler design problems which must be given special consideration when firing solid wastes are fouling of heating surfaces and potential corrosion. For chemical plant wastes, these design considerations are even more critical due to the diffuse nature of wastes incinerated and the increased possibility of acid corrosion. Despite these problems and the expense involved, a few waste heat boilers do operate successfully at chemical plants for both process and nonprocess waste incinerators.

A more common method of utilizing the heat from combustion of wastes is to fire the waste as a supplementary fuel into the main plant boilers. A number of plants were found to be using this disposal method for certain tars, solvents, and other organic liquids. Boilers were found with a waste fuel burning capability of up to 20% of capacity. Little information is available, however, on the design, operation, or applicability of this procedure to specific waste types.

VARIABLES AFFECTING PROPER WASTE COMBUSTION

Incinerators are generally classified by the form of waste that they burn—gas, liquid or solid. However, most incinerator systems, regardless of waste type, contain four basic components; namely, a waste storage facility, a burner and combustion chamber, an effluent purification device when warranted, and a vent or stack. It is the oxidation which occurs in the combustion chamber of the system which is of primary importance for proper hazardous waste disposal. It is here that the feed waste is converted to a less hazardous compound and that secondary pollutants which must be removed by further treatment are formed. The variables which have the greatest effect on the completion of the oxidation of the wastes are waste combustibility, dwell time in the combustor, flame temperature, and the turbulence present in the reaction zone of the incinerator (1).

Waste Combustibility

Combustibility is a measure of the ease at which a material can be oxidized in a combustion environment. A material's combustibility is characterized by its upper and lower flammability limits, its flash point, ignition temperature and autoignition temperature. In general materials with a low flammability limit, low flash point, and low ignition and autoignition temperatures may be combusted in a less severe oxidation environment (lower temperature and less excess oxygen) than those materials with a high flammability limit, high flash point and high ignition and autoignition temperatures.

Combustion Temperature

Of the "three T's" of good combustion, time, temperature and turbulence, only the temperature may be readily controlled after the incinerator unit is constructed. This can be done by varying the air/fuel ratio. If solid carbonaceous waste is to be burned without smoke, a minimum temperature of 1400°F must be maintained in the combustion chamber. Upper temperature limits in the incinerator are dictated by the refractory materials available. Whenever a temperature of 2400°F is exceeded, special refractories are required. A design range of 1800° to 2000°F is usually desirable, unless thermodynamic equilibrium considerations dictate some other temperature requirement. The rates of most combustion reactions increase rapidly with increases in temperature, while a few peak at relatively low values. The latter are rare but must not be overlooked when unusual fuels are burned.

There are four basic methods of controlling the combustion temperature:

[1] Excess Air Control: The adiabatic flame temperature is a function of both the type of fuel and the amount of air (oxygen) used. For example, the adiabatic combustion tem-

peratures of a cellulose waste that occur with variations in the amount of excess air vary dramatically, clearly indicating the importance of maintaining the proper air/fuel ratio when the maintenance of a specific temperature is required. In practice, the actual temperatures will be less because of thermal conduction losses through the furnace walls, thermal radiation to cold surfaces and possible incomplete combustion of the waste.

Control is obtained through automatically controlled or strenuously supervised operation. The designer can provide limiting orifices, nozzles, or pumps to prevent overfiring during continuous feeding of liquids and gases. For liquid fuels the problem is aggravated as the volatility of the fuel increases. If an incinerator is fed with discrete amounts at intervals, the hourly sum of these amounts must not exceed the rated capacity, and the intervals between feeding must be regulated. The greater the volatility of the waste, the smaller these discrete amounts must be and the more frequent the feed intervals. To illustrate, one may toss a teaspoon of gasoline every minute on the backyard grill without creating havoc, but a pint of gasoline tossed at once can have disastrous effects. All too often, this simple fact is overlooked in incinerator operation.

[2] Radiant Heat Transfer: Most combustion processes exist for the purpose of heat transfer, however, this is not usually the case for incinerators. Some large municipal waste incinerators use heat transfer surface as a means to control temperature and this design practice is growing although it is seldom economically feasible for industrial incinerators. It should be considered, however, when economics might make it attractive. Radiation to the sky is feasible, and several designs are capable of doing this. Where either heat transfer surface or radiation to the sky is used, surface area relationships can be used to estimate the combustion temperature.

[3] Two-Stage Combustion: When the combustion is divided into two distinct steps and the first stage is supplied with a deficiency of air, the first stage can act as a gasifier for certain fuels while burning incompletely at reduced temperatures. A second stage is necessary to burn the combustible products produced in the first stage, and its temperature must be limited by either the first method discussed (excess air control) or by heat transfer. Incinerators of this type must be carefully engineered to assure proper ignition and complete burning in the second stage. Additionally, feed rates must be carefully controlled since a decrease in the waste feed rate without a corresponding decrease in the air feed rate would result in the first stage progressively moving toward complete combustion with progressively higher combustion temperatures.

[4] Direct Heat Transfer: When heat-absorbing materials or other fuels are added to the waste fuel the temperature of combustion can be controlled. The most common method of achieving lower temperatures is to add water to the fuel since each pound of water added absorbs approximately 1,000 Btu for evaporation and ½ Btu for every degree Fahrenheit of temperature rise as sensible heat content. The water may be added with the fuel or sprayed into the combustion zone if carefully controlled to avoid quenching the fire or damaging the refractory. Temperature may be increased by burning other fuels with the waste stream.

Recommended minimum temperature requirements for complete combustion of those candidate waste stream constituents applicable to incineration are presented in the profile reports discussing the individual constituents in this volume.

Combustion Zone Turbulence

The degree of turbulence (intimate mixing) of the air for oxidation with the waste fuel will affect the incinerator performance significantly. In general, either mechanical or aerodynamic means are utilized to achieve the intimate scrubbing and mixing of the air and fuel. The completeness of combustion and the time required for complete combustion are significantly affected by the amount and the effectiveness of the flame turbulence. There is no accepted parameter that will quantitatively define an amount of turbulence, therefore it is judged by the combustion results that are produced.

Turbulence may be created in the combustion zone mechanically and aerodynamically. Turbulence can be induced mechanically through the use of fixed and moving grates, rotary kilns, mechanical pokers and hand pokers. There are two design factors which must be considered before applying mechanical turbulence producers. First, they depend principally upon natural factors to clear away gaseous combustion products and bring in fresh oxygen. The primary function of mechanical turbulence producers is the removal of noncombustible coverings (ash) to expose unburned material. Second, if mechanical devices are metal, they must be protected from elevated combustion temperatures. This protective cooling can usually be achieved by circulating either air or water through the device.

Aerodynamic turbulence can be defined as the creation of turbulence by gases in motion. High velocity jets of forced air can create a degree of turbulence that approaches perfection. The turbulence may be produced with convergent nozzles using air or steam, or it may be produced by air registers. Air registers are vane arrangements, usually circular, that surround a fuel injection nozzle. They serve as multiple nozzles both to increase the incoming air velocity and to create a forced vortex around the fuel nozzle. It should be pointed out that for fixed nozzle and vanes, and sometimes for variable ones, the degree of turbulence is almost always at its maximum with the maximum air flow. For this reason best results are often obtained if the incinerator is fired at its maximum rating. Shorter firing periods at maximum rate may produce better results than continued operation at low firing rates.

Another form of aerodynamic turbulence is achieved in the fluidized bed. Air is forced vertically through a bed of solids (usually cylindrical) at a rate that expands the bed without excessive solids carryover. If the bed container is lined with refractory and the bed material is [1] uniform and [2] does not fuse at the operating temperatures, excellent turbulence is created. Uniformity can be helped by utilizing a permanent bed of sand. By injecting new material at the rate of consumption, the bed is held level and the air-flow resistance held constant. Since solid fuels without appreciable ash (coke) can be added, the fluidized bed has found some application with waste solids that require auxiliary fuel.

Residence Time in the Combustion Zone

The third major requirement for good combustion is time. Sufficient time must be provided to the combustion process to allow slow-burning particles or droplets to completely burn before they are chilled by contact with cold surfaces or the atmosphere. The amount of time required depends on the temperature, fuel size and degree of turbulence achieved. Although it is customary to specify certain furnace volumes for heat releases in an attempt to obtain proper combustion times, this method is now generally recognized as inadequate. In the absence of specific data, combustion chambers with heat releases of 20,000 to 60,000 Btu/ft^3/hr are common.

These values are very conservative for high performance burners, and the use of small compact incinerators means lower investment and lower maintenance. The evaluation of the factor of time can be made only by tests of individual burners and furnaces. Important information on this subject can be obtained directly from burner manufacturers. When slow-burning items are present, such as carbon particles or carbon monoxide, additional chambers (secondary combustion chambers) may be needed to allow time for complete combustion. Recommended minimum residence time requirements for complete combustion of those candidate waste stream constituents applicable to incineration are presented in the profile reports discussing those individual constituents in later sections of this volume.

TYPES OF INCINERATORS

Afterburners

Afterburners are secondary combustion devices, in themselves incinerators, for cleaning up flue gas from primary incinerators. A device developed by J. Greenberg (6) accomplishes the removal of normally unburned pollutant products of combustion such as carbon, carbon monoxide and hydrocarbons, by contacting such materials with a molten salt. The molten salt acts as a catalyst in inducing relatively complete oxidation of the materials at temperatures below their normal combustion temperatures, and with substantial decrease in the unburned pollutant products. In one form, the salt may be a neutral salt with which the material, in the presence of oxygen, is contacted. In another form, the catalytic salt also contains a chemical oxidizer which continuously releases nascent oxygen and retakes ambient oxygen, thereby maintaining an equilibrium pressure of oxygen gas which aids the oxidation process. Figure 17 shows a suitable form of apparatus.

FIGURE 17: MOLTEN SALT AFTERBURNER DEVICE

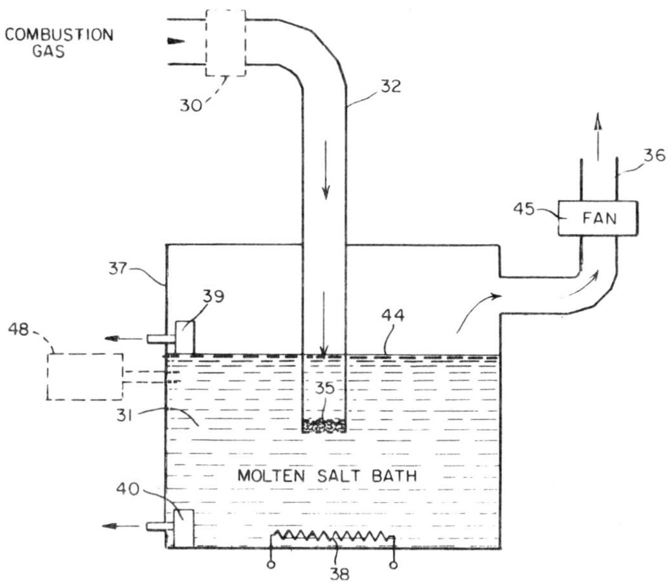

Source: U.S. Patent 3,647,358

The pollutant-containing combustion gas is introduced through an inlet **32** at a pressure which is sufficient such that the draft conditions of the combustion chamber, from which it came, are not disturbed. If necessary, a pump **30** may be provided to achieve a proper interface with the combustion chamber. The combustion gas normally contains a sufficient proportion of oxygen. For example, when the combustion takes place in a power plant where fuel is burned in the presence of air, the stack gases contain a proportion of oxygen which varies with the amount of air introduced into the combustion chamber, as well as with the degree of oxidation within the chamber. If desired, oxygen may be introduced into the bath from a separate source **48**.

It is to be noted that oxygen may be introduced into the salt electrolytically. For example, conventional electrolysis may be employed with a nitrate bath, with oxygen being produced at the positive electrode as a result of decomposition of nitrate anions. Further, while the oxidation reactions most often referred to in this patent are the combination of oxygen with carbon, carbon monoxide and/or hydrocarbons, it is also noted that, in broader terms, oxidation could occur by combination with chlorine or other similar gases, to produce innocuous oxidation products. A molten salt bath **31** is provided in a container

37 constructed of a suitably noncorrosive material, the molten salt bath filling the container to a level indicated at **44**. In the embodiment shown, inlet **32** carrying the gases of combustion extends below level **44**, the combustion gas passing through a filter **35** and into the molten liquid, where the desired catalytically induced oxidation takes place. Filter **35** is designed both to catch certain particulate matter such as fly ash and other solids, and to disperse the combustion gas into the liquid. The oxidized combustion gas evolves from the molten salt bath and is drawn through outlet **36** by conventional blower or fan **45** which, by itself or in combination with pump **30**, maintains a sufficient pressure differential from input to output to overcome the pressure head presented by the salt bath.

In some applications of the process the heat of the incoming combustion gas is sufficient to maintain the salt bath in a molten form and at a desired temperature. The catalytic nature of the bath insures that oxidation takes place at a temperature below the combustion temperature at which the combustion gas is produced and, accordingly, in many instances, the heat transfer from the combustion gas to the bath is sufficient to maintain it at the desired temperature. Further, the oxidation reactions which take place in the bath are generally exothermic, such that the bath temperature may rise above the temperature which would be maintained solely by heat transfer from the combustion gas.

In some applications, the temperature of the bath rises excessively unless the feed rate of combustion gas is properly controlled. However, in most applications involving power plant stack gases, automotive exhaust and the like, the feed rate is well within the desired limits, obviating the need for any feed control mechanism. In some applications, however, it is necessary or desirable to provide the bath with an external source of heat, as through a conventional heat source **38**. Heat source **38** may be in the form of conventional electrical heating coils immersed within the bath, or a conventional flame source (burner) located below and external to the bath container **37**. In practice, some residue is sometimes formed, which may be removed by a conventional skimmer device **39**. Also, certain products settle to the bottom of the bath from time to time; they may be removed by conventional dredging apparatus **40**.

A device developed by R.W. Foster-Pegg (7) is an afterburner including a regenerator disk having passages therethrough disposed over the combustion chamber, means for slowly rotating the disk, an inlet duct having a first rigid seal extending completely about the periphery of the upper surface of the disk, and an exhaust duct extending through the inlet duct, the duct chamber having a second seal extending about a sector of the upper surface of the disk within the first seal. The purpose of this device is to reduce the requirement for fuel of an afterburner system by recycle of exhaust heat to preheat the incoming gas.

A scheme developed by B.R. Hutchinson (8) involves coupling afterburner-centrifugal separator devices to existing incinerators. By means of exercising control over incinerator exhaust stack draft and afterburner temperature, complete oxidation and particulate separation can be achieved in an afterburner unit which may be coupled to an existing incinerator. The device further encompasses automatic safety features including provision of a bypass for incinerator effluents in the case of any failures in the afterburner system.

A device developed by M.W. Ehrlichmann et al (9) is an antipollution or smoke control device comprising an afterburner for the output of an incinerator that has means controlling combustion to meet pollution control standards of the various pollution control agencies.

An apparatus developed by E.M. Van Raden (10) is a fume incinerator for afterburning fumes containing air polluting elements that includes a casing having a burner disposed in a combustion chamber. The casing includes an outer compartment which is sealed from the combustion chamber and provides a dead air space between the combustion chamber and the outer walls of the casing. A blower receiving fumes from an industrial process or apparatus directs the fumes into the combustion chamber. A conduit in communication with the dead air chamber is in communication with the suction conduit of the blower so that any fumes which may leak into the dead air space or chamber are immediately removed and returned into the combustion chamber, thus preventing leakage of the same through the outer walls of the casing and into the atmosphere.

A device developed by R.H. Wieken et al (11) is a smoke eliminator which is connected to a chimney usually leading from an incinerator or the like. The eliminator includes a chamber capable of withstanding high temperatures. A power burner is provided to preheat the chamber. The temperature is sufficient to burn out the major portion of the impurities caused by the firing of the incinerator as the gases travel a tortuous path.

An apparatus developed by D.K. Longley (12) is an afterburner apparatus for oxidizing combustible materials such as vaporous gases and effluent gas from an incinerator or the like. The apparatus is fitted into the stack of an incinerator or other primary combustion apparatus so that the effluent gas from the incinerator flows upwardly into the interior of the afterburner apparatus. The afterburner includes a spherical chamber having a refractory lining, an inlet port at the bottom of the afterburner through which the effluent gas is introduced into the chamber and an outlet port at the top of the afterburner diametrically opposite the inlet port. A flame and stream of air are introduced into the lower portion of the chamber at a position near the inlet port and at an angle slightly upward from the horizontal and tangential to the interior surface of the chamber to cause the effluent gas to swirl upwardly. The flame and stream of air cause oxidation of the combustible materials in the effluent gas and the oxidation products then pass out the outlet port at the top of the chamber.

A device developed by A. Taeymans et al (13) is one in which the fumes and particle waste produced by a furnace are passed through a fluidized bed of granules wherein they are further incinerated and they then pass through a filter of granules where the particles are removed; when the filter becomes contaminated, the pressure drop caused thereby is sensed and contaminated granules are ejected therefrom into the fluidized bed; granules in the fluidized bed are simultaneously transferred back to the filter.

An apparatus developed by K.H. Hemsath et al (14) is an incineration system for incinerating flue gases which contain varying quantities of combustibles therein. Means is provided for controlling combustion air added to the flue gases when they contain combustible matter and means is provided for sensing the temperature of the combusting waste gases. When a preset temperature is exceeded, additional air is added to reduce the temperature of the gas, but when the temperature falls below the preset point, the burner firing rate is increased and the added air is cut off to insure proper incineration and achieve fuel efficiency.

Catalytic Incinerators

Catalytic incinerators are devices which are used to dispose of low concentration combustible materials in the gaseous state. Catalytic incineration, or as it is more commonly called, catalytic combustion or catalytic oxidation, has been successfully used in the chemical process industries for incineration of paint solvents, odors from chemical manufacture and food processing and many other functions that help offset the cost of air pollution control equipment. There are five steps in any solid catalyzed vapor phase reaction (1). These five basic steps are:

[1]   Diffusion of the reactants through the stagnant fluid around the surface
        of the catalyst.
[2]   Adsorption of the reactants on the catalytic surface.
[3]   Reaction of the adsorbed reactants to form products.
[4]   Desorption of the products from the catalytic surface.
[5]   Diffusion of the products through the pores and surface film to the bulk
        vapor phase outside the catalyst.

Therefore, given the identical support, the rate of steps [1] and [5] would be approximately equal regardless of the dispersed catalytic metal which is present. The criteria which govern the effectiveness of operation for various catalytic metals must therefore fall into steps [2], [3] or [4]. These criteria include reaction temperature, waste material concentration, excess oxygen available, the chemical composition of the catalyst and the geometric configuration of the individual catalysts.

The operating efficiency (percent of the waste organic combusted to carbon dioxide and water) of a catalytic system is strongly dependent upon the catalyst temperature. Increased catalyst temperatures generally result in increased removal efficiencies. A tabulation of removal efficiencies of various solvents as a function of catalyst temperature is presented (Table 12). The data summarizes the type and quantity of coating applied, the type and quantity of solvent evolved, the oven temperature, the catalyst temperature and the removal efficiency for a coating oven during tests with a catalytic incinerator.

TABLE 12: CATALYTIC OXIDATION OF SOLVENT VAPORS EVOLVED FROM A COATING OVEN*

Coating Type	Quantity of Coating (gal/hr)	Evolved Solvent	Solvent Quantity Evolved (lb/hr)	Average Oven Temp. ($^\circ$F)	Catalyst Temp. ($^\circ$F)	% Solvent Removal
Vinyl	19	Xylol and isophorone	120	350	800	28
					920	54
					1050	77
					1200	93
Vinyl	43	Methyl isobutyl ketone	271	340	840	77
					890	79
					930	85
					990	88
Epoxy	18.6	Xylol and butyl Cello-solve	86	350	800	79
					900	81
Phenolic	18.5	Mineral spirits	88	414	730	65
Oleoresinous	17.5	Mineral spirits	77	425	800	80
					920	89
					1050	94
					1200	95
Alkyd	8	Mineral spirits	30	290	690	41
					700	52
					800	80
					920	89
					1050	94
					1200	95

*The catalysts discharge temperature was held constant for each test and superficial gas velocities were constant within ±15% for all data points.

Source: PB 224 582

The waste compound concentration as well as the oxygen concentration in the gas stream have marked effects upon catalytic combustion efficiency. The combustible waste concentration in the gas stream have marked effects upon catalytic combustion efficiency. The combustible waste concentration in the stream to be treated should never exceed the lower flammability limit in order to guard against explosion and fire. However, it is desirable to operate with as high a concentration of combustible (below the lower flammability limit) as possible since conversion efficiencies are generally proportional to combustible compound concentrations when other operating variables are held constant. The conversion efficiency of catalytic combustors is also effected by the oxygen concentration present in the feed gas stream.

It has been found that increased oxygen concentration results in increased efficiency when other parameters are held constant. There is, however, a trade-off, as oxygen concentration is increased in a gas stream (through air injection), the contaminant concentration is decreased. That is, the efficiency of the system will be increased by oxygen addition while simultaneously being decreased by contaminant dilution. There is generally an optimum which must be determined for individual disposal systems. The normal effective range for catalytic oxidation extends from a very few parts per million of combustible up to a heating value of 20 Btu/std ft^3.

The chemical composition of a catalyst affects conversion efficiencies. The various noble metals such as platinum, palladium, rhodium, etc., as well as copper chromite and the oxides of copper, chromium, manganese, nickel, and cobalt, in varying concentrations have been applied successfully to the catalytic oxidation of various combustible compounds. However, in air pollution control it has not been practical to undertake research programs to develop specific catalysts for each problem. Therefore, the goal of commercial manufacturers has been to make available universal catalysts which are effective in oxidizing the entire range of organic materials over an extended period of exposure time with minimum maintenance and replacement.

When selecting a catalyst material care must be taken to make sure that there are no poisoning agents, activity suppressants or fouling agents present which inhibit the catalysts' effectiveness. With platinum family catalysts, contaminants to look out for are:

Poisons:	Heavy metals
	Phosphates
	Arsenic compounds
Suppressants:	Halogens (elemental and compounds)
	Sulfur compounds
Fouling Agents:	Alumina and silica dusts
	Iron oxides
	Silicones

Other materials that may reduce catalytic effectiveness in general are the vapors of metals such as mercury, zinc and lead. Catalyst manufacturers can generally specify which materials are detrimental to catalysts which they market.

The geometric configuration of the individual catalysts can also influence the extent of the oxidation of a gaseous waste. There are many commercial catalysts available in pellet, spherical or ring form. However, in air pollution work, which is generally operated at or near atmospheric pressure, usually the pressure drop through catalyst beds of these types is so large that the horsepower of the waste gas fan becomes too great for economical operation.

Therefore, three general commercial catalysts with a very low pressure drop ($\frac{1}{4}$ to $\frac{1}{2}$ inch w.g.) have been developed (Figure 18). The first is a mat type catalyst similar to an air filter. It consists of a ribbon type Nichrome or stainless steel wire to which the catalytic material has been applied, randomly packed between screens and mounted in a stainless steel frame. The second is a porcelain assembly consisting of two end plates which are secured by a center post and a number of teardrop-shaped rods to which the catalyst is applied. In this case the carrier is porcelain which is first coated with activated alumina and then with an active metal coating. The third is honeycomb type ceramic material to which the catalytic material is applied. The geometric configuration best suited for specific applications can usually be suggested by the catalyst manufacturer.

There are certain prerequisites required for an efficient catalytic operation. The basic requirements include intimate mixing of the combustibles in the stream to be treated. The stream must be brought up to the catalytic ignition temperature for the combustible to be burned, and good temperature distribution through the catalyst bed is essential. Sufficient oxygen must be present in the waste effluent or must be added to it to ensure oxidation of the materials. The system must be designed so that proper velocities and retention times through and within the catalyst bed are maintained.

The influent gas stream should be free of particulate in order to ensure against catalyst fouling. Therefore, if the stream contains high particulate loadings, a pretreatment of filtration or electrostatic precipitation would be required prior to catalytic oxidation. In such cases, it is generally more economical to treat the waste gas stream with thermal incineration.

FIGURE 18: COMMERCIALLY USED CATALYST CONFIGURATIONS

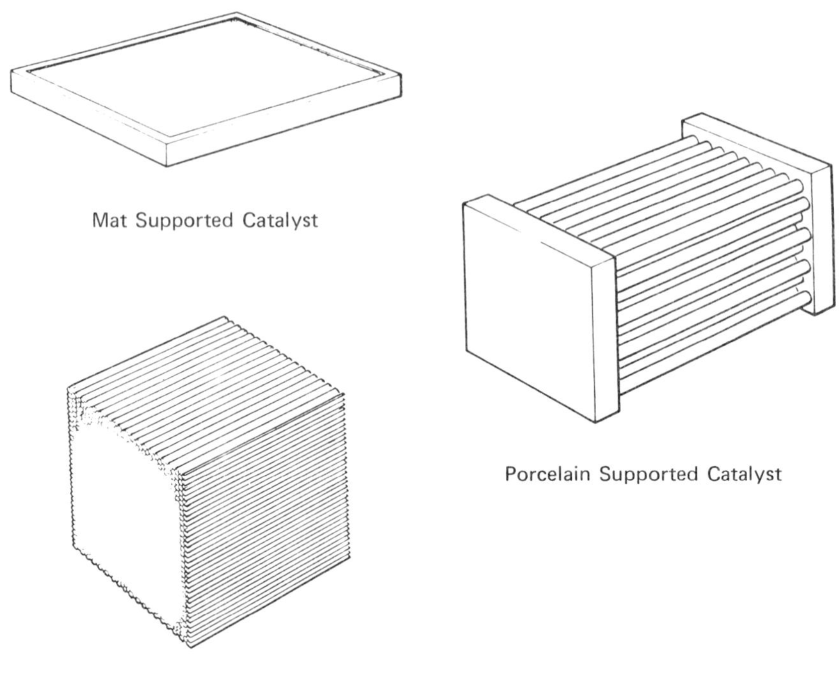

Source: PB 224 582

Basically, a catalytic incinerator consists of an afterburner housing containing a preheating section (if one is necessary) and a catalyst section. A gas burner preheats the contaminated gases before they flow to the catalyst section. Drawings of two catalytic incinerator installations are presented (Figures 19 and 20). An arrangement for the recovery of heat from the incinerator gases is illustrated (Figure 20).

Frequently, the contaminated gases are delivered to the afterburner by the fan exhausting process equipment. In one type of catalytic incinerator, the exhaust fan is located within the afterburner housing between the preheat burner and the catalyst bed. This fan also mixes the gases and distributes them evenly over the catalyst. Condensates do not occur in the fan since it operates above condensation temperature. Of course, the fan must be constructed of materials that can withstand the maximum temperature of the gases being handled.

The interior chamber of the afterburner may be constructed of 11 to 16 gauge black iron, heat-resisting steel, stainless steel, or refractory materials. Heat-resisting steel should be used for operating temperatures between 750° and 1100°F; stainless steel is recommended for operating temperatures exceeding 1100°F. Refractory materials are recommended for temperatures exceeding 1300°F. A thickness of 4 to 6 inches of similar thermal insulation is required unless refractory materials are used. The exterior sheet is usually fabricated from 16 to 20 gauge mild steel. The framework is usually fabricated from standard structural steel. Gas velocities through the chamber of about 20 fps have been found satisfactory.

FIGURE 19: CATALYTIC INCINERATION WITHOUT HEAT RECOVERY

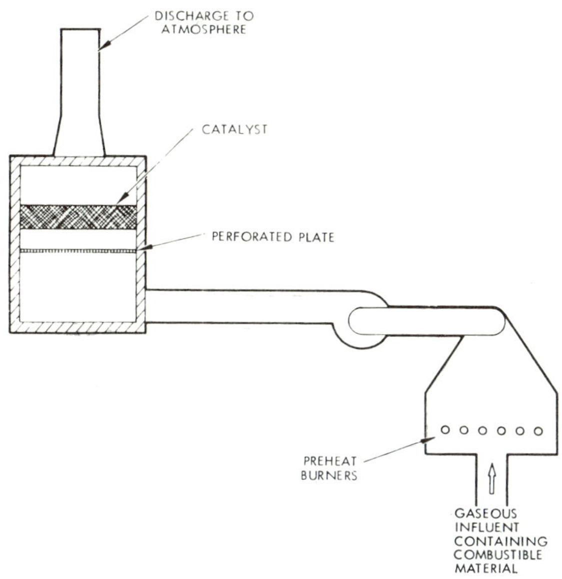

FIGURE 20: CATALYTIC INCINERATION WITH HEAT RECOVERY

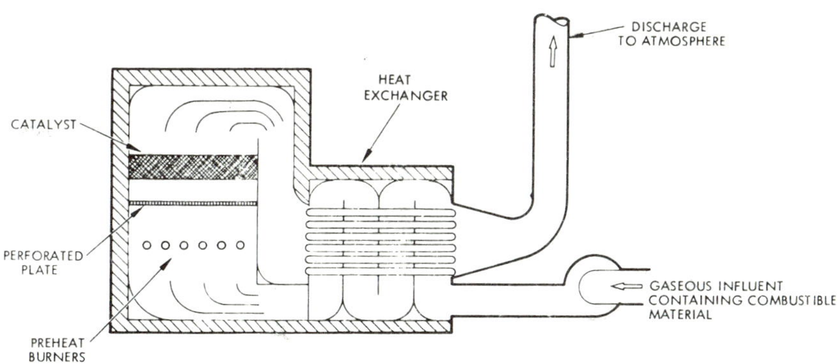

Source: PB 224 582

The contaminated gases are preheated to the reaction temperature by a gas burner before passing through the catalyst. When the preheat burner is on the discharge side of a fan, a premix gas burner is normally used because of the positive pressure in the combustion chamber. When the fan is between the preheat burner and the catalyst bed, an atmospheric burner may be used since a negative pressure exists in the preheat section of the combustor. Sizing the preheat burner to increase the temperature of the contaminated gases to the required catalyst discharge temperature without regard to the heating value of the combustible materials is advisable especially if considerable variation in concentration occurs.

The concentration of combustibles from process equipment is normally 25% of the lower explosive limit or less to meet the requirements of the National Board of Fire Underwriters. Experience indicates that the preheat burner should have sufficient capacity to heat the contaminated gas stream to 950°F minimum to obtain adequate catalytic combustion of the compounds that are more difficult to burn. Some burning of contaminants usually occurs in the preheating zone. The preheated gases then flow through the catalyst bed where the remaining combustible contaminants are burned by catalysis.

A direct relationship is believed to exist between the autoignition temperature of an organic vapor and the temperature at which catalytic oxidation will occur. In other words, the higher the autoignition temperature of a compound, the higher the expected temperature required for catalytic oxidation.

Catalytic incinerators possess an inherent maintenance factor not present in other types of incinerators, namely, that usage of the catalyst produces a gradual loss of activity through fouling and erosion of the catalyst surface. Occasional cleaning and eventual replacement of the catalyst are therefore required.

Modulating controls on the burner regulated by the catalyst discharge gas temperature are usually used. This allows the fuel gas input to the preheat burner to be reduced as the rate of heat released in the catalyst bed increases as a result of larger concentrations of combustible vapors. The sensing instrument commonly used is a type employing a fluid-filled bulb for detecting gas temperature with capillary and bellows. Movement of the bellows is amplified and transmitted to the preheat burner gas valve and combustion air blower blast gate. Electronic instruments are used less frequently because of considerably greater cost.

When operating conditions do not vary greatly, an improved means of ensuring maximum combustion efficiency seems to be the firing of the preheat burner at a fixed input capable of heating the contaminated gases to the temperature required for complete oxidation at the maximum rate of influx. Installation of a high-temperature-limiting control on the downstream side of the catalyst may be necessary to prevent overheating of the catalyst section.

Finally, the burned gases are discharged through a stack to the atmosphere, to a process that may use the sensible heat of the exhaust gases, such as a bake oven or dryoff oven, or they may be passed through an exchanger for heating the gases entering the combustor, which thereby reduces the amount of fuel required by the preheat burners.

The installed capital investment cost of a catalytic incineration system is a function of the difficulty of the combustion reaction, inclusion of auxiliary equipment, materials of construction and the extent of heat recovery. Operating costs generally reflect fuel consumption in the preheater (and are therefore a function of influent gas temperature, catalyst temperature, and heat recovery) as well as periodic catalyst replacement of activation. Estimates from various literature sources of both capital and operating costs are presented (Table 13).

Basically, the only process modifications utilized in catalytic incineration is waste heat recovery through heat exchange with or without recirculation of the hot effluent gas exiting the catalyst bed. Where the combustible concentration is relatively high, i.e., where the temperature rise through the bed approaches 400°F or more, it is often desirable to take a portion of the hot gases after passage through the catalyst and recirculate them, combining them with relatively low temperature influent prior to introducing them into the catalytic system. In this way, a stream at approximately 300°F could be combined with a portion of the effluent from the catalyst at 700°F, bringing the inlet temperature up to 500°F, thereby reducing the preheat to a minimum.

Another method of reducing the preheat required, even when the combustible content is relatively low, is to incorporate a heat exchanger. The influent to the process is passed through the cold side of the heat exchanger, then passed over the preheat burner and through the catalyst bed. The effluent from the catalyst is passed through the hot side of the heat exchanger, thereby increasing the temperature of the influent up to the approximate ignition temperature of the catalyst. These two alternatives along with that of not utilizing heat recovery are presented (Figure 21).

TABLE 13: PROCESS ECONOMICS FOR CATALYTIC INCINERATION

	Amount of Gas treated (scfm*)	Influent Gas Temp. (°F)	Capital Cost ($/scfm)	Annual Fuel Cost ($/year as of 1973)
Basic catalytic unit A	10,000	350	2.30	14,600
		550	2.30	4,800
Catalytic unit A with heat exchanger	10,000	350	2.85	7,100
		550	3.39	(**)
Basic catalytic unit B	10,000	300	2.00–2.50	(***)
Catalytic unit B with heat exchanger	10,000	300	3.50–4.50	(***)
Basic catalytic unit C	5,000	400	3.00	3,800

*Contaminants are at less than 25% of L.E.L.
**No fuel is required. The preheat exchanger saves $14,600 annually in fuel costs.
***No data available.

FIGURE 21: HEAT RECOVERY OPTIONS

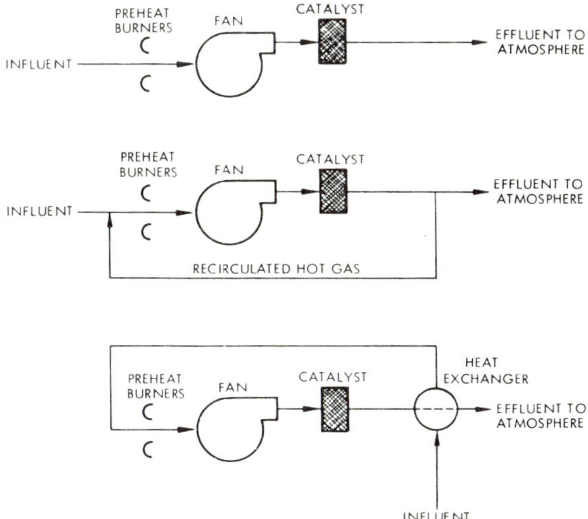

Source: PB 224 582

Due to the form of the waste material to be treated (dilute and in the gaseous state) catalytic incineration is best suited for use at the processing site where the waste material is generated. A listing of some of the typical industrial applications of catalytic incineration is presented (Table 14).

TABLE 14: TYPICAL INDUSTRIAL APPLICATIONS OF CATALYTIC INCINERATION SYSTEMS

Automotive paint baking	Vitamin manufacturing
Coil and strip coating	Rice browning
Metal parts finishing	Corn popping
Metal decorating	Nut roasting
Wire enameling	Coffee roasting
Hardboard coating and curing	Smoke houses
Resin manufacture	Potato chip cooking
Oil bodying	Rendering of animal by-products
Varnish cooking	Carbon baking
Oil sulfurization	Metal chip drying
Acrylate polymerization	Foundry core baking
Oil hydrogenation	Brake shoe bonding and burnoff
Oil quenching	Paper coating
Asphalt blowing	Printing
Tar and asphalt coating and saturating	Fabric finishing and curing
Phthalic and maleic anhydride manufacture	Paper mill digesters
Nitric acid plants	Fertilizer processing
Etching and dissolving metals with nitric acid	Wastewater stripping incinerators
	Investment casting and mold burnout
Fungicide manufacture	Synthetic rubber manufacture
Pharmaceutical manufacture	Sewage treatment

Source: PB 224 582

Catalytic incineration would find use at a National Disposal Site only as a secondary treatment (i.e., afterburner) on primary treatment processes evolving varying amounts of miscellaneous hydrocarbons, alcohols, amines, acids, esters, aldehydes and many other contaminants which are basically hydrocarbon in nature. These materials have varying degrees of toxicity and different odor levels; however, they all lend themselves to catalytic oxidation. Generally, the commercial catalysts available for installation in operations which emit compounds of this kind are not specific. That is, they tend to oxidize all combustible organic compounds in the stream regardless of their type and concentration. Catalysts are also effective in the reduction of oxides of nitrogen and in burning sulfur bearing compounds such as hydrogen sulfide and carbon bisulfide.

A process developed by L.E. Ravich (15) comprises utilization of a catalytic combustion surface for inducing and causing complete burning of combustible materials passed through flues, stacks and the like. The catalytic surface may comprise a perforated, porous refractory plate which carries a substantially flameless surface combustion on the face thereof, the surface combustion being supported independently of air from inside the flue. Other types of surfaces capable of supporting combustion may also be employed. They include the following: [1] a properly supported layer of granulated refractory material; [2] a diaphragm of bonded granular material; [3] a perforated ceramic plate. The face of the refractory plate is heated by the substantially flameless surface combustion to a temperature of from 1500° to 1650°F and higher to emit intensive infrared heat energy.

A scheme developed by H. Volker (16) is a cylindrical form thermal-catalytic incinerator unit for treating waste gas streams. The catalytic bed traverses the downstream end of the combustion chamber to effect contact with the mixture of waste gases and burner products from a burner axially positioned at the opposing inlet end of the combustion chamber. An annular heat exchange section around the combustion zone provides for cooling the resulting oxidized gas stream and preheating the incoming noxious stream which is then fed circumferentially around the burner to admix with the flame and hot gases therefrom.

A device developed by F. Tabak (17) (18) is a recuperative form of thermal-catalytic incinerator.

Direct-Flame Thermal Incinerators

Gas combustors, or as they are more commonly called, direct-flame thermal incinerators, are utilized to dispose of low concentration (usually less than 25% of the lower flammability limit) combustible gaseous waste. They have found wide application in the chemical and food processing industries. In direct-flame incineration combustible emissions are destroyed by exposure, under the proper conditions, to temperatures of 900° to 1500°F in the presence of a flame. The actual temperature required to do an effective job depends on the specific pollutants involved and the design of the combustion chamber.

The basic components of a direct-flame thermal incinerator are presented schematically (Figure 22). They are the combustion chamber, gas burner, burner controls, and temperature indicator. Operation of the unit is relatively simple. The contaminated gases are delivered to the combustor from the process equipment by an exhaust system. The combustion chamber must be designed for complete mixing of the contaminated gases with the flames and burner combustion gases. The presence of a flame is important for contaminant removal.

FIGURE 22: DIRECT-FLAME THERMAL INCINERATOR

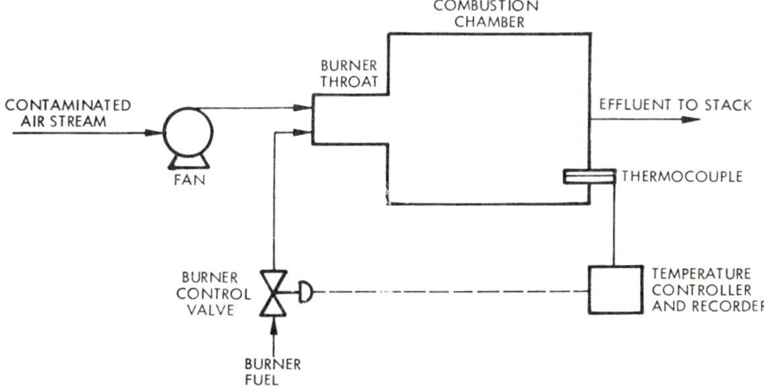

Source: PB 224 582

Evidence indicates that when using electric heat energy, much higher temperatures are required, 1500° to 1800°F, to obtain the same efficiency achieved with a direct-flame system at 1000° to 1400°F. One satisfactory method of achieving proper mixing is the admission of the contaminated gases into a throat where the burner is located. Sufficiently high velocities may be obtained here for thorough mixing of the gases with the burner combustion products in the region of highest temperature. Next, the gases pass into the main section of the combustor where velocity is reduced somethat by the larger cross-sectional area. Here the combustion reactions are completed and the incinerated air contaminants and combustion gases are discharged to either heat recovery equipment, scrubbers or directly to the stack.

Direct-flame incineration can be highly effective. Experience has shown that direct-flame incineration systems can operate continuously at efficiencies of 90 to 99+%. Such systems

can be readily adapted for automatic temperature control. This is an important consideration in processes where 100% continuous pollution abatement is required in spite of changing conditions. Temperature control can be obtained through the installation of modulating
gas burner controls. These controls may effect considerable savings in fuel where the volume
of gases or the amount of combustible material delivered to the combustor varies appreciably during the process cycle, or where both vary. A constant temperature in the afterburner chamber can be maintained through a gas temperature sensing element that actuates
the burner input control. When, however, the volume of contaminated gases and the amount
of combustible materials remain relatively constant, the firing of the burner at a fixed rate
is preferable.

An indicating- or recording-type temperature-measuring device is usually installed to show
the combustor operating temperature at all times. A bare thermocouple is normally used
because of low cost and rapid response to temperature changes. The thermocouple should
be located near the end of the combustion chamber to avoid large errors produced by direct
radiations from the burner flames. The thermocouples may be installed in a thermocouple
well for protection. A safety pilot is usually provided to shut off the burner gas supply
if the main burner malfunctions or the flow of contaminated gases to the combustor is
interrupted. It may also be advisable to install a high-temperature-limiting control to shut
off the gas burner fuel supply when combustion temperatures exceed safe operating levels.

In order to properly design an effective direct-flame fume incineration system, the following
information is required: flow to be handled (scfm); temperature and pressure of gases to
be handled; list of contaminants involved, type and concentration; deposit problem, if any;
fuel available, natural gas or oil; cost of fuel; number of hours of plant operations; and an
indication if heat energy can be used elsewhere in the plant.

There are basically two configurations of direct-flame thermal combustors; vertical (Figure
23) and horizontal units. The type of unit utilized is usually dictated by heat recovery requirements. That is, vertical units are usually used when no heat recovery is desired while
horizontal units are well suited for this application.

FIGURE 23: VERTICAL DIRECT-FLAME COMBUSTOR WITHOUT HEAT RECOVERY

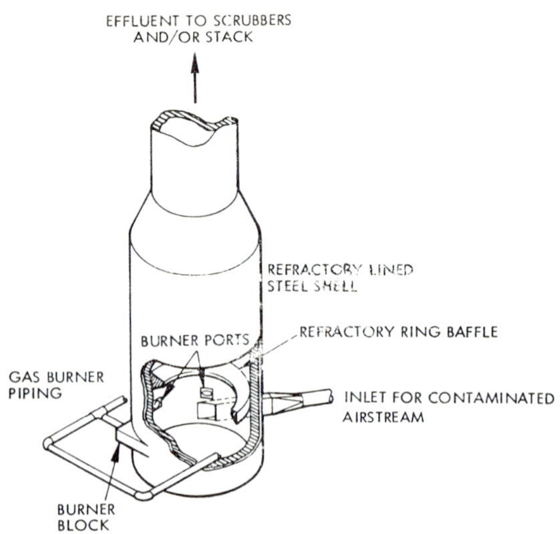

Source: PB 224 582

Direct-flame combustors, regardless of form, usually operate with throat velocities ranging from 15 to 25 fps, and combustion chamber residence times between 0.3 and 0.5 second. Operating temperatures are usually between 850° and 1500°F depending upon the waste being combusted. Standard industrial units are available with capacities ranging from 2,000 to 30,000 scfm of contaminated waste gas. Most direct-flame incinerators are constructed of firebrick or castable refractory with a sheet iron shell. Several types of gas burners have been successfully utilized in direct-flame combustors. Among these are: atmospheric, nozzle mixing, pressure mixing, premixing, and multijet gas burners.

The installed capital cost of a direct-flame thermal incinerator is a function of the difficulty of the combustion reaction, materials of construction and the extent of heat recovery operating costs generally reflect fuel consumption and are therefore dependent upon inlet gas temperatures and the required combustion temperature. They also reflect maintenance and labor. Estimates from various literature sources of both capital and operating costs are presented (Table 15).

TABLE 15: PROCESS ECONOMICS FOR DIRECT-FLAME INCINERATION

	Amount of Gas Treated, scfm*	Influent Gas Temp., °F	Capital Cost $/scfm	Annual Fuel Cost $/Year
Basic unit A	10,000	350	2.00	48,000
		550	1.95	30,700
Basic unit A with heat exchanger	10,000	350	2.40	33,500
		550	2.39	12,200
Basic unit B with heat exchanger	5,000	400	2.60	8,600
	25,000	150	3.20	35,400
Basic unit C	10,000	300	1.50-2.00	**
Basic unit C with heat exchanger	10,000	300	3.00-4.50	**
Basic unit D	**	**	5.00-10.00***	**

*Contaminants are at less than 25% of L.E.L.
**No data available.
***Installed capital cost.

Source: PB 224 582

The primary process modifications utilized in direct-flame thermal combustion are waste heat recovery options. The principle heat recovery options are influent preheat through heat exchange with the hot gaseous effluent and effluent heat exchange with other process streams (Figure 24).

FIGURE 24: HEAT RECOVERY OPTIONS

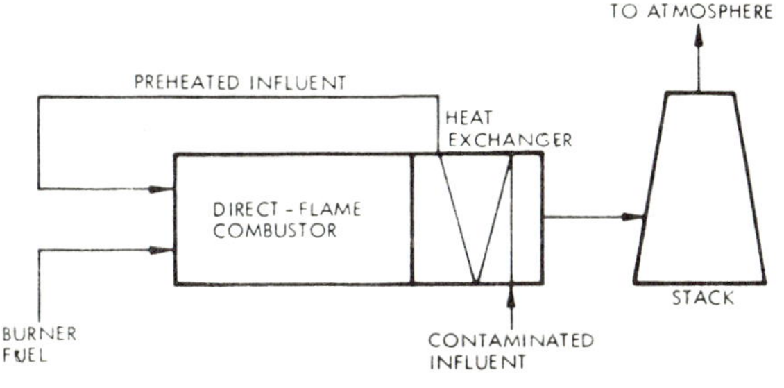

(continued)

FIGURE 24: (continued)

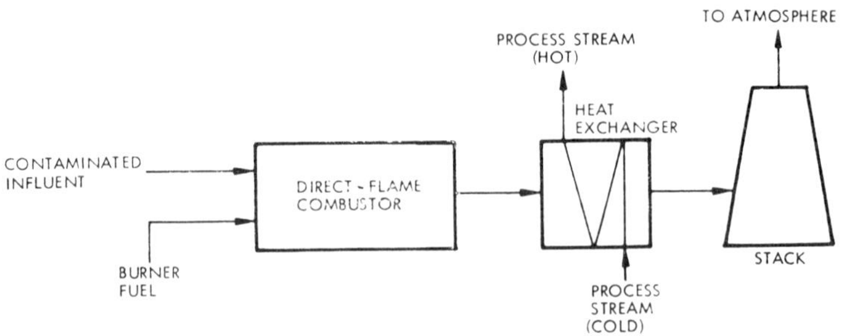

Source: PB 224 582

Additionally, secondary scrubbers may be utilized to further decrease concentrations of pollutants such as chlorides, fluorides, sulfur containing compounds and nitrogen oxides when they are present. This practice is usually very expensive since the contaminant levels in the combustor effluent stream are usually very low.

Due to the form of the waste material being treated (dilute and in the gaseous state) direct-flame combustors are best suited for use at the processing site where the waste is generated. A listing of some of the typical industrial applications of direct-flame combustion systems is presented (Table 16).

TABLE 16: TYPICAL DIRECT FLAME INCINERATION APPLICATIONS

Resin manufacturing	Phthalic and maleic anhydride
Paint and varnish cooking	manufacture
Wire enameling	Food processing
Metal decorating	Rendering of fats
Coil and strip coating	Bonding and burn-off
Carbon baking ovens	Grain dryers
Tar and asphalt blowing	Plastic curing
Fish meal processing	Sewage treatment
Printing press ink drying	Air sterilization

Source: PB 224 582

Direct-flame combustors would find use at a National Disposal Site as a secondary treatment (i.e., afterburner) on primary treatment processes evolving varying amounts of combustible contaminants. They are also well suited to the purification of ventilation air or any air which is monitored for pollutant control.

A device developed by R.J. Ruff et al (19) is an incinerating apparatus which utilizes a short, wide, high temperature flame pattern which will preclude flame from contacting or sweeping the interior walls of the furnace housing and which will have waste gas distributing means entirely circumscribing the flame pattern entering the furnace housing.

A device developed by W.A. Denny et al (20) is an improved direct flame incinerator which effects a spiral flow heat exchange preheating of the waste gas stream and a multiple stage spiralling discharge of such stream into an axially flowing flame pattern and resulting hot combustion gas stream.

An apparatus developed by L.C. Hardison et al (21) is an improved thermal incinerating system for contaminated air streams which uses the air content of the stream to provide all the oxygen requirements for the burning of the fuel. In this way handling and heating the primary air for the burner is eliminated.

A device developed by F. Tabak (22) is a recuperative form of direct flame incinerator unit providing a contaminated gas stream inlet plenum section which fully surrounds an axially positioned burner means so as to provide efficient mixing with the fuel within and downstream from the burner. The internally located combustion section of the unit is encompassed with an elongated annular form heat exchange section which in turn is provided with multiple tubes so as to have an elongated confined passageway for hot combustion gases and a separated adjacent passageway for the incoming gas stream. The passageways are in heat exchange relationship with one another to give a highly efficient recuperative heat transfer to the incoming gas stream and a uniform full 360° annular flow to the plenum section.

An apparatus developed by G.L. Brewer (23) is a fume incinerating unit having a multiple section, tubular-form combustion chamber, where the sections are connected in a serpentine manner and positioned within a fume receiving housing in a manner whereby the fumes will pass over and around the plurality of sections in a heat exchange relationship therewith. The unit is particularly adapted for use with wire enameling-drying ovens where the fumes from the plurality of coated wires can pass laterally into the housing for the incinerator unit and transversely across the plurality of tubular sections comprising the combustion chamber.

A device developed by J.H. Hirt (24) is a fume incinerator apparatus for purifying a continuous flow of waste gas that contains combustible material by incinerating such material. The incinerator includes a chamber, a fuel-operated burner in the chamber, and flow control means for rapidly and thoroughly mixing waste gas and the hot gaseous products from the burner for highly effective and efficient incineration of the combustible material.

A device developed by K. Zenkner (25) is a device for the thermal incineration of exhaust air from an industrial plant, such as a drying chamber, which air contains oxidizable foreign bodies, fluid particles or gases. The exhaust air is passed through a pair of heat exchangers in series, then through a burner while supplying air to the burner, and the resultant high combustion gases are led back outside the burning zone in a direction opposite to the original flow direction through the heat exchangers in heat exchange relation with the incoming exhaust air.

An apparatus developed by H.O. Ebeling et al (26) is an apparatus for incinerating waste gas which includes a hollow outer shell having a hollow inner liner supported therein so that an air passageway is formed therebetween. Front and rear walls are provided secured to the outer shell, and an intermediate wall having an opening disposed therein is secured within the outer shell. An air inlet is formed in the outer shell adjacent to the forward end thereof for admitting combustion air into the air passageway and means for injecting fuel gas into the outer shell at a position between the rearward end of the inner liner and the intermediate wall are provided. A waste gas inlet is formed in the outer shell for admitting waste gas thereto between the rear wall and the intermediate wall.

A direct flame incinerator unit developed by F. Tabak (27) makes use of heat exchange tubing in two different end-to-end heat exchange sections. Large diameter, self-supporting tubes are used to surround the flame-combustion zone and eliminate the conventional use of an expensive, interior cylinder positioned within and around the flame zone so as to separate the incoming gas stream from the combustion product stream. A multiplicity of small diameter tubes are used in the downstream part of the unit, with respect to combustion gas flow, and these tubes connect through a redistribution section to the larger diameter tubes around the flame-combustion section, whereby the incoming gas stream passes entirely countercurrently to the combustion gases and feeds directly into a plenum section around the burner means.

A device developed by B.G. Altmann (28) is a device for the complete combustion of gaseous waste products comprising a refractory chamber divided by means of a refractory wall into interconnecting first and second compartments which have an increasing internal cross-section from the point of gas entry to gas exit. Air is introduced in controlled amounts into the stream of gaseous wastes as it traverses the length of the chamber via the first and second compartments. Supplementary heating means are provided in the first compartment to raise the temperature above the point of combustion of the gas/air mixture whenever required.

Flares

The flare type burner has been utilized in many petroleum refineries and petrochemical plants to incinerate relatively large volumes of combustible gases and aerosols. Flare burners are of two basic types: the ground flare and the elevated or tower flare. The ground flare, as its name implies, is used at ground level where there is sufficient space around the flare for safety purposes to burn waste gas from an oil field operation or similar source. The tower flare, usually found in refineries, is elevated to keep the flame well above the level of surrounding process equipment protecting the refinery against possible fires.

Flares are basically open pipes which discharge a combustible gas directly to atmosphere with the end of the pipe containing a flame device and a continuous pilot or pilots to ignite the waste gas. Air for combustion is supplied by the surrounding atmosphere. Steam injection is often supplied to the flame of the flare to prevent smoking when burning waste hydrocarbon gases which have more than two carbon atoms. Flares are affected by atmospheric conditions, especially high winds. They cannot be considered an infallible method of waste gas disposal because unburned waste gases often escape from a flare system, but they are expedient and economical for high-volume discharges of combustible waste gases.

From a pollution viewpoint, the ideal flare is a combustion device that burns waste gases completely and smokelessly. But, in actual practice, flare utilization introduces the possibility of smoke and other objectionable gases such as carbon monoxide, sulfur dioxide, and nitrogen oxides. Some types of flares have been developed that ensure that combustion is smokeless and in some cases nonluminous. Luminosity, while not an air pollution problem, does attract attention to the operation and in certain cases can cause bad public relations. There is also the consideration of military security which nonluminous emergency gas flares would be desirable.

Smoke, when present, is the result of incomplete combustion. Smokeless combustion can be achieved by: [1] adequate heat values to obtain the minimum theoretical combustion temperatures, [2] adequate combustion air, and [3] adequate mixing of the air and fuel. An insufficient supply of air results in a smoky flame. Combustion begins around the periphery of the gas stream where the air and fuel mix, and within this flame envelope the supply of air is limited. Hydrocarbon side reactions occur with the production of smoke. In this reducing atmosphere, hydrocarbons crack to elemental hydrogen and carbon, or polymerize to form hydrocarbons. Since the carbon particles are difficult to burn, large volumes of carbon particles appear as smoke upon cooling.

Side reactions become more pronounced as molecular weight and unsaturation of the fuel gas increase. Olefins, diolefins, and aromatics characteristically burn with smoky, sooty flames as compared with paraffins and naphthenes. A smokeless flame can be obtained when an adequate amount of combustion air is mixed sufficiently with the fuel so that it burns completely and rapidly before any side reactions can take place. This rapid burning may be achieved by providing adequate mixing of fuel and air by means of jets. Air jets would be ideal, but the cost of providing high pressure air is prohibitive. Steam jets are the most satisfactory method of inspirating air into the combustion zone while giving good turbulence at the same time.

Steam also provides some fringe benefits: it reacts with the hydrocarbons, forming oxygenated compounds that burn readily at low temperatures, and it lowers the partial pressure of the fuel causing greater separation of the molecules. This inhibits polymerization, one

of the mechanisms likely to cause smoke. The ratio of steam to hydrocarbon is important.
The required ratio for smokeless operation increases with increased molecular weight and
fraction of unsaturates in the waste gas. Incineration of hydrocarbons in a typical steam-
inspirated-type elevated flare results in incomplete combustion of the feed gases. The results of a field test on a flare unit were reported in the form of ratios as follows: CO_2:hy-
drocarbons, 2,100:1; CO_2:CO, 243:1. These results indicate that the hydrocarbons and
carbon monoxide emissions from a flare can be much greater than those from properly operated boilers or furnaces.

Other contaminants that can be emitted from flares depend upon the composition of the
gases burned. The most commonly detected emission is sulfur dioxide, resulting from the
combustion of various sulfur compounds (usually hydrogen sulfide) in the flared gas. Toxicity, combined with low odor threshold, make venting of hydrogen sulfide to a flare an
unsuitable and sometimes dangerous method of disposal. Materials that tend to cause health
hazards or nuisances should not be disposed of in flares. Compounds such as mercaptans
or chlorinated hydrocarbons require special combustion devices with chemical treatment
of the gas or its products of combustion.

There are, in general, two types of flares for the disposal of waste gases: elevated flares
and ground-level flares. The essential parts of a flare are the burner, stack, seal, liquid
trap, controls, pilot burner, and ignition system. Smokeless combustion may be attained
through the use of elevated stack flares which utilize steam injection to provide turbulence
and inspirate air. Three main types of steam-injected elevated flares are in use. These
types vary in the manner in which the steam is injected into the combustion zone.

In the first type, there is a commercially available multiple nozzle which consists of an
alloy steel tip mounted on the top of an elevated stack (Figure 25). Steam injection is
accomplished by several small jets placed concentrically around the flare tip. These jets
are installed at an angle, causing the steam to discharge in a converging pattern immediately
above the flare tip.

FIGURE 25: STACK FLARE EQUIPPED WITH MIXING NOZZLE

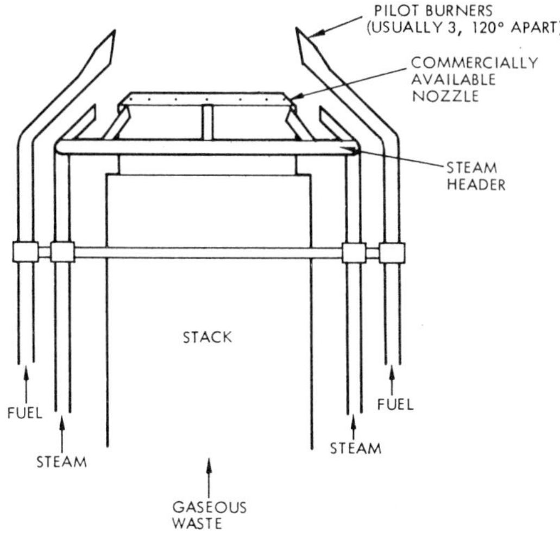

Source: PB 224 582

A typical refinery waste gas flare system utilizing a multiple steam jet burner is presented
(Figure 26). All relief headers from process units combine into a common header that conducts the hydrocarbon gases and vapors to a large knockout drum. Any entrained liquid
is dropped out and pumped to storage. The gases then flow in one of two ways. For
emergency gas releases that are smaller than or equal to the design rate, the flow is directed
to the main flare stack. Hydrocarbons are ignited by continuous pilot burners, and steam
is injected by means of small jet fingers placed concentrically about the stack tip.

FIGURE 26: WASTE GAS FLARE SYSTEM USING A MULTIPLE STEAM JET BURNER

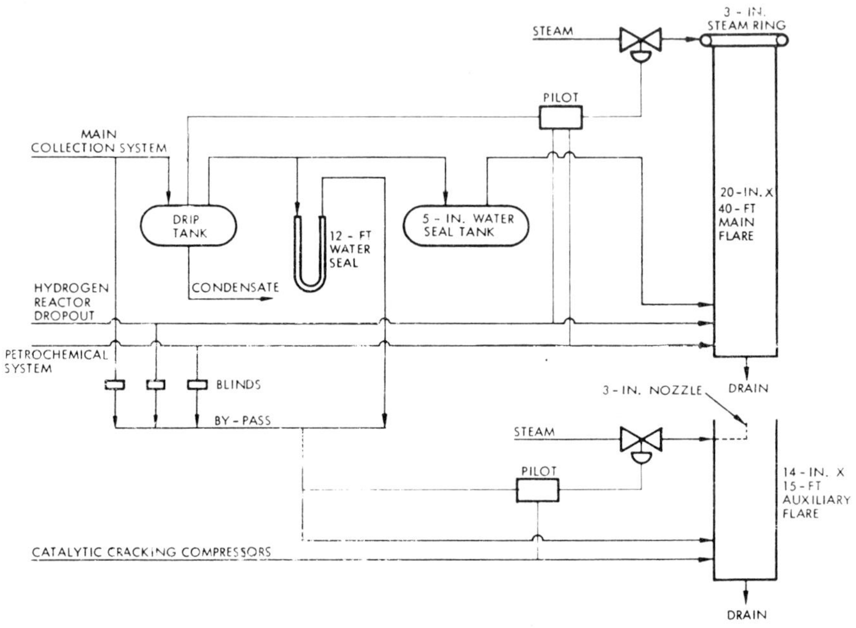

Source: PB 224 582

The steam is injected in proportion to the gas flow. The steam control system consists
of a pressure controller, having a range of 0 to 20 inches water column, that senses the
pressure in the vent line and sends an air signal to a control valve in the steam line. If the
emergency gas flow exceeds the designed capacity of the main flare, backpressure in the
vent line increases, displacing the water seal and permitting gas flow to the auxiliary flare.
Steam consumption of the burner at a peak flow is about 0.2 to 0.5 lb of steam per pound
of gas, depending upon the amount and composition of hydrocarbon gases being vented.
A small amount of steam (300 to 400 lb/hr) is allowed to flow through the jet fingers at
all times. This steam not only permits smokeless combustion of gas flows too small to
actuate the steam control valves but also keeps the jet fingers cooled and open.

A second type of elevated flare has a flare tip with no obstruction to flow, that is, the
flare tip is the same diameter as the stack. The steam is injected by a single nozzle located
concentrically within the burner tip. In this type of flare, the steam is premixed with the
gas before ignition and discharge. This configuration flare is generally referred to as an
Esso type flare (Figure 27).

FIGURE 27: ESSO TYPE FLARE

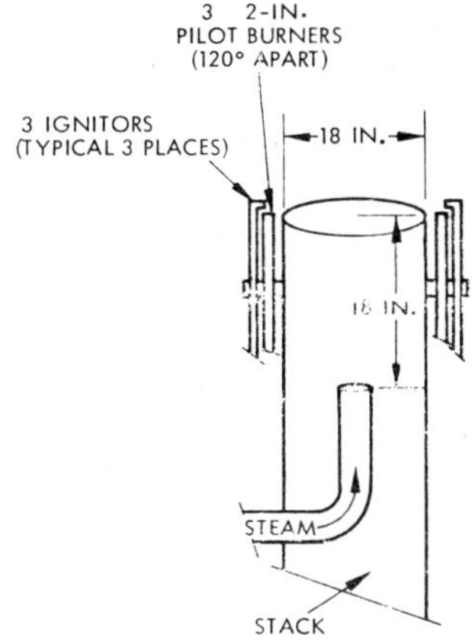

Source: PB 224 582

A typical flare system serving a petrochemical plant using this type burner is shown (Figure 28). The type of hydrocarbon gases vented can range from a saturated to a completely unsaturated material. The injection of steam is not only proportioned by the pressure in the blowdown lines but is also regulated according to the type of material being flared. This is accomplished by the use of a ratio relay that is manually controlled. The relay is located in a central control room where the operator has an unobstructed view of the flare tip. In normal operation the relay is set to handle feed gas which is most common to this installation.

FIGURE 28: WASTE-GAS FLARE SYSTEM USING ESSO TYPE BURNER

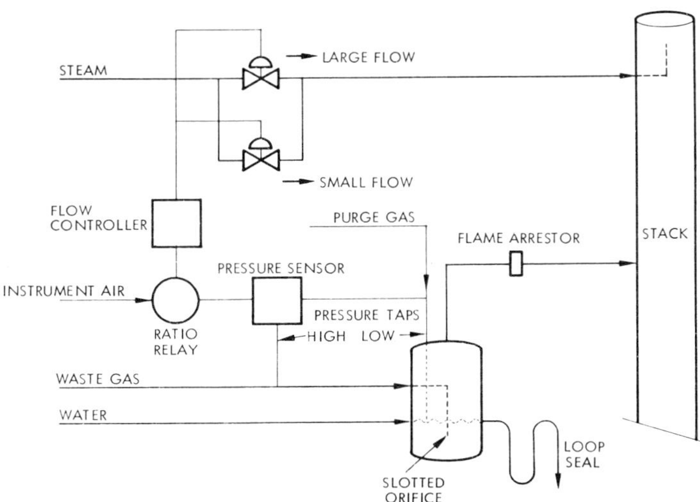

Source: PB 224 582

In this installation, a blowdown header conducts the gases to a water seal drum and the end of the blowdown line is equipped with two slotted orifices. The flow transmitter senses the pressure differential across the seal drum and transmits an air signal to the ratio relay. The signal to this relay is either amplified or attenuated, depending upon its setting. An air signal is then transmitted to a flow controller that operates two parallel steam valves.

The 1-inch steam valve begins to open at an air pressure of 3 psig and is fully open at 5 psig. The 3-inch valve starts to open at 5 psig and is fully open at 15 psig air pressure. As the gas flow increases, the water level in the pipe becomes lower than the water level in the drum, and more of the slot is uncovered. Thus, the difference in pressure between the line and the seal drum increases. This information is transmitted as an air signal to actuate the steam valves. The slotted orifice senses flows that are too small to be indicated by a pitot-tube-type flow meter. The water level is maintained 1 to 1½ inches above the top of the orifice to take care of sudden surges of gas to the system.

A 3-inch steam nozzle is so positioned within the stack that the expansion of the steam just fills the stack and mixes with the gas to provide smokeless combustion. This type of flare is probably less efficient in the use of steam than some of the commercially available flares but is desirable from the standpoints of simpler construction and lower maintenance costs.

A third type of flare, the Sinclair elevated flare (Figure 29) is equipped with a flare tip constructed to cause the gases to flow through several tangential openings to promote turbulence. A steam ring at the top of the stack has numerous equally spaced holes about ⅛-inch in diameter for discharging steam into the gas stream. The injection of steam in this latter flare may be automatically or manually controlled. In most cases, the steam is proportioned automatically to the rate of gas flow.

FIGURE 29: SINCLAIR TYPE FLARE

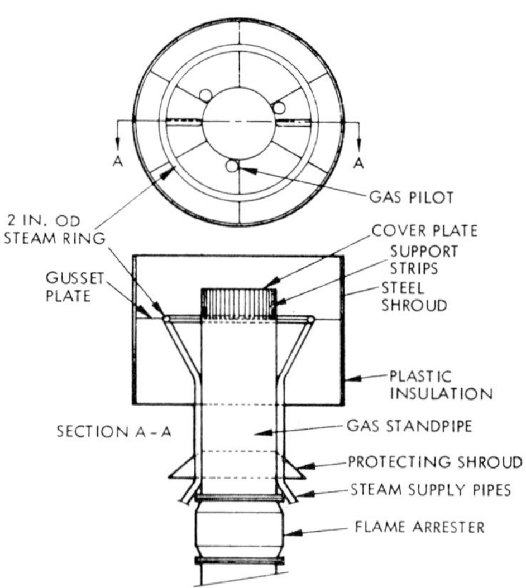

Source: PB 224 582

However, in some installations, the steam is automatically supplied at maximum rates, and manual throttling of a steam valve is required for adjusting the steam flow to the particular gas flow rate. There are many variations of instrumentation among various flares, some designs being more desirable than others. For economic reasons, all designs attempt to proportion steam flow to the gas flow rate.

There are four principal types of ground level flare: horizontal venturi, water injection, multijet, and vertical venturi. A typical horizontal venturi-type ground flare system is shown (Figure 30).

FIGURE 30: TYPICAL VENTURI GROUND FLARE

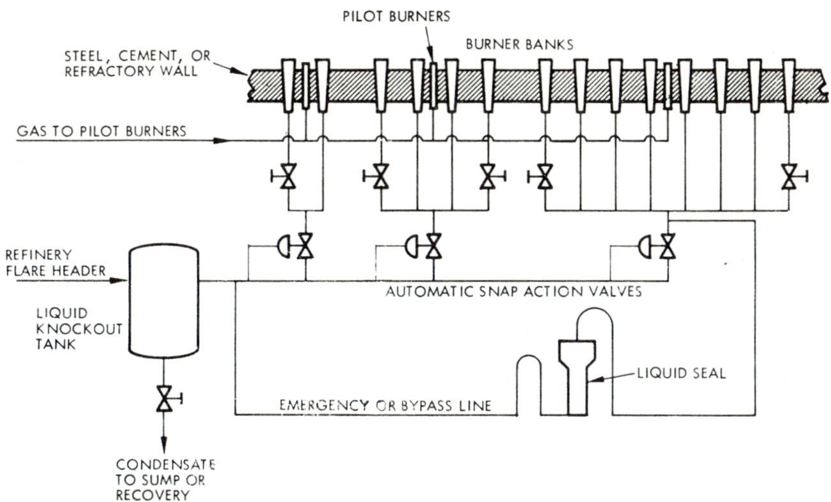

Source: PB 224 582

In this system, the refinery flare header discharges to a knockout drum where any entrained liquid is separated and pumped to storage. The gas flows to the burner header, which is connected to three separate banks of standard gas burners through automatic valves of the snap-action type that open at predetermined pressures. If any or all of the pressure valves fail, a bypass line with a liquid seal is provided (with no valves in the circuit), which discharges to the largest bank of burners.

Another type of ground flare used in petroleum refineries has a water spray to inspirate air and provide water vapor for the smokeless combustion of gases (Figure 31). This flare requires an adequate supply of water and a reasonable amount of open space. The structure of the flare consists of three concentric stacks. The combustion chamber contains the burner, the pilot burner, the end of the ignitor tube, and the water spray distributor ring. The primary purpose of the intermediate stack is to confine the water spray so that it will be mixed intimately with burning gases. The outer stack confines the flame and directs it upward. Water is not as effective as steam for controlling smoke with high gas flow rates, unsaturated materials, or wet gases. The water spray flare is economical when venting rates are not too high and slight smoking can be tolerated. A recent type of flare developed by the refining industry is known as a multijet. This type of flare was designed to burn excess hydrocarbons without smoke, noise, or visible flame.

FIGURE 31: TYPICAL WATER SPRAY TYPE GROUND FLARE

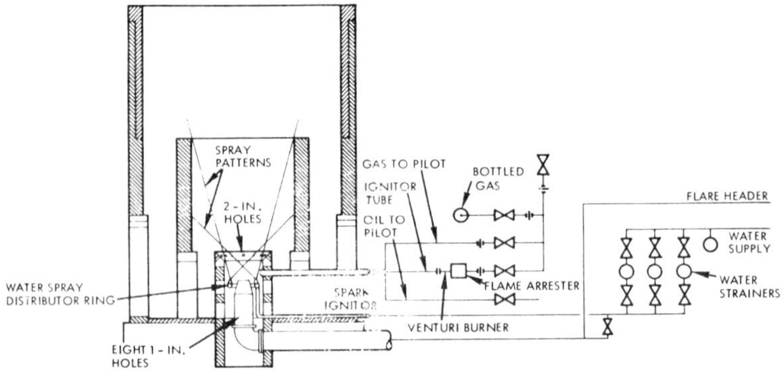

FIGURE 32: MULTIJET FLARE SYSTEM

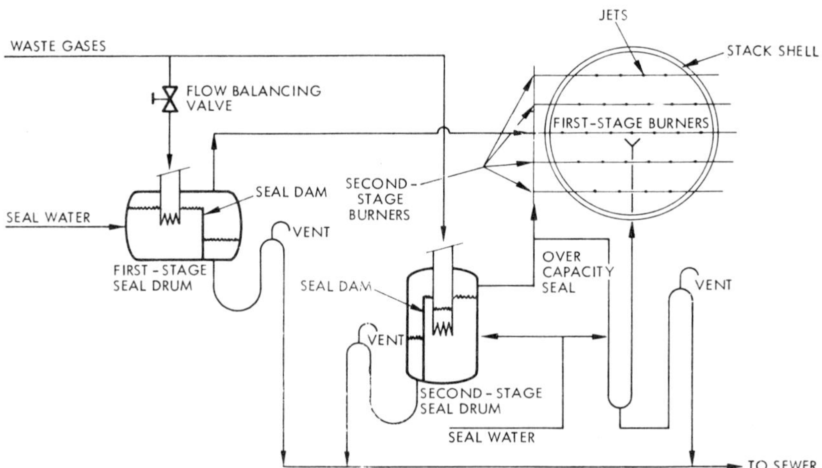

Source: PB 224 582

A sketch of a multijet flare installation is shown (Figure 32). The flare uses two sets of burners; the smaller group handles normal gas leakage and small gas releases, while both burner groups are used at higher flaring rates. This sequential operation is controlled by two water-sealed drums set to release at different pressures. In extreme emergencies, the multijet burners are by-passed by means of a water seal that directs the gases to the center of the stack. This seal blows at flaring rates higher than the design capacity of the flare. At such an excessive rate, the combustion is both luminous and smoky, but the unit is usually sized so that an overcapacity flow would be a rare occurrence. The overcapacity line may also be designed to discharge through a water seal to a nearby elevated flare rather than to the center of a multijet stack. Similar staging could be accomplished with automatic

valves or backpressure regulators; in this case, the water seal drums are used because of re-
liability and ease of maintenance. The staging system is balanced by adjusting the hand
control butterfly valve leading to the first-stage drum, which locks after its initial setting.

FIGURE 33: VERTICAL VENTURI TYPE FLARE

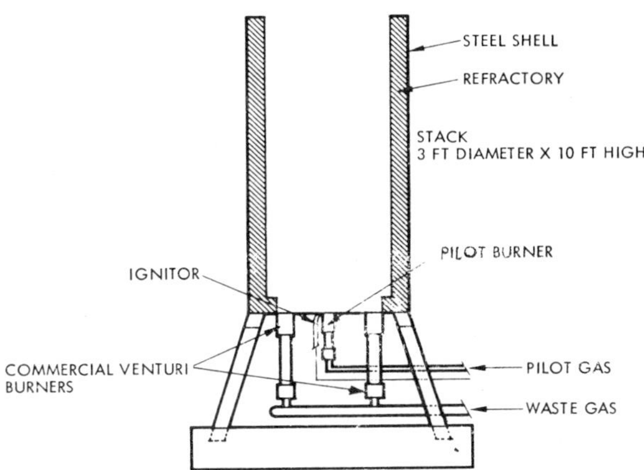

Source: PB 224 582

The fourth type of flare, based upon the use of commercial-type venturi burners, is presented
(Figure 33). This type of flare has been used to handle vapors from gas-blanketed tanks, and
vapors displaced from the depressuring of butane and propane tank trucks. Since the commercial venturi burner requires a certain minimum pressure to operate efficiently, a gas blower
must be provided. Generally, burners operate at a pressure of ½ to 8 psig. This type of flare
is suitable for relatively small flows of gas of a constant rate. Its main application is in situations where other means of disposing of gases and vapors are not available.

Most refineries and petrochemical plants have a fixed schedule for inspection and maintenance
of processing units and their auxiliaries. The flare system should not be exempted from this
practice. Removal of a flare from service for maintenance requires some type of standby
equipment to disperse emergency gas vents during the shutdown. A simple stack with pilot
burner should suffice for a standby. Coordinating this inspection to take place at the time
when the major processing units are also shut down is good practice. Flare instrumentation
requires scheduled maintenance to ensure proper operation. Most of the costs and problems
of flare maintenance arise from this instrumentation. Maintenance expenses for flare burners can be reduced by constructing them of chrome-nickel alloy. Because of the inaccessi-
bility of elevated flares, the use of alloy construction is recommended.

The capital investment will vary significantly depending upon the complexity of the overall
system, the type and quantity of waste gas being combusted and the materials of construction. Operating costs are mainly a function of steam, water and pilot fuel requirements,
maintenance and labor. Flares are generally applicable to the ultimate disposal of large
volumes of combustible gases and aerosols. They have found application in most petroleum
refineries and petrochemical plants. However, flares are not recommended for use at National Disposal Sites because of the associated lack of effluent control. This lack of control
might result in emissions to the surroundings of harmful combustion products. Additionally,
the form of waste handled by industrial flares (concentrated gases in large volumes) suggests
that flares are best suited for use at processing sites where the waste gas is generated.

A device developed by D.J. Bergman (29) is a flare stack which causes the flames issuing from it always to pass upwardly or horizontally from the discharge end thereof. It is its object to prevent flames from a flare from being sucked down behind the flare stack, thereby severely reducing its life and possibly creating a dangerous condition wherein burning may take place within the confined portion of the stack due to air passing into the stack through holes burned through its side.

An apparatus developed by C.A.H. Williams (30) comprises a water injection system for uoo on clevated flare stacks, comprising in combination (a) a burner head unit fitted with a nozzle adapted to direct an atomized spray of water into the combustion zone of the burner head unit, the nozzle being fitted with a cowl and a drainage tray, the cowl being adapted to return water to the drainage tray, (b) a storage tank adapted to receive water from the drainage tray, (c) a pump adapted to take suction from the storage tank and to deliver at the nozzle, and (d) a wind deflecter adapted to be set about the burner head unit in such manner that air is deflected to the leeward side of the stack under windy conditions.

A device developed by A.J. Turpin (31) is a flare stack combustion tip having a centrally disposed gas conduit having an opening in the lower end thereof for communication with a flare stack and the upper end thereof being at least partially closed, a plurality of spaced apart gas conducting channels extending outward from the gas conduit and in open communication along one end thereof with the gas conduit, and a gas emission orifice in an upper segment of each of the gas conducting channels and extending substantially the entire length of the upper segment, the gas emission orifices being disposed radially with respect to the axis of the gas conduit.

A device developed by A.E. Proctor (32) is an improved flare stack tip which will overcome the disadvantages of the known flare stack tips and achieve an improvement in smokeless and noiseless combustion. There is provided a flare stack tip including means for introducing ambient air into the gas stream prior to its leaving the outlet of the stack and combusting. Further, there is provided a method of combustion in a flare stack tip wherein air is introduced into the gas stream prior to its leaving the outlet of the stack and combusting.

The means for introducing ambient air into the gas stream can be a series of venturis or preferably one or more air-inducing devices based upon the phenomenon known as the Coanda effect. There is also provided a gas stream diverter in the flare stack tip the function of which is to divert the gas stream radially outwards so as to improve the properties of air/gas collision.

A scheme developed by V. Jasinsky et al (33) is one whereby flare stack smoke is reduced or eliminated by controlling the temperature of the flare. The temperature is sensed by a sensing element, e.g., one or more thermocouples connected in parallel and located in the vicinity of the flare; the sensing element produces a signal which directly or indirectly activates steam control valve to adjust the flow of steam to the tip of the flare stack so as to raise or lower the flare temperature to the temperature of optimum combustion of flare stack effluent.

A device developed by D.J. Frey et al (34) is an automatic ignition system for flaring waste combustible gases being exhausted through a stack, which utilizes an ignitor burner which is automatically energized when gases are passed to the stack. A main flame scanner, ignitor burner scanner, and a timer are so interrelated that a safe, economical, and efficient operation is performed.

A device developed by R.D. Reed (35) is a burner assembly for the upper end of a flare stack designed for the combustion of gases at significant elevations above the surrounding terrain wherein a smoke suppressant is thoroughly mixed with the gas prior to the burning reactions and to a structural arrangement which distributes the smoke suppressant in volumes proportional to the quantity of the gases to be burned.

A device developed by R.D. Reed et al (36) is a nonpolluting waste-gas disposal system for processing plants or other operation subject to variable quantities of waste–gas disposal. The system includes a low-level burner normally adapted to handle the usual volumes of plant waste gas, required to be disposed, without visible flame, smoke or noise pollution. An elevated flare can be used in combination to consume gases in excess of the normal capacity of the low-level flare.

A system developed by E.C. Eubanks (37) is one in which waste gases from refinery units are delivered into a vessel having a plurality of outlet lines connected into the vessel at different elevations. The outlet lines are connected into a waste-gas flare line for delivery of waste gas to a smokeless flare. A sealing liquid in the vessel can be displaced from the vessel through a line opening from the bottom of the vessel to a sealing liquid reservoir. The level of sealing liquid in the vessel, which is dependent upon the rate waste gases are discharged through the outlet lines, produces a signal to control the rate of steam delivery to the flare to maintain smokeless flare. As the rate of waste gas flow into the vessel increases, the sealing liquid level in the vessel is lowered to uncover successively lower outlet lines and allow increased rate of flow of waste gas to the flare.

A system developed by J.S. Zink et al (38) is a system for smokeless burning of hydrocarbon gases in which liquid water is used as input in place of the customary use of steam to provide hydrogen necessary to accomplish complete smokeless combustion. A particular design of atomizer is used in which the pressure of gas provides the energy for atomizing the water into droplets of such small size that they will instantly evaporate in the flame, and provide the water vapor to reform the hydrocarbons and provide the necessary hydrogen-to-carbon ratio for smokeless combustion.

A smokeless gas flare developed by J.J. Stranahan et al (39) includes a steam control valve responsive to a new very low gas flow detector utilizing a knockout drum water seal, a by-pass line with an orifice run, and a water level switch for supplying an empirically set fixed flow of steam to the flare for ensuring a smokeless flame prior to the gas flow rate reaching a measurable rate. A full range flare including all other gas flow includes also several steam flow valves, each valve being empirically set to provide the proper steam-to-gas ratios throughout its respective range responsive to several corresponding gas flow detectors for ensuring a smokeless flare as the flare gas flow varies from the immeasurable ranges through the measurable ranges of the detectors.

A process developed by R.W. Evans et al (40) involves passing waste fluids into a flare stack for combustion, passing air substantially continuously into the stack while passing waste fluids thereinto, and controllably passing additional air into the flare stack in response to the flow rate of waste fluids into the stack.

A design developed by W.L. Buchanan et al (41) is a waste-gas flare which provides two separate sequential burner arrays within a combustion chamber and which is enclosed at the sides but open at the top and bottom. The lower one of the two burner arrays handles the first 25% of the waste gas which the flare is designed to receive. A substantial excess of air is induced in order to produce low temperature combustion.

The combustion products and excess air from the lower burner array pass to an upper array which burns all the remaining waste gas (above the first 25% sent to the lower array). The flare is substantially smokeless without the need for steam injection at most gas rates, although at extremely low rates small quantities of steam are added. Each of the burner arrays produces, by means of slotted openings, a series of fan-shaped flames which require less steam for smokeless operation than is needed when gas is discharged from circular nozzles.

Fluidized Bed Incinerators

Fluidized bed incinerators are versatile devices which can be used to dispose of solid, liquid and gaseous combustible wastes. The technique is a relatively new method for ultimate disposal of waste materials. It was first used commercially in the United States in 1962 and has found limited use in the petroleum industry, paper industry and for processing nuclear wastes. In addition, applications of fluidized bed combustion to the disposal of sanitary sludge have been reported. A typical fluidized bed incinerator is shown schematically (Figure 34). Air driven by a blower enters a plenum at the bottom of the combustor and rises vertically through a distributor plate into a vessel containing a bed of inert granular particles. Sand is typically used as the bed material. The upward flow of air through the sand bed results in a dense turbulent mass which behaves similar to a liquid. Waste material to be incinerated is injected into the bed where combustion occurs within the fluidizing media.. Air passage through the bed produces strong agitation of the bed particles. This promotes rapid and relatively uniform mixing of the injected waste material within the fluidized bed.

FIGURE 34: SCHEMATIC OF A FLUIDIZED BED COMBUSTOR

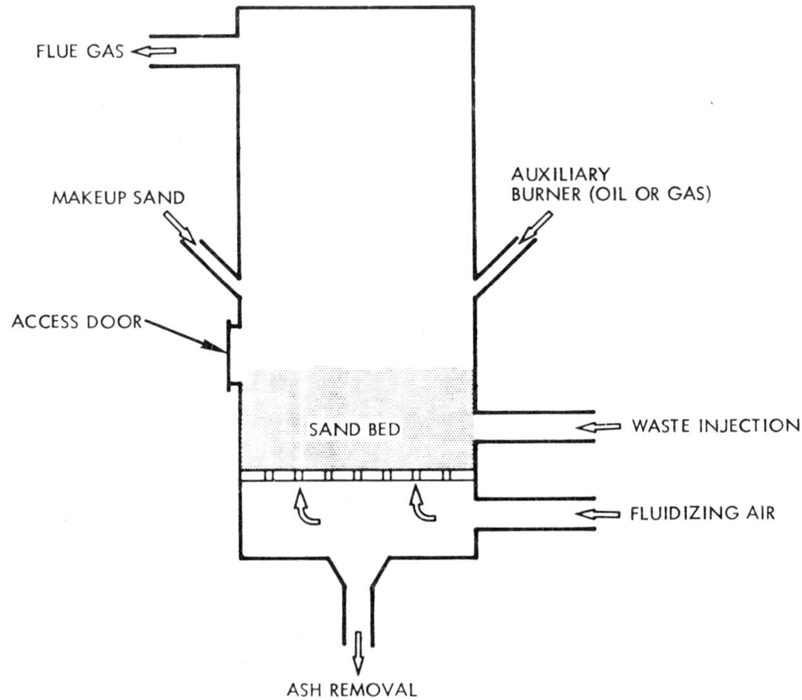

Source: PB 224 582

The mass of the fluidized bed is large in relation to the injected material. Bed temperatures are quite uniform and typically in the 1400° to 1600°F range. At these temperatures, heat content of the fluidized bed is approximately 16,000 Btu/ft^3 thus providing a large heat reservoir. By comparison, the heat capacity of flue gases at similar temperatures is three orders of magnitude less than a fluidized sand bed.

Heat is transferred from the bed into the injected waste materials to be incinerated. Upon reaching ignition temperature (which takes place rapidly) the material combusts and transfers heat back into the bed. Continued bed agitation by the fluidizing air allows larger waste particles to remain suspended until combustion is completed. Residual fines (ash) are carried off the bed by the exhausting flue gases at the top of the combustor. These gases are subsequently processed and/or scrubbed before atmospheric discharge. In specifying or designing a fluidized bed combustor, primary factors to be considered are: gas velocity; bed diameter; bed temperature; and, the type and composition of waste to be incinerated.

Gas velocities are typically low, in the order of 5 to 7 ft/sec. Maximum gas velocity is constrained by the terminal velocity of the bed particles and is therefore a function of particle size. Higher velocities result in bed attrition and an increased particulate load on downstream air correction equipment. Relatively low velocity reduces pressure drop and therefore lowers power requirements. Present fluidized bed design technology limits the bed diameter to 50 ft or less. At nominal values of gas velocity and temperature, the maximum volumetric flow would be approximately 2.5×10^6 acfm.

Bed depths range from about 15 inches to several feet. Variations in bed depth affect waste particle residence time and system pressure drop. One therefore desires to minimize bed depth consistent with complete combustion and minimum excess air. Bed temperatures are restricted by the softening point of the bed material. If sand is used, temperatures should be maintained below 2000°F to avoid softening and consequent agglomeration of the particles.

The type and composition of the waste is a significant design parameter in that it will impact storage, processing and transport operations (prior to incineration), as well as the combustion. If the waste is heterogeneous mixture such as municipal refuse and has a relatively low (<8,000 Btu/lb) heating value, processing (shredding, sorting, drying, etc.) operations will be more complex and auxiliary fuel addition to the combustor will be required. Homogeneous wastes which can be injected and uniformly dispersed in the bed should facilitate overall system design and minimize the bed volume.

Installation and operating costs will vary significantly depending upon the type of waste to be processed and the quantity and sophistication of water and air correction equipment required. Investment costs and operating costs have been estimated at approximately $20 and $5 per ton respectively. The fluidized bed combustor will normally be incorporated in an overall material handling, processing and disposal system to simultaneously cope with solid, liquid and gaseous waste or by-products. This is illustrated schematically in a block diagram (Figure 35) which has the following elements:

[1]   Receipt and storage of waste materials.
[2]   Processing or conditioning waste materials prior to incineration.
[3]   Waste material transport and handling.
[4]   Waste incineration.
[5]   Air correction of off-gases from combustion.
[6]   Disposal and/or recovery of residual solid and liquid by-products.

Incineration systems incorporating fluidized bed combustors vary depending upon the application and economic desire to utilize waste heat. Usually, systems will incorporate most if not all of the functional activities illustrated in Figure 35.

Most of the fluidized bed incineration application reported in the literature involve the disposal of sludges or slurried wastes. This may necessitate a dewatering step in processing the waste prior to incineration if combustion gases are to be used for steam-electric or gas turbine power generation. If power generation is a desired by-product of the incineration process, then waste moisture content values less than approximately 60% are required. Moisture values in excess of this value, or heavy concentrations of inert matter will require auxiliary fuel burners to preheat the waste and ensure sufficient heat content in the flue gases. Predrying of the sludge may be accomplished by aeration or more sophisticated mechanical systems involving the addition of heat. Waste material is penumatically, mechanically or gravity fed into the fluidized bed. Normally, inhomogeneous waste material must be reduced in size (shredded, pulverized, etc.) to facilitate the feed system operation and permit injection, distribution and combustion within the fluidized bed.

FIGURE 35: FUNCTIONAL DIAGRAM FOR FLUIDIZED BED INCINERATION DISPOSAL OF COMBUSTIBLE WASTE MATERIAL

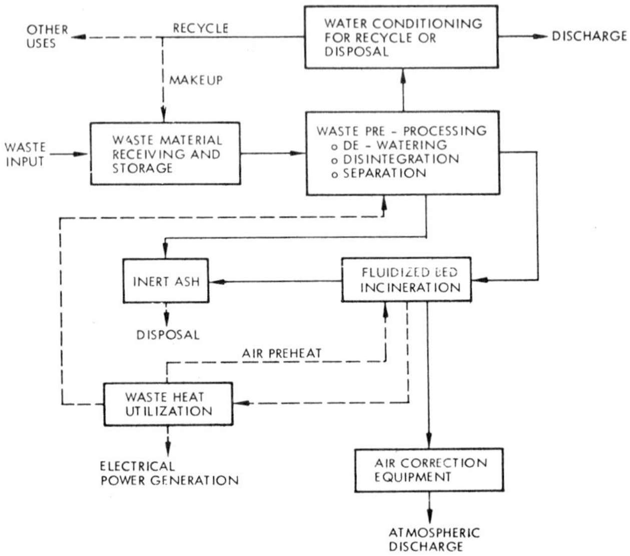

Source: PB 224 582

In addition to reducing moisture content and waste material size, separation of noncombustible material such as ferrous and nonferrous metals may be required. The former may be removed using magnetic separators. Nonferrous metals are commonly removed using ballistic-type separators. The tasks of receiving, storing, and transport of more hazardous wastes may often require a completely closed system. In this case, aeration in the conventional manner will be unacceptable. Enclosed hot air dryers using recycled combustion gases may be considered; however, the addition of gas or oil fuel burners to the incinerator (to accomplish waste drying) will probably result in higher initial equipment costs. An auxiliary burner system will be required in any event for startup and bed temperature conditioning.

Combustion of any waste which results in particulate, odors or gaseous stack emissions (other than water vapor and CO_2) may require air correction equipment to meet emission standards. Particulate emissions may be controlled using one or more of the following general categories of collectors; dry collectors, wet scrubbers, electrostatic precipitators and fabric filters.

Auxiliary air correction equipment for odor control should not be required with the fluidized bed incinerator process. In this instance, odors will be eliminated by oxidation in the combustor. The operating temperature of the fluidized bed combustor is 1400° to 1500°F which is adequate for most odor producing compounds. If odor control is a problem with certain hazardous wastes, then an afterburner can be added to the incineration process as a means of control. Control of gaseous pollutants will depend upon the waste and its combustion products. Because of its relatively low and controlled temperature environment, fluidized bed incinerators should produce little or no nitric oxide, a distinct advantage for this type of equipment. The fluidized bed incinerator is generally applicable to the ultimate disposal of combustible solid, liquid and gaseous wastes; a significant advantage over most other incineration methods. For that reason, it is probable that this type of incineration unit would find application at a National Disposal Site, especially considering its suitability to the disposal of sludges generated in any biological treatment facilities which would be present at the site. It has the following advantages and limitations:

Advantages

[1] The combustor design concept is simple and does not require moving parts in the elevated temperature regions of combustion.
[2] Designs are compact due to high volumetric heating rates (100,000 to 200,000 Btu/hr-ft^3) resulting in lower capital investment.
[3] Comparatively low gas temperatures and excess air requirements minimize formation of nitric oxide. For example, excess air requirements as low as 5% have been reported in the combustion of coal in fluidized bed reactions. Low excess air requirements reduce the size and cost of gas handling equipment.

Limitations

[1] Bed diameters are limited with design technology; therefore, maximum volumetric flow rates per unit are limited.
[2] Removal of inert residual material from the bed is a potential problem area.

A process developed by P.G. Marsh et al (42) is one in which waste sludges, particularly sludges resulting from the treatment of municipal sewage, but also including industrial waste sludges, are mixed with the residue obtained by the pulping of garbage, trash and other municipal refuse for the purpose of aiding the conjoint ultimate disposal of both types of wastes. The sludge is mixed with the pulped slurry of municipal wastes in order to take advantage of the ease with which the resulting mixed waste materials can be dewatered to a consistency appropriate for convenient ultimate disposal, especially by incineration in a fluidized bed reactor or other incinerating apparatus.

An apparatus developed by L.C. Hardison (43) carries out the continuous catalytic oxidation of a waste gas stream by providing for the cocurrent flow of subdivided catalyst particles therewith upwardly in the lower part of a reactor-stack unit, separating the catalyst particles from the contacted gas stream at the upper end of the reactor section and returning them to a collecting and flow regulating means for reintroduction into a waste gas inlet zone to the unit. Burner means, mounted in combination with the inlet zone, is utilized to provide a hot gas stream to contact the recirculated catalyst particles and thus heat and regenerate them for use in the continuous catalytic oxidation system.

A device developed by R.R. Muirhead et al (44) is a fluidized bed incinerator suitable for burning combustible refuse. The refuse is deposited on the surface of one region of a bed of hot particulate refractory material contained in a vessel, the bed is fluidized in such a manner to cause the bed material to circulate so that the refuse deposited on the surface of the first region is drawn into the bed where the combustible content of the refuse is burnt and the noncombustible content of the refuse is displaced while submerged in the bed to a second region spaced horizontally from the first region and from which region the noncombustible content is removable from the bed.

An apparatus developed by R. Menigat (45) is an apparatus for the combustion of sludge wherein the sludge is caused to fall freely through a fluidized-bed chamber from above through at least a sufficient distance to permit volatile components to be released by the cascading sludge and afterburning of volatile components. The exhaust gases are removed at least three meters above the sludge inlet to the chamber.

A device developed by S.G. Hibbert (46) is one in which the fluidizing gas is admitted so as to set up a recirculation of solid matter in a chamber through successive zones in which the material is successively unfluidized, fluidized, suspended, and finally recirculated by a secondary air stream across the top of the unfluidized zone.

A system developed by C.S. Miller, Jr. et al (47) is one for combusting or reacting mixtures of combustible and noncombustible waste matter. The improvement resides in a distributor plate having sides sloping toward the bottom of the apparatus and having an inlet for solid feed in the side of the distributor plate. A pipe extends downwardly from the base of the distributor plate and functions to capture noncombustibles in the solid feed. A lock is provided which permits noncombustibles to be removed from the apparatus without interruption of operation.

A process developed by E.J. Roberts (48) is one in which combustible waste materials containing chlorides are incinerated in a fluid bed reactor by a process in which formation of low-melting eutectics is avoided. Relatively refractory sulfate compounds are instead produced by the introduction of sulfur into the fluid bed and the chlorides pass off as HCl.

Liquid Waste Combustors

Liquid waste combustors are versatile units which can be used to dispose of virtually any combustible liquid waste with a viscosity less than 10,000 SSU. There are a wide variety of liquid waste combustors presently marketed, however they are generally classified as being either horizontal or vertical incineration units. These units have found wide usage throughout the manufacturing industries.

Before a liquid waste can be combusted, it must be converted to the gaseous state. This change from a liquid to a gas occurs inside the combustion chamber and requires heat transfer from the hot combustion product gases to the injected liquid. In order to effect a rapid vaporization (i.e., increase heat transfer), it is desirable to increase the exposed liquid surface area. Most commonly the amount of surface exposed to heat is increased by finely atomizing the liquid to small droplets of $40\,\mu$ size or smaller. This atomization can be achieved mechanically, by two phase flow, or by a combination of both methods. It is usually achieved in the liquid burner directly at the point of fuel and air mixing.

Atomization is the heart of any good liquid incinerator. Mechanical means of atomization include rotary cup and pressure atomization. The rotary cup consists of an open cup mounted on a hollow shaft. The cup is spun rapidly and liquid admitted through the hollow shaft. A thin film of the liquid to be atomized is centrifugally torn from the lip of the cup and surface tension reforms it into droplets. To achieve conical shaped flames an annular high velocity jet of air (primary air) must be directed axially around the cup. If too little primary air is admitted the fuel will impinge on the sides of the incinerator. If too much primary air is admitted the flame will not be stable, and will be blown off the cup. For fixed firing rates, the proper adjustment can be found and the unit operated long periods of time without cleaning.

Pressure atomizing may take many forms. The familiar garden hose nozzle is one example. Most commonly the liquid is given a direction by internal tangential guide slots to the center of the nozzle and then released axially through an orifice. Good atomization can be achieved at moderate pressures (100 to 150 psi). Disadvantages include a limited variable flow range at low pressures and, especially in the smaller sizes, a tendency to plug with foreign matter. Large sizes are reasonably free from this problem. Two-fluid nozzles may be used to impinge a compressible gas on a liquid to tear it into small particles. The compressible gas may be air, nitrogen, steam, etc. Steam is quite commonly used as a low cost source of compressed gas. These nozzles may take three forms: internal mix, external mix or sonic.

As the name implies, internal mix nozzles impinge the gas and liquid before it is sprayed from the nozzle. External mix nozzles impinge jets of gas and liquid together outside of the nozzle body. Sonic nozzles (Figure 36) use the compressed gas to create high frequency sound waves which are directed on the liquid streams. The liquid passage is large in diameter and requires little pressure drop. It can handle slurries or large particles without pluggage. Most two-phase nozzles can operate long periods without difficulty and without cleaning. A common consumption figure is ½ scfm of compressed gas per gph of fluid atomized.

FIGURE 36: TYPICAL SONIC NOZZLE

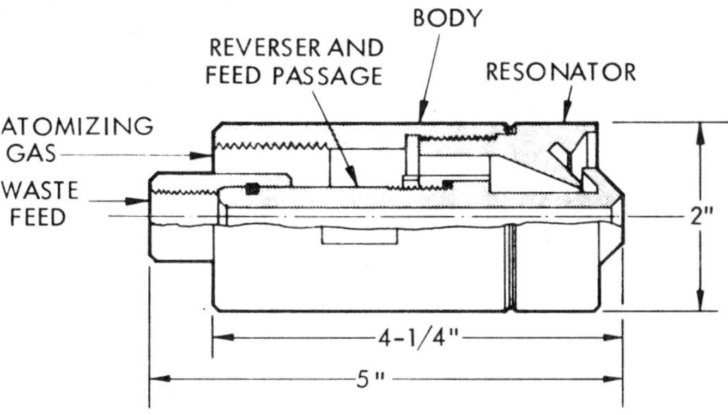

Source: PB 224 582

Liquid burners require considerably more turbulence and time to complete combustion than do gas burners. To complete combustion violent turbulence of the droplets is desirable, and the larger the particles, the greater distance they will go before being completely vaporized and burned. For this reason, forced draft units, if well designed, will have better combustion characteristics than natural draft units. Burners must also be located to prevent flame impingement on walls and, in the case of multiburner units, interference with one another. While multiple atomizers can be located within a single air register, the performance will suffer, and combustion volume must be added to offset this characteristic. Whenever possible, the number of liquid streams should be minimized.

Liquid streams can carry impurities of every sort. Furthermore, they may be highly viscous, which makes handling and atomizing difficult. Liquids should generally have a viscosity of 10,000 SSU or less to be satisfactorily pumped and handled in pipes. For atomization, they should have a maximum viscosity of 750 SSU. If the viscosity exceeds this value the atomization may not be fine enough, and the resultant droplets of unburned liquid may cause smoke or other unburned particles to leave the unit. Viscosity can usually be controlled by heating with tank coils or in-line heaters. Should gases be evolved in any quantity before the desired viscosity is reached, they may cause unstable fuel feed and burning. If this occurs, the gases should be trapped and vented safely, either to the incinerator or elsewhere. If preheating is not feasible, a lower viscosity and miscible liquid may be added to reduce the viscosity of the mixture.

Prior to heating a liquid waste stream, a check should be made to insure that undesirable preliminary chemical reactions such as polymerization, nitration, oxidation, etc., will not occur. Should these occur, it may be more desirable to fill disposable containers with the liquid and treat them as solids. Other preparatory steps may include filtration, degassing, pressurizing, neutralizing, storage, mixing, etc. In every one of these steps care must be employed to see that undesired and harmful results do not occur. Pump and piping materials of construction must be suitable for the liquids encountered. Heated liquids that can solidify or become too viscous should have jacketed or traced piping. Provision should be made to clean out the piping and equipment when long shutdowns occur. This is usually done by purging with steam. Certain atomizing nozzles should always be blown clear with steam whenever flow is stopped. If not, the residual heat in the incinerator may cause thermal cracking of the liquid remaining in the nozzles, resulting in partial or complete pluggage.

There are several basic considerations in the design of an incinerator for a partially combustible waste. First, the waste material must be atomized as finely as possible to present the greatest surface area for mixing with combustion air. Second, adequate combustion air to supply all the oxygen required for oxidation or incineration of the organics present should be provided in accordance with carefully calculated requirements. Third, the heat from the auxiliary fuel must be sufficient to raise the temperature of the waste and the combustion air to a point above the ignition temperature of the organic material in the waste.

Unlike the combustible waste, which sustains combustion by itself, this waste may not always be injected through the combustor or burner but may rather be atomized into the secondary chamber. If the waste material is marginal in combustibility, it may be fed directly through the burner or combustor along with the auxiliary fuel. Temperatures of 2200° to 2700°F will result, complete combustion of the organic in the waste will occur, and the products of combustion can be vented to the atmosphere.

The equipment for handling this type of waste is usually a horizontal or vertical refractory lined cylindrical furnace with an auxiliary fuel burner firing at one end or tangential to the cylindrical shell. The size of the incinerator depends upon the heat release in the system and the amount of combustion air to be used. Mixing is accomplished by baffles or a checker wall, and the temperature of the incinerator should vary, depending on the type and the amount of the waste. In most cases, it is possible to incinerate most organic aqueous mixtures below 1800°F and many in the range of 1200° to 1500°F. As with gaseous wastes, the autoignition temperature of the waste should first be determined, and the incinerator should be operated at a controlled temperature several hundred degrees above this point.

There are basically two forms of liquid waste incinerators; vertically and horizontally fired units. Units, regardless of form, usually operate at temperatures ranging from 1200° to 3000°F (most units operate around 1600°F) and residence times ranging from 0.5 to 1.0 second. Most units have combustion chamber volumes which provide for a heat release of approximately 25,000 Btu/hr-ft^3, however, the vortex type liquid combustor has an unusually high heat release of about 100,000 Btu/hr-ft^3.

A typical horizontally fired liquid waste incineration system is the one operated by the Dow Chemical Company at their Midland, Michigan facility. The unit is a 81 million Btu/hr incinerator which has a combustion chamber 35 ft long and 10 ft square in cross section. Residue are fed to the unit through a combination of four dual-fired nozzles. Combustion gases are quenched in a spray chamber, followed by a high-pressure-drop venturi scrubber, and a cooler/mist-eliminator. About 1,000 gpm of water is recycled from the primary tanks to the wastewater treatment facilities to furnish scrubbing water. This water flows back to the wastewater plant for treatment. About 1,100 hp is required for this unit.

The majority of the liquid wastes treated in the Dow unit are solids at room temperature and must be kept hot in order to remain liquid. Many residues are chlorinated and can contain as high as 50% chlorine, plus several percent of ash in the form of Fe, Ca, Mg, Na, oxides and chlorides. A drawing of the Dow unit is presented later in the volume in the

chapter on Pesticide Industry Wastes. A typical vertically fired liquid waste incinerator
is presented (Figure 37). This particular unit is designed and marketed by the Prenco
Division of Pickands Mather and Company. It is a versatile system in that it can be brought
up to operating temperature (1600° to 3000°F, depending on type of waste material to be
destroyed) in one to two hours with minimal fuel requirements. This quick warm-up
permits periodic rather than continual operation.

FIGURE 37: TYPICAL VERTICALLY FIRED LIQUID WASTE INCINERATOR

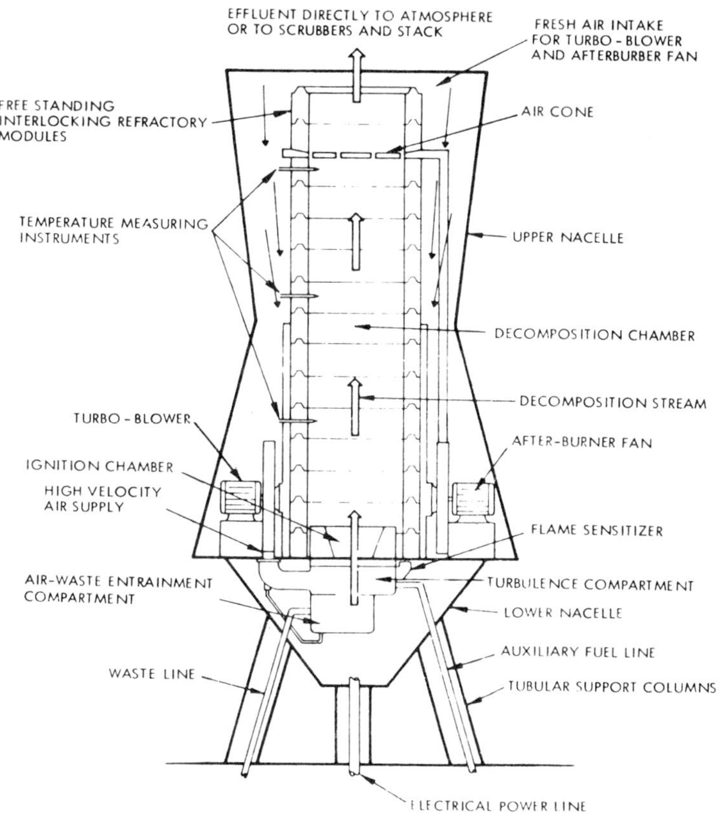

Source: PB 224 582

The Prenco vertical combustor operates in the following manner. A mixture of auxiliary
fuel (usually natural gas) and high pressure air are first fed into the vertical retort to bring
it up to proper waste decomposition temperature. When the retort reaches the correct
temperature, as determined by the temperature measuring instruments, fuel flow is mod-
ulated and waste is admitted to the air-waste entrainment compartment. From there the
aerated waste is fed into a turbulence compartment where it is mixed with more high pressure air and injected into the high-temperature vertical retort. Here the process breaks
down the waste by molecular dissociation, oxidation, and ionization. The gases and any
inert particles produced flow vertically through the air cone and out of the top of the
retort. Decomposition efficiency is greatly increased through the injection of pressurized

air at a point near the top of the retort through ports in a specially designed refractory module. The air cone, which serves as a fuel saver, increases decomposition efficiency by increasing heat retention. It also provides additional air for an after-burner effect. In addition, the air cone reduces the temperature of the decomposed effluent to about 650°F. As a result, scrubbers and effluent test equipment can be utilized if desired. The Prenco unit utilizes air pretreatment. Intake of air from the top of the upper nacelle causes it to be preheated as it travels down the outer wall of the decomposition chamber to both the turbo-blower and after-burner fans. The use of preheated air signficantly increases decomposition efficiency and economy of operation.

A fairly unique form of liquid waste incinerator is the vortex combustor (Figure 38). Its unusual characteristic is its high heat release capability (about 100,000 Btu/hr-ft^3). The vortex combustor is a cylindrical furnace which is tangentially fired with a modified oil burner. In operation, the furnace is preheated for 1 hr to a temperature of 800° to 1000°F. Operating temperature is 1200° to 1600°F, with 20% excess air. Liquid wastes, fuel gas and primary air pass through a hot ignition tunnel.

FIGURE 38: VORTEX LIQUID WASTE INCINERATORS

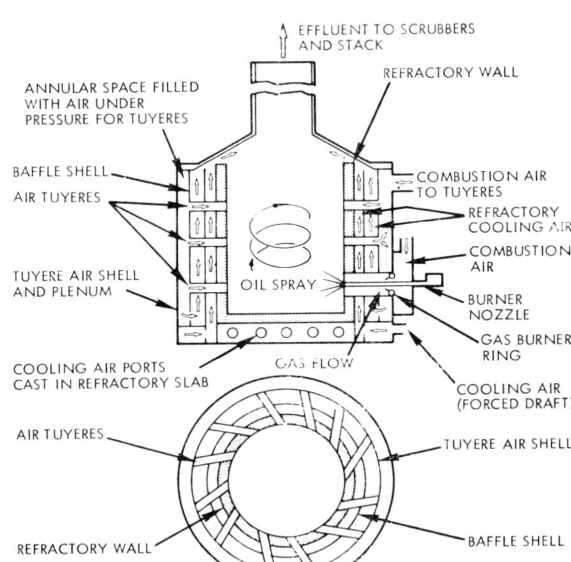

Source: PB 224 582

Tangential firing creates a vortex of hot gas and primary air that flow upward through the hot combustion chamber. As the gases rise, preheated high-velocity secondary air is introduced from tangential tuyeres, maintaining the vortex. The high heat release of this unit has resulted in some slagging and erosion of the refractory. Installation and operating costs will vary significantly depending upon the type and quantity of waste to be processed, the amount and the sophistication of any water and/or air correction equipment required, waste pretreatment requirements, materials of construction and the extent of heat recovery. Liquid waste incineration costs are reported to range between $1 and $100 per 1,000 gal of waste incinerated depending upon system complexity.

The primary process modifications utilized in liquid waste incineration are waste heat recovery options and burner design options. The principle heat recovery options are primary combustion air preheat through heat exchange with the hot gaseous effluent and waste heat boiler utilization (Figure 39). Waste heat boilers are generally used only when there is a demand for steam elsewhere on the industrial site and furthermore, heat recovery in any form is usually only considered economical for large installations.

FIGURE 39: HEAT RECOVERY OPTIONS IN LIQUID WASTE INCINERATION

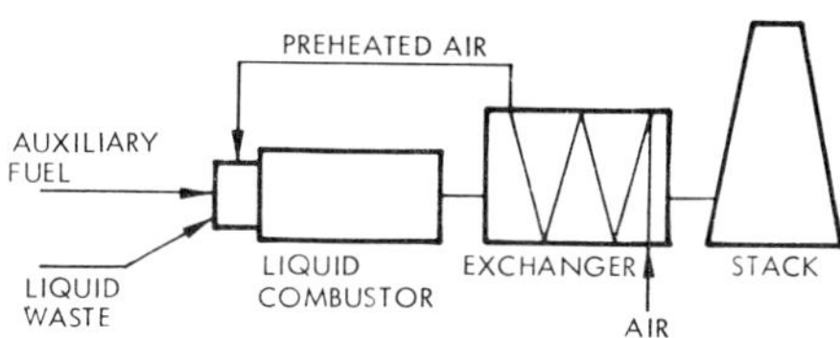

(A) COMBUSTION AIR PREHEAT

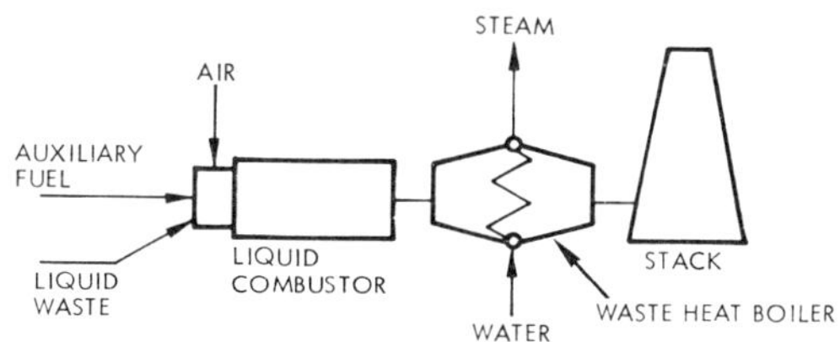

(B) WASTE HEAT BOILER UTILIZATION

Source: PB 224 582

The other major process modification is that of utilizing various types of burners for specific applications. Some liquid incineration units utilize very simple burners which are no more than a set of pipes; one injecting the combustible liquid waste under pressure and the other injecting air. This configuration is generally not very efficient and is usually very specific in application (i.e., designed for a specific waste material with specific characteristics and feed rate).

More sophisticated burners such as the John Zink Series DB-0 liquid waste burner, are designed for combination firing of auxiliary fuel gas with waste liquid and waste gas or waste liquid only. This type of burner uses steam to atomize the waste liquid and can handle liquids and slurries with viscosities up to 1,000 SSU. Inlet liquid pressures are variable (40 to 300 psi) as are atomizing steam requirements (0.15 lb to 0.4 lb of steam per pound of waste) depending upon the characteristics of the waste liquid. These burners may be equipped with pilot burners and electric ignitors.

Another type of complex burner is the TRW Combustor, designed and marketed by TRW Systems. It utilizes a central element injection technique which was first developed for rocket engines and has subsequently been adapted to the disposal of hazardous liquid wastes on a pilot scale. The burner has been used to combust a wide range of materials including solid waste, liquids and gaseous fuels. The basic injector design is a single central element configuration wherein air is injected as a continuous cylindrical sheet which impinges with fuel jets injected radially outward. The air and fuel mixture is then further mixed by means of a deflector on the control element. This deflector serves both to complete the mixing process and as a flame holder. This burner design produces an externally premixed flame with a short reaction zone. Due to the premixed flame and short reaction zone, the TRW burner has added feature of reducing NO_x formation. It has been demonstrated while using fuel oil as a fuel, that NO_x concentrations are cut by as much as 75% when the proper amount of excess air is used. Liquid waste incinerators are generally applicable to the ultimate disposal of most forms (including dilute) of combustible liquid waste materials and represent proven technology. Some of the materials currently being disposed with this type incinerator are presented in Table 17.

TABLE 17: LIQUID WASTES CURRENTLY BURNED IN LIQUID WASTE INCINERATION

Separator sludges	Animal oils and rendering fats
Skimmer refuse	Cyanide and chrome plating wastes
Oily waste	Lube oils
Detergent sludges	Soluble oils
Digester sludges	Polyester paint
Cutting oils	PVC paint
Coolants	Latex paint
Strippers	Thinners
Phenols	Solvents
Wine wastes	Polymers
Potato starch	Resins
Vegetable oils	Cheese wastes
Washer liquids	Dyes
Still and reactor bottoms	Inks
Soap and detergent cleaners	

Source: PB 224 582

Tar incinerators are specially designed for burning tar wastes, contaminated solvents, and slurries such as wastewater sludge. The design of the incinerator depends to a large extent on the combustion products of the tar to be burned. Tars vary considerably in chemical composition, and thus in combustion characteristics and products. The four basic types of tar burners are shown in Figure 40. The primary combustion zone and firing mechanism may vary (for some units) from those shown, but the basic sequence of operation is essentially the same.

Units 1 through 4 provide for progressively greater control of combustion. Burner **1** is usually used to incinerate highly volatile waste tars with negligible ash content and no gaseous pollutant combustion products. The combustion flame burns completely in the open, and when properly operated is smokeless. Burner **2** provides for greater control of the combustion process by enclosing the combustion zone and providing a stack for discharge of gases. A secondary mixing and combustion chamber and a settling chamber, as well as a higher stack, are provided in burner **3**. This burner would be used for tars which might generate small quantities of particulate matter on combustion. The settling chamber would remove some particulate, and the taller stack would assist in dispersion of particulates to the atmosphere.

For tars which on combustion generate high quantities of particulates or noxious gases, burner **4** is required. These tars require a highly refined burner system and air pollution control equipment sufficient to control pollutant emissions. Types of tars which contain halogens require equipment for incineration similar to that of burner **4**. Burning this material might generate the elemental halogen or the acid gas. Tar incinerators for this material are usually designed to generate the acid gas which is then scrubbed. An example of tars containing halogens are those from the production of chlorinated organic chemicals. Large volumes of this waste, containing some residual chlorine, are generated. The chlorine atom in the waste is not combustible, but carbon and hydrogen are. Hydrogen will react with chlorine to form the acid gas HCl, but unless there is sufficient hydrogen to combine with all the chlorine, unreacted chlorine will be released to the atmosphere in its elemental form.

FIGURE 40: SCHEMATIC DIAGRAMS OF TAR BURNERS

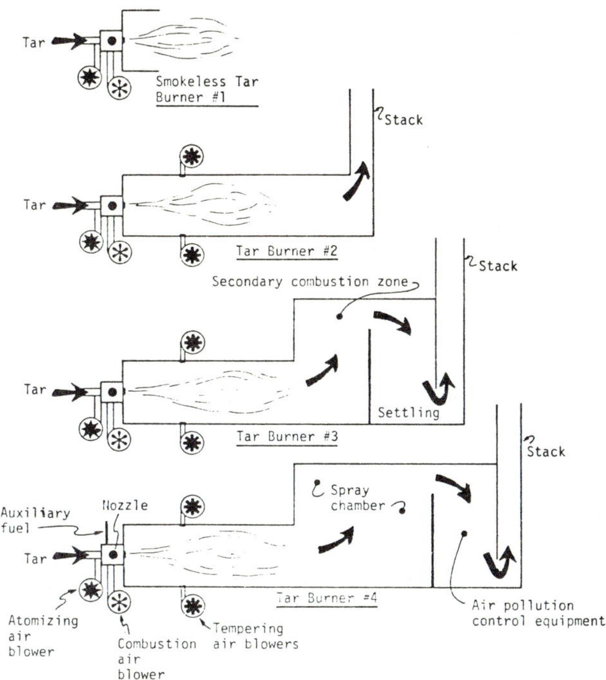

Source: PB 226 420

In addition, with low combustion temperatures, intermediates such as methyl chloride and phosgene may be formed and released. Elemental chlorine is difficult to scrub from a gas stream and requires special caustic scrubbing solutions. Incinerator designs for this tar provide for excess hydrogen through auxiliary fuel such as natural gas or steam at high temperatures. The acid gas can then be scrubbed from the exhaust using water in either low-energy-packed bed-type systems or high-energy venturi scrubbers. The scrubbing water becomes a weak HCl acid solution. It may either be reclaimed to produce various grades of hydrochloric acid or neutralized and sent to receiving waters, the usual practice.

Tar burners operate over a wide range of temperatures from about 1800°F to as high as 3500°F. At these high temperatures, and with the possible presence of acid gases, corrosion and damage to fire brick can be a serious problem. Tar burners can handle a variety of tars, slurries, gases, and liquids including wastes with a high percentage of noncombustible matter or highly aqueous wastes. Because the wastes are fired through a nozzle, there are limitations of particle sizes in slurries and on certain characteristics of tar wastes to prevent clogging of the nozzle. The waste tars which cannot be incinerated must be treated by an alternate method. These practices usually result in inadequate disposal.

Some of the engineering fundamentals involved in the design of incinerators for aqueous wastes have been presented by E.S. Monroe, Jr. of Du Pont (49). The use of fluid-bed incinerators to burn salty sludges such as NSSC paper mill waste liquor and oil refinery waste sludges has been described by C.J. Wall et al of Dorr-Oliver, Inc. (50). An apparatus developed by D.N. Garver et al (51) is one for thermally decomposing liquid waste material normally having some combustible constituents. The apparatus includes a high temperature retort which may be fired by a dual fuel burner.

An apparatus developed by E.S. Monroe, Jr. (52) is an apparatus for the oxidative disposal of a liquid waste stream made up primarily of organic materials diluted with water. Its operation involves spraying the waste stream counter to a combustion flame predominantly as an envelope external thereof but in close proximity to the combustion flame, thereby vaporizing the waste stream, and exhausting the waste stream in heated vaporous form together with products of the combustion flame and air past an oxidation catalyst effective to oxidize a substantial portion of the organic material to equilibrium products.

An incinerator developed by T.J. Wittmann (53) for incinerating liquid waste materials has a heated combustion bed of aggregate particles contained in a combustion vessel. The vessel is agitated by mechanical means to provide a constant motion to the particles in the bed. The aggregate particles are of relatively large size to eliminate the likelihood of solidification of the aggregate on shut down. Liquid waste materials, which may vary widely in viscosities and may contain widely varying concentrations of solid matter, are introduced into the combustion vessel and incinerated therein. Exhaust means are provided.

A process developed by R. Ragot (54) involves treating industrial residues such as chemical waste which are in a liquid form but which are substantially uncombustible. The process provides for the combination of the residues with at least 30% by weight combustible substances, preferably substances which are also waste materials, and the burning of the combination at the base of a chimney. The chimney is designed for the introduction of air into the burning zone, and a mechanism for producing a spray of the residue and substances introduced is utilized.

A cyclonic incinerator design has been developed by E.S. Monroe, Jr. (55). According to Monroe, such cyclonic incinerators can be improved in operation by the use of two rows of air injection nozzles of different sizes, one row supplying high velocity air, tangentially and continuously to the combustion zone. To further improve efficiency the flue from the incinerator should have a reduced diameter and provide a sharp cutoff to centrifugally retain small unburned particulates in the combustion zone. Such an incinerator is particularly useful for the combustion of industrial wastes, e.g., chemical wastes. It can handle solid, liquid or gaseous wastes such as aqueous wastes, plastics, soaps, sludge from sewage treatment plants, or halogenated compounds.

A process developed by H.M. Katz (56) is one in which incineration of waste products such as hydrocarbon oils or similar products which are difficult to combust is accomplished in a two-stage incinerator including a first stage vaporization zone with burner and forced air supply and a catalyst material. Communication is provided between the first stage vaporization zone and a second stage combustion zone with burner and forced air supply.

Molten Salt Incinerators

One of the newest types of incinerators is one incorporating a body of molten salt held at high temperature. The reader is also referred back to the earlier section of this chapter on Incineration to the section on Afterburners where a molten salt afterburner design is described and illustrated (Figure 17).

A process developed by S.J. Yosim et al (57) is one in which unwanted explosives and propellants are destroyed in a safe nonpolluting manner by being contacted with a hot molten salt containing as essential reactive component alkali metal carbonate or hydroxide or mixtures thereof. Some of the resulting decomposition products that would ordinarily be released as atmospheric pollutants react with or are retained in the melt. Any combustible matter present in the melt may be further reacted by use of oxidative molten salts alone or in combination with oxygen added to the melt. Formed gaseous products may be treated in a second reaction zone to complete oxidation of any combustible matter present before discharge of the gases to the atmosphere. The reactive component of the treated melt may be regenerated, or the melt may be disposed of, preferably by reaction with lime in an aqueous or molten medium to form a water-insoluble calcium salt residue. Particularly effective and the preferred reactive molten salts are the eutectic mixtures $NaOH$-KOH and Li_2CO_3-Na_2CO_3-K_2CO_3.

Another process developed by S.J. Yosim et al (58) is a process for the ultimate disposal of organic pesticides with negligible environmental pollution by feeding the pesticide and a source of oxygen into a molten salt containing an alkali metal carbonate and preferably also an alkali metal sulfate to pyrolytically decompose and at least partially oxidize the pesticide. Some of the resulting decomposition products react with and are retained in the melt; remaining gaseous products pass through the melt to the atmosphere or are conducted to a second reaction zone where oxidation of any combustible matter present is completed. Certain organic pesticides may be completely combusted in the first reaction zone by using an excess of oxygen or air in this zone during the decomposition reaction. Thereby the final gases vented to the atmosphere as a result of oxidative treatment in only the first zone or in both zones include only such gases as carbon dioxide, water vapor, oxygen and nitrogen. Figure 41 shows a suitable form of apparatus for the conduct of the process.

FIGURE 41: MOLTEN SALT INCINERATOR DESIGN FOR PESTICIDE DISPOSAL

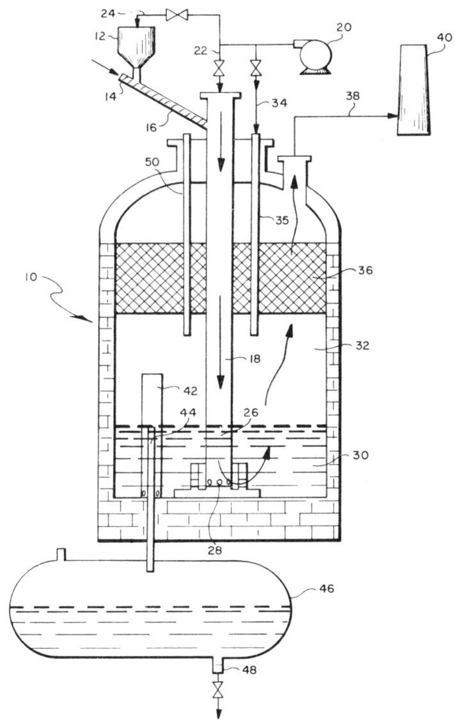

Source: U.S. Patent 3,845,190

Referring to the drawing, a refractory-lined pesticide disposal furnace **10** is shown. An organic pesticide as a free-flowing powder or an adherent component of a pulverized packaging or container material is contained in a feed hopper **12** attached to an auger-type screw conveyor **14** and is fed by way of a conduit **16** to a first zone **18** of furnace **10**. Alternatively, liquid pesticides or pesticides dissolved or dispersed in a liquid medium may be directly sprayed into zone **18**. Alumina is a suitable material of construction for this zone.

At the same time a stream of air from a blower **20** is fed by way of a valved conduit **22** to zone **18**. Air is also fed by way of a valved conduit **24** to hopper **12** so as to prevent any back pressure through conduit **16** due to evolved gases formed during the thermal decomposition or combustion reaction. The pesticide and the air stream impinge upon a pool **26** of molten salt disposed in the bottom of zone **18**. Preferably this molten salt pool consists of molten sodium carbonate containing from 1 to 25 weight percent sodium sulfate. The thermal decomposition reaction is effectively and preferably accomplished in this pool at a temperature between 900° and 950°C at which the salt is molten. Because of the exothermic reaction occurring when the char-containing salt is contacted with air, sufficient heat is internally generated to maintain the salt in the molten state.

At the same time, partial or complete oxidation of the organic pesticide present occurs, depending on the nature of the pesticide and the amount of air fed with the pesticide to zone **18**. The carbonaceous products formed and the gases evolved because of the decomposition reaction bubble through the molten salt, acidic gases formed such as sulfur dioxide and hydrogen chloride being immediately neutralized. Most of the carbonaceous material formed is consumed in molten salt pools **26** and **30**, which freely intermix, by reaction with the sodium sulfate therein to form carbon dioxide and sodium sulfide.

The effluent gases, which also generally contain carbonaceous particulate matter, bubble through openings **28** at the bottom of zone **18** into pool **30** of molten salt which also constitutes part of zone **18** because of the intermixing of these pools. Completion of combustion of combustible gases evolved from pool **30** takes place in a second reaction zone **32**. A stream of air is fed by way of a valved conduit **34** to a tube **35** communicating with second reaction zone **32**. For some applications tube **35** may be extended below the surface of the melt in pool **30** to cool the molten salt and at the same time preheat the air so that more rapid combustion occurs in zone **32**.

A corrosion-resistant wire mesh **36**, suitably an aluminized or alumina-coated stainless steel mesh, is contained within reaction zone **32** and serves to ignite and promote combustion of oxidizable gases and at the same time demist any molten salt particles carried to this point from the gas stream. The effluent gases are removed from furnace **10** by way of a conduit **38** and evolved to the atmosphere by a stack **40**. The evolved gases consist essentially of carbon dioxide and water vapor, formed by combustion, and oxygen and nitrogen present from the air.

The process may be operated as a batch, semicontinuous or a continuous operation. In furnace **10**, means have been provided suitable for continuous operation. An overflow chamber **42** is provided with a level regulator tube **44** so that as the liquid level is increased because of the presence of added pesticide and the reaction products formed, excess molten salt will be drained from the overflow chamber **42** by way of level regulator tube **44** into a soil pit (not shown) where it is solidified. The salt is then removed by a scoop conveyor for further disposal. Alternatively, the spent molten salt is treated with air to oxidize any residual sulfide to sulfate, and then the molten salt is drained into a body of water contained in a receiving vessel **46**

The spent salt solution in vessel **46** is then drained by way of a valved exit tube **48** for further processing and ultimate disposal. Also, in order to maintain the process in a continuous manner, additional sodium carbonate and sodium sulfate is conveniently added by way of hopper **12** in admixture with added pesticide. Alternatively, additional salt is added to the molten pool in zone **18** by way of tubular conduit **50**.

Multiple Chamber Incinerators

The multiple chamber incinerator has been employed by both municipal and industrial facilities for solid waste disposal. The configuration of modern multiple chamber incinerators falls into two general types (Figures 42a and 42b).

FIGURE 42: MULTIPLE CHAMBER INCINERATORS

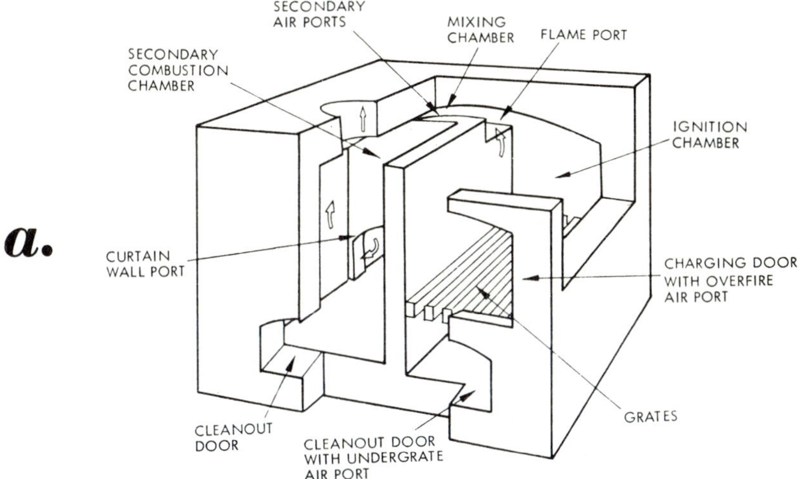

Retort Multiple Chamber Incinerator

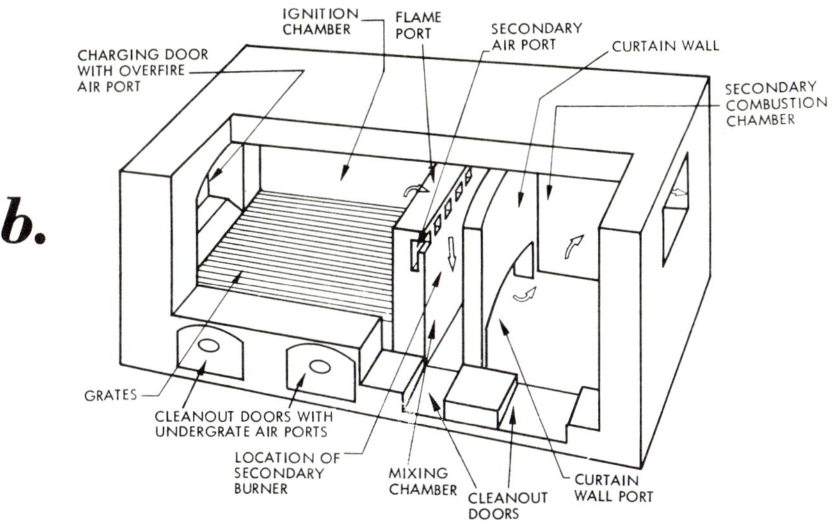

In-Line Multiple Chamber Incinerator

Source: PB 224 582

These are the retort type, named for the return flow of gases through the "U" arrangement of component chambers, and the in-line type, so-called because the component chambers follow one after the other in a line. Essential features that distinguish the retort type of design are as follows:

[1]   The arrangement of the chambers causes the combustion gases to flow through 90° turns in both lateral and vertical directions.

[2]   The return flow of the gases permits the use of a common wall between the primary and secondary combustion stages.

[3]   Mixing chambers, flame ports, and curtain wall ports have length-to width ratios in the range of 1:1 to 2.4:1.

[4]   Bridge wall thickness under the flame port is a function of dimensional requirements in the mixing and combustion chambers. This results in construction that is somewhat unwielding in the size range above 500 lb/hr.

Distinguishing features of the in-line-type design are as follows:

[1]   Flow of the combustion gases is straight through the incinerator with 90° turns only in the vertical direction.

[2]   The in-line arrangement is readily adaptable to installations that require separated spacing of the chambers for operating, maintentance, or other reasons.

[3]   All ports and chambers extend across the full width of the incinerator and are as wide as the ignition chamber. Length-to-width ratios of the flame port, mixing chamber, and curtain wall port cross sections range from 2:1 to 5:1.

Each style has certain characteristics with regard to performance and construction that limit its application. The combustion process in a multiple chamber incinerator proceeds in two stages—primary or solid fuel combustion in the ignition chamber, followed by secondary or gaseous-phase combustion. The secondary combustion zone is composed of two parts, a downdraft or mixing chamber and an up-pass expansion or combustion chamber.

The two-stage multiple chamber incineration process begins in the ignition chamber and includes the drying, ignition, and combustion of the solid refuse. As the burning proceeds, the moisture and volatile components of the fuel are vaporized and partially oxidized in passing from the ignition chamber through the flame port connecting the ignition chamber with the mixing chamber. From the flame port, the volatile components of the waste material and the products of combustion flow down through the mixing chamber into which secondary air is introduced. The combination of adequate temperature and additional air, augmented by mixing chamber or secondary burners as necessary, assists in initiating the second stage of the combustion process.

Turbulent mixing, resulting from the restricted flow areas and abrupt changes in flow direction, furthers the gaseous-phase reaction. In passing through the curtain wall port from the mixing chamber to the final combustion chamber, the gases undergo additional changes in direction accompanied by expansion and final oxidation of combustible components. Fly ash and other solid particulate matter are collected in the combustion chamber by wall impingement and simple settling. The gases finally discharge through a stack or a combination of a gas cooler (for example, a water spray chamber) and induced-draft system. Either draft system must limit combustion air to the quantity required at the nominal capacity rating of the incinerator.

The basic factors that tend to cause a difference in performance in the two incinerators are [1] proportioning of the flame port and mixing chamber to maintain adequate gas velocities within dimensional limitations imposed by particular type involved, [2] maintenance of proper flame distribution over the flame port and across the mixing chamber, and [3] flame travel through the mixing chamber, into the combustion chamber.

A retort incinerator in its optimum size range offers the advantages of compactness and structural economy because of its cubic shape and reduced exterior wall length. It performs more efficiently than its in-line counterpart in the capacity range from 50 to 750 lb/hr. In these small sizes, the nearly square across sections of the ports and chambers function well because of the abrupt turns in this design. In retort incinerators with a capacity of 1,000 lb/hr or greater, the increased size of the flow cross section reduces the effective turbulence in the mixing chamber and results in inadequate flame distribution and penetration and in poor secondary air mixing.

No outstanding factors favor either the retort or the in-line configurations in the capacity range of 750 to 1,000 lb/hr. The choice of retort or in-line configuration in this range is influenced by personal preference, space limitations, the nature of the refuse, and charging conditions. The in-line incinerator is well suited to high-capacity operation but is not very satisfactory for service in small sizes. The smaller in-line incinerators are somewhat less efficient with regard to secondary stage combustion than the retort type is. In in-line incinerators with a capacity of less than 750 lb/hr, the shortness of the grate length tends to inhibit flame propagation across the width of the ignition chamber.

This, coupled with thin flame distribution over the bridge wall, may result in the passage of smoke from smoldering grate sections straight through the incinerator and out of the stack without adequate mixing and secondary combustion. In-line models in sizes of 750 lb/hr or larger have grates long enough to maintain burning across their width, resulting in satisfactory flame distribution in the flame port and mixing chamber. The shorter grates on the smaller in-line incinerators also create a maintenance problem. The bridge wall is very susceptible to mechanical abuse since it is usually not provided with a structural support or backing and is thin where the secondary airlanes are located. Careless stoking and grate cleaning in the short-grate-in-line incinerators can break down the bridge wall in a short time.

The upper limit for the use of the in-line incinerator has not been established. Incinerators with a capacity of less than 2,000 lb/hr may be standardized for construction purposes to a great degree. Incinerators of larger capacity, however, are not readily standardized since problems of construction, material usage, mechanized operation with stoking grate, induced-draft systems, and other factors make each installation essentially one of custom design. Even so, the design factors discussed herein are as applicable to the design of large incinerators as to the design of smaller units.

Control of the combustion reaction, and reduction in the amount of mechanically entrained fly ash are most important in the efficient design of a multiple-chamber incinerator. Ignition chamber parameters are especially critical since solid contaminant discharges can be functions only of the mechanical and chemical processes taking place in the primary stage. Other important factors include the ratios of combustion air distribution, supplementary draft and temperature criteria, and portion factors. Some of these factors are functions of the desired hourly combustion rate and are expressed in empirical formulas, while others are assigned values that are independent of incinerator size.

The requirement for supplementary gas burners must be determined. In general, when the moisture content of the waste is less than 10% by weight, burners are usually not required. Moisture contents of from 10 to 20% normally necessitate installation of mixing chamber burners, and moisture content of over 20% usually necessitate inclusion of ignition chamber burners. The design and construction of multiple chamber incinerators are regulated in several ways. Ordinances and statutes that set forth basic building requirements have been established by most, if not all, municipalities. Air pollution control authorities have also set some limitations in material and construction that must be met, and manufacturers' associations have established recommended minimum standards to be followed. The most important element in construction of multiple chamber incinerators, other than the design, is the proper installation and use of refractories. High-quality materials are absolutely necessary if a reasonable and satisfactory service life is to be expected. Manufacturers must use suitable materials of construction since faulty construction may well offset the benefits of good design. In the choice of one of the many available materials, maximum service conditions should dictate the type of lining for any incinerator. Minimum specifications of materials in normal refuse should include high-heat-duty firebrick or 120 lb/ft^3 castable refractory. These materials, when properly installed, have proved capable of resisting the abrasion, spalling, slagging, and erosion resulting from high-temperature incineration. As the incinerator's capacity and severity of duty increase, superior refractory materials such as super duty firebrick and plastic firebrick should be employed. A recent improvement in standard construction has been the lining of all stacks with 2000°F refractory of 2-inch minimum thickness.

The grates commonly used in multiple chamber incinerators are made of cast iron in "Tee" or channel cross section. As the size of the incinerator increases, the length of the ignition chamber also increases. In the larger hand-charged incinerators, keeping the rear section of the grates completely covered is difficult because of the greater length of the ignition chamber. The substitution of a hearth at the rear of the ignition chamber in these units has been accepted as good practice. Since surface combustion is the primary combustion principle, the use of a hearth has little effect upon combustion rate.

Installation of a sloping grate, which slants down from the front to the rear of the ignition chamber, facilitates charging. A grate such as this also increases the distance from the arch to the grates at the rear of the chamber and reduces the possibility of fly ash entrainment. Stacks for incinerators with a capacity of 500 lb or less per hour are usually constructed of a steel shell lined with refractory and mounted over the combustion chamber. A refractory-lined reinforced, red brick stack is an alternative method of construction when appearance is deemed important. Stacks for incinerators with a capacity of more than 500 lb/hr are normally constructed in the same manner as those for smaller units but are often free standing for structural stability. Stack linings should be increased in thickness as the incinerator becomes larger in size.

Capital and operating cost data on multiple chamber incineration units are scarce. The installed capital investment will vary depending upon the type and quantity of waste being incinerated, the quantity and sophistication of air correction equipment, and materials of construction equipment, and materials of construction. The relative costs (as of 1968) of incinerator and air pollution control equipment of various capacities, exclusive of foundations are presented in Table 18.

TABLE 18: APPROXIMATE COSTS OF MULTIPLE CHAMBER INCINERATORS*

Capacity, lb/hr	Incinerator Cost, dollars	Wet Scrubber Cost, dollars
100	1,700	3,000
150	2,000	3,600
250	2,700	4,400
500	5,000	6,200
750	9,500	7,600
1,000	12,500	8,800
1,500	20,000	11,200
2,000	25,000	13,200

*Based on 1968 costs.

Source: PB 224 582

Operating costs are mainly a function of labor, power, fuel and refractory repair and replacement. Multiple chamber units generally can be operated by one to two men. As of 1968, overall processing costs were reported as being $15 to $16 per ton of waste incinerated. Multiple chamber incinerators are generally applicable to the ultimate disposal of most forms of combustible solid waste and represent proven technology. Some of the materials currently disposed of in this type of unit are general refuse, paper, garbage, wood, phenolic resins, rubber, wire coatings, acrylic resins, epoxy resins, and polyvinyl chloride. Although the multiple chamber incinerator is capable of handling various types of solid wastes, its inability to process liquids, gases, sludges and tars limits the application. There are other types of incineration units available which are much more diverse in application (i.e., rotary kiln fluidized bed and multiple hearth incinerators).

Multiple Hearth Incinerators

The multiple hearth incinerator (commonly called a Herreshoff furnace) is a versatile unit which has been utilized to dispose of sewage, sludges, tars, solids, gases, and liquid combustible wastes. This type of unit was initially designed to incinerate sewage plant sludges in 1934. In 1968, there were over 125 installations in operation with a total capacity of 17,000 tons per day (wet basis) for this application alone. There are currently numerous industrial installations in operation which are primarily utilized for chemical sludge and tar incineration as well as activated carbon regeneration. The multiple hearth furnace consists of a refractory-lined circular steel shell with refractory hearths located one above the other (Figure 43).

FIGURE 43: MULTIPLE HEARTH INCINERATION SYSTEM

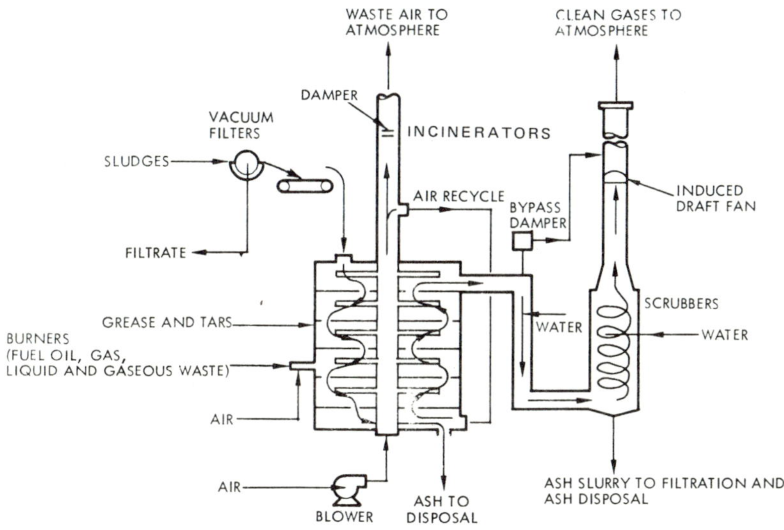

Source: PB 224 582

Sludge and/or granulated solid combustible waste feeds through the furnace roof by a screw feeder or belt and flapgate. A rotating air-cooled central shaft with air-cooled rabble arms and teeth plows the waste material across the top hearth to drop holes. It falls to the next hearth and then the next until ash discharged at the bottom. The waste is agitated as it moves across the hearths to make sure maximum surface is exposed to hot gases. Waste grease and tars are generally fed into the furnace through side ports. Liquid and gaseous combustible wastes may be injected into the unit through auxiliary burner nozzles. This utilization of liquid and gaseous waste represents an economic advantage since the secondary fuel (e.g., natural gas, fuel oil) requirements will be reduced thus lowering operating costs. The system has three operating zones: the top hearths where feed is dried to about 48% moisture; the incineration/deodorization zone, which has a temperature of 1400° to 1800°F; and the cooling zone, where the hot ash gives up heat to incoming combustion air. Exhaust gases exit at 500° to 1100°F.

Incinerator ash is sterile and inert. Volume discharged from the bottom hearth is about 10% of the furnace feed, based on sludge cake with 75% moisture and 70% volatile content in the solids. The ash usually has less than 1% combustible matter, which is normally fixed carbon. Discharge can be moved hydraulically, mechanically, or pneumatically, and used as landfill or roadfill. Current systems include gas cleaning devices on exhaust air. A number of multiple hearth incinerators are operating without difficulty in areas with strict air pollution codes. Although the exhaust does not violate opacity codes, existence of steam plumes has on occasion caused adverse public reaction.

Most multiple hearth incinerators are primarily designed for sludge disposal. The other forms of waste which are simultaneously fed to the system are usually considered a heat source to be utilized during sludge incineration. A heat balance across a multiple hearth furnace must consider the heat abosrbed by: latent heat in free moisture and combustion moisture, sensible heat in combustion gases, excess air, ash, radiation and shaft cooling. These quantities are balanced against the heat evolved from the combustibles in sludge solids and the fuel. A typical analysis of sludge combustibles is: C, 59.8%; H_2, 8.5%; O_2, 27.5%; and N_2, 4.2%. Calorific value of this sludge is 10,000 Btu/lb.

Sludge parameters that have the most influence over incineration are moisture content, percent volatiles and inerts, and calorific value. Moisture is the principal one over which the plant operator has some control. Minimum moisture is important because of its thermal load on the incinerator. Volatiles and inerts, which affect the Btu value of the sludge, can be controlled to some extent by treatment processes such as degritting, mechanical dewatering and sludge digestion. Almost all combustibles are present as volatiles, much in the form of grease. Volatile percentage can vary a great deal, so equipment must be designed to handle a range of values.

The sizing of a multiple hearth incinerator is dependent upon waste combustion characteristics and water content. Incinerator burning rates vary from 7 to 12 lb/ft²-hr for sewage plant sludges with the value 7.5 lb/ft²-hr generally accepted as typical. The area referred to in the burning rate is the total hearth area of the unit. Standard multiple size hearth incinerator sizes range from 85 ft² of hearth to greater than 3,000 ft² of hearth. The secondary fuel requirement is dependent upon the water content of the waste being incinerated. For instance, a waste sludge with a heating value of 10,000 Btu/lb of volatile solids which is composed of 60% volatile solids, will require about 100 ft³ of natural gas per ton of wet feed when the moisture content of the sludge is 75%. This same sludge will require about 1,200 ft³ of gas per ton of wet sludge when the moisture content is 82.5%.

The multiple hearth incinerator is usually operated so that the top hearth temperature is in the 600° to 1000°F range, the combustion hearths are in the 1400° to 1800°F range, while the cooling hearths are maintained in the 400° to 600°F range. Capital and operating costs for multiple hearth incinerators will vary significantly depending upon the type and quantity of waste being incinerated, the sophistication of water and air correction equipment, waste pretreatment requirements, materials of construction, secondary fuel requirements, and labor. Total disposal cost per ton of dry solids fed are reported to range between $8 and $15 while operating costs generally run between $0.50 and $5 per ton of dry solids (Table 19) depending on the size of the unit.

The multiple hearth incinerator is generally applicable to the ultimate disposal of most forms of combustible wastes and represents proven technology. It can incinerate combustible sludges, tars, granulated solids, liquids and gases and is especially well suited to the disposal of spent biological treatment facility sludges.

TABLE 19: MULTIPLE HEARTH INCINERATION COSTS

Sludge incinerated, tons/wk. (wet basis)	28.0	56.0	139.0	278.0	2,780.0
Sludge incinerated tons/wk. (dry basis)	7.0	14.0	34.75	69.5	695.0
Operating schedule, hr/wk.	35	35	70	70	168
Furnace feed, lb/hr	1,600	3,200	3,960	7,920	33,000
Furnace required	10 ft-9 in. OD 5 Hearth	14 ft-3 in. OD 5 Hearth	14 ft-3 in. OD 6 Hearth	18 ft-9 in. OD 6 Hearth	Two-22 ft-3 in. OD 8 Hearth
Installed Cost, $	120,000.00	185,000.00	200,000.00	310,000.00	750,000.00
Weekly fuel cost, $	27.50	45.00	30.00	45.00	50.00
Weekly power cost, $	9.00	12.00	13.00	25.00	165.00
Total utility cost, $	36.50	57.00	43.00	70.00	215.00
Operating cost-$/ton dry solids	5.20	4.06	1.24	1.01	0.31
Filtration cost-$/ton dry solids	8.00	8.00	8.00	8.00	8.00
Maintenance cost-$/ton dry solids	.70	.60	.60	.50	.40
Total disposal cost-$/ton dry solids	13.90	12.66	9.84	9.51	8.71

Cake moisture	75%
Volatile content	65%, 10,000 Btu/lb volatile solids
Solids removal	90%

Source: PB 224 582

An incinerator developed by J.C. Eck (59) is a furnace of the multihearth type, comprising a lower combustion zone having a plurality of vertically spaced annular hearths and an upper drying zone having a plurality of vertically spaced annular hearths. The central area of the drying zone hearths is open and provides a passage for rubbish and garbage to fall directly from a hopper at the top of the furnace upon the uppermost hearth in the combustion zone.

Rabbling apparatus moves the rubbish and garbage progressively from hearth to hearth and thence in ash form to a removal chute. The sewage sludge is deposited on the uppermost of the drying zone hearths and moved progressively downward by rabbling apparatus in a zig-zag path from hearth to hearth and then falls in a dry state to the uppermost hearth of the combustion zone, where it is incinerated along with the garbage and rubbish, and then removed in ash form via the removal chute. A forced air system for cooling the rabbling apparatus is also provided.

Open Burning

Open burning is the burning of waste materials on open land without the use of combustion equipment. This form of incineration is utilized mainly for the disposal of waste or excess high explosives. It is generally unacceptable for the disposal of other forms of waste because of the associated lack of combustion product effluent control. A current disposal technique utilized for many high explosive wastes such as TNT, as well as wet explosive machining waste is open burning. A typical open burning operation (conducted at the rate of about 2,000 lb per week) is to place the waste explosive and explosive contaminated waste on an asbestos pad covering a flat gravel base in a remote open area of the plant grounds. The wastes are thoroughly wet down with fuel oil and ignited from a safe distance by the use of a bridgewire and lead. Considerable black smoke along with NO_x, CO and HF are evolved during operation and are emitted directly to the atmosphere.

These emissions are the result of uncontrolled combustion temperature, incomplete combustion due to the inability of oxygen to efficiently mix with the waste, and the inability to effect sufficient residence time of the generated particulate at elevated temperature. Because of the emission problem, there is currently an effort to develop combustion units

which are applicable to explosives and incorporate effluent scrubbers. Open burning is
not considered to be an adequate form of waste disposal because of the associated loss
of gaseous effluent control. Although open burning is currently utilized for the disposal
of explosives and explosive wastes, it is anticipated that this practice will cease when new
technology is developed for this application.

Open Pit Burners

An additional type of incinerator used to handle both process and nonprocess waste is the
open pit incinerator. It is presently used to incinerate trash, tars, and sludges. The incinerator is a box-shaped pit with no top, permitting maximum radiation of the flame to
the sky. Depth of the pit ranges from 10 to 30 feet. Normally, all air is supplied from
an overfire position to produce maximum turbulence and recirculation of the combustion
gases.

For solids with high calorific value and solids that tend to melt, the open pit incinerator
combustion rate and performance have been reported as high, particularly for wastes
with less than 1 or 2% ash. However, such an incinerator provides little control over emission of particulate matter or noxious gases which are generated by combustion of some
wastes. Open pit incinerators have been used to dispose of high heat content solids and
liquids. These incinerators solve the problem of high heat flux by eliminating enclosure.
Their chief drawbacks are the lack of confinement of combustion product effluents and
relatively high particulate emissions.

Open-pit incinerators vary from the pedestal-mounted oil burner used by Union Carbide
to the Du Pont pit (Figure 44). The Union Carbide installation burns organic liquids
containing up to 25% water without visible smoke. The installation consists of burners
firing horizontally and mounted about 5½ ft above grade. The firing area is surrounded
by earthworks for personnel protection. Heat is dissipated by direct convection and radiation. By the very nature of the design, there is always excess air.

FIGURE 44: DU PONT OPEN-PIT INCINERATOR

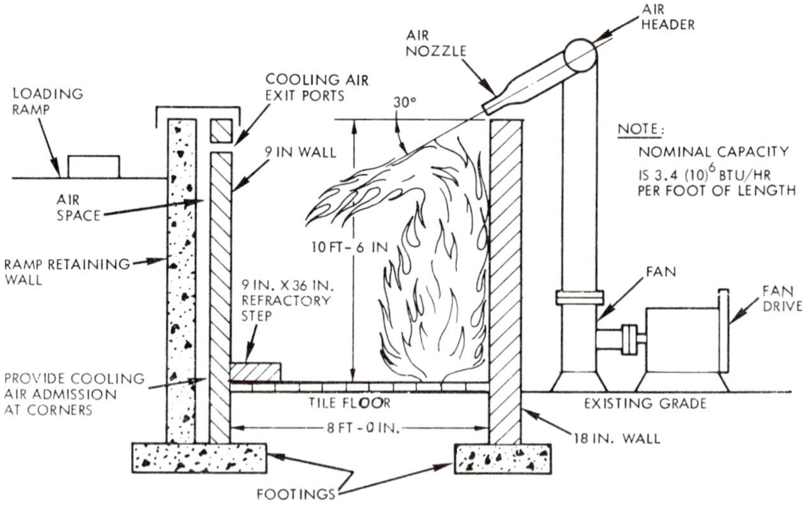

Source: PB 224 582

The open-pit incinerator was originally developed at Du Pont for the safe destruction of nitrocellulose that presents an explosion hazard in a conventional closed incinerator. The incinerator has an open top and an array of closely spaced nozzles that create a rolling action of high-velocity air over the burning zone (Figure 44). Very high burning rates, long residence times leading to complete combustion, and high flame temperatures are achieved. The visible smoke is readily eliminated and the smuts are contained by proper screening. Oversized wastes and plastics that create problems in conventional incinerators are easily destroyed in the open-pit incinerator. It should be noted that the concentration of particulates is slightly higher than conventional incinerators, and there is no way to clean the exit gases. Although these pit incinerators are used for liquid wastes, they are more efficient for solid wastes, especially rubber and plastic.

The incinerator consists of a rigid shell of either reinforced-concrete or steel, lined with refractory on the floor and walls. Empirical and theoretical calculations indicate the optimum width to be 8 ft between refractory walls. The capacity is determined by length usually between 8 and 16 ft. The pit is about 10 ft with cleanout doors located at either end. Normally, a screened enclosure is placed over the pit to contain large airborne particles and for insect and rodent control when burning garbage.

The over-fire air is supplied from a manifold running along one edge of the pit, with alternating 2- and 3-inch nozzles directed downward at an angle of 25° to 35° across the incinerator. Charging is from the opposite side of the nozzle from a leading ramp. The pit should be oriented so that the loading ramp is located upwind. The high-velocity air jets create turbulence in the burning zone, and the excess air aids complete combustion. When the equipment is properly operated, the air pattern creates a sheet of flame under the air manifold on the back wall, rolling the flame across the top of the pit. Particulates and unburned gases are largely returned to the burning-zone more or less eliminating smoke. Smoke intensity rarely exceeds Ringelmann 1 when properly operated.

Operating capacity of the pit depends on the lower heating value of the feed, combustion characteristics, quantity of overfire air, size and configuration of the pit, and method of charging. Not less than 100% excess air is required, and 300% is usual. Du Pont's criteria are: 850 ft^3/min of overfire air at 11 inches of water column per foot of length for standard trast (5,000 Btu/lb). The incinerators are charged intermittently by dump trucks, although hydraulic rams and skip hoists have been used. The rate of charge depends on the material being burned, in order to meet the heat-release capability of the pit. High-heating-value materials such as plastics are fed in small quantities at frequent intervals. The operator's skill is the major factor in minimizing emissions while maintaining a high burning rate.

Direct operating costs are low. Two men can operate two pits. The only other costs are for energy to drive the blower and operate the loader and any cleanout device. No auxiliary fuel is used, as lighting off a small amount of combustible material will ignite the pit. Maintenance is slight and largely consists of repair of the refractory. Capital cost is low. An average price for a commercial unit 16 ft long, 8 ft wide by 10 ft deep, completely installed, including a covered storage building for the waste and screened enclosure for the pit is about $65,000. The capacity is 5,000 lb/hr of low heat-release material, and about half that for high heat-release material.

A variety of wastes have been burned in the pit incinerator. It readily accepts heavy timbers, cable reels and construction wastes. It burns plastics and similar high heat-release materials that might detonate, or erode the refractory in a closed unit. It effectively handles numerous types of manufacturing and process wastes both liquid and solid, plant trash and rubber wastes. Although the open-pit incinerator is currently used industrially, it is not recommended for use because of the associated lack of effluent control. This lack of control might result in emissions to the surroundings of harmful combustion products such as chlorides, fluorides, cyanides, sulfur compounds, carbon monoxide, or any partially combusted waste material. An open pit burner for industrial wastes which has been described by C.H. Boll et al (60) consists of a combustion chamber having vertical side walls, an open top and inclined hearth at the bottom. A plurality of nozzles are positioned across the top of one side wall of the combustion chamber to direct high velocity air into the combustion chamber. The surface area of the material being incinerated is a critical factor in obtaining smokeless and odor free combustion. The surface area is controlled by controlling the rate of feeding of the material to be burned. The surface area control is accomplished by the inclined bottom. The construction is such that the higher the level of the material being incinerated on the hearth, the greater is the exposed surface area. Conversely, the lower the surface level of the material being burned is on the hearth, the narrower is the exposed surface area of the material. This incinerator is claimed to be an improvement over the DuPont design which has a constant area hearth.

A process developed by R.P. Lanyon et al (61) is one in which viscous liquid wastes having a high ash content are incinerated in an open pit incinerator having a hearth inclined up-wardly from the point of introduction of the liquid wastes to the hearth. The liquid wastes are spread into a thin film by means of a reciprocating grid, and air for maintaining combustion is introduced into the open pit over one of its walls through downwardly directed jets. An ash pit is provided adjacent the highest point of the hearth. Hot ashes may be quenched therein and removed from the pit by a conveyor.

Rotary Kiln Incinerators

Rotary kiln incinerators are versatile units which can be used to dispose of solid, liquid and gaseous combustible wastes. They have been utilized in both industrial and municipal installations. In addition, applications of rotary kiln incineration to the disposal of obsolete chemical warfare agents and munitions have been reported.

The rotary kiln incinerator is a cylindrical shell lined with firebrick or other refractory and mounted with its axis at a slight slope from the horizontal. It is a highly efficient unit when applied to solids, liquid, sludges and tars because of its ability to attain excellent mixing of unburned waste and oxygen as it revolves. Its use as a concentrated waste gas combustor is considered a secondary application. This is due to the fact that although proper conditions are present for efficient gas combustion (i.e., long residence time at elevated temperatures) there is no need for the cylinder to be rotating. Therefore rotary kiln incinerators are used for gaseous waste combustion only in conjunction with solid or liquid waste incineration.

Rotary kiln incinerators used in municipal applications are generally designed to handle large volumes of solid combustible waste (refuse) along with any entrained liquid. In this type of facility, the kiln actually serves as a secondary combustion unit since the waste material is ignited on traveling grates prior to entering the kiln (Figure 45). In this In-stance, the kiln serves mainly as an efficient mixer of the burning waste with combustion air.

Rotary kiln incinerators when applied to industrial (includes military) applications are generally designed to accept both solid and liquid feed. A typical major industrial in-stallation is operated by the Dow Chemical Company at Midland, Michigan (Figure 46). This particular unit consists of a 65 million Btu/hr kiln that is used for the incineration of solid chemical refuse, liquid residues, paper, wood and other solids of varying Btu per content. A pack-feed mechanism is used to feed packs and drums of solid waste chemicals into the incinerator.

Liquid wastes transported to the incinerator are transferred to designated receiving tanks that contain compatible wastes. All drums of liquid wastes are also transferred to the receiving tank by the way of a drum-dumping dock. The waste is strained as it is pumped from the receiving tank into a burning tank, where it is blended for optimum burning characteristics. All liquid residues are burned in suspension by atomization with steam or air. Drum quantities of solid tars are destroyed by feeding them into the rotary-kiln incinerator via a hydraulically operated drum and pack-feeding mechanism. All refuse, ex-cept full drum and packs of material, is dumped into the refuse pit.

FIGURE 45: MUNICIPAL ROTARY KILN INCINERATION FACILITY

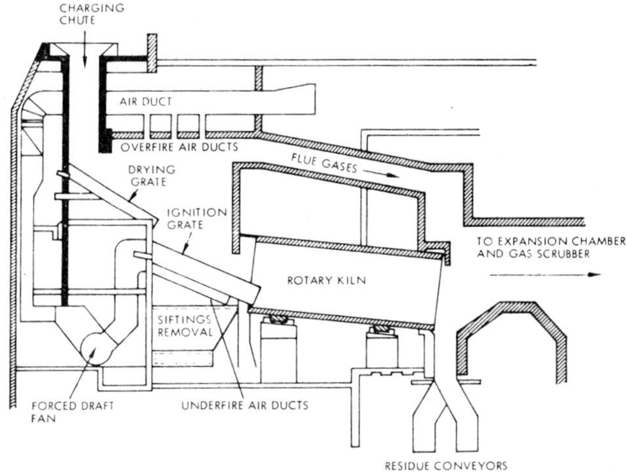

Source: PB 224 582

FIGURE 46: TYPICAL MAJOR INDUSTRIAL ROTARY KILN INCINERATION FACILITY

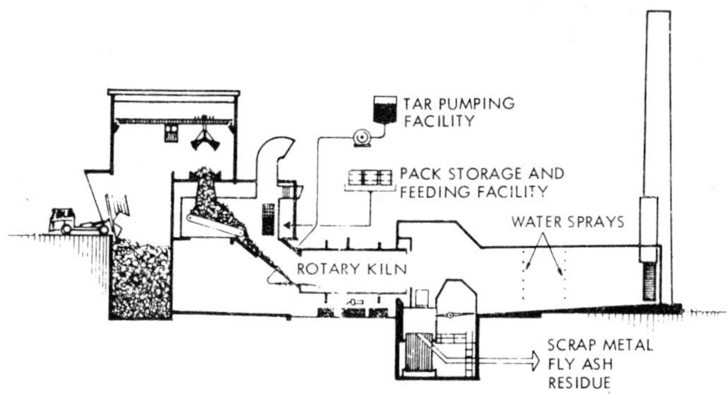

Source: PB 224 582

An overhead crane is used to mix the refuse and raise it to the charging hopper of the rotary kiln (see Figure 46). While the solid refuse is being fed, liquid tars are fired horizontally into the rotary kiln. As the refuse moves down the kiln, organic matter is destroyed, leaving an inorganic ash. This ash is made up primarily of slag, and other nonburnables such as drums and other metallic material. The ash discharges from the end of the kiln into a conveyor trough that contains water. After quenching, the material is conveyed into a dumping trailer, and then to a landfill.

After leaving the kiln, the products of combustion enter the secondary combustion chamber and impinge on refractory surfaces that cause a swirling action. No secondary fuel or afterburners are used. Downstream of the secondary combustion chamber, the gases pass through several banks of water sprays in which the fly ash is knocked down and sluiced onto the ash-conveyor floor. Cooled gases pass under a stack damper and then to a 200-ft stack. There are a variety of small relatively portable and inexpensive rotary kiln incinerators currently marketed. These units usually are not as versatile as large installations because they generally lack the auxiliary equipment utilized in pretreatment of heterogeneous feeds.

An example of this type of unit is the completely packaged compact Thumble-Burner, designed by Bartlett-Snow (Figure 47). These units have capacity ranging from 100 lb to 2 tons per hour with corresponding overall system dimensions ranging from 5 x 5 x 15 ft to 14 x 15 x 34 feet. This type of system will efficiently incinerate properly sized solid waste material with heat content ranging from 1,000 to 15,000 Btu/lb. It is also capable of burning gaseous and liquid wastes when they are injected through the auxiliary burner which is used for incinerator temperature control.

FIGURE 47: PORTABLE ROTARY KILN INCINERATION UNITS

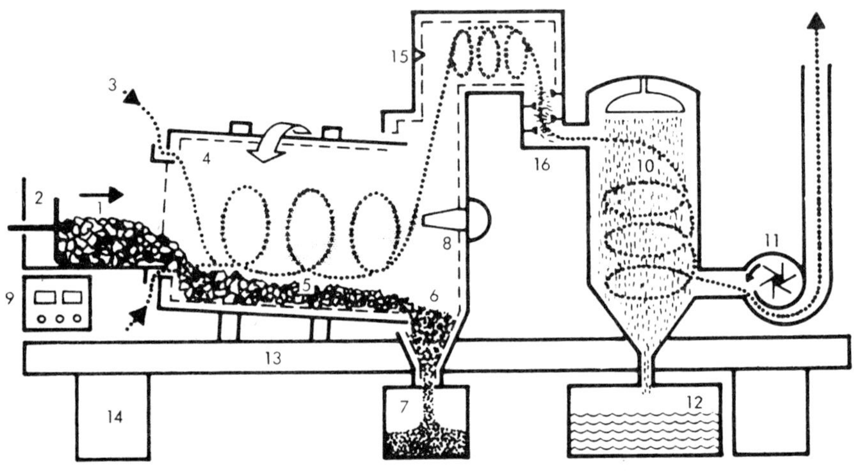

[1] Waste to Incinerator
[2] Auto-Cycle Feeding System:
    Feed Hopper, Pneumatic Feeder, Slide
    Gates
[3] Combustion Air In
[4] Refractory-Lined, Rotating Cylinder
[5] Tumble-Burning Action
[6] Incombustible Ash
[7] Ash Bin
[8] Auto-Control Package:
    Programmed Pilot Burner

[9] Self-Compensating Instrumentation-
    Controls
[10] Wet-Scrubber Package:
    Stainless Steel, Corrosion-Free Wet
    Scrubber, Gas Quench
[11] Exhaust Fan and Stack
[12] Recycle Water, Fly Ash Sludge Collector
[13] Support Frame
[14] Support Piers
[15] Afterburner Chamber
[16] Precooler

Source: PB 224 582

The rotary kiln has been successfully applied to the incineration of obsolete or excess chemical warfare agents (GB, VX and mustard). In this case, the waste is carefully fed to the unit at a relatively slow flow rate and supplemented by fuel oil flame. The liquid chemical warfare agents are fed cocurrent to the fuel oil flame. All incineration controls are fail-safe and the agent feed is equipped with a fast acting cutoff valve in the event of loss of flame. The combustion gases leaving the incinerator are water quenched before being processed through a cross-flow packed scrubber to remove objectionable constituents of the combustion products. The scrubbing media are $Ca(OH)_2$ and NaOH for both GB and mustard and HNO_3, NaOH and $Ca(OH)_2$ for VX incineration products. In all cases, the products recovered and dried by the pollution control system are calcium salts. These solids must then be sent to ultimate disposal.

Rotary kilns have also been used to incinerate explosives such as obsolete munitions. In this case, the explosive is fed countercurrent to a fuel oil flame. The kiln is equipped with an internal spiral to convey materials through the furnace. Feed and discharge is accomplished with metal conveyers. The effluent fume scrubbing system consists of a packed bed scrubber, utilizing a sodium carbonate solution, followed by a hydroclone or venturi scrubber. The scrubber liquor containing fly ash and a sodium nitrate-nitrite mixture is then dried and sent to ultimate disposal. Specific data on rotary kiln incinerator design parameters are scarce. This is due to the fact that incineration is a relatively new application for rotary kilns. Additionally, information of this type is generally considered proprietary by manufacturers.

Information sources indicate that rotary kiln incinerators generally have a length to diameter ratio (L/D) between 2 and 10. Smaller L/D ratios result in less particulate carry over. Rotational speeds of the kiln are usually much slower than those for kilns which are utilized as calciners or dryers and are on the order of 1 to 5 ft/min measured at the kiln periphery. Both the L/D ratio and the rotational speed are strongly dependent upon the type of waste being combusted. In general, larger L/D ratios along with slower rotational speeds are used when the waste material requires longer residence times in the kiln for complete combustion.

The residence time and combustion temperature required for proper incineration is totally dependent upon the waste materials combustion characteristics. Combustion temperatures usually range from 1600° to 3000°F. Required residence times vary from seconds to hours. For instance, a finely divided propellant may require 0.5 seconds while wooden boxes, municipal refuse, and railroad ties may require 5, 15 and 60 minutes respectively. When it is desired to increase the capacity of an existing kiln incinerator, consideration should be given to the following changes:

[1]   Increase charge to the kiln.

[2]   Increase temperature and quantity of combustion gases.

[3]   Decrease quantity of air in excess of combustion needs.

[4]   Increase speed of rotation of kiln.

[5]   Increase capacity of feeding and discharge mechanisms.

[6]   Decrease moisture content of feed material.

[7]   Increase temperature of feed material.

[8]   Preheat all combustion air.

[9]   Reduce leakage of cold air into kiln.

[10]   Increase stack draft by increasing height or by use of jets.

[11]   Install instrumentation to control the kiln at maximum-capacity conditions.

Efficient air seals are essential for the controlled and economical operation of kiln incinerators. They reduce outside air entrance; certain types effectively prevent entrance of all outside air. The inflow of air is the result of the kiln incinerator operating under reduced pressures which are caused by downstream induced draft fans and thermal lift from the stack. This reduced pressure is necessary to ensure against any leakage of undesirable material to the surroundings.

The simplest type of air seal is a floating T-section ring mounted on a wearing pad around the feed end of the kiln shell. The web of the T-ring is confined within circular retainer plates. The floating-type discharge-end air seals consists of a circular bar which floats on a wearing pad and which can be moved to provide the desired operating clearance between air seal and support. The floating ring and the fixed portion of these seals can be furnished with renewable wearing surfaces. Air infiltration through this type of seal is usually less than 10%. For further reduction of air infiltration, lantern-ring-type floating seals, pressurized with inert gas or stack gases, are employed. Capital and operating cost data on rotary kiln incineration systems are scarce. The installed capital investment will vary significantly depending upon the type and quantity of waste being incinerated, the quantity and sophistication of water and air correction equipment, waste pretreatment equipment, materials of construction and the extent of heat recovery equipment. Operating costs are mostly dependent upon the amount of secondary fuel required, replacement of refractory linings (usually about once per year), heat recovery and labor.

Uninstalled costs of the rotary kiln itself are reported to run between $30 and $60 per cubic foot of kiln. Installation is generally about 200% of the purchased cost. Kiln maintentance average 5 to 10% of the total installed cost per year but is dependent largely on the life of the refractory lining. These costs do not take into account secondary combustion chambers, heat recovery equipment or air correction equipment. Installed costs for large municipal type rotary kiln incineration systems are on the order of $10,000 per daily ton of feed capacity. This cost includes waste heat broilers utilized for steam generation. The installed cost of relatively small industrial type rotary kiln incinerator systems range from $2,500 to $5,000 per daily-ton of feed capacity, depending on the specific application. Basically, the only process modification utilized in rotary kiln incineration is waste heat recovery (Figure 48).

FIGURE 48: HEAT RECOVERY OPTIONS

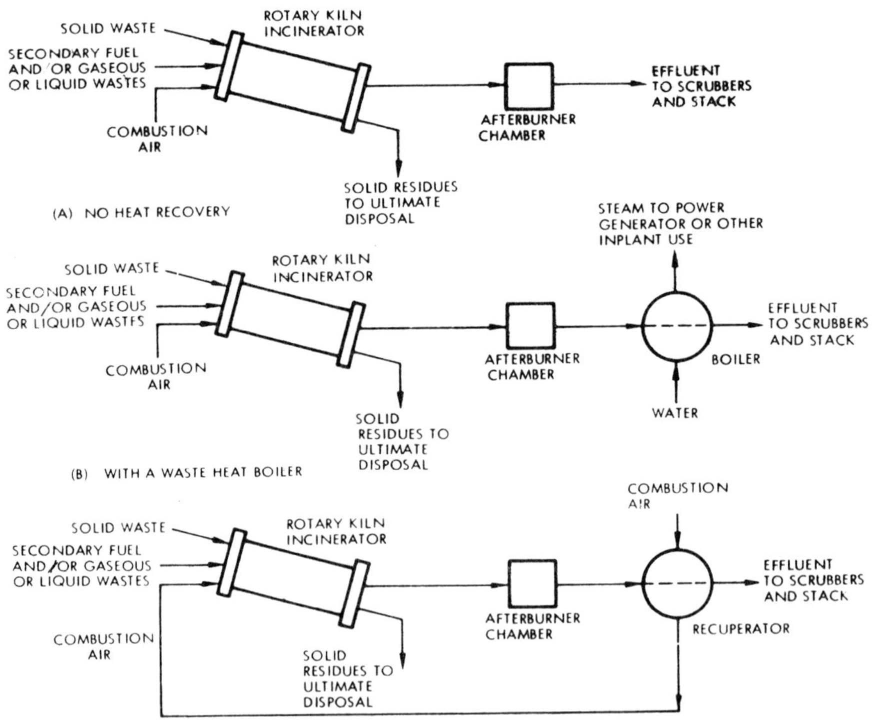

Source: PB 224 582

This practice is seldom followed in industrial and military applications due to the expense of heat recovery equipment and the fluctuations incurred in both waste feed quantity and composition. There have been instances however, when waste heat boilers have been used to recover heat from gaseous effluents where there is need for steam elsewhere on the industrial site. In these cases, the incinerators also function as boilers and constant heat output must be maintained through the use of auxiliary fuels.

Large municipal installations generally utilize heat recovery either for power generation or preheating of combustion air. In the latter case, a significant increase in incineration capacity can be realized. These alternatives are generally economically attractive because of the large volumes of waste (refuse) and the relatively constant heat content of the waste (usually 4,800 to 6,500 Btu/lb). The rotary kiln incinerator is generally applicable to the ultimate disposal of any form of combustible waste material and represents proven technology. It can incinerate combustible solids (including explosives), liquids (including chemical warfare agents), gases, sludges and tars.

A device developed by S.M. Porter et al (62) is a solid waste disposal apparatus that includes a rotary furnace arranged upstream of a rotary dryer in end-to-end relation to the latter but separated therefrom by a burning chamber and stationary ash removal subassembly interposed therebetween. The raw feed is mixed with a portion of the dried feed from the discharge end of the dryer and introduced into the intake end of the dryer as a moist mixture containing between approximately 40 to 60% water. The remaining dried material is introduced into the furnace. A forced draft burner at the entrance to the furnace is canted in the direction of furnace rotation so as to direct the hot products of combustion spirally along the wall thereof in concurrent flow relation to the moist mixture.

Combustion takes place within the upstream end of the furnace inside a separate hollow combustion chamber that cooperates with the furnace wall to define an annulus into which the dryer gas is recycled and mixed with the products of combustion as they leave the aforementioned combustion chamber to reduce the temperature and axial velocity of the latter preparatory to delivering same to the burning chamber and ash removal subassembly complete with exhaust gas system, and dryer inlet therebeyond. The dryer is intricately baffled with a short fall fill that materially increases its efficiency. The ash removal subassembly includes a hopper into which the ashes fall and are sucked out of the top thereof and used to preheat the primary combustion airstream. The exhaust gases are exhausted through a heat recovery system and then scrubbed prior to release to the atmosphere.

An apparatus developed by H. Tsuruta et al (63) is a rotary kiln type solid waste incinerating system and a method for continuously and efficiently incinerating a large amount of solid waste in which a vortex is established within the rotating cylinder of the kiln by injection of air to vigorously agitate the atmosphere within the kiln and thereby ensure uniform distribution of oxygen and efficient combustion. In one embodiment, the system includes an air induction duct which is approximately concentric with the rotary cylinder and which has a plurality of air injection nozzles at suitable intervals along its length for injecting air into the rotary cylinder in a direction tangential to its inner surface to establish a vortex.

An apparatus developed by E.A. Sargent et al (64) is an apparatus for disposing of solid wastes having a kiln sealed at its feed and discharge ends. A compacted slug of waste material is collected at the feed end and when the slug is of sufficient size and density, it is pushed into the kiln combustion chamber where it is burned. The combustion gases enter a secondary combustion chamber which is provided with an afterburner.

After complete combustion, the gases exhaust through a stack. The kiln is provided with a unique air distribution system and is formed of a steel outer jacket and a spaced inner refractory jacket which has a plurality of openings to permit the feeding of oxygen or other combustion supporting gas from outside the kiln to the inside of the kiln. The kiln rotates about its axis to improve the combustion of the waste material. The burning is in a controlled atmosphere since the discharge end of the kiln is water sealed. The ash from the combustion chamber drops onto a water submerged conveyor for automatic removal from the system while maintaining the seal of the controlled combustion furnace.

A process developed by J.W. Bolejack, Jr. et al (65) is one in which waste propellant or explosives in particulate form is mixed with water forming an aqueous suspension. The aqueous suspension is burned in a rotary incinerator under conditions which permit sequential evaporation of water from the suspension, drying of the particulate propellant or explosive, and then ignition of the propellant or explosive. The combustion gases are scrubbed with water prior to passing into the atmosphere. Figure 49 shows a suitable form of apparatus.

FIGURE 49: ROTARY KILN INCINERATOR FOR DISPOSAL OF WASTE PROPELLANTS AND EXPLOSIVES

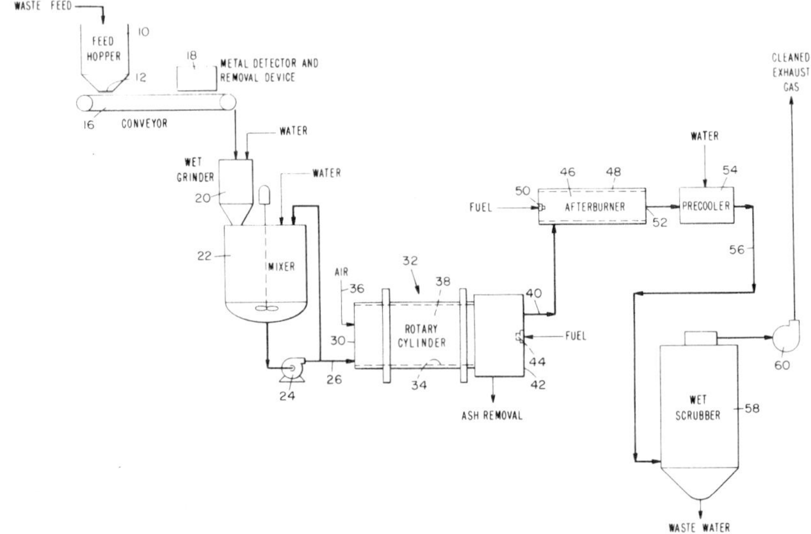

Source: U.S. Patent 3,848,548

As shown, waste solid propellant or solid explosive, is fed into feed hopper **10**. The feed hopper **10** is equipped with vibratory means (not shown) for uniformly feeding the waste solids through a bottom opening **12** in the feed hopper **10** into a fiber glass conveyor **16**. Conveyor **16** is used to convey the waste solids through a metal detection and removal device **18** wherein any extraneous metal, i.e., metal which is not associated with or part of either the solid propellant or solid explosive waste feed is detected and after detection the metals are removed from the feed to wet grinder. Extraneous metal-free waste solid propellant or solid explosive is then conveyed into a wet grinder **20** at a controlled rate. In the wet grinder **20**, the solid waste is ground in the presence of water. The resulting solid particles have a length or maximum dimension averaging 0.25 inch or less.

The ground water-wet solid waste particles are washed into mixer **22**. In mixer **22**, the particles are admixed with additional water, as necessary, to form a suspension of solid waste particles in water. This suspension is pumped by pump **24** through feed line **26** into the inlet-end **30** of kiln **32**. Flow through the feed line **26** must be maintained in the turbulent range, i.e., at a Reynolds Number of at least about 6,400 to maintain the solid particles in suspension. Kiln **32** is lined with refractory brick. Kiln **32** rotates about its longitudinal axis through use of drive means (not shown). The aqueous suspension of waste solids is deposited on the wall of the kiln **32** and immediately flows to a thin layer of suspension. Movement of the wall **34** of kiln **32** constantly provides continual contact of a hot-dry surface to the layer of suspension. Air is continuously introduced into the kiln **32** at the inlet-end **30** through air line **36**. Air passes through the inner chamber **38** of kiln **32** and through exhaust outlet **40** at the burner-end **42** of kiln **32**. A burner **44** is used constantly during incineration to maintain the temperature in the inner chamber **38** of the kiln **32**. Exhaust gases from combustion of the waste solid propellant or explosive within inner chamber **38** of kiln **32** pass through exhaust line **40** into a second incinerator **46** consisting of a refractory lined wall **48** and burner **50**.

The second incinerator **46** functions as an afterburner for further combustion of combustible products contained in the exhaust gas from kiln **32**. The resulting exhaust gas passes out of the exit-end **52** of second incinerator **46** through a precooler **54** and through feed line **56** into a wet scrubber **58**. In wet scrubber **58** particulate matter and noxious gases are removed from the exhaust gas. An exhaust fan **60** draws the exhaust gas through the wet scrubber **58** and exhaust the resulting product gas to the atmosphere.

Seagoing Incinerators

Some of the aspects of incineration and of ocean disposal are combined in the use of special shipboard-mounted incinerators for the disposal of refractory and toxic wastes. The first incineration at sea was practiced by Stahl Und Blech-Bau GmbH (SBB) of Hamburg, Germany (66). They outfitted a former coastal tanker named Matthias I in 1968 and built a second ship, Matthias II in 1970.

All of the chlorinated hydrocarbon residues from the German chemical industry that have burned at sea were performed by SBB. SBB has also burned similar wastes from Holland, Belgium, Scandinavia, and France. The other practitioner of ocean incineration is Ocean Combustion Services (OCS), a subsidiary of the Hansa Line, the West German steam shipping company (Bremen, Germany). OCS outfitted the Vulcanus with incinerators and can accommodate 4,200 metric tons of such wastes. It became operational in 1972.

Wastes from Shell Chemical Co. have been burned aboard the Vulcanus about 190 miles from the U.S. shore in the Gulf of Mexico. As of mid-November 1975, the U.S. Environmental Protection Agency gave a limited seal of approval to ocean incineration of organochlorine wastes, based on experience with the Vulcanus (67). EPA's limited endorsement of the technique may relieve waste-disposal headaches for makers of vinyl chloride, chlorinated solvents, epoxy resin, glycerin intermediates, pesticides, and similar chlorinated hydrocarbon derivatives. But EPA is hedging on the technology's viability for treating other potent chemical wastes.

Incineration at sea boasts two key advantages for the chemical process industries. First, they wouldn't incur any capital costs. Companies would simply contract for the services of an incinerator ship, such as the Vulcanus. While this certainly doesn't come cheap, about $50/ton in Shell's case, sources report, it may be used on an interim basis, e.g., while developing new treatment technology, or installing land-based incinerators. And it might be cost-effective for small companies that don't have the waste volume to justify their own equipment.

In operation, fuel oil preheated the about 17-ft diameter furnace to at least 2700°F. Then, 22 to 27.5 tons/hr of the 6,000-Btu/lb wastes (see Table 20 for analysis) were fed to the incinerator, and the unit's flame temperature was adjusted to an average of about 2460°F. Excess air varied from 90 to 160%. The 2000° to 2200°F stack gases generally contained from 25 to 75 ppm carbon monoxide, 9.0 to 12.5% oxygen, and 5.2 to 6.2% hydrochloric acid, and less than 200 ppm of chlorine—all well within permit specifications, reports EPA. Overall, more than 99.9% of the toxic wastes were oxidized to relatively innocuous forms.

An invisible plume trailed at a 20° angle from the Vulcanus stack, reaching a maximum altitude of 2,800 ft. Hydrochloric acid concentration in the 3,900-ft wide plume only hit a high of 3 ppm. Followup marine-environment tests showed no ill effects caused by un-oxidized chlorinated hydrocarbons, trace heavy metals, or hydrochloric acid, says EPA.

TABLE 20: TYPICAL ANALYSIS OF SHELL CHEMICAL'S ORGANOCHLORINE WASTES

Compound	Concentration, % by weight
1,2,3-trichloropropane	27 – 28
Dichloropropenes and lighter	20 – 22
1,1,2-trichloroethane	13
1,2-dichloroethane	10 – 11
Dichlorobutanes and heavier	10 – 11
Dichlorohydrins	8 – 9
Tetrachloropropyl ether	6
Allyl chloride	3

Source: *Chemical Engineering,* (January 5, 1976)

REFERENCES

(1) R.S. Ottinger, J.L. Blumenthal, D.F. Dal Porto, G.I. Gruber, M.J. Santy and C.C. Shih; *Recommended Methods of Reduction, Neutralization, Recovery or Disposal of Hazardous Waste,* Vol. III, Report PB 224 582; Springfield, Va., Nat. Tech. Information Service (August 1973).

(2) J.I. Frankel; *Chem. Eng.* 73, 91-96, (Aug. 29, 1966).

(3) R.S. Ottinger, J.L. Blumenthal, D.F. Dal Porto; G.I. Gruber, M.J. Santy and C.C. Shih; *Recommended Methods of Reduction, Neutralization, Recovery or Disposal of Hazardous Waste,* Vol. I, Report PB 224 580, Springfield, Va., Nat. Tech. Information Service (August 1973).

(4) J.K. Holcombe and P.W. Kalika; *Solid Waste Management in the Industrial Chemical Industry,* Report PB 226 420, Springfield, Va., Nat. Tech. Information Service (1973).

(5) M. Sittig; *Pollution Control in the Organic Chemical Industry,* Park Ridge, N.J., Noyes Data Corp. (1974).

(6) J. Greenberg; U.S. Patent 3,647,358; March 7, 1972; assigned to Anti-Pollution Systems, Inc.

(7) R.W. Foster-Pegg; U.S. Patent 3,718,440 ; February 27, 1973.

(8) B.R. Hutchinson, B.H. Hunter and R.N. Brooks; U.S. Patent 3,730,112; May 1, 1973; assigned to Silent Glow Corp.

(9) M.W. Ehrlichmann and J.R. Bjorklund, Jr.; U.S. Patent 3,749,032; July 31, 1973; assigned to West Creek Co., Inc.

(10) E.M. Van Raden; U.S. Patent 3,754,869; August 28, 1973; assigned to Mahon Industrial Corp.

(11) R.H. Wieken and H.C. Potter; U.S. Patent 3,832,144; August 27, 1974; assigned to Temperature Control, Inc.

(12) D.K. Longley; U.S. Patent 3,843,329; October 22, 1974.

(13) A. Taeymans, G. Dumont and W. Balleux; U.S. Patent 3,847,904; November 12, 1974; assigned to Belgonucleaire.

(14) K.H. Hemsath and A.C. Thekdi; U.S. Patent 3,898,317; August 5, 1975; assigned to Midland-Ross Corp.

(15) L.E. Ravich; U.S. Patent 3,056,467; October 2, 1962; assigned to Hupp Corp.

(16) H. Volker; U.S. Patent 3,690,840; September 12, 1972.

(17) F. Tabak; U.S. Patent 3,806,322; April 23, 1974; assigned to Universal Oil Products Co.

(18) F. Tabak; U.S. Patent 3,898,040; August 5, 1975; assigned to Universal Oil Products Co.

(19) R.J. Ruff and W.C. Verner; U.S. Patent 3,090,675; May 21, 1963; assigned to Universal Oil Products Co.

(20) W.A. Denny and W.C. Verner; U.S. Patent 3,311,456; March 28, 1967; assigned to Universal Oil Products Co.

(21) L.C. Hardison and W.L. Hable; U.S. Patent 3,484,189; December 16, 1969; assigned to Universal Oil Products Co.

(22) F. Tabak; U.S. Patent 3,549,333; December 22, 1970; assigned to Universal Oil Products Co.

(23) G.L. Brewer; U.S. Patent 3,706,533; December 19, 1972; assigned to Universal Oil Products Co.

(24) J.H. Hirt; U.S. Patent 3,738,816; June 12, 1973; assigned to Hirt Combustion Engineers.

(25) K. Zenkner; U.S. Patent 3,827,861; August 6, 1974.

(26) H.O. Ebeling and R.D. Smith; U.S. Patent 3,837,813; September 24, 1974; assigned to Black, Sivalls and Bryson, Inc.

(27) F. Tabak; U.S. Patent 3,838,975; October 1, 1974; assigned to Universal Oil Products Co.

(28) B.G. Altmann; U.S. Patent 3,875,874; April 8, 1975.

(29) D.J. Bergman; U.S. Patent 2,976,922; March 28, 1961; assigned to Universal Oil Products Co.

(30) C.A. H. Williams; U.S. Patent 3,162,236; Dec. 22, 1964; assigned to The British Petroleum Co., Ltd.

(31) A.J. Turpin; U.S. Patent 3,547,567; December 15, 1970; assigned to Smoke-Ban Manufacturing, Inc.

(32) A.E. Proctor; U.S. Patent 3,554,681; January 12, 1971.

(33) V. Jasinsky and A.T. Upfold; U.S. Patent 3,667,408; June 6, 1972; assigned to Polymer Corp., Ltd.

(34) D.J. Frey and W.P. Opp; U.S. Patent 3,697,229; October 10, 1972; assigned to Combustion Engineering, Inc.

(35) R.D. Reed; U.S. Patent 3,697,231; October 10, 1972; assigned to John Zink Co.

(36) R.D. Reed, J.S. Zink, R.E. Schwartz, H. Glomm, J.C. Corble and H.F. Koons; U.S. Patent 3,779,689; December 18, 1973; assigned to John Zink Co.

(37) E.C. Eubanks; U.S. Patent 3,782,880; January 1, 1974; assigned to Gulf Oil Corp.

(38) J.S. Zink, R.D. Reed and H.E. Goodnight; U.S. Patent 3,814,567; June 4, 1974; assigned to John Zink Co.

(39) J.J. Stranahan, J.C.L. Hollier and H.C. Deloney; U.S. Patent 3,829,275; August 13, 1974; assigned to Texaco, Inc.

(40) R.W. Evans and D.M. Simmons, Jr.; U.S. Patent 3,837,785; September 24, 1974; assigned to Phillips Petroleum Co.

(41) W.L. Buchanan and A.C. Worley; U.S. Patent 3,859,033; January 7, 1975; assigned to Exxon Research Engineering Co.

(42) P.G. Marsh and E.T. Blakley; U.S. Patent 3,549,010; December 22, 1970; assigned to The Black Clawson Co.

(43) L.C. Hardison; U.S. Patent 3,632,304; January 4, 1972; assigned to Universal Oil Products Co.

(44) D.R. Muirhead and D.A. Michell; U.S. Patent 3,702,595; Nov. 14, 1972; assigned to Power-Gas Corp.

(45) R. Menigat; U.S. Patent 3,736,886; June 5, 1973; assigned to Metallgesellschaft AG

(46) S.G. Hibbert; U.S. Patent 3,745,940; July 17, 1973; assigned to Sprocket Properties, Ltd.

(47) C.S. Miller Jr., H.K. Staffin and R. Staffin; U.S. Patent 3,772,999; November 20, 1973; assigned to AWT Systems, Inc.

(48) E.J. Roberts; U.S. Patent 3,864,458; February 4, 1975; assigned to Dorr-Oliver, Inc.

(49) E.S. Monroe, Jr., "Burning Waste Waters", *Chem. Eng.* (September 23, 1968).

(50) C.J. Wall, J.T. Graves and E.J. Roberts, "How to Burn Salty Sludges," *Chem. Eng.* (April 14, 1975).

(51) D.N. Garver, L.A. Davis and D.F. Davis; U.S. Patent 3,489,108; January 13, 1970; assigned to Garver-Davis, Inc.

(52) E.S. Monroe, Jr.; U.S. Patent 3,611,954; October 12, 1971; assigned to E.I. du Pont de Nemours and Co.

(53) T.J. Wittmann; U.S. Patent 3,814,568; June 4, 1974; assigned to Systems Technology Corp.

(54) R. Ragot; U.S. Patent 3,828,700; August 13, 1974; assigned to Speichim.

(55) E.S. Monroe, Jr.; U.S. Patent 3,865,054; February 11, 1975; assigned to E.I. du Pont de Nemours and Co.

(56) H.M. Katz; U.S. Patent 3,881,430; May 6, 1975; assigned to Phillips Petroleum Co.

(57) S.J. Yosim, L.F. Grantham and D.A. Huber; U.S. Patent 3,778,320; December 11, 1973; assigned to Rockwell International Corp.

(58) S.J. Yosim, D.E. McKenzie, L.F. Grantham and J.R. Birk; U.S. Patent 3,845,190; October 29, 1974; assigned to Rockwell International Corp.

(59) J.C. Eck; U.S. Patent 3,777,680; December 11, 1973; assigned to Wilputte Corp.

(60) C.H. Boll and R.P. Lanyon; U.S. Patent 3,483,832; December 16, 1969; assigned to Solvents Recovery Service of New Jersey, Inc.

(61) R.P. Lanyon and C.H. Boll; U.S. Patent 3,592,150; July 13, 1971; assigned to Separation Processes Corp.

(62) S.M. Porter, E.C. Weimer and H.W. Shideler; U.S. Patent 3,716,002; February 13, 1973; assigned to Stearns-Roger Corp.

(63) H. Tsuruta and M. Makiguchi; U.S. Patent 3,827,379; August 6, 1974; assigned to Nittetu Chemical Engineering, Ltd.

(64) E.A. Sargent and A.J. Doner; U.S. Patent 3,842,762; October 22, 1974; assigned to Grumman Ecosystems Corp.

(65) J.W. Bolejack, Jr., T.D. Daniel, Jr. and D.E. Rolison; U.S. Patent 3,848,548; November 19, 1974; assigned to Hercules, Inc.

(66) *Environmental Science and Technology* 9, No. 5, 412-13 (May 1975).

(67) *Chemical Engineering,* 83, No. 1, 86, 88 (January 5, 1976).

LAND APPLICATION

Land disposal of solid wastes by chemical plants consists of either dumping the waste in piles on the ground or burying it. All the major types of process solid wastes are sometimes disposed of by this method, including sludges, tars, off-quality product, filter residues, and fly ash (1).

Wastes dumped on the ground are principally dry chemicals, filter residues, and heavy sludges. They are generally inert and insoluble inorganic chemicals which do not generate odors on decomposition or pollute surface and ground waters through leaching of pollutants. Other effects of some solid wastes (such as emission of noxious gases, dusting, or esthetic problems) may prevent disposal by this method. A large percentage of the solid wastes dumped on the ground are sludges dredged from settling ponds or lagoons.

The disposal of liquid digested sludge on open land surfaces is quite common among smaller wastewater treatment plants. In England, the disposal of liquid sludge to farmland is very popular and in the arid and semiarid parts of the United States, the reclamation of water from municipal sewage is becoming increasingly recognized as an important water conservation measure. Liquid digested sludge and supernatant have been applied to lands for final disposal to fertilize grass or agricultural crops for soil conditioning (2).

Digestion, aerobic or anaerobic, is almost always required before spreading liquid sludge on land. Sludge is distributed on the land and processed in a variety of ways; it may be injected into the subsoil under pressure or pumped or gravity fed through a pipeline to agricultural fields or land to be reclaimed. A common technique is disposal directly to the land by spraying from tank wagons. Design considerations and process variables involved are: proximity of surface waters and distance to groundwater table, toxic constituents of the sludge, nutritional value of the sludge, availability of disposal sites, transportation costs, suitability of soil for sludge disposal, nature of any esthetic nuisance, effects on vegetation, application rates, atmospheric and climatic conditions, and method of sludge application. Design criteria can only be developed through the use of fairly large demonstration sites and extensive time-involving studies are required. The effects of land disposal on crops and ground and surface waters is of paramount importance.

REFERENCES

(1) J.K. Holcombe and P.W. Kalika, *Solid Waste Management in the Industrial Chemical Industry,* Report PB 226 420, Springfield, Va., Nat. Tech. Information Service (1973).
(2) M. Sittig, *Pollution Control in the Organic Chemical Industry,* Park Ridge, N.J., Noyes Data Corp. (1974).

LAND BURIAL

This is to be differentiated from land application (to the land surface) or landfill (very near the surface). Land burial is adaptable to those hazardous materials that require permanent disposal. Disposal is accomplished by either near-surface or deep burial. In near-surface burial the material is deposited either directly into the ground or is deposited in stainless steel tanks or concrete lined pits beneath the ground. The standard procedures for deep burial are disposal in salt mines or hard bedrock, or in shale formations by using hydraulic fracturing. Hydraulic fracturing is not covered here but is covered under deep well disposal.

In land burial the waste is transported to the selected site, where it is prepared for final burial. Transportation of the wastes to the burial site can be accomplished in three ways: by common carriers with the waste packaged along with ordinary shipments of wastes, by contract carriers that handle only the hazardous materials to be buried but collect from various sources, or by private carriers that transport their own wastes from the point of origin to the burial site.

Either solid or liquid wastes can be received at the burial site. To reduce the mobility of the wastes before burial all liquid wastes should be converted to a solid form. This requires that special solidification equipment be located at the burial site. Coupled with this special solidification equipment heavy equipment for excavation and lifting and special monitoring instruments and stations will also be required.

At the present time near-surface burial of both radioactive and chemical wastes is being conducted at several Atomic Energy Commission (AEC) and commercially operated burial sites (1). These wastes are buried in unlined trenches approximately 20 feet in depth. The trenches are filled to within 2 to 5 feet of the surface and are covered with either asphalt or soil and vegetation to reduce infiltration of water. Radioactive wastes are stored in either liquid or solid form in steel tanks enclosed in concrete (2)(3).

Pilot plant studies have been conducted for deep burial in salt formations (4)(5) and hard bedrock (6). These wastes would be buried approximately 1,000 to 1,500 feet beneath the ground in unlined tunnels. The wastes are lowered into these tunnels by means of a central access shaft. After the filling operation is complete in a tunnel, it is sealed off by backfilling with salt and using a positive seal, for example, concrete.

SITE SELECTION

The selection of a site for the disposal of hazardous materials is dependent upon several factors. These include physical characteristics of the wastes to be buried, environmental characteristics of the area, operating equipment and waste handling procedures required, and the geographic characteristics of the surrounding area. The types of hazardous wastes to be buried at a particular disposal site are important in determining whether the site is owned and operated by a private concern or by the government. If only short-lived materials are to be disposed of then private ownership and operation with state or federal licensing and regulation can be considered. For long-lived materials it is imperative that the disposal site be located on state or federally owned land to ensure that perpetual monitoring and care can be maintained. Even though government ownership of the site is required, on-site operation can be performed by a private concern.

In selecting the location of a disposal site the environmental characteristics of the area are important. The environmental factors of principal concern are meteorology, geology, hydrology, and geoseismology. Detailed meteorological data are required since if a particle or gas escapes to the outside environment, its fate is determined by the prevailing meteorological conditions. The frequency of wind direction toward any given sector determines the degree of possible risk to the population within that sector from material emitted upwind. Besides wind direction, wind speed affects the dilution rate of the material. The amount and rate of rainfall are significant factors in determining the amount of material that can be leached from the wastes.

The geology and hydrology of the area determine whether waste is dispersed or confined. The factors which influence the movement of the waste are: main formations in the area, such as gravel, clay, sand and shale; permeability and ion exchange capacity of the soil; and depth of the water table.

Since water represents the main vehicle of transporting any significant quantities of wastes from the burial site, the site should be as far as possible from any important groundwater sources. Since the groundwater can convey the wastes to the surface streams, it is necessary to determine the possible movement of groundwater from the burial site into the streams, springs, and water sources. The points of groundwater discharge must be established and the dilution capacity of the surface streams determined.

In near-surface burial the trenches or concrete and stainless steel lined pits should be constructed to hold the wastes above the water table. This is to prevent leaching of the waste by the groundwater. The wastes should also be buried as far as possible from any surface stream or water wells in order to maximize the retention time in the soil if leaching of the wastes does occur. In this way the waste can be retained by the natural processes of absorption, filtration, and ion exchange.

The trenches or pits should be covered with an impermeable material to prevent infiltration of rainfall. Also, infiltration of rain can be reduced by covering with grass or other vegetation. This latter method is less desirable than the other two since some infiltration of water can occur especially during periods of heavy rainfall.

The geoseismology data such as faults, vibrations and tsunamis are the major earthquake phenomena that must be considered. Since there is a general lack of knowledge about earthquakes, it is necessary to make conservative estimates and evaluations of the critical geoseismological data. A seismic probability map of the United States is shown in Figure 50 depicting zones of no, moderate and major damage. The largest zone of possible major seismic damage lies along the west coast of the United States.

In locating the disposal site it is necessary to provide sufficient distance between the site and the surrounding population to minimize the danger to the general public by either normal operation or accidental releases. For nuclear reactor plants federal regulations (10CFR100) specify that the reactor plant be surrounded by a zone of low population. This same regulation should also apply to a disposal site of hazardous materials. A population density map of the United States is shown in Figure 51. The major areas with a population density of less than 30 persons per square mile are located in the midwest, southwest and northwest regions of the United States.

FIGURE 50: SEISMIC PROBABILITY MAP

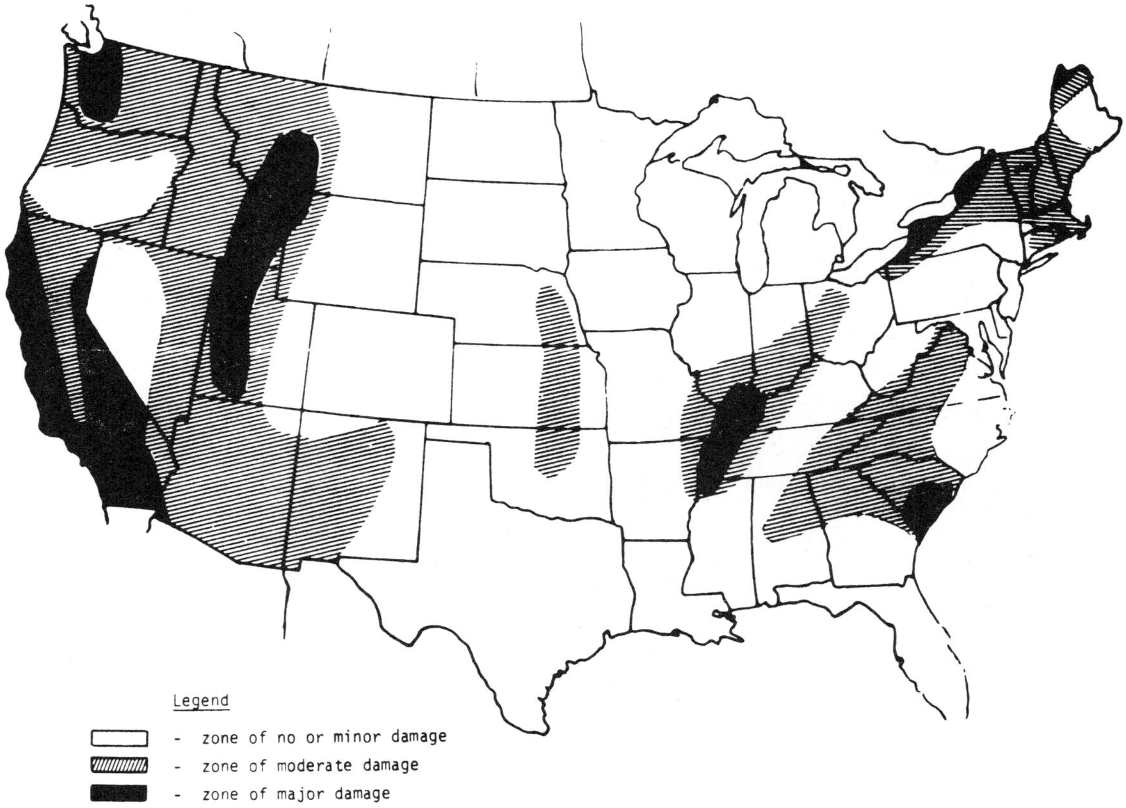

Source: ORNL-4451

FIGURE 51: POPULATION DENSITY IN THE UNITED STATES

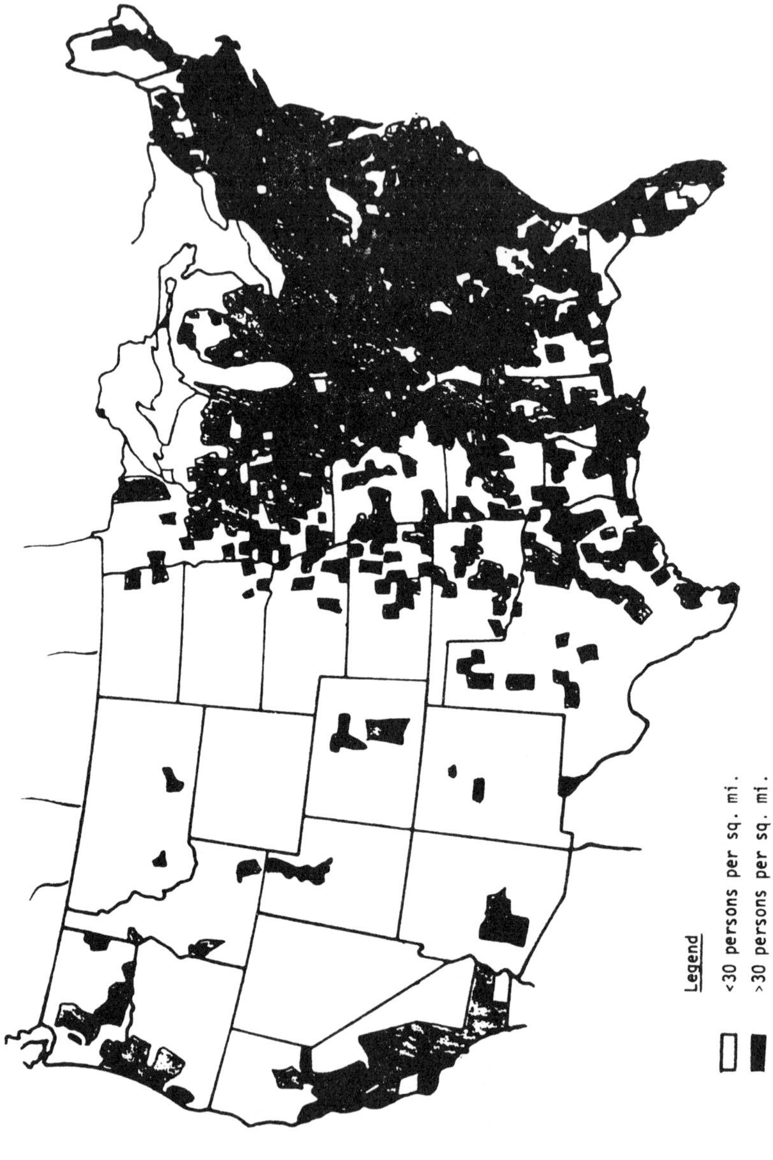

Source: ORNL-4451

In addition to selecting an area of low population density the site should be located to minimize the distance required to transport the hazardous materials to the site.

MONITORING SYSTEM

For each waste received at the burial site inventory records should be kept identifying the type of waste received, its activity and toxicity, and the source and quantity of the waste. Also the form (liquid or solid), type of container and date received should be recorded. A coding or permanent marking system should be devised to record the location of all buried wastes. These data should then be recorded on a map. A monitoring system is also required to measure the amount and location of any discharged wastes. This should include direct monitoring of the wastes in each burial site and monitoring of test wells, surfaces, streams, and lakes in the general area of the burial site.

In deep burial in either salt mines or hard bedrock a waste retrieval plan should be devised. This plan should call for the development of systems capable of retrieving the wastes. This plan should also be coupled with a worst case hazards analysis to determine what happens if the integrity of the site is destroyed or the waste retrieval system does not perform according to design. A continuous monitoring system is not only needed to measure the discharge of any wastes but to also measure any changes in the geology of the area and in the location of the buried wastes.

PRESENT DESIGNS

At the present time most hazardous materials are disposed of by near-surface burial. These materials are either buried directly in the ground or in stainless steel tanks or concrete lined pits beneath the ground. Research and pilot plant studies are being conducted for deep burial in salt mines or hard bedrock.

Near-surface burial of radioactive wastes is being conducted at several AEC sites and also at six commercial burial sites. In addition to radioactive wastes some commercial burial sites also handle certain chemical wastes. These commercial burial sites are regulated by the AEC or by an AEC agreement state.

A complete description of the operation and facilities at several of the commercial burial sites is provided by R.J. Morton, AEC (1). At each of these sites the wastes are buried in trenches approximately 20 feet in depth. These trenches vary in width from 25 to 60 feet and vary in length from 300 to 700 feet. The design of the trenches at each site is fairly similar. The trenches are designed not to intercept the groundwater table and are constructed with a bottom drain and sump for water monitoring. The trenches are unlined, so that the extent of leaching is dependent on the permeability of the soil. At each site liquid wastes are solidified by mixing with various additives, such as concrete, which absorb and solidify the wastes. These commercial facilities also offer packaging and transportation services.

Radioactive wastes have also been stored as liquids in stainless steel encased in concrete and buried underground. These tanks range in size from 0.33 to 1.3 million gallons. The tanks are equipped with devices for measuring temperatures, liquid levels, leaks and for agitating the contents. These tanks are considered as an interim storage technique due to a general lack of confidence in their long-term integrity (2).

Stainless steel bins buried beneath the ground have also been used at the Idaho Chemical Processing Plant, Idaho Falls, for solid radioactive wastes. The life of these bins has been estimated at 500 years. The bins are constructed of one-quarter inch thick stainless steel and each bin is 12 feet in diameter and 42 feet high. Six bins are enclosed in a concrete vault. The vault is constructed of 2 foot thick reinforced concrete and is 46 feet in diameter and 69 feet high. The vault is 45 feet below ground level and rests on bedrock.

The vault is provided with a cooling air system to provide convective cooling of the bins. A detailed design of these storage facilities is available (3). Pilot plant studies have been conducted for deep burial in salt mines. Detailed designs of a salt mine disposal facility for solid radioactive wastes (4) and for solid chemical wastes (5) have been prepared. At the salt disposal facility the wastes would be received at a surface facility and lowered down a steel-lined shaft into the working area of the mine. The working area would be located approximately 1,000 feet below the surface. The wastes would then be transported to the disposal area in the mine by either a specially designed underground waste transporter or by a conveyor belt. After the waste disposal operations are complete in a particular area, this area is then shut off by backfilling with salt.

Studies have been conducted at the Savannah River Plant (6) near Aiken, South Carolina for the disposal of radioactive wastes in vaults excavated in crystalline rock 1,500 feet beneath the surface. Access to the vault would be provided by a 15 foot diameter shaft. The wastes would be stored in tunnels extending from the central shaft. These tunnels would be approximately 30 feet wide and 18 feet high. Each tunnel is provided with a 2 foot diameter service shaft. For the disposal of liquid wastes the tunnel is isolated from the main shaft by two concrete bulkheads, each 10 feet thick. The tunnel is then filled via the main shaft with the service shaft serving as an air vent. After the tunnel is filled, it is sealed by two concrete bulkheads.

PROCESS ECONOMICS

Typical rates charged at the six commercial burial sites for near-surface burial of radioactive or chemical wastes are included (Table 21). The wastes received at each burial site must be enclosed within containers that are in accordance with AEC, U.S. Department of Transportation (DOT), or U.S. Bureau of Explosives regulations. The minimum rate charged for unloading and burying these containers is $0.75 per cubic foot. Special surcharges are also made for containers weighing >15 tons and for containers that require special handling.

TABLE 21: TYPICAL RATES CHARGED AT COMMERCIAL BURIAL SITES

(1) Basic rate for containers less than 15 tons total weight: $0.70 per cubic foot plus state charge of $0.05 per cubic foot.

(2) Surcharge for containers in excess of 15 tons:

| | ------------- Surcharges ($) ------------ | | |
Weight, Tons	Per Shipment		Per Container
0–15	0.00	plus	0.00
15–30	130.00	plus	200.00
30–50	260.00	plus	330.00
50–60	520.00	plus	475.00
60–80	1,600.00	plus	1,200.00
80–130	3,200.00	plus	2,500.00

(3) Surcharge for special handling of containers consisting of two or more parts. This is for removing and burying inner containers which have been shipped inside a shielded coffin, cask or container:

Primary Containers with Surface Dose Rates	$ Per Shipment		$ Per Hour*
0.2–10 r/hr	25.00	plus	26.00
10–50 r/hr	50.00	plus	26.00
50–100 r/hr	100.00	plus	26.00
100–500 r/hr	250.00	plus	26.00

*The $26.00 per hour referred to includes consulting and preparation of proper procedures

(4) Minimum charge for any shipment is $20.00.

Source: R.J. Morton, *Land Burial of Solid Radioactive Wastes: Study of Commercial Operations and Facilities*

Cost studies have been conducted (2) for tank storage of radioactive wastes. The waste management and storage costs were based on an economics model using a discounted cash flow technique. This type of model requires that the income received must provide for the recovery of investment, the desired return on investment, all cash expenses, and the establishment of a reserve account to pay all waste management operations that remain to be completed after all income has ceased. Using this model, costs were determined for perpetual tank storage of high-level liquid radioactive wastes.

For a 50 year tank life and a 1,000,000 gallon capacity the costs varied from $4,100 to $8,200 per ton of fuel depending on the type of waste (acid or alkaline) and type of ownership (government or private). Costs for interim solid storage of radioactive wastes in water-filled canals were presented as a function of age of the waste. The costs range from $1,275 per ton of fuel for 1 year storage of 30 year old waste in 6 inch diameter pots to $4,100 per ton of fuel for 30 year storage of 1 year old waste. Costs were also presented for solidifying these wastes using the pot calcination technique. For calcination in 6 inch diameter pots these costs ranged from $4,200 per ton of fuel for 1 year old waste to $800 per ton of fuel for 30 year old waste.

For deep burial of hazardous materials in salt mines cost studies for radioactive wastes and for chemical wastes (5) have been included. For radioactive wastes the costs were based on the same economic model described above. The wastes were assumed to be buried in vertical holes in the floor of the salt mine at a depth of 1,000 feet. The burial costs vary with the heat generation rate and age of the waste at burial. For burial in 6 inch diameter pots the disposal costs range from $2,800 per ton of fuel for 1 year old waste to $260 per ton of fuel for 30 year old waste.

Detailed cost estimates for constructing and operating a chemical waste storage facility in bedded salt have been derived (5). The facility was located in Boco County, Colorado and had a storage space of 43.6 million cubic feet mined in bedded salt at a depth of 1,330 feet. The total cost of the facility was estimated at $41 million which gives an average cost of $0.96 per cubic foot of waste stored.

Cost for disposal in a solution mined facility were also presented. This facility would consist of four caverns having an average diameter of 67 feet and a height of 4,000 feet. The total volume of the four caverns would be 43.6 million cubic feet and the average cost of disposal was estimated at $0.32 per cubic foot of waste stored.

PROCESS APPLICABILITY

Land burial is a possible choice for those hazardous materials that require complete containment and permanent disposal. This includes radioactive wastes as well as highly toxic chemical wastes. Disposal can be accomplished by either near-surface or deep burial. Deep burial is more applicable to the highly toxic or dangerous materials since better isolation from the biosphere is afforded.

The important criterion in evaluating a particular land burial process is determining the integrity of the site. Sites with a life expectancy of a few hundred years are not applicable to wastes with a life expectancy of a few thousand years. In addition, before any land disposal methods can be selected, it must be determined if eventual retrieval of the wastes is required. This could be required if new reprocessing techniques are devised or under emergency conditions.

At the present time only near-surface burial is used for the disposal of most wastes. Low-level radioactive wastes and some chemical wastes are buried in unlined trenches 20 feet in depth. High-level radioactive wastes are stored as liquid in steel tanks located near the ground surface. For deep burial in salt formations or hard bedrock only pilot plant studies are being conducted at this time.

REFERENCES

(1)   R.J. Morton, *Land Burial of Solid Radioactive Wastes: Study of Commercial Operations and Facilities,* Atomic Energy Commission, Washington, WASH-1143 (1968).

(2)   Staff of the Oak Ridge National Laboratory, *Siting of Fuel Reprocessing Plants and Waste Management Facilities,* Oak Ridge National Laboratory, ORNL-4451 (July 1970).

(3)   C.L. Bendixsen, *Storage Facilities for Radioactive Calcined Waste Solids at the Idaho Chemical Processing Plant,* Idaho Nuclear Corp., Idaho Falls, IN-1155 (July 1968).

(4)   R.L. Bradshaw et al, *Evaluation of Ultimate Disposal Methods for Liquids and Solid Radioactive Wastes. VI: Disposal of Solid Wastes in Salt Formations,* Oak Ridge National Laboratory, ORNL-3358 (March 1969).

(5)   C.S. Dunn et al, *Feasibility of Permanent Storage of Solid Chemical Waste in Subsurface Salt Deposits,* Fenix and Scission, Inc., Tulsa, Oklahoma, F&S-196 (October 1971).

(6)   *Proceedings of the Symposium on the Solidification and Long-Term Storage of Highly Radioactive Wastes,* sponsored by Atomic Energy Commission, Richland, Washington, CONF-660208 (Feb. 14-18, 1966).

LANDFILLS

Some of the primary findings of EPA's Report to Congress on Hazardous Waste Disposal (1) are that usage of the land for hazardous waste disposal is increasing due to the implementation of air and water pollution controls, and the limitation of disposal methods such as ocean dumping. The Clean Air Act (as amended), the Federal Water Pollution Control Act (as amended), and the Marine Protection, Research, and Sanctuaries Act (as amended), are curtailing the discharge of hazardous pollutants into the nation's air and water. Increasing volumes of sludges, slurries, and concentrated liquids will therefore find their way to land disposal sites.

Few economic incentives exist to encourage waste generators to utilize environmentally acceptable disposal methods. Current methods frequently result in contamination of groundwaters from leachates; surface waters from runoff and leachate; and air from evaporation, sublimation, or dust dispersal. For example, toxic heavy metals create a chronic hazard when deposited in the land environment. As a result of arsenic buried more than 30 years ago, several people in Perham, Minnesota, had to be hospitalized in 1972 due to arsenic poisoning of drinking water from a groundwater supply source contaminated by leachate from the buried deposit (1).

Because of the lack of effective controls, many hazardous wastes are currently being disposed of in dumps and conventional sanitary landfills. As an example, for several years a large municipal land disposal site in Delaware accepted both domestic and industrial wastes. In 1968, this disposal site had to be closed because chemical and biological contaminants had leached into the groundwater. By 1974, two major groundwater supply fields which had provided water for about 40,000 households in the area were contaminated. The cleanup costs are expected to be over $10 million. Although this situation has not directly been linked to the hazardous nature of any of the industrial wastes constituents, this example serves to point up the potential problem caused by disposing of any wastes in an unacceptable land disposal site (2).

DEFINITIONS

Common landfill disposal methods include the following: mixing with soil, evaporation and infiltration, and/or shallow burial. Combinations of these methods can be involved in a disposal process. For example, in the spreading of a slurry on land, the liquid content may either evaporate or infiltrate into the subsoil. Solid wastes will normally be incorporated in a landfill and buried. Liquids, slurries, and sludges might also be incorporated into a landfill; however, due to the large quantity of moisture contained in these wastes, disposal practices may involve spreading them on land or placing them in ponds to maximize evaporation or infiltration (3). Thus, there are some rather vague boundaries between land application (spraying or discharge onto land surface), landfill (commonly covering waste with a few inches to a few feet of soil by bulldozer application) and land burial (in steel or concrete containers or under many feet of soil).

Definitions of landfills (4) are as follows:

> General landfill: A site which is limited to disposal of inert solid wastes which should not pose a threat to water quality. The site may contain water (for example, marshy areas, gravel pits, or periodically flooded areas) with no threat to water quality from the wastes.
>
> Approved landfill: A site which is suitable for the disposal of inert solid wastes and decomposable organic materials. The site must provide separation of the wastes from underlying or adjacent usable water because of leachate possibilities.
>
> Secured landfill: A site suitable for disposal of all wastes, including liquid and/or solid hazardous wastes. The site must allow no discharge of these materials or their by-products to usable ground or surface waters by leaching, percolation or any other means. Another feature which may be included is inventory control on the wastes buried in the secured landfill. The prime requisite for such disposal is that the hazardous contents of the landfill be isolated from the surrounding environment. Water quality of surface and groundwater must not be compromised. Air quality must also be maintained.

Landfills may be classified as follows: general purpose landfills; general purpose approved landfills for small volume hazardous wastes; approved landfills for large volume hazardous wastes; general purpose secured landfill operations for small volume, extremely hazardous wastes; and secured landfill for specific wastes for intermediate or large volume extremely hazardous wastes. It should be noted that this classification of landfills is not that of the Environmental Protection Agency but is by an EPA contractor (4).

General Purpose Landfills: It is estimated that 55% of concentrated hazardous wastes of the land-destined hazardous waste from the inorganic chemicals industry currently finds its way into general purpose landfill sites. General purpose landfills are characterized by their acceptance of a wide variety of wastes, including garbage and other organic materials, and by the usual absence of special containment, monitoring, and leachate treatment provisions for hazardous wastes.

The potential for environmental damage by landfilled hazardous wastes differs depending on both the composition and quantity of that waste. Many general purpose landfills will accept small quantities of hazardous wastes, particularly if they are in drums or plastic containers, but refuse large amounts. Cyanides, arsenic compounds and some heavy metal compounds are examples of such materials. On the other hand, general purpose landfills are usually wary of even small amounts of beryllium and mercury compounds. When the hazardous level is relatively low, due either to the inherent characteristic of the compound or its low concentration in the overall waste mass, even large quantities of hazardous wastes may be accepted.

The preferable disposal technique for many wastes (2) such as municipal solid wastes, is the conventional sanitary landfill which may be defined as "a land disposal site employing an engineered method of disposing of solid wastes on land in a manner that minimizes environmental hazards by spreading the solid wastes in thin layers, compacting the solid wastes to the smallest practical volume, and applying cover material at the end of each operating day" (5).

The potential for leachate generation exists even in a well-designed and operated sanitary landfill (6). However, good site selection and design and careful attention to operating procedures minimizes this potential, and, in many instances prevents its occurrence. Other potential problems include escape of hazardous vapors and gases and possible explosive reactions within the fill.

Thus, additional precautions over and above those taken during sanitary landfilling of municipal solid wastes are required for land disposal of many hazardous wastes. The conventional landfill might be used, however, in those instances where the wastes contain a hazardous substance but in a form which is not particularly hazardous, that is, insoluble salts, or in a concentration so low as to be innocuous. Certain other wastes should probably never be land disposed because of extreme hazards posed by escape of even small quantities.

General Purpose Approved Landfills: Each general purpose landfill has its own ambience—geologically, hydrologically, and environmentally. Ideally, a general purpose landfill would be located in an isolated, dry part of the country with a thick layer of impermeable soil between the waste and the water table. Such areas are plentiful in the western United States, but not in the east. However, many existing and future landfill sites throughout the United States can approach ideal conditions. The degree of approach is differentiated here as approved landfills and secured landfills. Secured landfills are defined and discussed below.

Approved general purpose landfills are defined to meet the following criteria: the composition and volume of each hazardous waste is known and approved for site disposal by pertinent regulatory agencies; the site should be environmentally suitable for hazardous wastes; and provision is made for monitoring wells and leachate control and treatment if required. The advantages of approved landfill sites include the following: many hazardous wastes may be disposed of in a controlled and environmentally safe fashion; selection of landfill sites and disposal technology for environmental suitability still leaves a great number of available landfill sites; and disposal costs, for both transporting the waste to the site and the landfilling itself, are kept to levels close to those for general purpose sites and still much lower than for secured landfill.

From a practical standpoint many local regulatory agencies and landfill site owners are informally practicing much of this discrimination by selective acceptance of waste materials. Sites with known high potential for surface and groundwater contamination are thereby avoided.

Approved Landfill for Large Volume Hazardous Wastes: Whenever the volume of a potentially hazardous waste is large, general purpose landfill operations are no longer appropriate. These wastes warrant, and because of their small number can be given, special attention. In the inorganic chemicals industry these large volume wastes usually come from ore residues and process by-products. Examples are chrome ore residues from sodium dichromate production and calcium fluoride-calcium sulfate residues from hydrofluoric acid production. Hazardousness for these wastes is generally of a lower level than for more concentrated and/or toxic compounds.

On the basis of this selection, a rough estimate is that 5% of the total wastes are disposed of in this fashion. Large volume hazardous wastes normally have their own landfill site so that interaction with other wastes is not a factor. Also, since transportation costs are high, disposal is usually either on-site or within a few miles of the plant. In view of their small number, large size, and transportation restrictions, each of these wastes can be given in-depth disposal analysis for environmental safety.

General Purpose Secured Landfills: The inorganic chemicals industry has a number of small volume wastes of extremely hazardous potential. For these wastes landfilling involves additional safeguards beyond those described for approved landfills.

Criteria for these secured landfills include the following: the composition and volume of each extremely hazardous waste is known and approved for site disposal by pertinent regulatory agencies; the site should be geologically and hydrologically approved for extremely hazardous wastes (included in the criteria would be a soil or soil-liner permeation rate of less than 10^{-7} cm/sec, a water table well below the lowest level of the landfill and adequate provision for diversion and control of surface water); monitoring wells are provided; leachate control and treatment (if required); records of burial coordinates to avoid any chemical interactions; and registration of the landfill site for a permanent record of its location once filled.

A number of landfills which meet the physical requirements (if not all the regulatory criteria) are located around the country. California has a number of Class 1 impermeable landfills which accept extremely hazardous materials. Texas has similar sites. A number of low level radioactive waste landfill sites accept industrial hazardous wastes. In addition to the radioactive waste sites, various other private secured landfills also take extremely hazardous wastes. Currently, secured landfills are scattered and not fully utilized. Part of the lack of utilization stems from the fact that the majority of the sites are in isolated western areas away from inorganic chemicals industrial centers. Another reason for the lack of utilization is the high cost as compared to other available disposal methods. The present utilization of secured landfills is estimated at 5% (4).

Relatively isolated impermeable soil conditions exist in many areas of the country. If impermeable soil is not available then clay, special concrete, asphalt, plastic and other liners and covers are available to accomplish similar containment and isolation of wastes. Once a landfill area has been isolated from surface and groundwater contact and leachates are being handled satisfactorily, almost any nonflammable, nonexplosive and nonair-polluting hazardous waste can theoretically be disposed of safely.

The following are a number of practical restrictions to this approach, however; in wet climates the impervious landfills are flooded with heavy rainfall and dumping of liquids or sludges into the landfill only accentuates the problem; some inorganic hazardous wastes create hazards for landfill personnel and/or give air pollution problems; and chemical interactions with both other materials and the liner can cause undesirable side effects.

Methods have been developed to modify the conventional sanitary landfill to make it acceptable for receipt of hazardous materials. Taken together, these modifications result in what is called above a "general purpose secured landfill" or otherwise designated as a "chemical waste landfill." In general terms, such operations provide complete long-term protection for the quality of surface and subsurface waters from hazardous waste deposited therein, and against hazards to public health and the environment. Such sites should be located or engineered to avoid direct hydraulic continuity with surface and subsurface waters.

Generated leachates should be contained and subsurface flow into the disposal area eliminated. Monitoring wells should be established and a sampling and analysis program conducted. The location of the disposal site should be recorded in the appropriate local office of legal jurisdiction. A special operating permit will most likely be required under the terms of future regulations. Of course, these requirements are also desirable in standard sanitary landfills. The primary difference involves the degree of concern and care which must be exercised where hazardous materials are involved. If there is potential for hazardous wastes to percolate or leach to groundwater, then the use of barriers and collection will be necessary.

Due to potentially hazardous reactions, wastes must be segregated and records kept of disposal areas. Neutralization, chemical fixation, encapsulation, and other pretreatment techniques are often necessary. Because of the high concentrations of hazardous wastes, attenuation capacity may be reached relatively quickly. Leachate treatment may be more complex due to the wide variety of waste types and constituents. Due to volatility or for other reasons, hazardous materials may require immediate cover. Due to these reasons, land disposal of hazardous wastes normally requires a greater degree of care and sophistication in design and operation at a given site than would normally be necessary with municipal refuse.

PRINCIPALS OF OPERATION

Landfills operate on two principles: [1] utilization of the absorptive capacity of the soil and, perhaps, some biological degradation of the wastes by soil microorganisms and [2] storage of wastes such that they are isolated from direct contact with man and the surface environment. Some liquid wastes are currently discharged to infiltrate and percolate into the underlying porous sediments where there is no possibility of ground waste impairment. In other cases simple, shallow burial of solid wastes in a geologically dead area is the ultimate method of disposal. It must be stressed that the usability of any landfill site is basically determined by the site's characteristics and investigation is of utmost importance to site selection.

Very little information is available on the mechanism and rate of decomposition of most chemical wastes when landfilled (7). The principal soil transformation process is biological decomposition. Microorganisms are capable of biological oxidation or reduction of both inorganic and organic chemicals, resulting in a broad array of chemical and biochemical products.

It has been reported that most aliphatic hydrocarbons are rapidly decomposed in soil. Hydrocarbons which are unsaturated, branched, and of high molecular weight are generally more susceptible to degradation than their saturated, unbranched, low-molecular weight analogues. Aromatic materials, on the other hand, are generally considered quite resistant to microbial degradation in soil and water, and carbon in aromatic forms constitutes a major portion of the relatively stable soil organic fraction. Once the aromatic ring is cleaved, however, the resulting straight-chain hydrocarbons are subject to relatively rapid degradation and oxidation to carbon dioxide and water.

The rate, extent and direction of microbial transformations in soil, however, are commonly dependent upon the type and quantity of available energy sources, availability of essential nutrients, degree of aeration, temperature, moisture, pH, and the presence of toxic substances. Large variations in these factors exist in chemical landfills depending on the procedures used and the chemical wastes disposed.

In some cases, when disposing of solid chemicals or sludges, the material will be mixed with the soil. In other cases no mixing is practiced, resulting in large slugs of chemical wastes in the fill. Waste tars or liquids contained in drums are often buried, and when the barrels corrode, the contents leach into the landfill. More information is necessary on the effect of these practices and on the fate of landfilled chemical wastes before land disposal procedures can be formulated which adequately protect the environment.

The chemical landfill operation conducted by Union Carbide Corporation at Institute, West Virginia has been described in great detail by E.E. Slover and J.D. Hainley (8). Sanitary landfills have also been discussed extensively by B.G. Liptak (9) with regard to site selection and site preparation, groundwater pollution, disposal costs, and landfill utilization. Landfill design, construction and operation (albeit with primary emphasis on municipal wastes rather than industrial wastes) has been reviewed in detail by S. Weiss (10).

QUALITATIVE EVALUATION OF LANDFILLS

The parameters resulting from the investigation of potential sites outlined above must be compared against standards designed to protect man and the environment from the hazards associated with the various wastes. In California a set of standards for selecting landfill sites which is based on contamination of usable water supplies has been defined and used by the State Department of Public Health, the Department of Water Resources, and the various California Regional Water Quality Control Boards.

According to a paper by L.A. Burch of the State Department of Health (11), three classes of wastes are recognized as requiring distinct levels of control of site effluents (surface or subsurface): [1] water-soluble materials that constitute hazards of high toxicity or special water pollution potential; [2] decomposable organic materials; and [3] relatively inert, nondecomposable materials.

Correspondingly, three classes of landfill disposal sites are recognized and are described by Burch as follows. Class 1 sites are those sites located over nonwater-bearing sediments or with only unusable groundwater underlying them. The site location must provide complete protection from flooding, surface runoff or drainage, and waste materials and all internal drainage must be restricted to the site. In essence, a Class 1 site is a large container providing safe, ultimate storage of toxic or hazardous materials; a secondary function of the site might be the processing of the waste such as evaporation to reduce the volume of the material to be disposed of. These sites can accept almost any type of materials, liquid or solid. These are the only sites where the first group of wastes, such as toxic materials, oily sludges and soluble industrial chemicals may be placed.

It should be noted that possible public health hazards must be recognized at the Class 1 sites in addition to water quality protection. Certain very toxic chemicals such as pesticides or tetraethyllead may require special handling techniques to protect site personnel and to provide long-term protection of public health and the environment. There are nine criteria developed by California which must be met by Class 1 facilities (Table 22).

TABLE 22: CALIFORNIA CLASS 1 SITE CRITERIA

A	Geological conditions are naturally capable of preventing hydraulic continuity between liquids and gases emanating from the waste in the site and usable surface or groundwaters.
B	Geological conditions are naturally capable of preventing lateral hydraulic continuity between liquids and gases emanating from wastes in the site and usable surface or groundwaters, or the disposal area has been modified to achieve such capability.
C	Underlying geological formations which contain rock fractures or fissures of questionable permeability must be permanently sealed to provide a competent barrier to the movement of liquids or gases from the disposal site to usable water.
D	Inundation of disposal areas shall not occur until the site is closed in accordance with requirements of the regional board.
E	Disposal areas shall not be subject to washout.
F	Leachate and subsurface flow into the disposal area shall be contained within the site unless other disposition is made in accordance with requirements of the regional board.
G	Sites shall not be located over zones of active faulting or where other forms of geological change would impair the competence of natural features or artificial barriers which prevent continuity with usable waters.
H	Sites made suitable for use by man-made physical barriers shall not be located where improper operation or maintenance of such structures could permit the waste, leachate, or gases to contact usable ground or surface water.
I	Sites which comply with A, B, C, E, F, G and H but would be subject to inundation by a tide or a flood of greater than 100 year frequency may be considered by the regional board as a limited Class 1 disposal site.

Source: California State Water Resources Control Board, *Disposal Site Design and Operation Information*

Class 2 sites are underlain by usable groundwater and may be located adjacent to streams. To protect the underlying groundwater quality, a distance of separation must be maintained between the bottom of the fill and the water table. Any surface water must also be restricted from the site to preclude water from contacting the wastes. The second group of wastes (decomposable materials such as refuse) is the acceptable material at this class of site, along with the third group materials.

Class 3 sites are those sites which intercept groundwater or where wastes will be dumped directly into water. Examples of these sites are deep gravel pits with groundwater ponded in the bottom and swampy areas where filling operations commence without construction of levees and removal of the water. Only the third group of wastes is allowed to be disposed of in this class of site. These nonwater-soluble, nondecomposable inert materials such as concrete and bricks will not adversely affect the quality of water that they may contact. This type of classification together with geologic consideration such as faults location, etc., described earlier is necessary for proper management of wastes.

FACTORS IN SITE SELECTION AND EVALUATION

Chemical waste landfills should be sited to take advantage of geologic factors responsible for optimum attenuation of the wastes and any decomposition products, and designed to overcome the disadvantages posed by less favorable sites (2). The factors to be considered in the selection of a site include: waste characteristics, topography, geology (rock type, geologic structure, weathering characteristics), hydrology (permeability, depth to water table, direction and rate of groundwater flow), climate, and composition of soils (which affect pH and sorptive capacity).

Design factors to be considered include: waste preparation, construction of impermeable liners, leachate collection systems, and monitoring equipment. The objectives of an engineering design are to overcome the natural drawbacks of the site and to control and monitor the release of hazardous wastes into the environment. In selecting and evaluating a chemical waste landfill site, some of the general criteria to be considered are as follows (6):

Chemical waste landfills ideally should be located in areas of low population density, low alternative land use value, and low groundwater contamination potential.

All sites should be located away from flood plains, natural depressions, and excessive slopes.

All sites should be fenced, or otherwise guarded to prevent public access.

Wherever possible, sites should be located in areas of high clay content due to the low permeability and beneficial adsorptive properties of such soils.

All sites should be within a relatively short distance of existing rail and highway transportation.

Major waste generation should be nearby. Wastes transported to the site should not require transfer during shipment.

All sites should be located an adequate distance from existing wells that serve as water supplies for human or animal consumption.

Wherever possible, sites should have low rainfall and high evaporation rates.

Records should be kept of the locations of various hazardous waste types within the landfill to permit future recovery if economics permit. This will help facilitate the analysis of causes if undesirable reactions or other problems develop within the site.

Detailed site studies and waste characterization studies are necessary to estimate the long-term stability and leachability of the waste sludges in the specific site selected.

The site should be located or designed to prevent any significant, predictable leaching or runoff from accidental spills occurring during waste delivery.

The base of the landfill site should be a sufficient distance above the high water table to prevent leachate movement to aquifers. Waste leachability and soil attenuation and transmissivity characteristics are important in determining what is an acceptable distance. Evapotranspiration and precipitation characteristics are also important. The use of liners, encapsulation, detoxification, and/or solidification-fixation can be used in high water or poor soil areas to decrease groundwater deterioration potential.

All sites should be located or designed so that no hydraulic surface or subsurface connection exists with standing or flowing surface water. The use of liners and/or encapsulation can prevent hydraulic connection.

In arid regions where the cumulative precipitation is less than the evapotranspiration, water will not likely accumulate in the landfill or migrate through the soil. Under such conditions, leachate containment precautions (liners, etc.) will not be necessary unless the water table is high or large quantities of liquid wastes are disposed.

Unless leachate generation or escape is prevented in some manner, such as by encapsulation, location in arid regions or naturally impermeable basins, or by immediate cover with an impermeable membrane to prevent infiltration, it will be necessary to line the basin with an impermeable membrane, collect the leachate in headers, and recycle it through the fill or pump it to an appropriate treatment facility.

All liners, cover materials, and encapsulating materials must be tested or have known chemical resistance to the materials they will contain or might otherwise come in contact with. Ideally, such materials should have an effective life greater than the toxic life of the wastes they contain.

Studies will be necessary to determine general site monitoring requirements. Hydrogeological monitoring will be required to detect routine and accidental releases of liquid effluents. A system of observation wells should be installed in aquifers around the site and concentrated in potential water and waste movement paths downgradient from the site. A monthly sampling frequently has been suggested by one source. Downstream monitoring stations and a bimonthly sampling frequency were suggested for surface streams in the site vicinity.

As part of the selection investigations of a proposed hazardous waste disposal site the basic meteorology of the site must be investigated. The two primary elements of this investigation are the determination of the average rainfall in the area and the construction (from available historical data) of a wind rose for the site (13). Demographic data for the area consist of a plot of the population distribution within a 25 mile radius of the site which can be compared with the direction of the prevailing winds.

The geological and groundwater conditions should be investigated through a program of field inspection and testing that involve soil and rock examination and the boring of test holes (14). The investigation should study the depth and occurrence of groundwater, its natural quality, and the existence of natural impervious barriers. The soil types, permeability, depth and thickness of impervious layers, extensiveness of their lateral continuity, and occurrence of dip and strike of the layers should also be determined. The investigation should indicate either that geologic and hyrologic conditions will prevent migration of hazardous material onto adjacent properties or that appropriate design features are feasible to preclude such migration. Hydrogeologic conditions of the disposal facility should be described in the report.

The number of test holes required to indicate underlying geologic conditions should be related to the adequacy of detailed information from other sources. Information should be provided on underlying geology to confirm rock types and groundwater conditions (absence of groundwater and/or its occurrence and quality). Shallow zone exploration should involve drilling a minimum of three test holes on the site to a depth determined by the geologist in charge of the investigation.

More test holes may be necessary depending on the size of the property and the potential for variable geologic conditions. A rough guideline is one test hole per each five acres of the actual area to be used for waste disposal. Drilling logs should be included in the report for the test holes and any wells constructed. The area used for any hazardous waste disposal facility should be free from potential geological hazards, such as known earthquake faults and land slippage or slide zones. In areas of major subsidence, this hazard should also be evaluated.

Land slippage or settlement can result in rupture of levees surrounding industrial waste ponds, exposure of buried hazardous materials, or slippage of earth masses into large ponds which can result in liquids breeching or overtopping pond walls. The effect of waste liquids percolating through soils on slope or levee stability of other zones of weakness must be considered in the design of waste disposal areas.

If the method of operation relies on the infiltration of large quantities of liquids, the natural soils on the property should be relatively permeable to allow infiltration to occur, and sufficient subsurface storage capacity for the liquids should exist. Conversely, if impervious basins are desired and the native soils are not suitable for that purpose, impermeable materials may have to be imported or artificial linings installed.

Soil and rock types should also be suitable for the type of excavation work anticipated. Excavations made to allow location of the disposal facility should not create hazards of slope instability or problems of erosion. The degree of slopes should be consistent with good engineering practice for the particular soil or rock type. Erosive soils should be protected such as by use of mulches or hydroseed applications.

Finally, if artificial barriers are to be installed, a report should be submitted indicating the long-term competence of such a barrier. Response to seismic activity and possibility of destruction through shrinking and cracking due to drying or action of the hazardous wastes should be evaluated. Pretests should be made on all prospective liners to determine compatibility with the material being disposed.

FACTORS IN SITE DESIGN AND PREPARATION

Although the criteria used by the State of California and such industrial users as Union Carbide and Bayer vary, all have incorporated site design and preparation requirements considerably more stringent than those normally required for a standard sanitary landfill.

Liners: The use of liners is becoming more widespread, and is being incorporated even in some conventional sanitary landfills. When impervious basins are desired at a landfill site and the existing soil is not suitable, artificial liners are a potential solution to the problem. All prospective liners should be pretested for strength and compatibility with the expected wastes. Due to relatively few applications and recent emergence of various liner materials, the long-term effects of different hazardous wastes in a landfill upon the liner's life cannot be determined in a definitive manner.

In addition, the use of liners for environmental protection may require collection and treatment of leachate if rainfall is significant. Common types of liner materials include clay, rubber, asphalt, concrete and plastics such as Hypalon (a chlorinated polyethylene plastic) and PVC (polyvinyl chloride). The leachate collection process usually requires plastic pipes, risers, and pumps. Leachate treatment methods are not well defined but may require neutralization, biological treatment, evaporation or precipitation.

DuPont, the manufacturer, claims that 30 mil Hypalon sheeting is essentially impermeable to water. The material is also said to resist tearing and puncturing, but may be readily patched if an accident occurs. Also, it is claimed that the liner resists aging, weather, ozone (a chief enemy of rubber), and a wide range of hydrocarbons and chemicals. It is reportedly not adversely affected by soil chemicals and microorganisms.

The cost of rubber and Hypalon liners varies between $0.25 and $0.50 per square foot, while certain other plastic liners cost $0.15 to $0.25 per square foot. The plastic pipes and risers for leachate collection range between $3 and $7 per linear foot. A sanitary landfill on Long Island which uses a 20-mil thick Hypalon liner with sand cover has the following costs. The liner costs were $20,000 per acre installed and the leachate collection system adds an additional $6,000 to $7,000 per acre.

To protect groundwater, a Pennsylvania firm has lined a 52-acre sanitary landfill with one-half inch thick asphalt covered with a one-foot thick layer of sandy loam. It is being developed in five-acre sections. The low point of the landfill is two feet above the groundwater level. The base is excavated, graded and rolled. The asphalt is applied in several coats. Sandy loam is applied on top of the asphalt to protect the liner and allow a flow path for leachate. The depth of each 5-acre fill is 22 feet. Leachate is collected from a manhole at the low point of the fill. Laboratory tests have indicated that the asphalt liners resist normal leachate. Solvents cannot be accepted, however, since tests have indicated that dissolution of the asphalt will result.

Lab tests indicate that the life of an asphalt liner is at least 50 years. Asphalt liner costs, including installation, vary between $6,000 and $12,000 per acre. The higher cost applies when the sand cover must be trucked long distances. The asphalt used is a special flexible type, and not the normal paving grade. It is applied at the rate of 2 gallons per square yard. An estimated 65 gallons per minute of leachate is expected upon completion of this facility. A leachate treatment plant will be constructed, though process details are not currently available.

Another approach is to collect the landfill leachate and circulate it back through the wastes. This reportedly recycles the successful flora and nutrients which may improve and speed waste degradation. Further research is needed regarding the interactions between appropriate types and concentrations of microorganisms and different hazardous wastes.

The Hypalon liner was introduced commercially in 1951. Primary uses of Hypalon include the lining of pits, ponds, lagoons, and landfills. At least two of the larger regional hazardous waste processing firms have begun using this material in their operations. Rollins Environmental Services, Inc. (RES), experienced holding basin failures using rubber liners and clay liners (8 to 12 inches thick), and have switched with apparent success to a concrete base with a Hypalon liner. Rollins estimates that construction of a 500,000 gallon holding basin, square in cross section and 9 feet deep in the center, costs approximately $19,000, including 4 to 8 inches of concrete. The Hypalon liner adds an additional 20 to 25 cents per square foot (or approximately $4,500) to the cost. Company officials indicated that initial difficulties were experienced with the adhesive used to bond the liner to the concrete.

The EPA-sponsored Kansas City Model Sanitary Landfill demonstration project is operated on a 46-acre site. The cost of installing an 18-inch clay liner was $54,500, or approximately $1,185 per acre. A summary of available cost data for the liner types discussed above is presented in Table 23.

TABLE 23: LINER COSTS

Liner Type	Cost per Acre (1973)
Clay	$1,185
Asphalt	$ 6,000 - $12,000
Rubber	$11,000 - $22,000
Hypalon	$11,000 - $22,000
Polyvinyl chloride (PVC)	$ 4,840 - $9,680

Source: Report EPA/530/SW-165

The Lindenmaier-Precision Company of West Germany is promoting the use of polyurethane foam to line and seal landfills. A top layer of the same foam material is used to cover the compressed waste, and a final earth cover is applied over the foam. Complete containment of the waste reportedly results. There is no infiltration of water into the landfill, and no contamination of air and water resources. Recent controlled research evaluated the stability of concrete, asphalt, rubber, and plastic in contact with selected acids and organic solvents (benzene, ethyl alcohol, acetone, chloroform).

The relative durability was found to be in the following decreasing order: (a) concrete, (b) plastic, (c) rubber, and (d) asphalt. Asphalt was the least suitable, according to the tests, reacting with all of the reagents tested and completely dissolving in benzene and chloroform.

The above study also investigated the effect of pH on soil attenuation capabilities. A low pH, apart from inhibiting the growth of beneficial microorganisms, reportedly increases the solubility of metals and affects the ion exchange and absorption properties of the colloidal fraction of soils. Clays are most effective absorbers of metals at higher pH's while most organics are more effectively absorbed under more acid conditions. As a general principle, maintaining the soil pH at 7.0 to 8.0 is encouraged to reduce leaching potential of heavy metals and promote biological activity. The effectiveness and longevity of most liners is also improved.

Cover Materials: A sufficient supply of suitable cover material is a necessary item. Ideally, the cover material will minimize or eliminate infiltration of water, and prevent sublimation or evaporation of harmful pollutants into the air. A recent EPA study indicated that a good cover for a chemical waste landfill in arid regions of the United States might consist of a one foot layer of sand topped by a four foot layer of silty loam or clay. However, in other regions of the country, more stringent requirements may be necessary. If infiltration of water to the fill can be minimized sufficiently, very little leachate will form, and collection and treatment might not be necessary. It is apparent that landfilling wastes on a one shot basis, as opposed to semicontinuously, has advantages since the site can immediately be sealed to infiltration, eliminating the need for leachate collection.

The importance of adequate cover materials is demonstrated by the following case history (2). In early 1973, excess levels of hexachlorobenzene (HCB) were detected in slaughtered cattle from the Ascension Parish area of Louisiana. A quarantine was imposed on food animals in an area of over 100 square miles surrounding this area. Studies conducted by State and EPA Region VI personnel confirmed that the problem was associated with chlorinated hydrocarbon manufacture in the vicinity. The HCB transfer mechanism from manufacturing operations to cattle is believed to be sublimation from two dump sites receiving wastes from the manufacturing facilities. No cover was provided at the sites.

To rectify the situation, land disposal of the HCB wastes has been halted and one of the dumps has been sealed with a sheet of 10-mil polyethylene covered with two feet of silty sand material dredged from river banks. The polyethylene sheet is separated from the wastes by a one to two foot layer of soil material. Air monitoring by state Department of Health officials indicates a marked decline in HCB concentrations over the dump site.

When a top liner is used at a landfill to provide a waterproof covering, care must be exercised to avoid potential gas problems. Gas venting mechanisms must be provided, since even minimal accumulations can cause ballooning and rupture. A recent journal article mentioned one instance where the application of a clay soil cover forced migrating methane gas into an adjacent farm, ruining crop production (15).

The primary factors affecting the rate at which gas is produced in a landfill are: moisture—the greater the moisture, the greater the rate of decomposition; temperature—increased temperatures tend to increase bacterial productivity and resulting gas production; amount of organic matter—greater amounts of organic material increase the amount of substrate material from which the microorganisms can produce gas; and pH—a pH of 6.5 to 7.5 is optimum for methane gas production.

The possibility of recovering gases from sanitary landfill operations is beginning to be examined. The Solid and Hazardous Waste Research Laboratory, EPA, is planning a case study of the methane recovery method in effect at the Palos Verdes sanitary landfill operated by the Los Angeles County Sanitation Districts. The objectives of the study will be to determine whether such methods of methane recovery are feasible, and how the economics and techniques of such methods might be exploited.

Additional work is being sponsored by OSWMP and Pacific Gas and Electric (PG & E) at the Mountainview, California municipal landfill. This work will investigate gas withdrawal rates in relation to stability of gas quality over time. PG & E will build a facility to dehydrate the gas but plans for ultimate use have not been finalized.

The possibility that biodegradation of materials within landfills may be able to yield pipeline gas is thus giving new impetus to landfill studies as described by L.J. Ricci (16). Wastes already disposed of at landfill sites produce methane by natural decomposition. Tapping of the Palos Verdes, California dump rates as the world's first commercial recovery operation. Gas output runs about 1 million cubic feet per day and the 172 acre site reportedly can sustain a 5 million cubic feet per day level of production for as long as 15 to 20 years, according to Ricci (16).

Not all landfills satisfy key criteria of volume, depth, age and refuse composition needed to assure commercial viability of tapping, but Palos Verdes does not rank as all that isolated a case. The Los Angeles Department of Water and Power, for instance, has been capturing gas experimentally for two years from a 100-foot-deep well at the Sheldon-Arleta dump in Sun Valley, California, to fuel a 300-horsepower internal combustion engine that drives a 200-Mw generator.

Observation and Monitoring Wells: Prior to the deposition of hazardous wastes, observation and monitoring wells should be installed around the periphery of the site. Locations should be determined by the appropriate regulatory authorities based on the site topography and hydrogeological conditions. A OSWMP documented case history (1) illustrates the importance of monitoring wells. A company in the north central United States had utilized the same dump site for laboratory waste disposal since 1953. More than half of the waste dumped was arsenic. Although the monitoring wells around the site were superficial in nature, arsenic concentrations greater than 175 ppm were detected.

The United States Public Health Service drinking water standard for arsenic is 0.05 ppm. The dump site is located above a limestone bedrock aquifer which supplies about 70% of a nearby city's residents with drinking and crop irrigation water. Indications are that this water is in danger of being contaminated by arsenic seepage through the bedrock. Without monitoring wells, this waste transport would not have been detected, and serious illness could have resulted.

FACTORS IN WASTE PREPARATION FOR LANDFILL DISPOSAL

Most land-destined hazardous wastes from the inorganic chemicals industry are company land stored (dumped on the land, stored in steel drums or containers or on the bottom of settling ponds) or in company on-site landfills (4). Safeguards to avoid environmental contamination include the aforementioned steel drums and chemical fixation to change hazardous sludges and liquids into solids with low leaching rates.

Many of the hazardous wastes disposed of in chemical waste landfills should be prepared or treated in some manner prior to deposition to lessen potential environmental and health effects. Methods of hazardous waste preparation for chemical landfill disposal include chemical stabilization (fixation), volume reduction, waste segregation, detoxification/degradation, and encapsulation (2).

Chemical Fixation: Chemical fixation of industrial waste materials has been developed by several companies, including: the Chemfix Division of Environmental Sciences, Inc., I.U. Conversion Systems, Inc., Dravo, Inc., and Chicago Fly Ash Company. Although the environmental adequacy of these processes has not been evaluated by OSWMP, the resulting solidified waste sludges are less likely to cause environmental damage than if the wastes were deposited on land as is. Long-term leaching and defixation potentials are not understood at this time. In all fixation systems, proprietary chemicals are mixed with the waste sludges, and the resulting mixture is pumped onto the land, where solidification occurs between a few days and a few weeks (depending upon the process). Some of these processes result in the formation of a matrix in which wastes are entrapped; others claim that pollutants such as heavy metals are chemically bound in insoluble complexes. Processes such as these have been applied to many varied waste streams, including heavy metal sludges, oil refinery wastes, and lime/limestone wet scrubber sludges. Reduced leaching should result, but the permanence of the resulting structure and the absolute environmental adequacy of these techniques have not as yet been fully demonstrated. Typical costs quoted by one firm are in the range of 2 to 10 cents per gallon; however, certain waste types involve much higher treatment costs (2).

Except for the western United States where many aqueous hazardous wastes are being directly landfilled or evaporated, aqueous liquids are usually handled by chemical treatment as described. Hazardous sludges on the other hand are being increasingly treated either on-site or in collection areas by mixing them with inorganic chemicals and catalysts to set up the entire mass into solid structures with low leachability and good land storage or landfill characteristics. There are a number of such processes which produce solids ranging from crumbly soil-like materials to concrete to ceramic slags. There has been no reported instance of chemical fixation being used on inorganic chemicals industry wastes. These processes have been used in other industries, however, to treat inorganic chemical type wastes and will most likely be used in the inorganic chemicals industry in the near future (4).

Volume Reduction: Incineration is the most widely used hazardous waste volume reduction technique. Approximately 60% by weight of the hazardous waste generated in this country are organics and can normally be destroyed and/or detoxified by incineration. The potential for use of incineration as a hazardous waste management technique is apparent. Many wastes can be completely destroyed; others leave small amounts of solid residues which may or may not be hazardous. In any case, they must be disposed of, usually on the land. Several of the larger regional hazardous waste processing firms use incineration in combination with land disposal. Emission control devices are usually required for hazardous waste incineration since combustion by-products may also be hazardous.

Waste Segregation: Segregation by type and chemical characteristics of wastes is usually practiced to prevent undesirable reactions within the fill. A number of dangerous problems can develop from mixing. For example, acid wastes combined with cyanide-containing wastes produce extremely toxic hydrogen cyanide gas. Segregation prior to disposal may allow the acquisition of sufficient quantities of particular waste types to realize economies of scale in design of treatment facilities for detoxification or recovery. Also, it may be possible to use acidic wastes to neutralize high pH wastes, or perhaps to use waste sulfides to precipitate toxic heavy metals.

Detoxification: Detoxification prior to landfill disposal can often be accomplished by thermal, chemical, or biological processes. Included in this category are such techniques as ion exchange, neutralization, oxidation-reduction, pyrolysis, incineration, activated sludge, aerated lagoons, waste stabilization ponds, and trickling filters.

Degradation: Some chemical degradation methods being developed and/or utilized primarily for pesticides include hydrolysis, dechlorination, photolysis, and oxidation. No single chemical procedure for degrading the entire spectrum of hazardous materials is effective. Hydrolysis is the best method for destroying organophosphorus and carbamate pesticides. Chemical dechlorination can be used to degrade polychlorinated pesticides.

Photolysis may be applied to partially degrade 2,4-D and 2,4,5-T. The use of strong oxidants offers still another approach to destroying some pesticides and herbicides. However, the water insolubility of many of the compounds, particularly the chlorinated pesticides, makes the use of strong oxidants in aqueous solution impractical. The above methods are usually more expensive than alternative pesticide disposal methods (e.g., incineration, biodegradation, etc.), and for the most part have not been demonstrated on a full-scale basis. Economically, biological degradation of pesticides by soil incorporation may be a useful disposal method. Soil degradation requires that the soil microorganisms not be inhibited, and be capable of metabolizing the waste components. Also, the site must have minimum potential for pollution of groundwaters or via dust dispersal.

Encapsulation: Steel drums, alone or with plastic liners, not only provide some long-term containment but also are the most convenient storage and transportation for relatively small quantities of inorganic hazardous waste. As such they are used widely for hazardous wastes of approximately one ton per day or less. Drum costs rapidly mount as the volume of hazardous wastes increases much above the one ton per day range and bulk handling then becomes the preferred route for storage and transportation. A rough estimate of 10 to 20% of the off-site waste is handled in drums (4).

In wet climates, sections of or entire landfill areas are encapsulated by adding clay or asphalt caps or covers to impervious isolation cells or landfill liners. The impervious cover is necessary to protect the hazardous waste from rainfall flooding. Neutralizing or pH control ingredients such as lime may also be used to encase or surround the hazardous waste to avoid solubility, decomposition or other change in the character of the waste to increase its environmental damage. In dry climates, there is no need to encapsulate the entire landfill since rainfall and water buildup is not a problem. Isolation cells may still be constructed, however, for specific hazardous waste containment. Perhaps 10% of the off-site wastes are handled this way (4).

Those wastes which are not amenable to detoxification may be encapsulated in some permanent material prior to disposal. Available materials include concrete, molten asphalt, and plastics (polyurethane, polyethylene). Leachable heavy metal wastes are examples of wastes which may require encapsulation prior to land disposal. In some cases, the resulting encapsulated wastes will require casting in drums prior to deposition in the landfill. The purpose of encapsulation is to limit the leachability of the potentially toxic materials contained therein by physically keeping water from contacting the hazardous materials or their containers.

Direct encapsulation of hazardous inorganic chemicals industry wastes in concrete in 55 gallon drums is now practiced by at least one contract disposer. The practice is used for small quantities of containerized miscellaneous hazardous wastes. Plastic film encapsulation has been used for arsenic disposal. Direct bulk plastic encapsulation of solid wastes has not been encountered. Encapsulation is used for less than 0.1% of the off-site waste disposal (4).

A recent OSWMP study (17) provides some cost data regarding encapsulation of heavy metal sludges (20% solids by weight). For asphalt or polyethylene scrap encapsulation, it is assumed that still bottoms or other tar residues might be used at an average cost of one cent per pound. Off-standard polymers are available at the same price. It is further assumed that wastes are cast into used steel 55-gallon drums costing about $2 each. The study estimates the fixed capital expenditures for asphalt encapsulation of 115 cubic feet per day of chrome waste sludge at $21,000. The corresponding operating costs are $0.65 per cubic foot of sludge encapsulated, and an additional $0.12 per cubic foot of sludge landfilled.

In another process, dilute metal sulfide or hydroxide can be used as added water in mixing concrete, thus incorporating the wastes into the poured concrete. A portable cement mixer can be used to mix the cement and the water containing the insolubilized metal. The cement mixture is then cast into fiber drums, or used steel drums. It is estimated that cement encapsulation and burial on-site of volatile sludges cost about $0.10 per gallon of sludge. According to this report, cement is preferred over molten asphalt or plastics for metal sulfides or hydroxides since volatile heavy metal sludges may have high vapor pressures at the temperature of the molten asphalt or plastic polymers.

The Lindenmaier-Precision Company of West Germany has developed a unique encapsulation technique in which the waste sludge is placed in 55-gallon drums. An inch thick layer of polyurethane foam material is then sprayed completely over the drum's exterior, so that air and water can no longer reach this surface. If the inside of the drum is not resistant

to the sludge deposited therein, an inside liner of plastic or some other suitable material
might be necessary. The polyurethane foam prevents rusting of the steel, thus eliminating
deterioration of the capsule and ultimate release of the contents. Long-term testing of this
approach is continuing under actual landfill conditions. Encapsulating a 55-gallon drum
in polyurethane foam costs about $4 in Germany.

FACTORS AFFECTING LANDFILL USE

The following factors affect landfill use, according to R.G. Shaver et al (4).

Rainfall: In southwestern United States the annual rainfall is significantly less than the
evaporation. In the San Francisco area a net evaporation rainfall differential of four feet
exists. Therefore, in these dry climates, liquid and sludge hazardous wastes can be mixed
into either refuse or fill dirt without having the landfill area flooded. California private
and public Class 1 landfill areas accept liquid hazardous wastes as allowed under state requirements. Eastern United States, on the other hand, tends to handle these liquid wastes
by treatment, ponding and final discharge to surface water.

Personnel and Air Pollution Hazards: Some hazardous chemicals, particularly liquids and
sludges, are not usually landfilled even in dry climates because of danger to landfill personnel and/or air pollution. Strong acids and caustic solutions, reactive sulfides, volatile solvents and mercaptans are examples of such materials. One hazardous waste handling company is so concerned with hydrofluoric acid exposure that it has issued special treatment
cards to its waste handling personnel specific to this chemical.

Chemical Interaction: Indiscriminate dumping of hazardous chemicals into landfills leads
to serious interactions. The dumping of acid sludges on top of cyanides is a classic example which has actually occurred in California and perhaps other places. Acid similarly
attacks metallic sulfides and changes the solubility of heavy metal precipitates. Dumping
of liquids into landfill areas is particularly prone to give such interactions because of the
relative mobility of liquids in seeping throughout the wastes.

In addition to the chemical interactions between landfill components, liners can also be
attacked. Many plastic materials used in liners are attacked by organic solvents, oxidizing
agents and other waste components. Clay liners are attacked by a variety of chemicals and
become more porous. An old construction practice is to mix clay with lime to give a sand-
like soil. There have been instances in California where clay liners have failed to stop seepage of hazardous wastes such as chromates into groundwater. It is believed that chemical
impairment of the clay liner was responsible.

FACTORS IN LANDFILL OPERATION

Leachate Collection and Treatment: In wet climates particularly, both private and public
landfills are paying increasing attention to leachate collection, monitoring and treatment.
Landfill areas in the state of Pennsylvania are representative of those in a wet climate and
leaching treatment has been initiated in some public landfill areas. Leachate monitoring
and treatment is also practiced in an on-site inorganic chemicals plant landfill. The vast
majority of the landfill operations handling hazardous wastes, however, do not have any
leachate control and treatment provisions.

Coordinate Record Keeping: Landfilling of hazardous wastes as discussed earlier can lead
to undesirable chemical interactions. A few public and private landfill operations keep records of all hazardous waste burials by location and composition. By means of this record
undesirable interactions may be avoided and potentially reactive chemicals isolated from
each other. One necessary corollary to coordinate record keeping is prior knowledge of
hazardous wastes coming to the landfill area so that a satisfactory disposal section may be
selected. Prior written requests for the specific hazardous waste disposal are already required
for some public and private landfill areas.

It is obvious that there are many things not known, or that are known imperfectly, and thus, there are many technical questions which need to be answered if hazardous wastes are to be properly controlled in a secured landfill. More work is required to fully answer such questions as:

> Which hazardous materials can be satisfactorily landfilled?
> How must a hazardous waste material be prepared before deposition in a landfill?
> How must the landfill site be prepared before deposition of the hazardous waste material?
> What monitoring requirements are necessary for effective landfill site operation?
> How might a landfill site be prepared for reuse at a future date?
> What are the requirements for long-term surveillance of such sites?

EPA, in cooperation with other governmental agencies and the private sector, is endeavoring to find answers.

QUANTITATIVE EVALUATION OF LANDFILLS

The procedures outlined above are all necessary to the proper selection of a site utilizing landfill disposal. The data provided by the various procedures include both quantitative and qualitative information, but the evaluation of these parameters is currently handled on a totally subjective basis. This subjective evaluation does not provide the necessary methodology for comparing one site with another or for determining absolute suitability of a site for a particular waste material. A methodology providing the framework for quantitative evaluation has been proposed by Pavoni, Hagerty, and Lee (13). Considerable research is required to test and revise the quantification, but such methodology is necessary to ensuring that factors other than economics will receive consideration in site selection.

In order to evaluate the potential danger of depositing any hazardous material in a particular landfill site it is necessary to critically examine three general characteristics of that site: [1] the potential for precipitated surface waters to infiltrate the deposited waste material; [2] the potential for the waste material to be transported through fluid transmission from its deposit location through underlying bottom soils to groundwater systems; and [3] other mechanisms for the removal of hazardous materials from the site and their transport to other areas. A number of factors have been included quantitatively in the site rating procedure originally presented by Pavoni, Hagerty and Lee (13) which follows.

Soil Parameters

Infiltration Potential: The potential for water to enter a waste deposit may be quantitatively expressed as the ratio of the amount of water which may enter the top surface of the cover soil divided by the amount of water necessary within the cover soil to produce a full passage of moisture from the top of the layer to the bottom of the layer and out into the contained refuse.

The amount of water (i) which could theoretically enter the site or enter the cover soil at the site may be estimated as the total area under all of the rainfall intensity graphs for the site, beneath a horizontal line representing the infiltration rate of the cover soil (see Figure 52). The infiltration rate of the cover soil may be expected to vary from 0.01 in/hr for bare heavy clay soils to approximately 3 in/hr for loose sands. The probable range of (i) will be from 1 to 64 inches.

The amounts of water necessary for passage of moisture through the cover soil may be related to the volumetric field capacity of the cover soil layer. In other words, whereas the field capacity refers to the amount of water as a percentage of the dry unit weight of the soil required for passage of water through a unit volume, the volumetric field capacity in this instance would refer to the product of the thickness of the cover soil layer times the field capacity of the soil. Thus, let (FC)H be the denominator of the infiltration potential term, where FC is field capacity of the soil expressed as a decimal and H is thickness of cover soil layers (inches). The field capacity will vary from 0.05 for a clean sand to 0.40 for a clay, whereas H will vary from approximately 30 to 72 inches. The infiltration potential may be finally quantitated as:

$$I_p = \frac{2i}{(FC)H}$$

having a practical range of 0.02 to 20. This infiltration potential may be thought of as one of the most significant factors in determining the site potential for waste transmission.

FIGURE 52: PRECIPITATION/INFILTRATION CHART

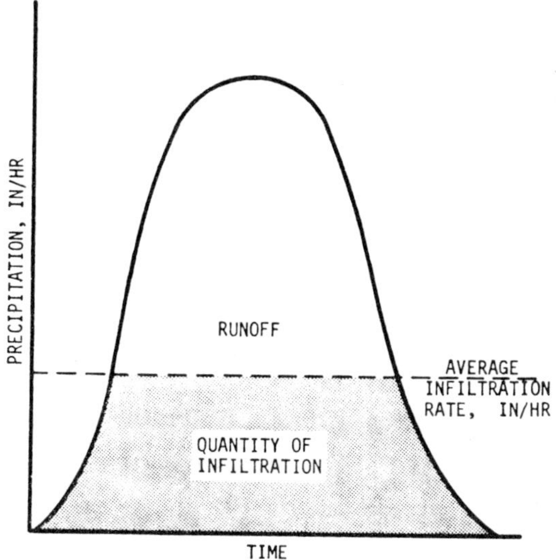

Source: PB 224,582

Bottom Leakage Potential: In addition to the problem of water entering the refuse cells and removing the contained hazardous material, consideration must be given to the action of a waste in suspension or solution in water, or in liquid form, passing through the bottom soil layer from its original location and entering the groundwater system.

The potential hazard for a waste to travel through a bottom soil from the bottom of the refuse cell through the containing soil layer and into a groundwater flow system may be evaluated in terms of the permeability of the bottom soil layer and its thickness. Since all natural geological materials possess some finite permeability it is fatuous to think in terms of an impermeable bottom in a landfill. Even in the situation where an artificial lining material has been applied to the bottom of a refuse cell, it is quite probable that the artificial liner is in truth not impermeable. For example, thin sheets of impervious polyvinyl chloride or polyethylene lining may easily be pierced and penetrated during placement or after placement by sharp-edged equipment or refuse items. Asphaltic liners likewise may crack because of distortions experienced when the bottom soils settle as a result of the applied loads of the landfill. Thus, in all cases, a certain finite permeability of the bottom confining layer must be anticipated. Therefore, in a true sense, the migration of materials from the landfill site into the substrate must always be anticipated and the only variable to consider is the time which will be required for such migration; in other words, the migration time for a hazardous substance through a bottom soil layer consisting of clay minerals may be sufficiently long so that the substance's half-life is greatly exceeded.

In such a case the virulence and hazardous nature of these substances will be diminished. For this reason this bottom leakage factor has been quite simply expressed in the form shown below to give a measure of the time factor for migration of a hazardous material in terms of permeability and thickness of the bottom soils

$$\text{Bottom leakage potential (Lp)} = \frac{1,000\ \sqrt[3]{K}}{T}$$

where K is bottom soil permeability (cm/sec), and T is bottom soil thickness (ft). The approximate range for K for all practical problems will be about 10^{-1} to 10^{-10} cm/sec, whereas T will vary from 5 to 50 feet. The overall range of Lp will therefore be from approximately 0.02 to 20.

Filtering Capacity: A less important characteristic of the bottom soils will be their ability to remove solid particles traveling downward (through the bottom soil layer) in a fluid suspension. In general, this filtering capacity is dependent upon the sizes of the pore spaces between individual soil grains. In other words, the physical filtering ability of the bottom soil will depend upon void-space size in that soil and may therefore be related to the size of the soil particles themselves. The physical filtering capacity may be considered proportional to the inverse of the average grain size in the soil stratum. Therefore, the filtering capacity of the bottom soil layer may be easily expressed as shown below:

$$\text{Filtering capacity (Fc)} = \frac{-8.75}{\log \phi}$$

where ϕ is average particle diameter (inches). The average particle diameter of various soils will vary from about 0.25 to 2.5×10^{-5} inches. Therefore the filtering capacity will vary between approximately 2.1 and 15.0.

Adsorptive Capacities: In addition to the removal of solid particles through physical filtering within the bottom soil layer, certain materials will be removed from suspension and solution in a migrating fluid by the physical-chemical attraction of the mineral constituents within the soils. Adsorption of materials both organic and inorganic in the migrating fluids will take place principally on colloidal-size particles consisting of clay minerals which describe the attracting of such minerals for the migrating particles. A general measure of such attraction is the cation exchange capacity of the clay mineral.

In this rating system the greater the danger of transmission of a hazardous material from a landfill site the greater the rating factor; therefore, the greater the ability of the bottom soil layer to adsorb migrating materials the smaller should be the adsorption factor. The ability of the soil is evaluated as an inverse quantity and a factor is obtained by dividing a numerator by cation exchange capacity in the denominator.

The cation exchange capacity alone will not reflect the potential for adsorption of a material on the minerals present in the soil. If the available adsorption positions on the soil mineral are already occupied then no further adsorption can occur. The occupancy of the adsorption sites in the soil are already occupied by organic compounds and complex organic ions. Therefore, the complete adsorptive capacity factor will consist of the organic content as the numerator and the cation exchange capacity as the denominator as shown on the following page.

$$\text{Adsorptive capacity (Ac)} = \frac{10\ (o)}{(\log CEC) + 1}$$

In the above, o is organic content expressed as a decimal and CEC is cation exchange capacity, me/100 g. The range in the numerator will therefore be from 0 to 10, whereas the log CEC will range between about 0.6 to 2.2. The adsorptive capacity (Ac) will therefore vary between approximately 0 and 16.

Groundwater Parameters

Organic Content: Transmutations of a hazardous material following contact with groundwater must also be evaluated. Assuming that a hazardous waste has reached the groundwater after disposal in a landfill, probably the most important single water parameter to be considered would be that of organic carbon content. The organic content of the groundwater may be quantitated in terms of the biochemical oxygen demand or BOD. The higher the organic content (BOD) of a groundwater, the higher the substrate potential, and consequently the higher the potential it may afford pathogenic organisms.

Groundwater organic content was assigned a third order of priority with regard to landfill ranking factors so that its range of values was fixed between 0 and 10 dependent upon BOD values as follows: Oc = 0.2 BOD where Oc is organic content rating (maximum value of 10) and BOD is biochemical oxygen demand of groundwater (mg/l).

Buffering Capacity: The buffering capacity of a groundwater is another important parameter when considering transmutations of hazardous wastes in groundwater systems. Any waste material having acidic or alkaline characteristics would be less hazardous to the groundwater ecosystem if the water it is entering possesses a high buffering capacity. In other words acidic or basic waste characteristics would be neutralized or moderated upon contact with a high buffering capacity water system.

Groundwater buffering capacity was assigned a third order of priority and was quantitated in relation to pH, acidity, and alkalinity. For the purposes of this study the buffering capacity ranking (Bc) will be equal to ten minus the smallest number of milliequivalents (maximum of ten) of either an acid or base required to displace the original groundwater pH below 4.5 or above 8.5. The buffering capacity ranking will therefore vary from 0 for a strong buffer to 10 for a weak buffer.

Potential Travel Distance: The potential for travel of a hazardous waste once it enters a groundwater system will determine how much of the immediate landfill environment it may affect. This potential travel distance was assigned a fourth order of priority and varied in value from 0 to 5 depending upon the greatest possible distance a molecule of water could travel from a point directly beneath the landfill through the groundwater system and surface water systems, and thence to the sea.

Potential Travel Distance	Travel Distance Ranking (Td)
0 to 500 feet	0
500 to 4,000 feet	1
4,000 feet to 2 miles	2
2 miles to 20 miles	3
20 miles to 50 miles	4
Greater than 50 miles	5

Groundwater Velocity: The groundwater velocity will determine how fast a hazardous material may spread into the environment. A groundwater system having a high velocity should therefore be assigned a higher ranking since the time of waste transmission would be reduced. The groundwater velocity, having a fourth order of priority, may be defined as: v = kS, where v is velocity, k is permeability and S is gradient. Values of k will vary between 10^{-1} and 10^{-9} cm/sec whereas values of S will usually range between 0 and 20 feet per mile.

Groundwater velocities were ranked according to the following formulation:

$$Gv = \frac{S}{\log[(1/k) + 1]}$$

where Gv is groundwater velocity rank, k is permeability (cm/sec), and S is gradient (feet per mile). The groundwater velocity rank will approximately range between 0 and 20.

Air Parameters

Prevailing Wind Direction: The third major site characteristic to be investigated is air. The hazardous potential of any toxin or pathogen escaping through the atmosphere from the landfill would depend upon the prevailing wind direction in relation to the distribution of population surrounding the site. Obviously the worst situation would be one in which a strong prevailing wind blew from the site to the center of a very dense population.

The following procedure was therefore developed to quantitatively evaluate the potential of the prevailing wind direction. Initially, a 25 mile radius circle was constructed with the landfill site as its center (see Figure 53). This circle was then divided into four quadrants by drawing two lines, one north-south, and one east-west. The population of each quadrant was determined (Pi) and a point representing the center of population (population node) was located in all four quadrants (PNi).

FIGURE 53: PREVAILING WIND ROSE

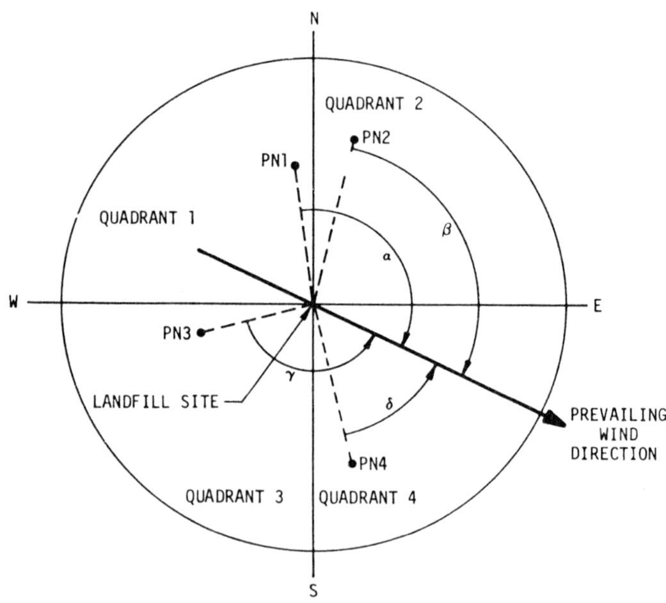

Source: J.L. Pavoni et al, paper presented at the Seventh American Water Resources Conference, Washington, D.C.

A radius was then drawn from the site to each quadrant's population node. The prevailing wind direction was determined and a radius drawn in this direction from the site (center of circle). The angles from the prevailing wind direction to each site-population node radius were determined (α, β, γ, δ) and incorporated in the following prevailing wind potential formula:

$$Wp = \sum_{i+1}^{4} [(5 - Ai/36) \log Pi] /15$$

where Wp is prevailing wind potential rank, Ai is the angle from the prevailing wind direction to each site-population node, and Pi is the population of each quadrant. Wp quantitatively interrelates the prevailing wind direction, site location, and population nodes of each quadrant. Wp has a practical range of 0 to 5.

Population Factor: The population immediately surrounding the landfill site will determine how many persons could be adversely affected by escaping hazardous materials. The higher the population within a specified radius of the landfill site the higher the population factor ranking as shown: Pf = log p, where Pf is population factor rank, and p is population within a 25 mile radius of the landfill site. The population factor rank will range between 0 and 7. The total landfill site ranking formula may not be assembled by uniting the various soil, water, and air parameters as follows:

Landfill site rank = Ip + Lp + Fc + Ac + Oc + Bc + Td + Gv + Wp + Pf

where Ip is infiltration potential; Lp is bottom leakage potential; Fc is filtering capacity; Ac is adsorptive capacity; Oc is organic content; Bc is buffering capacity; Td is potential travel distance; Gv is groundwater velocity; Wp is prevailing wind direction; and Pf is population factor. The first four parameters (Ip, Lp, Fc and Ac) describe the soil system, the next four factors (Oc, Bc, Td and Gv) delineate the groundwater characteristics, and the last two terms (Wp and Pf) depict air parameters. The total landfill rank may assume values from approximately 0 to 110, the lower the rank the better the landfill for hazardous waste disposal.

The following data have been accumulated concerning two existing landfill sites in Louisville, Kentucky so that a ranking comparison can be developed.

	Site #1*	Site #2**
Yearly rainfall	43 inches	43 inches
Soil type	clean sand	heavy clay
Infiltration rate (percent of rainfall)	75	10
Field capacity	0.05	0.40
Permeability	10^{-3}	10^{-8}
Soil cover (inches)	60	24
Bottom thickness (feet)	20	15
Average particle diameter (mm)	0.25	0.002
Organic content of soil	0.5	0
Groundwater BOD	10	10
Cation exchange capacity	0	80
Buffering capacity (meg)	7	4
Groundwater travel distance (miles)	750	750
Gradient (feet/mile)	5	5
Population within 25 mile radius	10^{6}	10^{6}
Prevailing wind direction	WNW	WNW

*Site #1 ranking parameters: Ip = 11.5, Lp = 5, Fc = 14.5, Ac = 5, Oc = 1, Bc = 7, Td = 5, Gv = 1.66, Wp = 4.05, and Pf = 6.
Total landfill rank (Site #1) = 60.71.

**Site #2 ranking parameters: Ip = 1.03, Lp = 0.145, Fc = 3.2, Ac = 0, Oc = 1, Bc = 4, Td = 5, Gv = 0.625, Wp = 2.9, and Pf = 6.
Total landfill rank (Site #2) = 23.9.

Landfill #2 having a much smaller rank than landfill #1 would be more condusive to land disposal of hazardous wastes.

REGULATORY CONSIDERATIONS

With the exception of radioactive and pesticide wastes, land-based hazardous waste treatment, storage, and disposal activities are essentially unregulated at the Federal level. The Atomic Energy Act of 1954, as amended (P.L. 703) and the Federal Insecticide, Fungicide, and Rodenticide Act, as amended (P.L. 92-516) do provide mechanisms for control of disposal of radioactive, and pesticide-containing wastes. Hazardous waste legislation has been enacted in a few states, of which the states of Oregon, California, New York and Minnesota are examples.

The disposal of the majority of hazardous wastes generated in the United States is not regulated by the State or Federal Government. Of those few states with some type of hazardous waste management controls, less than half have acceptable treatment/disposal facilities within their boundaries. Due to the generally spotty nature of federal, state and local solid waste and land protection legislation, regulation and enforcement, there has been little pressure applied to generators of hazardous residues to force disposal by environmentally acceptable methods.

ECONOMICS

The operating cost of a sanitary landfill depends on the cost of labor and equipment, the method of operation, and the efficiency of the operation. The principal items in operating costs are (18): personnel; equipment including operating expenses (gas, oil, etc.), maintenance and repair, rental, depreciation or amortization; cover material, including material and haul costs; administration and overhead; and miscellaneous tools, utilities, insurance, maintenance to roads, fences, facilities, drainage, features, etc. Wages ordinarily make up about 40 to 50%; cover material, administration, overhead, and miscellaneous amount to about 20%. The operating costs per ton versus the amount of solid wastes handled in tons and the population equivalent may be charted (Figure 54).

FIGURE 54: SANITARY LANDFILL OPERATING COSTS

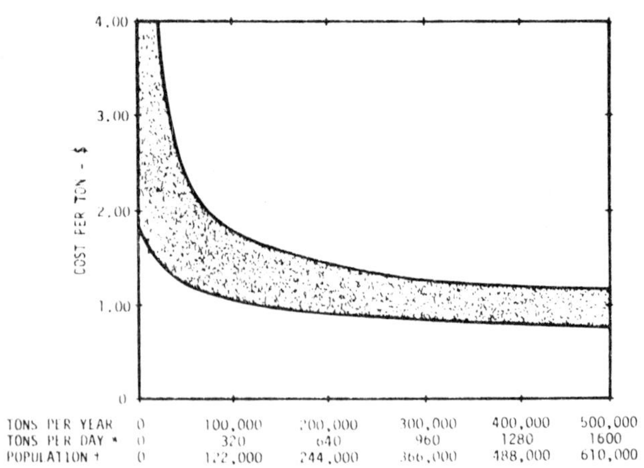

Source: PB 224,582

The operating cost of a small operation handling less than 50,000 tons per year varies from $1.25 to approximately $5.00 per ton. This wide range is primarily due to the low efficiency of the smaller operations which are usually operated on a part-time basis. Full-time personnel, full-time use of equipment, specialized equipment, better management, and other factors that lead to high efficiency are possible at large sanitary landfill operations. The increased efficiency results in lower unit cost of disposal. The unit cost of a large landfill handling more than 50,000 tons per year will generally fall between $0.75 to $2.00 per ton.

APPLICABILITY OF LANDFILLS

The utilization of landfill procedures for the disposal of certain hazardous waste materials in an industrial environment will undoubtedly be required in the future. In order to ensure that no damage to man or the environment results from this technique it is recommended that all sites currently used or proposed for the landfill disposal of hazardous wastes be subjected to the design procedures specified above. It is further recommended that any site considered as a disposal site be subjected to analyses whether it is expected that landfill will be a primary disposal mode at that site or not since account must also be taken of possible accidental spillage of materials which represents an unintentional but direct application of the landfill technique.

The waste constituents considered in this volume are primarily in the category described as Class 1. It is therefore recommended that any disposal facility handling these materials be required to meet the Class 1 site criteria. Finally, it is recommended that the landfill disposal model described by Pavoni, Hagerty, and Lee (13) be tested, modified, and applied to provide the best possible sites for all types of landfill disposal.

Most industrial waste landfill sites are those operated by the small private hazardous waste management industry. As an example, Chem-Trol Pollution Services, Inc., Model City, New York operates an industrial waste landfill which receives residues from its physical-chemical hazardous waste treatment plant. This plant receives a large variety of industrial wastes for treatment (2).

The Chem-Trol landfill consists of a series of clay lined pits or cells into which solid sludges, chemically stabilized or solidified liquids, and slurries are deposited. A sump at the bottom of each cell recycles leachate to the treatment plant. A three-dimensional inventory is kept of wastes buried in each cell to facilitate reclamation at a future date, should economics permit. The company estimates this landfill can be utilized for the next 150 to 200 years.

A few large United States chemical companies also have landfill facilities which are reportedly capable of handling hazardous waste materials (Table 24). The Union Carbide Corp., for example, has operated a state licensed chemical landfill at their Institute, West Virginia, plant since 1965 (19). The initial system experienced drainage problems, and resulted in a reengineered landfill which was completed in 1969. Wastes coming to the landfill are generated by the broad range plant production mix of some 200 or more chemicals (mostly organic).

TABLE 24: SOME U.S. COMPANIES WITH CHEMICAL WASTE LANDFILLS

Union Carbide Corporation
Institute, West Virginia

Dow Chemical Company
Midland, Michigan

American Cyanamid
Willow Island, West Virginia

Source: Environmental Protection Publication SW-115

According to the company, all leachate from the landfill is collected and treated either in the plant's five million gallon per day activated sludge wastewater treatment system, or burned as a source of heat for steam generation. Biological sludge from the treatment of wastewater is dried in special beds, cycled back into the chemical landfill, and mixed with soil and incoming chemical waste sludge. Dried chemical sludges are introduced into the landfill on a one-to-one basis by blending with soil. Blending the wastes with earth reportedly reduces the gas and fire hazards sometimes associated with conventional landfill cell construction techniques. It also tends to hasten biooxidation of the chemical wastes.

The landfill handles approximately 10 tons (20 cubic yards) per day of chemical waste sludges, and 28 tons (33 cubic yards) per day of wastewater treatment plant sludges. The chemical wastes make up only 3 to 6% by volume (2 to 4% by weight) of the total plant wastes, but are by far the most difficult and costly to manage. According to the company, the cost of chemical landfill disposal is $36.80 per ton ($9.27 per cubic yard), while the costs for disposing of municipal-type plant wastes (garbage, rubbish, metals, etc.) in a conventional sanitary landfill are $2.50 per ton ($0.63 per cubic yard).

The Union Carbide landfill has a two-foot-thick rolled clay liner to keep leachate from entering adjacent groundwaters. A 20-year life (based on a 4,000 tons, or 12,000 cubic yards per year waste disposal rate) has been projected for the landfill. An internal drainage system permits all-weather operation, and serves to collect the leachate for treatment. The basic operating procedures for hazardous waste disposal consist of strict segregation of in-plant wastes; deactivation before landfilling, where practical; continuous blending of wastes and soil, and daily earth cover.

Union Carbide indicates that not all chemical wastes are degraded in the landfill. Some liquid flows out of the landfill as the oil layer into a contaminated water basin, where it is skimmed for residue fuel. Other waste leaves as dissolved chemicals in the leachate and goes to wastewater treatment. The estimated costs for the expected 20-year life of the fill can be summarized as follows.

	1973
Study and design	$ 77,495
Land costs	100,000
Capital costs	250,000
Operating costs (20 year)	2,884,505
Total	$3,312,000

Some foreign companies also operate industrial waste landfill facilities. The Bayer Chemical Company's main plant in Leverkusen, West Germany, for example, has a large (150 acre) specially designed landfill (1). About 1,000 cubic meters (35,310 cubic feet) per day (about 5,000,000 metric tons, or 550,000 tons, per year) of solid waste are deposited in the landfill, which has a one-meter-thick clay bottom. Landfilling is done in 10-meter (33 feet) layers, which will lead ultimately to construction of a plateau 60 meters (197 feet) high. Estimated life is 70 years.

Approximately 35 monitoring wells are located around the landfill. Wastes accepted at the landfill include organic sludge from biological wastewater treatment, slag from the plant's high temperature (1200°C to 2190°F) incinerator, insoluble salts from titanium dioxide production, and heavy metal hydroxide sludges precipitated from inorganic production wastewaters.

Table 25 presents hazardous waste stream constituents for which landfill disposal is considered an acceptable waste disposal alternative. A brief summary of each applicable landfill disposal process is found in Table 26, including design and operating parameters where known. These processes are coded alphabetically in Table 25. Other equally acceptable or preferable treatment/disposal techniques are mentioned. By examining the disposal methods in Table 26, it is obvious that a great deal of additional detailed information on suitable operating parameters is needed.

TABLE 25: LAND DISPOSAL OF SPECIFIC MATERIALS

Hazardous Material	Recommended Disposal Method*
Aluminum fluoride	A
Aluminum oxide	B
Ammonium bifluoride	C
Ammonium fluoride	C
Ammonium perchlorate	D
Ammonium persulfate	D
Antimony pentafluoride	E
Antimony pentasulfide	A
Antimony sulfate	A
Antimony trifluoride	E
Antimony trisulfide	A
Barium fluoride	F
Barium nitrate	G
Barium sulfide	G
Benzene sulfonic acid	H
Beryllium carbonate	I
Beryllium chloride	I
Beryllium oxide	I
Beryllium (powder)	I
Beryllium selenate	I
Boron trifluoride	J
Cacodylic acid	K
Cadmium fluoride	F
Calcium arsenate	L
Calcium arsenite	L
Calcium fluoride	B
Calcium hypochlorite	D
Calcium phosphate	B
Chromic acid (liquids, chromium trioxide)	M
Chromic fluoride	N
Chromic sulfate	N
Cobalt chloride	G
Copper acetoarsenite	L
Copper acetylide	X
Copper arsenate	L
Copper nitrate	O
Copper sulfate	O
Diphenylamine (phenylaniline)	P
Hypochlorite (sodium)	D
Lead	Q
Lead arsenate	L
Lead arsenite	L
Lead oxide	R
Magnesium arsenite	L
Magnesium chlorate	D
Magnesium oxide	B
Manganese	B
Manganese arsenate	L
Manganese chloride	S
Manganese sulfate	S
Metallic mixture of powdered magnesium and aluminum	B
Nickel antimonide	T
Nickel arsenide	T
Nickel selenide	T
Nitrochlorobenzene (dilute)	A
Potassium arsenite	L
Potassium bifluoride	C
Potassium binoxalate	U

(continued)

TABLE 25: (continued)

Hazardous Material	Recommended Disposal Method*
Potassium fluoride	C
Potassium oxalate	U
Potassium permanganate	V
Selenium (powdered)	A
Silica	B
Sodium arsenate	L
Sodium arsenite	L
Sodium bifluoride	C
Sodium cacodylate	K
Sodium carbonate peroxide	D
Sodium fluoride	C
Sodium oxide	W
Sulfur	B
Tantalum	B
Thallium (dilute)	A
Thallium sulfate (dilute)	A
Vanadium pentoxide	B
Zinc arsenate	L
Zinc arsenite	L
Zinc chlorate	D
Zinc oxide	B

See Table 26 for definitions.

TABLE 26: DISPOSAL METHODS

A   Disposal in a chemical waste landfill.

B   Disposal in a sanitary landfill. Mixing of industrial process wastes and municipal wastes at such sites is not encouraged however.

C   Reaction of aqueous waste with an excess of lime, followed by lagooning, and either recovery or land disposal of the separated calcium fluoride.

D   Dissolve the material in water and add a large volume of concentrated reducing agent solution, and then acidify with H_2SO_4. When reduction is complete, soda ash is added to make the solution alkaline. Ammonia will be liberated and will require recovery. The alkaline liquid is decanted from any sludge formed, neutralized, diluted and discharged. The sludge is landfilled.

E   The compound is dissolved in dilute HCl and saturated with H_2S. The precipitate (antimony sulfide) is filtered, washed and dried. The filtrate is air stripped of dissolved H_2S and passed into an incineration device equipped with a lime scrubber. The stripped filtrate is reacted with excess lime, the precipitate (CaF-CaCl mixture) is disposed of by land burial.

F   Precipitation with soda ash or slaked lime. The resulting sludge should be sent to a chemical waste landfill.

G   Chemical reaction with water, caustic soda, and slaked lime, resulting in precipitation of the metal sludge, which may be landfilled.

H   Biological or chemical degradation of dilute streams using conventional wastewater techniques; treatment with lime to precipitate out calcium benzene sulfonate which can be disposed in a chemical waste landfill.

I   Wastes should be converted into chemically inert oxides using incineration and particulate collection techniques. These oxides may be landfilled.

J   Chemical reaction with water to form boric acid, and fluoroboric acid. The fluoroboric acid is reacted with limestone forming boric acid and calcium fluoride. The boric acid may be discharged into a sanitary sewer system while the calcium fluoride may be recovered or landfilled.

K   Long-term storage in concrete vaults or weatherproof bins; small amounts may be disposed in a chemical waste landfill.

(continued)

TABLE 26: (continued)

L   Long-term storage in large, weatherproof, and sift-proof storage bins or silos;
     small amounts may be disposed in a chemical waste landfill.

M   Chemical reduction of concentrated materials to chromium-III and precipitation
     by pH adjustment. Precipitates are normally disposed in a chemical waste land-
     fill.

N   Alkaline precipitation of the heavy metal gel followed by effluent neutralization
     and discharge into a sanitary sewer system. The heavy metal may be disposed
     in a chemical waste landfill.

O   Copper wastes can be concentrated through the use of ion exchange, reverse os-
     mosis, or evaporators to the point where copper can be electrolytically removed
     and sent to a reclaiming firm. If recovery is not feasible, the copper can be pre-
     cipitated through the use of caustics and the sludges deposited in a chemical
     waste landfill.

P   Wastes may be incinerated, or disposed in a chemical waste landfill.

Q   Recycle using blast furnaces designed for primary lead processing to convert
     waste into lead ingots. Small quantities may be disposed in a chemical waste
     landfill.

R   Chemical conversion to the sulfide or carbonate followed by collection of the
     precipitate and lead recovery via smelting operations. Landfilling of the oxide is
     also an acceptable procedure.

S   Chemical conversion to the oxide followed by landfilling, or conversion to the
     sulfate for use in fertilizer.

T   Encapsulation followed by disposal in a chemical waste landfill.

U   Ignite to convert it to a carbonate. The carbonates (nontoxic) may be sent to a
     landfill.

V   Chemical reduction in a basic media, resulting in manganese dioxide formation.
     The material may be collected and placed in a landfill.

W   Chemical neutralization followed by solids separation with deposit of solids into
     a chemical waste landfill.

X   Detonation (on an interim basis until a fully satisfactory technique is developed);
     the copper salts liberated may be disposed of in a chemical waste landfill.

The material in Tables 25 and 26 is drawn primarily from Reference (20) as supplemented
by OSWMP of EPA (1). Reference to these tables will provide the user with an indication
of whether a material in question is landfillable and in many cases, some of the operating
parameters and procedures required. These tables should be used in making preliminary
investigations to indicate the overall practicality of the landfill approach to specific hazardous waste problems.

PATENTED TECHNIQUES

A technique developed by M.T. Present (21) involves filling a land area with successive
layers of refuse and covering each layer of refuse with a layer of foamed plastic resin which
is applied as a liquid, is expanded in situ and solidifies to a rigid cell structure.

More specifically this method involves the steps of covering successive layers of refuse with
foamed polyurethane, polyether, or polyester based plastic resins. These materials are
easily handled in liquid form and easily spread over a layer of refuse by means of a gun or
nozzle having a mixing chamber. Two liquid chemical components of the desired plastic
resin are introduced under pressure into the mixing chamber where they are mixed together
and forced out of the nozzle. When the two components are mixed, there is an exothermic
chemical reaction which produces a gas and a stable plastic resin. The formation of the
gas, which may be Freon or carbon dioxide, occurs generally uniformly throughout the
reacting components and results in the foamed or cellular structure of the plastic resin.

Although the exothermic reaction begins in the mixing chamber of the nozzle, it takes place mostly after the components have been sprayed on the layer of refuse. In order to provide greater control over the thickness of the layer of foamed plastic resin spread over the refuse, the nozzle may be provided with a source of atomizing air. Although the atomizing air may contribute slightly to the forming of the plastic resin, the foaming is due primarily to the gas produced in the exothermic reaction; the atomizing air merely acting to break up the stream ejected from the nozzle into a wide angle spray so that the foamed plastic resin may be spread more uniformly.

When the layer of foamed plastic resin solidifies, which requires only a few minutes, it has a rigid cellular structure that is impervious to gas and water. It therefore seals in any noxious odors created by the decomposition of the refuse. In addition, the plastic resins mentioned above are ratproof. In fact, rats are actually repulsed by these materials and will not gnaw through them to get to the refuse.

The foamed plastic resin layer is also an effective insect barrier. Not only does it prevent adult insects from laying eggs in the refuse after it has been spread, it also prevents the escape of any insects from the refuse that should hatch from eggs already lain therein. The foamed plastic resin layer of this method does not adhere to the tires of dump trucks hauling refuse to the landfill area, and thereby eliminates the mess attendant the use of asphaltic oil, even on the hottest of summer days.

The advantages of this method over the conventional method of using earth fill are numerous. For example, it is claimed to cost less than earth fill, it takes up much less space and thereby makes more of the volume to be filled available for refuse rather than being occupied by earth, it is impervious to rats and insects, eliminates the need for heavy earth moving equipment and can easily be spread by one man.

A landfill scheme developed by G.P. Larson (22) is one in which an earth subbase is formed in a land area and the subbase is covered with a liquid impervious layer. All of the solid waste fill in successive layers is placed on the impervious layer. Water flowing through the landfill becomes contaminated by the solid waste, is collected by the impervious layers and is then drained off the landfill and treated. The treated water, or if not available, water from an external source, is distributed on the landfill which together with natural precipitation produces a substantially continuous flow of water through the landfill thereby accelerating the decomposition reaction of the solid waste.

An operation developed by T. Nieman (23) is a landfilling operation, utilizing city and industrial trash which is compacted, moistened, sprayed with insecticide and morticed with a clean earth layer to develop a landfill suitable for recreation or both residential and commercial construction activity. The sequence of steps involved in this operation is shown in Figure 55.

A process developed by C.C. Cook et al (24) is one in which waste slimes and tailings obtained from the beneficiation of various ores are utilized in the reclamation of otherwise useless land by formulating them into a reconstituted fertile soil having acceptable bearing strength. A principal generator of copious amounts of slimes and tailings is the phosphate rock processing industry, a substantial portion of which is located in the state of Florida.

A process developed by C.L. Smith et al (25) is one in which waste sludges containing small amounts of certain types of reactive materials are treated by adding to such sludges materials capable of producing aluminum ions, lime and/or sulfate bearing compounds to produce a composition having a sufficient concentration of sulfate ions, aluminum ions and equivalents thereof, and calcium ions and equivalents thereof.

Fly ash is a preferred source of aluminum ions for this purpose. Over a period of time such compositions harden by the formation of calcium sulfoaluminate hydrates. Hardening of the sludge facilitates its disposition and may permit the reclamation of the land now occupied by large settling ponds for such sludge.

FIGURE 55: SEQUENCE OF STEPS IN A SANITARY LANDFILL OPERATION

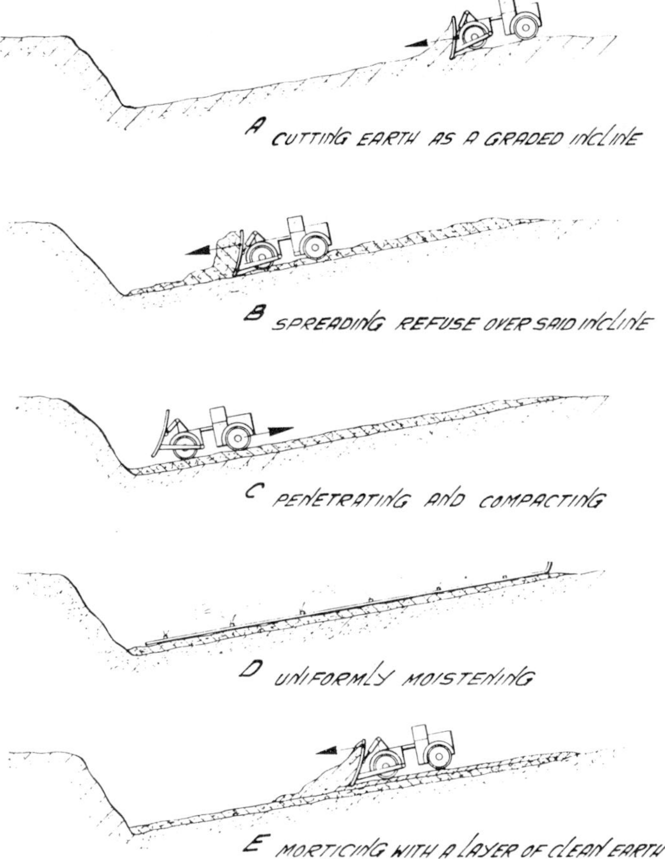

Source: U.S. Patent 3,614,867

Still further, the solidification of such settling ponds may provide permanent landfill which permits immediate use of the land without the necessity for removal of the sludge. Aggregate materials may also be incorporated in the solidified waste.

A scheme for disposal of liquid industrial wastes at a landfill site has been developed by R.F. Dickson (26). According to a first aspect of the scheme, a waste disposal facility is constructed by placing an artificial liner over the native soil, the liner being water impervious and having a vessel-shaped configuration. The purpose of the liner is to permit various industrial liquid wastes to be dumped within the liner and yet be prevented from contaminating the underlying and surrounding soil structures and water supplies.

The liner itself is constructed from sandy clay soil which is usually at or near the disposal site and which is mixed with water and then mechanically compacted to provide the desired water-impervious characteristic. The materials and procedures used in constructing the liner and the configuration of the liner are selected in such manner that the finished liner is not entirely rigid but has considerable flexibility and hence will not be readily cracked or fractured by ordinary stresses and strains or even by an earthquake and will generally heal itself of any tears that may occur. According to the second phase of the scheme, liquid waste material is prepared for disposal by mixing it with loose earth, the ratio of loose earth to liquid waste being sufficient so that the earth absorbs the liquid completely, and the earth is approximately at its optimum moisture content which thereby prevents uncontrolled flow of the liquid.

The resulting mix is then spread on the surface of a disposal site and is compacted by mechanical means. As the disposal process continues, additional quantities of liquid waste material are mixed with loose earth, spread on the disposal site in successive interlocked layers with each layer being mechanically compacted at optimum moisture with properly weighted rollers before the next layer is spread on the site.

A process developed by J.A. Franciscovich (27) is one in which fuel is injected into a refuse landfill surface where nitrogenous gases have been created by natural decay of the vegetable and animal refuse material. The fuel is ignited in the selected subsurface region to heat the nitrogenous gases to the point of combustion to effect self sustained burning of the refuse material beneath the landfill surface.

A landfill process for disposing of refinery waste sludge materials developed by W.K. Lorenz et al (28) comprises dewatering and deoiling refinery sludges by filtering through a filter press at a temperature of from about 100° to 200°F, washing the filter cake with hot water, and subjecting the solid filter cake thus obtained to soil biodegradation.

A process developed by J.R. Cinner et al (29)is one in which waste material is treated by mixing it with an aqueous solution of an alkali metal silicate and a silicate setting agent containing polyvalent metal ions to cause the silicate and setting agent to chemically react with each other and convert the mixture into a consolidated chemically and physically stable earth-like material substantially insoluble in water and with its pollutants entrapped. Landfill material and the treated waste material are disposed in contact with each other on land to be filled and in such relation that the leachate from the landfill material will trickle down through the treated waste material.

A scheme developed by C.M. Jaco, Jr. (30) involves operating a landfill for disposing of power plant wastes including sulfur dioxide removal system wastes and fly ash waste. The sulfur dioxide removal system wastes comprise a sludge containing varying amounts of finely divided alkaline-sulfur composition solids which must be stabilized to serve as landfill material. The fly ash is preferably included with the sludge to promote stabilization.

The sludge is admixed with a cementitious material and deposited in an initial holding basin. The admixture is permitted to settle to dewater somewhat and to partially stabilize the solids. The settled solids are removed from the bottom of the basin and deposited at a secondary dewatering and stabilizing zone whereat the water content may be further reduced and the solids stabilized. The stabilized solids are then moved to the landfill zone.

Figure 56 shows the essential elements of this scheme. The landfill operation **30** shown comprises an initial holding basin **34**, which is fed from fill pipe **28** which is used to transport the solids containing sludge to the landfill site. The fill basin can be established by building a small initial toe dam **32**. The sludge is admixed with cementitious material in an amount of about 1 to 10% by weight of the solids contained in the sludge.

The solids content of the sludge arriving via fill pipe **28** will be of the order of 30 to 50 weight percent. In general, the higher the solids content the less cementitious material required to stabilize the resultant product. The sludge developed by a typical sulfur dioxide removal system either directly in the scrubber or in a regenerative system contains substantial quantities of calcium sulfite, calcium sulfate, and other calcium compounds. These calcium-sulfur composition solids form particles of randomly attached crystal platelets which are finely divided and exhibit a low mass to volume ratio.

FIGURE 56: LANDFILL DISPOSAL OF POWER PLANT WASTES

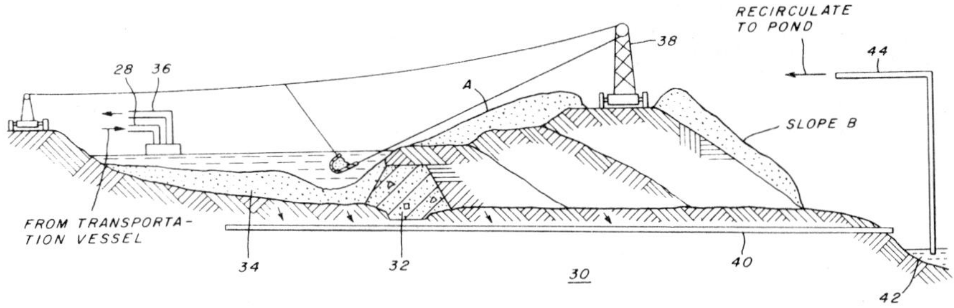

Source: U.S. Patent 3,859,799

These particles will not settle out normally to a water content of less than about 40 to 50% by weight. The sludge may contain coal fly ash as it is generated if the flue gases are passed through the wet scrubber to remove the fly ash. The fly ash can be collected by a dry precipitator and admixed with the sludge. It has been discovered that a fly ash content of at least 25 weight percent of the total solids weight in the sludge is desirable in achieving a stabilized landfill material.

The partially settled solids are moved from the bottom of the holding basin **34** by a drag cable material moving apparatus **38**, and deposited above the toe dam **32** to zone **A** which is a secondary dewatering and stabilizing zone. Several earlier landfill layers are shown below zone **A** from earlier operations. The material at zone **A** is substantially stabilized as a result of further drainage of water therefrom, and includes conversion of free water to hydrated water as the cementitious material cures. The substantially stabilized material can then be moved to landfill zone **B** as the start of another landfill zone.

The landfill site is prepared by providing an underdrain **40** which collects any fluid leached from the landfill and collects it in pond **42**, which is recirculated back via pipe **44** to the holding basin. Various cementitious materials have been found useful including lime, Portland cement, cement baghouse dust, and steel making residue slag. When the material is particularly high in silica and alumina an accelerating agent such as lime can be added to promote cementitious reaction.

These cementitious materials are preferably added to the sludge or slurry in an amount of 1 to 10% by weight of the solids weight, as the material is deposited in the holding basin. The admixed material will settle to some extent in the holding basin and supernatant fluid can be pumped back for reuse in slurry formation through pipe **36**.

A holding time of about 2 to 3 weeks in the initial basin **34**, can be followed by holding the material at zone **A** for another week. The material is then substantially stabilized and can be placed at zone **B**. The various holding times can, of course, be larger if the initial sludge has a high water content and if the percentage of cementitious material admixed is low. The holding times can likewise be shortened if the sludge has a lower water content and a higher percentage of cementitious material is added.

REFERENCES

(1) Office of Solid Waste Management Programs. *Report to Congress; Disposal of Hazardous Wastes.*
      Environmental Protection Publication SW-115, Washington, D.C., U.S. Government Printing
      Office (1974).

(2)   T. Field, Jr. and A.W. Lindsey, *Landfill Disposal of Hazardous Wastes: A Review of Literature and Known Approaches,* Report EPA/530/SW-165, Washington, D.C., U.S. Environmental Protection Agency (September 1975).

(3)   R.S. Ottinger, J.L. Blumenthal, D.F. Dal Porto, G.I. Gruber, M.J. Santy and C.C. Shin, *Recommended Methods of Reduction, Neutralization, Recovery and Disposal of Hazardous Waste,* Vol. III, Report PB-224,582, Springfield, Virginia, National Technical Information Service (August 1973).

(4)   R.G. Shaver, L.C. Parker, E.F. Rissman, K.M. Slimak and R.C. Smith, *Assessment of Industrial Hazardous Waste Practices—Inorganic Chemical Industry,* Report PB 244,832, Springfield, Virginia, National Technical Information Service (March 1975).

(5)   "Thermal Processing and Land Disposal of Solid Waste—Guidelines," *Federal Register* 39 (158):28334 (August 14, 1974).

(6)   Battelle Memorial Institute, *Program for the Management of Hazardous Wastes for Environmental Protection Agency,* Washington, D.C., Office of Solid Waste Management Programs (July 1973).

(7)   J.K. Holcombe and P.W. Kalika, *Solid Waste Management in the Industrial Chemical Industry,* Report PB 226,420, Springfield, Virginia, National Technical Information Service (1973).

(8)   E.E. Slover and J.D. Hainley, *Solid Waste,* C.L. Mantell, Ed., New York, John Wiley & Sons (1975).

(9)   B.G. Liptak, *Environmental Engineers Handbook,* Radnor, Pennsylvania, Chilton Book Co. (1974).

(10) S. Weiss, *Sanitary Landfill Technology,* Park Ridge, New Jersey, Noyes Data Corp. (1974).

(11) L.A. Burch, "Solid Waste Disposal and Its Effect on Water Quality," *California Vector Views* 16, No. 11.

(12) California State Water Resources Control Board, *Disposal Site Design and Operation Information,* Sacramento (March 1975).

(13) J.L. Pavoni, D.J. Hagerty and R.E. Lee, "State of the Art of Land Disposal of Hazardous Wastes," paper presented at the Seventh American Water Resources Conference, Washington, D.C., Oct. 24-28, 1971.

(14) California State Department of Public Health, *Tentative Guidelines for Hazardous Waste Land Disposal Facilities* (January 1972).

(15) R.W. Eldredge, "Minimizing Leachate at Landfills," *American Public Works Association Reporter,* p 22-23 (January 1974).

(16) L.J. Ricci, *Chemical Engineering* 82, No. 24, 128, 130, 132 (November 10, 1975).

(17) J.F. Funkhouser, *Alternatives to the Management of Hazardous Wastes at National Disposal Sites,* Washington, D.C. U.S. Environmental Protection Agency (May 1973).

(18) T.J. Sorg and H.L. Hickman, Jr., *Sanitary Landfill Facts,* Washington, D.C., U.S. Department of Health, Education and Welfare (1970).

(19) E.E. Slover, "Solid Waste Disposal in a Multi-Product Chemicals Plant, Institute Plant—Union Carbide Corp., Institute, West Virginia," presented at Symposium on the Textile Industry and the Environment—1973, Washington, D.C., May 22-24, 1973.

(20) R.S. Ottinger, J.L. Blumenthal, D.F. Dal Porto, G.I. Gruber, M.J. Santy and C.C. Shih, *Recommended Methods of Reduction, Neutralization, Recovery of Disposal of Hazardous Waste,* Vol. I, Report PB 224,580, Springfield, Virginia, National Technical Information Service (August 1973).

(21) M.T. Present; U.S. Patent 3,466,873; September 16, 1969.

(22) G.P. Larson; U.S. Patent 3,586,624; June 22, 1971; assigned to Warner Company.

(23) T. Nieman; U.S. Patent 3,614,867; October 26, 1971; assigned to Landfill, Incorporated.

(24) C.C. Cook and E.M. Haynsworth; U.S. Patent 3,718,003; February 27, 1973; assigned to American Cyanamid Company.

(25) C.L. Smith and W.C. Webster; U.S. Patent 3,720,609; March 13, 1973; assigned to G. & W.H. Corson, Inc.

(26) R.F. Dickson; U.S. Patent 3,732,697; May 15, 1973.

(27) J.A. Franciscovich; U.S. Patent 3,743,486; July 3, 1973; assigned to Fibre-Weld, Inc.

(28) W.K. Lorenz, E.C. Sebesta and C.L. McClellan; U.S. Patent 3,835,021; September 10, 1974; assigned to Sun Oil Company of Pennsylvania.

(29) J.R. Cinner and R.J. Polosky; U.S. Patent 3,841,102; October 15, 1974; assigned to Environmental Sciences, Inc.

(30) C.M. Jaco, Jr.; U.S. Patent 3,859,799; January 14, 1975; assigned to Dravo Corporation.

OCEAN DISPOSAL

BASIC TECHNIQUES

The oceans have always served both man and nature as the ultimate disposal sink for all
the waterborne waste material carried by the natural and man-made streams discharging
at their shores, and for all of the atmospheric pollutants scrubbed from the air by the rain
that falls on their surfaces. In addition, with increasing frequency in this century, hazardous waste materials have been deliberately shipped out to sea and dumped as either an
expedient or an economically attractive disposal technique. The hazardous waste materials
thus disposed of have varied widely in type, in quantity, and in frequency of disposal.
Three examples of this diversity may be cited as typical.

[1]   Spent sulfuric acid (7 to 10% H_2SO_4, and up to 30% $FeSO_4$) wastes from
steel pickling and titanium oxide pigment manufacture processes are shipped
daily to sea in specially designed barges, at the rate of 2.7 million tons
per year (1).

[2]   The U.S. Army program for deep sea disposal of obsolete chemical munitions was terminated in 1970 with the scuttling in the Atlantic of a stripped
cargo vessel laden with 418 concrete vaults which contained a total of
135,432 pounds of GB chemical warfare agent (nonpersistent nerve gas)
and 32,605 pounds of explosives (2).

[3]   Individual 55 gallon drums filled with sodium metal sludge (75% Na, 25%
Ca) are pierced and dropped from the decks of merchant vessels into the
Gulf of Mexico on an intermittent, unscheduled basis (1).

The three examples cited above illustrate the three basic techniques for ocean disposal of
hazardous waste materials. The first basic technique is bulk disposal of liquid or slurry-
type wastes. The waste materials are loaded into barging equipment; generally specially
designed tank barges. The barges are towed to sea, and emptied while underway at off-
shore distances that range from 10 to 125 miles.

The second basic technique involves scuttling of vessels filled with waste. In the past, the
U.S. Army and U.S. Navy have stripped obsolete or surplus World War II cargo ships, and
loaded the ships with obsolete munitions of all types. The explosive waste laden hulks
were towed out to the pelagic depths beyond the Atlantic and Pacific continental shelves,
and scuttled in predesignated sites. The third basic technique employed for deep sea disposal of hazardous materials is the sinking at sea of containerized hazardous/toxic wastes.
The individual containers, generally 55 gallon drums, are carried as deck cargo on merchant

vessels, and are discharged overboard at distances from shore that, dependent upon the contents, may be well over 300 miles. The operating principles involved in the three basic techniques employed for deep sea disposal differ in their use of the ocean. Seawater is used as a reacting, neutralizing medium and/or a diluent in the bulk disposal of industrial wastes from tank barges.

By contrast, obsolete munitions detonated in the deep sea employ the ocean as a cushioning, isolating medium, to protect the on-shore environment from the effects of the detonation. Similarly, disposal of concrete-encased obsolete chemical munitions and (undetonated) obsolete conventional ordnance items by burial in several thousand feet of water use the ocean as a means of isolation, to minimize or prevent both potential and actual impact upon the on-shore ecosphere.

The deep sea disposal of containerized hazardous wastes is based upon the principles cited above. Where the drums are deliberately ruptured at the surface, the ocean is used as reactant and/or a diluent. Those drums that are weighted and sunk intact beyond the continental shelf employ the thousands of feet of water for protective isolation of the barrels and their contents.

The District Offices of the U.S. Army Corps of Engineers have handled all applications for permission to engage in ocean disposal of hazardous (and other) waste materials. If the other Federal (and State) agencies to whom the Corps circulates the application are reasonably agreeable, the Corps issues a letter of no objection to the applicant tantamount to an authorization to proceed (1). The Corps may, in addition, specify regulatory procedures to be followed in connection with the disposal processes.

There were 281 ocean areas designated for the disposal of wastes of all types in 1969. Of these areas, 117 were employed for the disposal of hazardous wastes. The regional distribution of these locations is summarized by Smith and Brown (Table 27).

TABLE 27: MARINE DISPOSAL AREAS FOR HAZARDOUS WASTES (BY REGION AND WASTE TYPE)

Waste Type	Pacific	Atlantic	Gulf	Total
Industrial waste	9*	15*	16	40
Radioactive waste	10*	25*	2	37
Explosive and chemical munitions	19*	19*	11	49
Total	38	59	29	126

*Areas used for two or more types of wastes.

Source: D.D. Smith and R.P. Brown, *Ocean Disposal of Barge-Delivered Liquid and Solid Wastes from U.S. Coastal Cities.*

ENVIRONMENTAL EFFECTS

The report to the President on ocean dumping by the Council on Environmental Quality (3) makes a number of strong, broad-based recommendations "to ban unregulated ocean dumping of all materials, and strictly limit ocean disposal of any materials harmful to the marine environment." Specifically, legislation is recommended to:

Require a permit from the Administrator of the Environmental Protection Agency (EPA) for ocean, estuary, or Great Lake disposal of any waste;

Authorize the EPA Administrator to ban specific materials and specify safe sites; and

Provide for Coast Guard enforcement, and establish penalties for violations.

The Council recommended the use of the following principles in regulating ocean disposal:

Stop ocean dumping materials clearly identified as harmful to the marine environment or man.

Phase out ocean disposal where existing information on effects is inconclusive but where best indicators are that the materials dumped could create adverse conditions. If and when conclusive proof is obtained that disposal of the materials in question produces no damage to the environment (short term, cumulative, and long-term), permit dumping under regulation.

Include in the criteria for setting disposal standards and for urgency in stopping disposal operations:

Present and future impact on the marine environment, human health, welfare and amenities.

Irreversibility of the impact of dumping.

Volume and concentration of materials involved.

Location of disposal, such as, depth and potential impact of one location relative to others.

Give high priority to protecting the estuaries and shallow, near shore areas. Specifically, the Council made the following recommendations to discontinue, prohibit, or phase out the ocean dumping of the various categories of hazardous wastes:

Continue prohibiting the ocean disposal of high-level radioactive wastes.

Prohibit the ocean disposal of all other radioactive wastes, excepting only the federally regulated discharge from vessels and land-based nuclear facilities of low-level liquid wastes, or such other low-level radioactive wastes as have no alternative offering less harm to man and the environment.

Prohibit further ocean disposal of chemical warfare materials.

Continue prohibiting the ocean disposal of biological warfare materials.

Terminate as soon as possible ocean dumping of explosive munitions.

Terminate immediately dumping of toxic industrial wastes, excepting only those which have no alternative offering less harm to man and the environment.

Phase out ocean dumping of all industrial wastes.

A number of studies have been made of the environmental effects of ocean dumping of hazardous materials. Due to the (unique) requirement by the Galveston District of the Corps of Engineers, that laboratory and field studies of the effects of the wastes be filed in support of disposal applications, the majority of these studies have been carried out in the Gulf of Mexico (1). Inspection of the results of the various studies performed (see Table 28 for a summary of the key findings) indicates that the toxic effects of the hazardous chemical and pesticide wastes are generally limited to short time periods and areas in immediate proximity to the discharge or dump. The rate of dilution is, in general, so high that, after 12 hours, it is impossible to detect analytically chemical differences between contaminated and uncontaminated sea areas.

TABLE 28: SUMMARY OF ENVIRONMENTAL STUDIES ON INDUSTRIAL WASTES DISCHARGED AT SEA

Waste Type	Description of Disposal Area				Barging Characteristics				Waste Characteristics		Laboratory Toxicity Studies		
Industrial	Distance from shore (miles)	Water depth (feet)	Latitude	Longitude	Amount of waste/ trip	Depth of discharge	Rates of discharge tons/min.	Towing speed (knots)	Given description		Type of test organisms	TLM (range)	Remarks
Spent sulphuric acid	15 from New Jersey coast	80	40°20'N	73°40'W	3200 tons 5000 tons	15 feet	18 70	6 7	Fe_2SO_4 (10%) H_2SO_4 (8.5%)		Marine phytoplankton.	2.7-35.5 mg/l 6-16 day test duration.	Investigated effects of various concentrations of iron on growth of algae.
Chlorinated hydrocarbons	125 SE of Galveston, Texas	2400	27°36'N	94°36'W	1200 tons	12 feet	5	6	Beta-chloropropylene (22%), trichloropropane (5%), isopropylchloride (38%), allyl chloride (11%), misc. chlorides (33%), heavy ends (3%), pH - 9.8, specific gravity 0.9 - 1.34.		Marine phyto-, and zoo-plankton, anemones, brine shrimp, bacteria, fish.	0.02-2.5% saturated 0.0005-1g/l, 2-24 hr test duration.	Investigated acute toxicity and inhibition to photosynthesis.
Paper mill wastes (black liquor)	125 SE of Galveston, Texas	2400	27°36'N	94°36'W	1300 tons	10 feet	5	6	Paper mill wastes, 47% solids, BOD_5 -100,000 ppm Na_2CO_3, Na_2SO_4, NaOH etc., pH - 13, Specific gravity 1.27 at 60°C.		Marine phytoplankton zooplankton.	0.001-1.0g/l, 22 hr test duration.	Investigated acute toxicity and inhibition to photosynthesis.
Ammonium sulfate (mother liquor)	100 S of Freeport, Texas	2760	27°35'N	95°20'W	1700 tons		7	4	Ammonium sulfate (23%), nitrogen (8%), carbon (12%), organics (29%) (alcohols, esters, amides), IOD-90 mg/l., BOD_5 57,000 ppm, pH -4.3, S.G. 1.23.		Brineshrimp, top-water minnows.	1.25% by volume 200 ppm, 24 hr test duration.	Acute toxicity.
Waste liquor	70 S of Freeport, Texas	960	27°48'N	94°54'W	2400 tons		5-24	6	Na_2S (Na_2S_2) (6%) (Na_2S_3) NaHS 1% S (total) (6%), NaCl (21%), organic 2%, solids (dissolved and suspended), specific gravity 1.24 at 60°C.		Marine phytoplankton, top-water minnows, brine shrimp.	0.0005-0.18% by volume, 24 hr test duration.	Acute toxicity.
Chlorinated waste liquor	125 SE of Galveston, Texas	2400	27°36'N	94°36'W	1400 tons (proposed)		13-25 (recommended)	5 (recommended)	(Organic waste) chlorinated organics (10-15%), inorganic salts (Na_2SO_4) (5-6%); (acid) chlorinated organics (1%), sulfuric acid (10-15%), nitric acid (0.1%).		Marine phytoplankton, top-water minnows, brine shrimp.	0.02-0.64% by volume, 24 hr test duration.	Acute toxicity, waste acid more toxic to fish than brine shrimp.
Sodium sludge (containerized)	110 S of Galveston, Texas	2400	27°36'N	94°36'W	15 55-gal. drums (500-570 pounds per drum)	Surface	Variable	10	Metallic sodium (75%), calcium (24%), barium, magnesium, potassium (1%).				
Pesticides	95 SSE of Galveston, Texas	720	27°49'N	94°30'W	50 55-gal. drums per trip	1200 ft	1 barrel/2 mins. (600' intervals on bottom	3	Anilines (chloroaniline, monochlorobenzene), liquid organics (methanol, p-xylene, chlorobenzene), dry chemicals—insoluble (*thiram, thiram-E, thionex, zineb, ferbam, monuron*, carbon disulfide).		Minnows, largemouth bass, salmon, bluegills, mosquito fish, channel catfish, bluegill sunfish.	1-135 mg/l, 48-96 hr test duration.	Toxicity values obtained from literature.
Caprolactam wastes	35 S of Sabine Bank Light	60	29°10'N	93°40'W									

(continued)

TABLE 28: (continued)

| Waste Type | Reported Field Observations | | | Mixing Characteristics | | General study conclusions |
Industrial	Physical-Chemical	Biological	Observed effects	Initial dilution	Diffusion coefficient cm^2/sec.	
Spent sulfuric acid	pH, iron concentration (0-60'), wind, weather, sea state.	Plankton (0-60'), benthic organisms, pelagic fish (30').	Water discoloration, plankton temporarily immobile, iron settled rapidly from surface layer, no appreciable accumulation of iron found in bottom sediments.	1:5,000	2.9×10^3	Mixing and diffusion of wastes occurs rapidly in the wake of the barge. No evidence to indicate adverse effects. Each new proposed waste disposal operation needs careful study prior to allowing ocean disposal.
Chlorinated hydro-carbons	Temp. (0-900'), salinity oxygen, waste concentration (0-500'), surface currents, wind, weather, sea state, bottom mud.	Plankton, O_2 evolution, C_{14} uptake, chlorophyll-A, visual inspection of sargassum weed.	Water discoloration: fish, plankton killed on direct contact of waste, no harmful effects seen after 2-4 hrs. Bulk of waste sank. Low diffusion of waste at depth.		2.5×10^3 (average)	Disposal of toxic wastes at sea can be accomplished with only a slight effect on organisms in the biomass within a limited oceanic area. Each new waste disposal operation needs careful study prior to sanctioning it.
Paper mill wastes (black liquor)	Temp. (0-900'), salinity (0-600'), waste concentration surface samples, pH, oxygen (0-600'), wind, weather, sea state.	Plankton (0-100'), O_2 evolution, C_{14} uptake, chlorophyll-A.	Slight water discoloration. No mortality to marine life. Bulk of waste sank.	1:300,000	2.5×10^3 (average)	Disposal of "black liquor wastes" in the deep sea can be accomplished without determinable effects on marine biota. Ultimate disposal is expected to be accomplished by bacteria. Advisable to monitor each separate load of waste to determine toxicity in laboratory.
Ammonium sulfate (mother liquor)	Waste concentration (0-150')		No fish mortality. No floating oil. Bulk of waste sank. Maximum waste concentration at depth.		9×10^2	No undesirable effects were observed. Diffusion great enough to ensure good dispersion to minimize harmful effects to biota.
Waste liquor	Waste concentration (0-150'), temp., salinity pH, oxygen, (0-600').	Plankton, pelagic fish, sargassum weed communities.	No evidence of subsurface maximum waste concentration.	1:10,000-1:100,000 in 2 hrs		Disposal should produce no significant mortality in the biota, nor any prolonged effects.
Chlorinated waste liquor						No significant mortality would be expected from disposal of this waste in the open ocean. It is suggested that a full-scale disposal operation be properly monitored and continued to verify preliminary study results.
Sodium sludge (con-tainerized)	pH	Plankton, fish, sargassum weed.	No mortality to fish. Flying debris hazardous to disposal personnel. 30% mortality to plankton due to collection methods.			Explosions caused by reaction of sodium with seawater had no significant effects on sargassum and zooplankton populations. Absence of fish-kill was probably due to barrenness of disposal area.
Pesticides				100:1	0.002×10^3	Consideration of available toxicity and diffusion data from literature sources indicate that the zone of water containing toxic concentrations of waste surrounding each disposal drum will be limited in extent and duration and will not endanger motile aquatic life in the disposal area to a significant degree.
Caprolactam wastes	Temperature (0-60'), salinity (surface), wind, weather, sea state.	Plankton, bottom samples, C_{14}, chlorophyll-A,-B,-C.				Limited scale (one day) of survey precluded any significant results. Recommend future surveys be conducted on 3-day basis for better results.

Source: D.D. Smith and R.P. Brown, *Ocean Disposal of Barge-Delivered Liquid and Solid Wastes from U.S. Coastal Cities.*

The toxic effects of waste acid discharged from tank barges at sea are minimal; the zoo-plankton from samples in the immediate discharge zone were immobilized temporarily, but recovered rapidly in unpolluted water. Chlorinated hydrocarbon wastes discharged in Gulf waters killed fish and plankton in direct contact with the undiluted waste. In contrast, due to dilution, there was no effect observed on marine life at the surface two to four hours after discharge.

The general ocean surface and upper level effects of chlorinated hydrocarbon discharge range downward from the upper extreme noted above, fish and plankton kill, through laboratory-study-detected inhibition of photosynthesis and respiration (4) to total absence of observable effect. The possible effects of the discharged chlorinated hydrocarbons at deeper levels, and on the bottom, have not been determined.

Smith and Brown (1) report that unpierced barrels loaded with sodium sludge (75% metallic sodium, 24% metallic calcium) have on two occasions been retrieved in the Gulf of Mexico by fishermen. These barrels were not pierced prior to dumping as prescribed, nor were they dumped in the prescribed area. Pierced barrels of the sodium sludge exploded when dropped overboard, produced no significant effects on the microbiota, and, due to the probable barrenness of the area, produced no visible fish kill.

The probable effects of deep sea disposal in the Gulf of Mexico of weighted steel drums, designed to rupture on the sea floor, and containing herbicide and fungicide wastes (chloro-aniline, aniline, monochlorobenzene, methanol, para-xylene, theram, theram-E, Thionex, zineb, ferbam, and carbon disulfide) were studied on a theoretical basis. No field (at-sea) confirmation of the study findings, that dilution due to eddy diffusion and chemical degradation would reduce concentrations to below the median tolerance limits, have been reported. On the contrary, pesticides at sublethal dosages have been shown to reduce the size and strength of mollusk shells, and to reduce growth rate and reproduction activity in fish (3).

The National Academy of Sciences-National Research Council (NAS-NRC) Committee of Oceanography, on the basis that the understanding of most of the physical and biological processes in the ocean was too poor to permit precise predictions of the results of the introduction of a given quantity of radioactive materials at a particular location in the sea, has proposed attacks upon several basic problem areas. These were all oriented towards the potential hazards associated with deep sea disposal of radioactive wastes.

The most critical of these is the possibility of return of the radioactivity to humans. Second in criticality was marine organism damage due to exposure to radioactive waste. The two avenues cited for occurrence of such damage are transport of the radioactive wastes from the disposal sites to the coastal zone; and uptake of the radioactive wastes by one or more of the trophic levels in the marine biota, with possible return to man via commercially important fish and shellfish.

The NAS-NRC studies produced reports (5)(6)(7) which covered the key factors to be considered in disposal of low level radioactive wastes at sea, both containerized and liquid wastes from vessels and outfalls, and the disposal sites selections. Relatively little of the further research recommended in the NAS-NRC reports to fill in the wide gaps in factual data has been carried out. As of 1969 (1) only three off-shore sites employed for disposal of low-level containerized radioactive wastes had been resurveyed for the Atomic Energy Commission: [1] the site near the Farallon Islands, off San Francisco; [2] the Santa Cruz basin site, west of Los Angeles (70 miles); and [3] the site 130 miles east-southeast of Cape May, New Jersey.

The first two sites, based on beta-gamma counting of sediment samples, had no apparent radioactive waste leakage. There were some indications in 1961, based on similar counts, of possible leakage from the containers at the site off Cape May. There are no reports of further investigations. As noted earlier, low-level radioactivity liquids discharged from vessels and nuclear facilities, if as per federal regulations and international standards, have not

been recommended for termination; the ocean dumping of other radioactive wastes was (3).
The President's Council on Environmental Quality, in the report on ocean dumping (3) cited
earlier, recommended termination of the disposal of chemical warfare material and explosive munitions by dumping at sea. As support for this recommendation, a Department of
Defense calculation was cited which indicated that the 1,000 tons of explosives detonated
in Deep Water Dump in the Pacific off Washington on September 4, 1970 was capable of
generating a shock wave that would kill most marine animals within 1 mile of the explosion
and will probably kill those fish with swim bladders out to 4 miles from the explosion.
It should be noted that, of 15 explosive ordnance laden hulks scuttled, 12 detonated during
scuttling (8).

The President, in response to the above recommendation and the others cited earlier, stated
on October 7, 1970 that he would recommend legislation to stop ocean dumping. The
Secretary of the Navy immediately placed a moratorium on Deep Water Disposal (DWD)
operations. Subsequently, the Chief of Naval Operations directed that an investigative program for conventional explosives, be instituted to prepare a comprehensive environmental
condition report on representative past explosive ordnance DWD sites; develop criteria for
selection of future DWD sites; and determine the monitoring required at future DWD sites.
This program was started in the late spring of 1971.

The representative sites selected were: an area off Cape Flattery in the Pacific, where five
ships were sunk and exploded in 8,400 feet of water and an area 175 miles southeast of
Charleston where one ship was sunk and did not explode, in 6,300 ft of water. Results
as of the date of the report showed no evidence of cratering of the sea floor in the
Pacific. A substantial number of organisms typical to the general region are apparently
residing within the debris area.

There has been some fear that the ocean disposal of chemical warfare munitions by techniques similar to those used in CHASE involved the possibility that the explosive portions
of the munitions would detonate after sinking, and release the chemical warfare agent.
In this context, the chemical warfare material encased in concrete vaults that were disposed
of in 7,000 feet of water by scuttling the laden hulks off New Jersey in 1967 and 1968
did not explode and rupture the vaults. It was also stated that seawater, because of its
alkalinity, will both hydrolyze and dilute any chemical warfare agent that is released (2).

DESIGN OF OCEAN DISPOSAL SYSTEMS

Deep sea disposal bulk transport systems vary from modern, specialized tank barges to obsolete hulks. The majority of bulk liquid and slurry hazardous wastes dumped at sea are
transported in specially designed tank barges, from 1,000 to 5,000 short tons in capacity.
The tank barges are of double-skinned bottom construction, and must be certified for ocean
waters by the U.S. Coast Guard. The barge cargo is under U.S. Coast Guard regulations
covering the bulk shipment of chemicals from the U.S.

Industrial waste-laden barges are transported to the industrial waste disposal areas designated
in Figures 57, 58 and 59 for the Pacific, Atlantic, and Gulf of Mexico coasts, at off-shore
distances that depend upon the type of waste and the regulatory procedures. Typical distances are, for acid wastes, 15 miles from New York City; for toxic chemical wastes, 125
miles into the Atlantic; for Gulf of Mexico operations, 125 miles from the coast (at the
2,400 foot depth line).

In the disposal area, typical barge speeds from 3 to 6 knots are used; typical discharge is
at 6 to 15 feet submergence, at rates that vary between 4 and 20 tons per minute. A
characteristic prediction rate for chlorinated hydrocarbon concentration in the disposal
area is $C = $ (pumping rate)$/(0.063t)$ where C is concentration in parts per million; pumping rate is in grams per centimeter and t is time in minutes after pumping (4).

FIGURE 57: PACIFIC COAST DISPOSAL AREAS

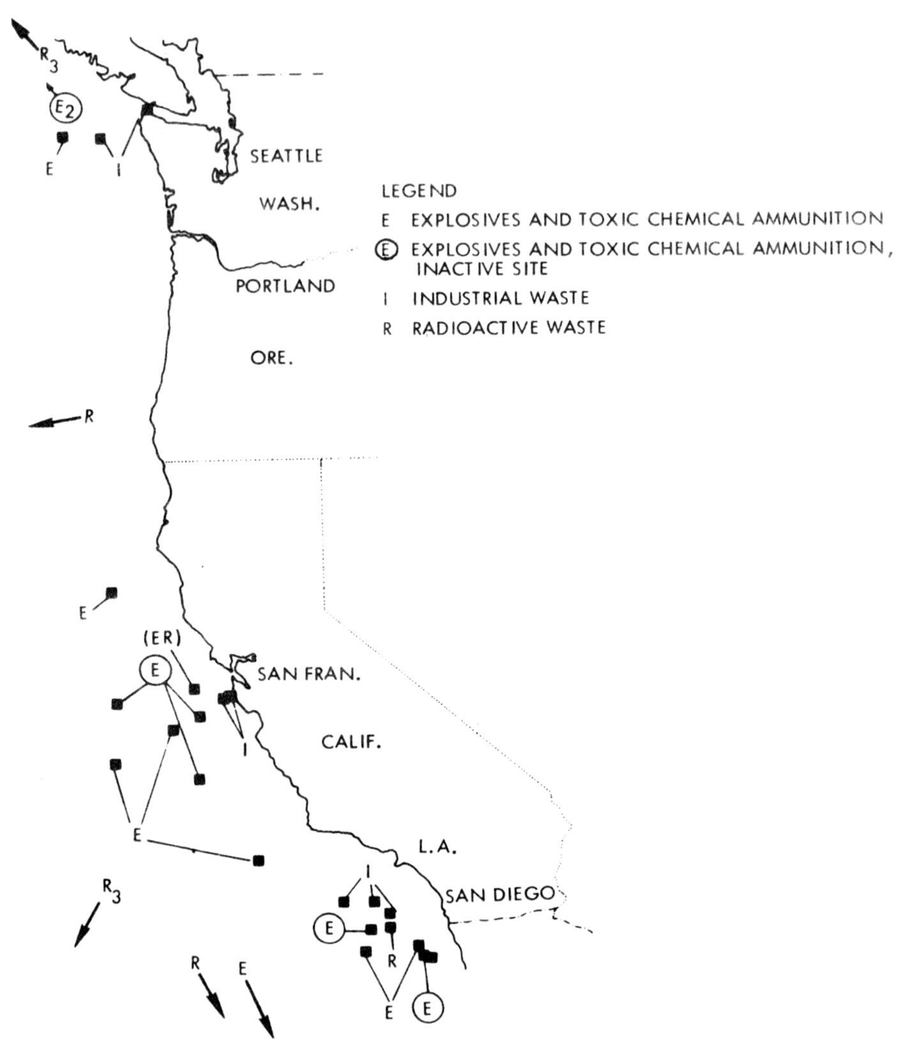

Source: D.D. Smith and R.P. Brown, *Ocean Disposal of Barge-Delivered Liquid and Solid Wastes from U.S. Coastal Cities*

FIGURE 58: ATLANTIC COAST DISPOSAL AREAS

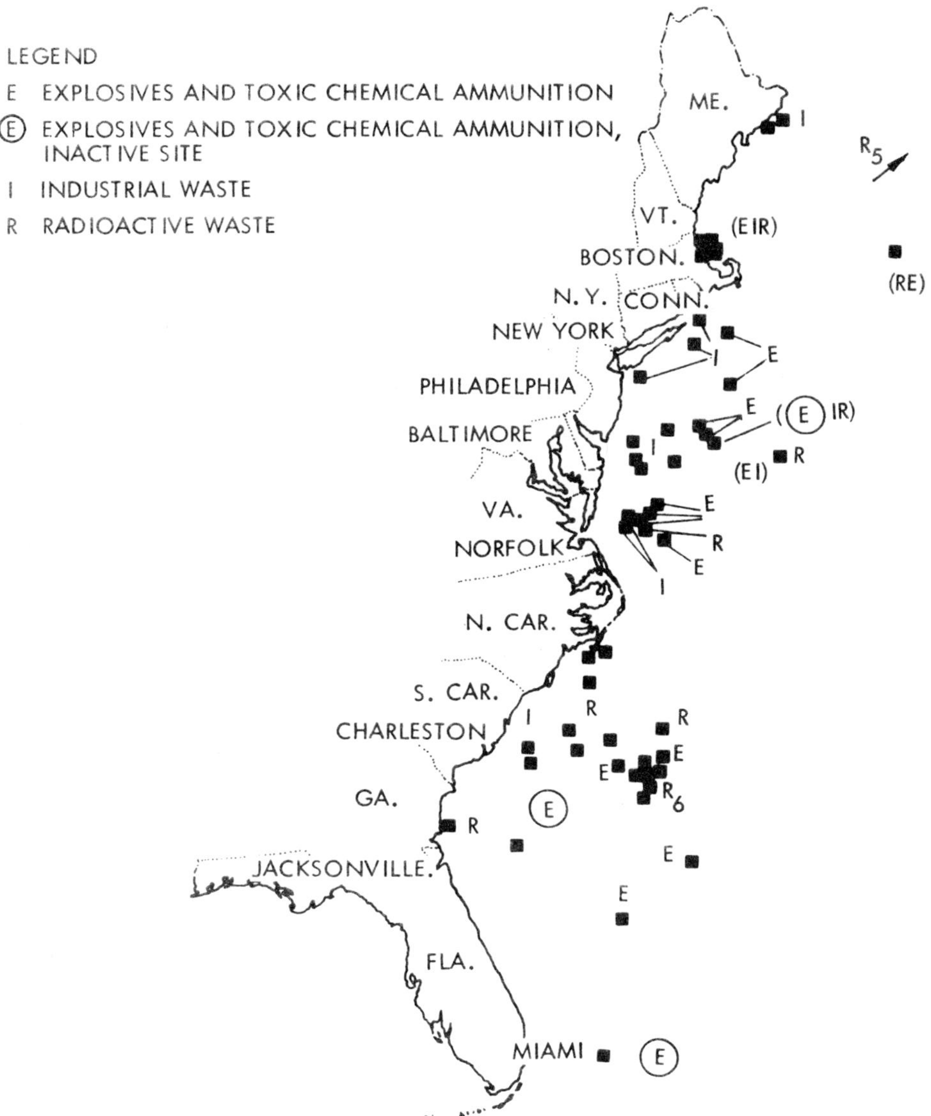

Source: D.D. Smith and R.P. Brown, *Ocean Disposal of Barge-Delivered Liquid and Solid Wastes from U.S. Coastal Cities*

FIGURE 59: GULF OF MEXICO DISPOSAL AREAS

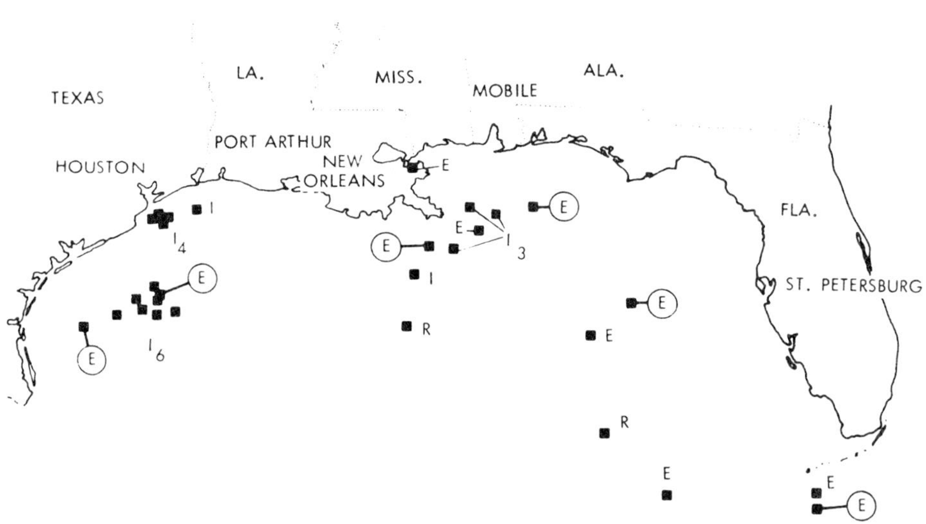

Source: D.D. Smith and R.P. Brown, *Ocean Disposal of Barge-Delivered Liquid and Solid Wastes from U.S. Coastal Cities*

The last practice employed in deep water disposal of obsolete explosive and chemical warfare ordnance was that of the U.S. Navy CHASE (Cut Holes and Sink 'Em) disposal program. The Navy, which handled deep water dumping of munitions for all of the services, obtained merchant hulks for this purpose from the U.S. Maritime Administration Reserve of surplus World War II merchantmen. The ship was stripped of anything readily removable or loose. The fuel tanks were cleaned to eliminate oil contamination and scuttling valves were installed to allow water to enter the ship.

To permit the water to spread evenly, soft patches were installed in the bulkheads between the holds. The material for dumping was made negatively buoyant (bulk density higher than seawater), to prevent it from floating to the surface, and loaded into the bulk at one of two Naval Depots (Earle, New Jersey, or the former Naval Ammunition Depot at Bangor, Washington. After the operation had been cleared with all responsible authorities, the munition-laden hulk was towed to the selected dumping site under naval escort, and scuttled. The selected sites have been at least 10 miles from any shore, and in waters at least 3,000 feet deep. The majority of sites employed before the moratorium on DWD were at sea depths in excess of 6,000 feet. Of the 15 explosive laden hulks scuttled, four were detonated deliberately (two at 1,000 foot depth, two at 4,000 foot depth) and eight detonated unintentionally. Low level radioactive wastes have in the past been encased in concrete contained in 55 gallon steel drums, which were required by AEC regulations to be of a minimum gross weight of 550 pounds to ensure sinking. Another packaging technique has been to encase 55 gallon drums, loaded with liquid low level radioactive wastes, in a concrete block. The concrete packaged wastes were then taken to designated disposal sites, and dropped overboard.

Most of the wastes disposed of in the Pacific were dumped at two sites. Disposal in the Atlantic, with one exception, has been at depths greater than 6,000 feet. The exception (in the area of Massachusetts Bay, about 12 to 15 miles offshore) was in 300 feet of water. Low level radioactive liquid wastes, resulting from the operation of U.S. Navy nuclear submarines, are discharged at sea in accordance with regulations on depth and rate of pumping. As of 1970, one commercial organization, two government agencies, and one university were the only entities authorized to dispose of radioactive wastes in the ocean.

Containerized toxic industrial wastes, as noted earlier, are dumped at sea after transport as deck cargo on either merchant vessels or contract disposal vessels. The individual containers are either ruptured at the surface, or weighted for sinking. There is no single operating design or operating practice that covers the wide variety of materials thus disposed of.

The question of permits for ocean dumping has been summarized neatly by R.A. Young and P.J. Cheremisinoff (9). As pointed out by the above authors, there are several varieties of permits: [1] general permits, [2] special permits, [3] emergency permits, [4] interim permits, and [5] research permits. Some of the conditions attached to these various types of permits are as follows:

[1] General permits may authorize the dumping of nontoxic materials disposed of in small quantities. These permits must be published in the *Federal Register* specifying types and amounts of material and the sites for dumping.

[2] Special permits cover materials not covered by general permits with the exception of toxic metals, oils, inorganic wastes and oxygen-consuming matter. These permits have a three year expiration date and are renewable upon application.

[3] Emergency permits allow for the dumping of prohibited materials listed under special permits, where there is a risk to human life and there are no other feasible solutions.

[4] Interim permits are issued for the prohibited materials listed under special permits in excess of the permissible levels under the following conditions:

An assessment of the environmental impact is required as a part of the application, along with alternatives to ocean dumping.

The development and implementation of a plan to eliminate or bring within agreeable limitations set by the act. The permit's expiration date will be set according to the proposed plan, but shall not exceed one year. However, it is not renewable. A new permit may be obtained upon application, provided certain phases of the proposed plan have been satisfactorily completed.

No interim permits may be granted for new facilities, or for expansions of existing facilities.

[5] Research permits may be issued if the scientific merit of the project outweighs the potential damage that may occur and only under the following conditions:

A detailed statement of the proposed project and its impact must be provided.

There must be a public hearing.

Research permits shall expire in 18 months, but are renewable.

All applications for ocean dumping permits must include the following: name and address of the applicant; name and location of the firm to be used for transportation and dumping; physical and chemical descriptions of the material and any containers to be dumped; quantity of material; means of conveyance and proposed dates and times of disposal; proposed dump site; method of disposal at site; identification of the process which produces the material; and descriptions of alternative methods and why these are not used.

Permit holders are liable for transporting and dumping material not conforming with the permit application (9). All permits for dumping at existing sites shall have a fee of $1,000. An additional $3,000 shall be charged for dumping at a site other than the designated one. A fee of $700 shall be charged for the renewal of a permit. No agencies of the Federal, State or local government shall be charged for dumping.

Under no circumstances will the EPA allow the dumping of the following materials: high-level radioactive materials, such as irradiated reactor fuels; materials of any form produced for radiological, biological or chemical warfare; any materials which are not sufficiently described to permit their impact on marine ecosystems; and persistent inert materials, synthetic or natural, which will remain in suspension, unless they are processed to sink and remain in place.

The EPA will not permit the dumping of the following materials as other than trace contaminants: organohalogen compounds or materials which will form these compounds in the marine environment; mercury and its compounds; cadmium and its compounds; and crude oil, fuel oil, heavy diesel oil, lubricating oil and hydraulic fluids. The trace contamination level is defined as follows:

> Organohalogen compounds—0.01 of a concentration shown to be detrimental to the marine environment.
>
> Mercury—solid phase, not greater than 0.75 mg/kg; and liquid phase, not greater than 1.5 mg/kg.
>
> Cadmium—solid phase, not greater than 0.6 mg/kg; and liquid phase, not greater than 3.0 mg/kg.
>
> Oils—there must not be a visible surface sheen in undisturbed water when it is added at 1 part oil to 100 parts water.

Finally, materials which are stated to require special care in ocean disposal are the following: arsenic, lead, zinc, copper, nickel, selenium, vanadium, beryllium and chromium, and any of their compounds; organosilicon compounds; inorganic processing wastes such as cyanides, fluorides, chlorine and titanium dioxide; petrochemicals, organic chemicals, and organic processing wastes; biocides; oxygen-consuming and/or biodegradable organics; and other radioactive wastes.

ECONOMICS

The average cost for ocean disposal of all types of wastes in 1968 (Table 29) was slightly over $0.60 per ton (1). The 1968 average cost for disposal of bulk industrial wastes was $1.70 per ton. Average ocean disposal costs for explosives in 1968 were $15 per ton. Since the quantity of containerized low level radioactive wastes dumped at sea in 1968 was zero, and only 4.2 tons per year were dumped in 1969 and 1970, costs were not calculated for this category.

The ocean disposal costs cited for industrial wastes and miscellaneous wastes represent transportation and dumping costs only, and do not include other costs incurred for treatment, storage and loading of the wastes. The costs reported for explosives include hull preparation, towing and loading costs, and are given as dollars per ton of total waste cargo. The costs for ocean dumping of industrial wastes are significantly lower than those of other disposal techniques currently employed.

TABLE 29: 1968 COSTS PER TON FOR MARINE DISPOSAL OF HAZARDOUS WASTES IN THE UNITED STATES COASTAL WATERS

Type of Waste	Total United States		Pacific Coast	
	Average Cost	Reported Range	Average Cost	Reported Range
Industrial Wastes				
Bulk	$1.70/ton	$0.60-9.50/ton	$1.00/ton	$0.60-9.50/ton
Containerized	$24/ton	$5-130/ton	$53/ton	$50-130/ton
Explosives	$15/ton	$15-90/ton	-	-
Miscellaneous*	$15/ton	$5-600/ton	$15/ton	$5-600/ton

Type of Waste	Atlantic Coast		Gulf Coast	
	Average Cost	Reported Range	Average Cost	Reported Range
Industrial Wastes				
Bulk	$1.80/ton	$0.60-7.00/ton	$2.30/ton	$0.75-3.50/ton
Containerized	$7.73/ton	$5-17/ton	$28/ton	$10-40/ton
Explosives	-	-	-	-
Miscellaneous*	-	-	-	-

Note: Although reference (1) quotes costs as on the basis of the weight of the volume of water in which the wastes were contained, marine costs are generally quoted on the basis of the weight of the volume of water displaced. It is believed that the costs cited in this table are so based.

*Includes barreled chemicals and sludges.

Source: D.D. Smith and R.P. Brown, *Ocean Disposal of Barge-Delivered Liquid and Solid Wastes from U.S. Coastal Cities*

The costs of minimum environmental impact disposal techniques for the bulk industrial wastes are very much higher than the costs of ocean dumping. As an example, material (lime) costs for neutralization of the waste acid from TiO_2 pigment manufacture are estimated at roughly $1.00 per ton; operating costs and equipment amortization would add at least an equal amount and land burial with its attendant costs would still be required for the solid calcium sulfate-iron hydrate produced as a product of neutralization, after recovery by lagooning or filtration.

In general, the predominant factors which have given rise to ocean disposal of hazardous wastes have been economic, avoidance of capital outlay, and/or a cheaper operating cost than the costs of other techniques. This economic incentive will be increased as federal and local regulations increase the requirements for the use of minimum impact disposal techniques, with their attendant higher costs.

Few of the hazardous waste materials disposed of by ocean dumping present an economically attractive recycle or by-product recovery picture. Recycling and reprocessing of waste acid (generally, waste sulfuric acid) which constituted 58% of the bulk industrial wastes dumped at sea, has been the objective of many economically fruitless steel and pigment company research and development projects. In fact, the major reason for the change over from sulfuric acid to hydrochloric acid as the preferred material for pickling of steel was the virtual impossibility of economic recycle via regeneration of the spent sulfuric acid.

Sludge may be transported by rail, truck, barge, pipe or conveyor to the embarkation point for ocean disposal. For hauls up to 50 miles, it is likely that truck transportation would be more economical than rail. Freight rate structures in the United States are extremely diverse, causing comparisons of transportation methods to be difficult at best. In addition to the cost of hauling, there will be an incurred cost for loading sludge into hopper cars in the case of rail haul or pumping sludges to barges in the case of barge haul. Barge rates are lower than rail rates since the tractive force involved is less and barging normally does not bear the cost of the waterway.

For coastal communities or communities adjacent to the well-developed inland waterway systems on the Mississippi, Ohio, Missouri, Illinois, and Tennessee Rivers, sea disposal is frequently the most economical and simplest method available for ultimate sludge disposal, although the entire concept is under review by the regulatory agencies. It should be recognized that present and future laws governing sea disposal practice will in large measure dictate its economics. Design consideration and process variables involved are:

[1]   proximity of final disposal site
[2]   pretreatment required
[3]   transportation unit costs
[4]   type of transportation available
[5]   distance to loading point
[6]   requirements for final disposal
[7]   sludge characteristics

The decision to use transportation in conjunction with sludge disposal is based upon an analysis of alternate dewatering and disposal methods. Laboratory analyses of the sludge are required for such an investigation. There are many factors that may rule out the possibility of sludge transportation, such as the unavailability of economical transportation or the distance to a final disposal site. Barging and sea disposal should be considered as a means for ultimate sludge disposal for plants located near rivers close to the sea.

SPECIFIC APPLICATIONS

The list of hazardous waste materials dumped in the ocean is almost endless. The broad classes of hazardous waste materials have been categorized as follows (1): industrial wastes; obsolete, surplus, and nonserviceable conventional explosive ordnance and chemical warfare material; radioactive wastes; miscellaneous hazardous wastes. The major types of industrial waste dumped at sea which are considered hazardous are: spent acids; refinery wastes; pesticide wastes; and "chemical" wastes. Other, lower hazard industrial waste types disposed of at sea are: pulp and paper mill wastes; oil drilling wastes and waste oil.

Conventional explosive ordnance and chemical warfare material which has received deep water disposal includes nonserviceable or obsolete shells, mines, solid rocket fuels, propellants, small arms ammunition, rockets, pyrotechnics, and mines and rockets containing HS, GB and VX lethal chemical warfare agents. The miscellaneous waste category covers, for the most part, materials disposed of in small lots, without sanction by any regulatory authority. Hazardous wastes covered under this catch-all heading include pesticides and complex chemical solutions. There were 4.7 million tons of all types (hazardous and nonhazardous, bulk and containerized) of industrial waste dumped at sea in 1968 (Table 30). Conventional and chemical warfare munitions subjected to ocean disposal in 1968 totaled 15,200 tons.

TABLE 30: SUMMARY OF QUANTITIES OF HAZARDOUS WASTES DISPOSED OF IN UNITED STATES COASTAL WATERS

Waste Type	Pacific Coast Annual Tonnage	Atlantic Coast Annual Tonnage	Gulf Coast Annual Tonnage	Total Annual Tonnage
Industrial Wastes	981,300	3,013,200	696,000	4,690,500
Bulk	981,000	3,011,000	690,000	4,682,000
Containerized	300	2,200	6,000	8,500
Munitions	--	15,200	--	15,200
Radioactive Wastes	--	--	--	--
Miscellaneous*	200	--	--	200
Total, All "Hazardous" Wastes[†]	981,500	3,028,400	696,000	4,705,900

*Rough Estimate

[†]Includes all categories of industrial wastes dumped at sea

Source: D.D. Smith and R.P. Brown, *Ocean Disposal of Barge-Delivered Liquid and Solid Wastes from U.S. Coastal Cities*

No containerized radioactive materials were ocean dumped in 1968 (average for 1969-1970 was 4.2 tons per year). Miscellaneous wastes amounted in 1968 to an estimated 200 tons. The hazardous industrial wastes which constitute by far the largest class of hazardous waste materials dumped at sea are waste products of pigment processing, steel production, petroleum refining, petrochemicals manufacture, insecticide-herbicide-fungicide manufacture, chemical manufacture, and metal finishing-cleaning-plating processes, among many others.

It is believed that the recommendations of the President's Council on Environmental Quality should be reconsidered in the following specific areas after the prerequisite research has been performed: [1] termination of the ocean dumping of explosive munitions; and [2] phase-out of ocean dumping of spent sulfuric and/or hydrochloric acid wastes from steel pickling, and from titanium pigment manufacture.

On the basis of data presented by Reed (8), the dumping in the ocean of conventional explosive munitions at selected, predesignated charted sites where depths are in excess of 6,000 feet presents no apparent hazard if the munitions are not detonated, and a minimal impact where the munitions detonate at the lower depths. Similarly, the evidence presented by Smith and Brown (1), if verified by additional laboratory and field test findings in the Atlantic, indicates a minimal acceptable impact on the ecosphere if proper current practice is followed for acid unloading at prescribed depths and rates while underway at usual tow speeds in designated deep sea disposal areas.

Further research with these specific wastes and with other selected materials is necessary to determine the necessary additional information on the effects of the selected materials on the ocean environment. Additionally, the effects of the ocean environment on the wastes to be dumped must also be determined to ensure that toxic materials are not formed as the result of reaction and interaction.

Finally, research is necessary to develop waste forms stabilized to ensure compatibility with the ocean environment on both short and long term bases. A review of ocean waste disposal ecology, economics, practice and monitoring has been presented by A.W. Reed (10).

REFERENCES

(1)　D.D. Smith and R.P. Brown, *Ocean Disposal of Barge-Delivered Liquid and Solid Wastes from U.S. Coastal Cities.* (Dillingham Corporation, La Jolla, California). Contract No. PH 86-68-203, U.S. Environmental Protection Agency, Solid Waste Management Office (1971).

(2)　*Ocean Disposal of Unserviceable Chemical Munitions.* Hearings, Subcommittee on Oceanography, Committee on Merchant Marine and Fisheries, House of Representatives, 91st Congress, Serial No. 91-31 (August 1970).

(3)　*Ocean Dumping; A National Policy.* Council on Environmental Quality, Report to the President (October 1970).

(4)　D.W. Hood, B. Stevenson and L.M. Jeffrey, "Deep Sea Disposal of Industrial Wastes," *Industrial and Engineering Chemistry,* 50(6):885-888 (June 1958).

(5)　National Research Council Committee on Oceanography. *Radioactive Waste Disposal into Atlantic and Gulf Coastal Waters; A Report from a Working Group of the Committee on Oceanography of the National Academy of Sciences-National Research Council.* National Research Council Publication No. 655, Washington, D.C., National Academy of Sciences (1959).

(6)　National Academy of Sciences Committee on Effects of Atomic Radiation on Oceanography and Fisheries. *Considerations of the Disposal of Radioactive Wastes from Nuclear-Powered Ships into the Marine Environment.* National Research Council Publication 568, Washington, D.C., National Academy of Sciences-National Research Council (1959).

(7)　National Research Council Committee on Oceanography. *Disposal of Low-Level Radioactive Waste into Pacific Coastal Waters.* National Research Council Publication 985, Washington, D.C., National Academy of Sciences-National Research Council (1962).

(8)　W.F. Reed, Jr., Captain, U.S.N. *Assessment of the Environmental Effects of Past Deep Water Dumping Operations.* Thirteenth Annual Explosives Safety Seminar, Minutes, Armed Forces Explosives Safety Board, San Diego, California (September 1971).

(9)　P.J. Cheremisinoff and R.A. Young, *Pollution Engineering Practice Handbook,* Ann Arbor, Michigan, Ann Arbor Science Publishers (1975).

(10)　A.W. Reed, *Ocean Waste Disposal Practices,* Park Ridge, New Jersey, Noyes Data Corporation (1975).

AUTOMOTIVE INDUSTRY WASTES

DISCARDED TIRES

See Rubber Industry Wastes, page 475.

STAMPING PLANT WASTES

Incineration

A process developed by J.W. Brophy (1) utilizes a unique incinerator design in which the atomization of the fluid material is greatly accelerated and instant combustion of its combustibles promoted by (a) preheating the fluid material, at superatmospheric pressure, to at least the vaporizing temperature of a vaporizable component of the fluid material and then (b) ejecting the so-heated stream through an orifice into the elongated space so as to effect atomization of the fluid material including the prompt vaporization of the vaporizable content of the fluid material. Figure 60 shows such an apparatus.

The retort **10** rests on a base plate **11**, which latter may and preferably is supported above the ground level on suitable legs (not shown). The retort walls are composed of an outer steel shell **12**, an intermediate annulus of heat-insulating material **13**, and an inner lining of high-temperature refractory material **14**, which lining is not tied to the steel shell, but rather is self-supporting. Preferably, it is monolithic in structure. At the base of the retort there is positioned a ceramic tile **16** formed of a ceramic refractory material capable of withstanding high temperature gases. The tile is in the form of an annular member the inner wall of which is divergent in the direction of flow of the gases passed through the tile. That is to say, the space bounded by the curved inner wall of the member is in the form of an inverted frustrum of a cone.

The ceramic tile **16** is positioned at the lower end of, and in coaxial relationship with respect to, the retort. An annular opening in base plate **16** surrounds the base of ceramic tile **16** and provides an annular space between the outer periphery of the tile and the opposed inner surface of lining **11** of the retort. Beneath the ceramic tile **16** there is disposed an inspirator **20**. This latter consists of generally vertically disposed cylindrical part **21**, having at its upper end an inwardly flared end portion **22**. The cylindrical part **21** is closed at its base by closure member **23**. An elongated ejection pipe **25** extends vertically from a point beneath the base **23** through the latter and coaxially upwardly through the cylindrical part **21**, terminating in an outwardly flared mouth portion **26**. The outwardly flared part **26** and the inwardly flared end portion **22** of the cylindrical part **21** are arranged and positioned so as to form between them an annular slit or nozzle **24** for the discharge of an annular stream of gas (air) under pressure.

FIGURE 60: SECTION OF INCINERATOR FOR AUTOMOTIVE STAMPING PLANT WASTE DISPOSAL

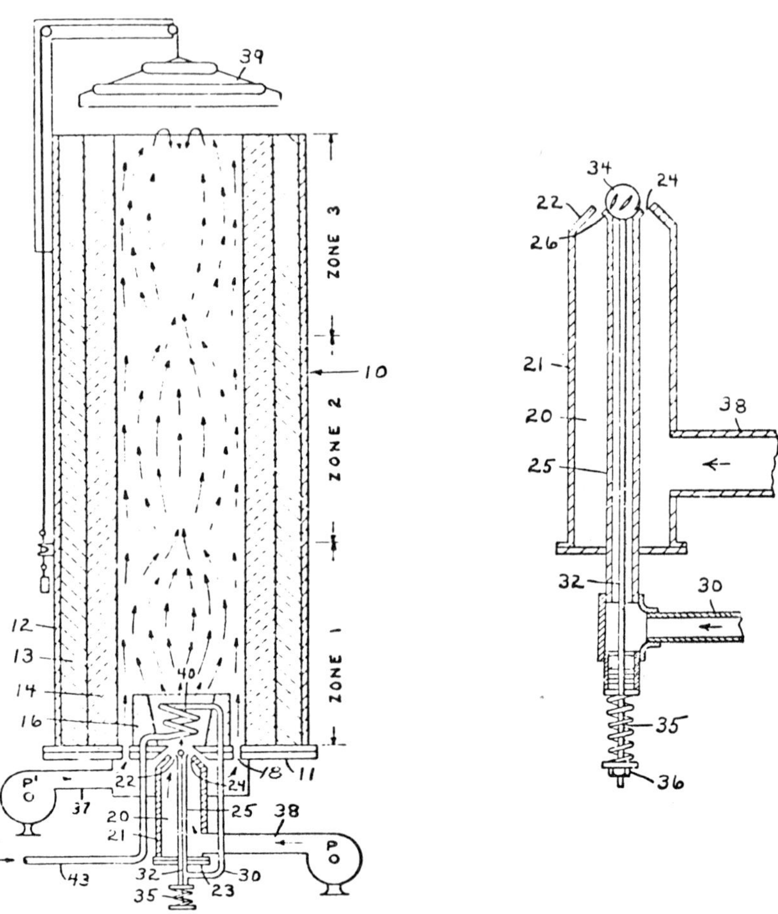

Source: U.S. Patent 3,357,375

Adjacent the lower end of the ejection pipe is a feed pipe **30** for the delivery, to the ejection pipe, of fluid (or fluidized) waste material to be incinerated. Extending through the ejection pipe **25** and to a point beneath the lower end of the latter is a wire or rod **32** at the upper end of which there is rotatably mounted a ball-like rotatable member **34**, which latter may and preferably is provided with spaced vanes which are angularly disposed so as to impart a spinning action to the ball-like member when struck by fluid waste material under pressure. At its lower end, the wire or rod **32** is associated with a biasing means, e.g., a compression spring **35**, tending to bias the ball-like member **34** toward the mouth **26** of the ejection pipe **25**. The biasing force of the spring can, as will be well understood, be adjustable by means of the nut **36** threaded to the base of the rod **32**. This rotatable ball-like part of the inspirator functions to break up and disperse the fluid forced upwardly through the mouth of the ejection pipe, and it also acts as a relief valve rod ridding the inspirator of gross pieces of solid material which may inadvertently be forwarded through feed pipe **30**. Adjacent the lower end of the cylindrical part **21** there is provided a conduit **38** for delivery of primary atomizing air, under pressure, from any suitable source, e.g., pump **P**, to the inspirator for eventual discharge through the annular slit **24**. Secondary air is supplied to the retort by a suitable source, e.g., pump **P'**, of air under pressure and its discharge conduit **37** leading to a chamber **27** surrounding the upper part of inspirator **20** and immediately beneath base plate **16**.

Annular opening **18** communicates between chamber **37** and the interior of the retort **10** through the aforementioned annular space between tile **16** and the retort lining **14**, for providing secondary air for use in the pyro-decomposition operation within retort **10**. Capping the incineration retort is a balanced, weighted cover member **39** which is so secured, with respect to the retort, that it serves to retard the ready egress of combustion gases thereby imposing upon the gaseous contents of the retort a desired superatmospheric pressure and providing a desired retention time of the gases in the incinerator for complete pyro-decomposition. As was noted hereinabove, fluid waste material is introduced into the inspirator through a feed pipe **30**. This feed pipe preferably is an extension of a helical coil **40** which lines the conical inner wall of the ceramic tile **16**. This coil is connected to the pressure side of an inspirator pump by means of a pipe **43**.

Its operation will now be described with reference to the disposal of the fluid waste (other than that from the sanitary system) accumulating in a stamping plant of an automobile manufacturing company. The materials to be disposed of consisted of watery wastes having an oily content which came from a variety of plant sources. Although total volume was fairly constant from day to day (about 12,000 gallons in a 24-hour day), the character and content of the waste varied widely. Although nominally the accumulated fluid material was made up of about 90% water and 10% oil, it sometimes varied from a 97 to 3 ratio one day to as much as 75 to 25 later in the same week. Components of the total accumulation were derived as follows.

In the blanking area, sheet metal was automatically washed and cleaned as it was uncoiled. The waste from this operation consisted essentially of water and cleaning fluid mixed with oil from the steel stock. Once a week the cleaning fluid supply was drained and replenished with fresh liquid, and at that time a higher-than-normal cleaning-fluid-to-water ratio obtained. Other waste was generated when the basement was cleaned (once or twice a week). This created a special problem because a detergent was used to assist in the cleaning. All manner of waste (dirt, grease, oil and the detergent itself) was swept and flushed into the main conveyor drain. Another source of waste were the steam cleaning booths which produced a high volume of water-base mixture of a variety of waste materials.

In the main assembly area of the plant, the welding equipment was water-cooled. This process water went directly to drain after passing through the welders. Also, there were machine leaks and occasionally machine breakdowns that injected small to large quantities of fire-resistant hydraulic fluid into the store of liquid waste. Sometimes the hydraulic fluid became contaminated and had to be pumped out and replaced. All of the above wastes were carried ultimately into the main plant sumps, from which they were pumped to the waste disposal system. In sum, the total wastes were watery mixtures, containing varying proportions of machine oil, coolants, hydraulic fluid, cleaning fluids, detergent and grease, plus an assortment of solids—string, bits of paper, cigarettes, metal chips and the like.

Formerly, the worst of the above wastes were accumulated in tank trucks and hauled to locations (in the country) of uncultivatable land where they were dumped. This mode of disposal had three serious drawbacks: [1] It was costly. [2] It did not dispose of waste from basement washdown or waste from steam-cleaning booths. [3] It did not solve the problem from the plant itself to the dump area. Because this mode of disposal was undesirable, an incinerator plant, and the disposal process were adopted. The incinerator was sized to dispose of 500 gal/hr of the mixed fluid waste, i.e., 12,000 gallons per day. The equipment included two surge tanks each of 15,000 gallons capacity, equipped with agitator means. Two such tanks were provided in order to operate the process on a batch basis. By filling a large holding tank and thoroughly agitating, several important purposes were served: [1] Thereby there was provided a 15,000-gal batch of a homogeneous mixture; [2] the correct amounts of air and auxiliary fuel for most efficient burning of a given batch could be quickly and easily determined and set; and [3] from the time the adjustment was made until the one tank was emptied, about 30 hours later, the system required no more attention than routine scheduled monitoring by power house personnel.

When one tank had been emptied and the valving had been switched over to admit waste from the other tank, the power house engineer needed only to observe burner performance for a short period of time as indicated on the control panel. A rise in temperature, for example, indicated that a higher oil-to-water ratio existed in the new 15,000 gallon batch of waste liquid. This called for a decrease in auxiliary fuel, and the resetting of fan and blower dampers to produce continued optimum incineration efficiency. Conversely, a temperature drop indicated a lower oil content and called for a proportionate increase in the rate of feed of auxiliary fuel.

In starting up the incinerator, the latter was progressively heated to an internal temperature of approximately 2200°F. Thereupon, liquid waste admixed with auxiliary fuel was pumped from the surge tank to and through the heating coil and thence into the inspirator whence it was atomized with primary air into the lower part of the vertical retort. The amount of auxiliary fuel (fuel oil) admixed with the fluid waste was so adjusted as to bring the operating temperature within the upper part of the retort to from about 2800° to 3000°F, the ratio of fuel oil to watery waste being about 15 gallons of fuel oil to each 100 gallons of the waste. Thereafter the operation was continued observing this same ratio of added fuel to waste, thereby maintaining the aforementioned maximum temperature in the upper part of the retort.

The rate of introduction of primary air was determined to be approximately 584 cfm and that of secondary air was approximately 1749 cfm. The gas pressure maintained in the retort was calculated to be about 1½ to 3 times atmospheric pressure. The Btu content of the waste was calculated to be approximately 7,000 Btu's per gallon, and the fuel oil used was rated to have a 130,000 Btu's/gal thermal value. High efficiency was achieved through precise control of waste atomization, combustion air, fuel-air mixture, ignition and temperature. Controlled incineration reduced all of the fluid waste material to simple elements and harmless compounds of microscopic particle size. All material was thermally decomposed without smoke, odor or visible ash, thus eliminating any and all possibility of pollution.

USED ENGINE OIL

The disposal of used engine oil by various techniques including incineration as well as disposal by land spreading and microbial assimilation has been reviewed by Kimball (7).

Incineration

A device developed by P. Widdig (2) is one for burning materials, especially waste oils, oil sludges and chemical waste products ordinarily difficult to burn. The device employs two combustion chambers connected with one another each having its coordinated supply of pressurized air so that the second combustion chamber has a greater volume than the first combustion chamber.

An incinerator developed by H. Appelhans and W. Schumann (3) is a muffle furnace designed to handle a wide range of waste substances of liquid or sludgelike character which are to be burned, such as emulsions, solvents, lubricating oils, cutting oils, electroplating muds and the like. The apparatus features means for feeding the material to be incinerated to the muffle, the feed means comprising a plurality of material feed tubes passing through the top side of the muffle and disposed to drop the material in free fall onto the hearth of the muffle, first air feed means mounted in the front end of the muffle, and at least one second air feed means arranged to feed air into the muffle from the top and both sides.

A device developed by S. Nakano (4) is an incinerator adapted to consume organic waste products of relatively high viscosity, such as used oil, tar, and pitch, hereinafter merely called waste oil, which would otherwise be difficult to completely burn because of its water and mineral content. The burning is accomplished by the mixing of air with the waste oil to be incinerated while automatically controlling the air supply in response to the burning condition.

The waste oil is atomized by the jet of primary air, filling a combustion chamber with oil particles which are burned in a state of being well mixed with secondary and tertiary air. The air supply is automatically controlled and the remaining waste oil on the bottom is stirred at an adequate speed in accordance with the burning condition of the waste oil, thus accomplishing the constant mixture of air with the incinerating substance. This enables the substance to burn in a continuous and complete manner. The dirt or particles contained in the waste oil settle and may be removed, thereby providing no difficulty in the combustion of the waste oil. Figure 61 shows such a device.

FIGURE 61: INCINERATOR FOR WASTE OIL

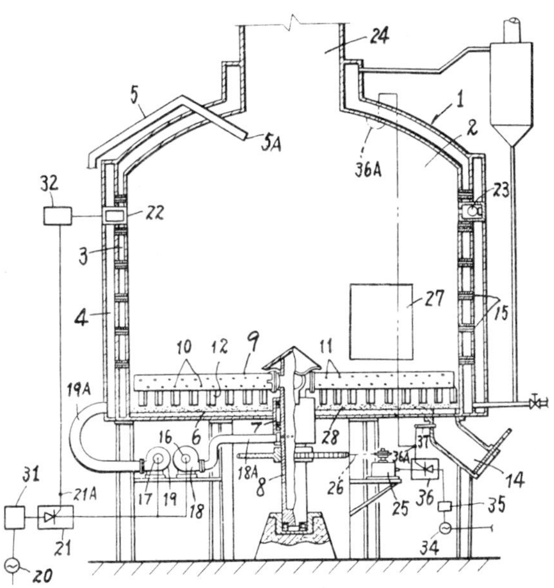

Source: U.S. Patent 3,671,167

The body of the incinerator 1 has a three-fold enclosure, i.e., an innermost combustion chamber 2, a middle water jacket 3 around the combustion chamber 2 to prevent possible overheating and an outermost third-air chamber 4. Waste oil to be incinerated is supplied into the combustion chamber 2 through an outlet 5A of a supply pipe 5 entering from outside the combustion chamber 2. Ash residue is discharged through an outlet 14 provided at the bottom of the combustion chamber, as near to its corner as possible. A bottom floor **6** of the combustion chamber is provided with a bearing **7** at its center, which rotatably supports a rotary cylinder **8** having its upper end extending into the combustion chamber **2**. The upper end of the rotary cylinder **8** is provided with a number of radially extending cylindrical arms **9** positioned parallel to the bottom floor **6** of the combustion chamber **2**. The rotary cylinder **8** and the cylindrical arms **9** communicate internally with one another, providing passages of primary and secondary air. Two types of air are blown into the passages by means of a pair of blowers **18** and **19**.

In the example illustrated, the blower **18** is connected to the rotary cylinder **8**. The cylindrical arms **9** have vanes **12** secured to their underfaces. The vanes are spaced from each other and are positioned adjacent to the bottom floor **6**. In addition, the arms **9** are provided with primary air outlets **11**. Preferably, the former are directed diagonally downward, and the latter are directed diagonally upward. The side wall of the combustion chamber **2** is provided with third air outlets **15** communicating with the third air chamber **4**, to which the tertiary air is transferred by means of a blower **18** or **19**. In the example illustrated, it is blown by a blower **19**. The blowers **18** and **19** are driven DC motors **16** and **17**, respectively, all of which are mounted at convenient locations under the combustion chamber **2**. Furthermore, in the example illustrated, the blower **18** is connected to the rotary cylinder **8** through a duct **18A**, whilst the blower **19** is connected to the third air chamber **4** through a duct **19A**.

Thus, air is blown under pressure into the combustion chamber **2** through its respective route as primary, secondary and tertiary air. The DC motors **16** and **17** are connected to an AC source **20** through a silicon controlled rectifier **21** and a full-wave rectifier **31**. The gate **21A** of the silicon controlled rectifier **21** is electrically connected to a photodiode **22** through a phase-reversing circuit **32**. The photodiode **22** is mounted in the side wall of the combustion chamber **2**. Mounted opposite diode **22** is a light projector **23**, diode **22** thereby electrically detecting the burning condition of the waste oil inside the combustion chamber **2** through changes in the strength of the light beam received from the projector **23**.

The rotary cylinder **8** is rotated by a motor **25** by means of a chain-and-sprocket unit **26**. The motor **25** is a DC speed-reduction motor, and is connected to an AC source **34** through another all-wave rectifying circuit **35** and a silicon controlled rectifier **36**, the gate **36A** of which is connected to a thermoelectric thermometer **37** provided at the ash discharging outlet **14**. Hence, when the discharged ashes are still red-hot, the condition is sensed by the thermoelectric thermometer, and the motor **25** is caused to slow down, reducing the rotating speed of the rotary cylinder **8**.

In this way, adequate stirring of the waste oil is ensured so as to lead to its complete combustion. In the reverse case, the motor **25** is caused to speed up, avoiding unnecessary build-up of the waste oil on the bottom floor **6**. In place of the thermoelectric thermometer **37**, a resistance thermometer or a bimetal can be of course used. Alternatively, instead of detecting the temperature of ash residue, it is also possible to electrically examine the color of the flames brought about through the burning of the waste oil. This may be accomplished by placing a photoconductive cell or a photoelectric tube thermometer in the top wall of the combustion chamber **2** as shown by the imaginary line, connected to the gate **36A** of the silicon rectifier **36**. Smoke, if any, is discharged from the combustion chamber **2** through a chimney **24**. The burning condition of the waste oil is visually viewed through a window **27**. The bottom floor of the combustion chamber **2** is covered with sand or ash **28**.

The operation of this device will be described as follows. First, the motors **25**, **16** and **17** are driven to actuate the rotary cylinder **8**, and the blowers **18** and **19**, respectively. Subsequently, a combustible agent, such as light oil, is initially supplied through the outlet **5A** of the supplying pipe **5** only for igniting purposes. After it has been ignited, a valve (not shown) is switched to stop its supply, and to start the supply of waste oil to be incinerated. The waste oil falls, drop by drop, around the center of the bottom floor **6**, where the waste oil is stirred by the vanes **12** of the rotating cylindrical arms **9**, and is atomized by the jet of primary air. In the course of rising in the combustion chamber **2**, the atomized waste oil is burned. On the other hand, the waste oil remaining on the bottom floor **6** is gradually conveyed to the corner of the bottom floor **6**, creating eddies thereon. In the course of its conveyance it is substantially burned out, leaving solid carbon residue on the bottom floor **6**, which is broken down into pieces by the vanes **12**, and which is taken away from the oil surface. Finally, the residue is discharged out of the bottom floor **6** through the outlet **14**. In this case, the temperature of the ash is examined by the thermoelectric thermometer **37**, which is transmitted to the motor **25**, thereby controlling its rotating speed to ensure the adequate stirring of the waste oil by the vanes **12**.

In consequence of the combustion of the waste oil, gas is created in the combustion chamber **2**, which gas is completely consumed as it is burned in its admixture with secondary and tertiary air, without the possibility of air-pollution when it is discharged out of the chimney **24**. When incomplete combustion takes place due to the shortage of oxygen, with smoke filling the combustion chamber **2**, the beam of light from the projector **23** is only weakly received by the photodiode **22**, which transmits a signal to the silicon controlled rectifier **21** through the phase-reversing circuit **32**, thereby speeding up the motors **16** and **17** to increase the air supply into the combustion chamber **2**. When smoke disappears or is weakened, in which case the beam of light from the projector **32** is thus strongly received by the photodiode **22**, the motors **16** and **17** return to their respective normal speed. In this way the adequate amount of air is automatically adjusted in accordance with the burning condition of the waste oil in the combustion chamber **2**, ensuring the complete combustion of the waste oil.

In place of the photodiode **22** mentioned above, a phototransistor or a phtocell can be employed. Under the system thus constructed, in which waste oil successively supplied into the combustion chamber is stirred by the vanes of the rotating cylindrical arms on the bottom floor of the combustion chamber, during which primary air is jetted to atomize it, waste oil otherwise hard to burn is easily burned in its well admixture with air, under automatic control of air supply into the combustion chamber. The stirring of the waste oil, in response to the burning condition thereof, thus provides labor-saving equipment without any problem of air pollution and public hazard by harmful gases.

An incinerator design developed by R.A. Dingwell (5) is designed to handle waste oils, oily wastes, oil-carrying and oil-saturated materials, and various other liquids, solids, and semisolids that must or can be incinerated and materials that even if they cannot be completely burned can be broken down by heat into forms more suitable for disposal. One example of such disposal problems is that represented by used crankcase oil of which substantial volumes regularly accumulate at garages and service stations and for which there is presently no market except in those few areas where rerefineries exist. Collection services are usually available but the disposition of such oil presents problems since, if conventionally burned, an oily smoke results.

The device consists of a primary incinerator in communication with a waste collector having a stack. The primary incinerator provides in its combustion chamber a spinning gas stream and the plant includes means to deliver into the combustion chamber materials to be incinerated and a specific objective is to provide that such materials be introduced into the combustion chamber as a stream to cause or assist the spinning of the hot gas stream therein and supplementing the air supply.

Instead of using special incinerators, another option is to pretreat used oils by

Setting	Solvent extraction
Centrifugation	Vacuum distillation
Acid/clay treatment	

so that they can be handled in conventional boilers.

S. Chansky et al (6) evaluate the technical, economic and environmental feasibility of waste oil reuse as a fuel. The supply and potential marketability of waste oil fuel is considered in relationship to existing and projected fossil fuel usage in the United States. Although the total automotive waste oil generated annually represents less than 0.5% of the total U.S. fossil fuel production, it is concluded that waste oil can serve as an economically advantageous supplement to present domestic fuel supplies. Moreover, its use will alleviate a serious waste oil disposal problem.

The physical and chemical properties of waste oil are presented and serve as the basis for subsequent assessment of waste oil usage options. Options considered are the use of untreated waste oil as a blended fuel oil or as a supplement to coal combustion and the use of waste oil following treatment to alleviate technical and environmental impacts. Although the use of untreated waste oil blends appears feasible for large utility and industrial boilers, some treatment will be required for smaller boilers. Various treatment methods are discussed and their cost and effectiveness assessed. The reduction of environmental impacts by the use of particulate emission control systems also is considered in relationship to the cost and effectiveness of control equipment, and utility and industrial utilization of fuel and control equipment.

REFERENCES

(1)   J.W. Brophy; U.S. Patent 3,357,375; December 12, 1967; assigned to Prenco Mfg. Co.
(2)   P. Widding; U.S. Patent 3,552,331; January 5, 1971.
(3)   A. Appelhans and W. Schumann; U.S. Patent 3,559,595; February 2, 1971; assigned to Polyma Maschinebau, Dr. Appelhans GmbH.
(4)   S. Nakano; U.S. Patent 3,671,167; June 20, 1972; assigned to Iwatani & Co., Ltd.
(5)   R.A. Dingwell; U.S. Patent 3,838,651; October 1, 1974.
(6)   S. Chansky, J. Carroll, B. Kincannon, J. Sahagian and N. Surprenant, *Waste Automotive Lubricating Oil Reuse as a Fuel,* Report EPA-600/5-74-032, Washington, D.C., U.S. Environmental Protection Agency (September 1974).
(7)   V.S. Kimball, *Waste Oil Recovery and Disposal,* Park Ridge, N.J., Noyes Data Corp. (1975).

ELECTRICAL INDUSTRY WASTES

INSULATED WIRE SCRAP

Incineration

A process developed by W.D. Jones (1) is a system for recovering copper from insulated copper scrap containing electrical insulation of the type comprising plastic, rubber and the like which, upon being thermally decomposed and partially combusted, forms a waste gas which may pollute the atmosphere. The scrap is thermally treated in a cupola to form blister copper and the thermally decomposed insulation is separated as a waste gas and passed through an after burner where it is substantially completely combusted and the effluent gas therefrom then scrubbed and passed through an electrostatic precipitator following which it is vented to the atmosphere as relatively clean gas.

The scheme of operations in this process is shown in Figure 62 which depicts a pile of insulated scrap wire **10** which is sheared (**11**) and the sheared scrap fed to a vertically standing cupola **12** as shown. In starting the cupola, a bed of carbon is prepared at the bottom thereof and the carbon ignited and the combustion thereof maintained by injecting air **13** via tuyeres disposed around the cupola. The alternate layers of scrap and carbon are fed into the cupola to the top thereof.

As the temperature increases due to the combustion of the carbon, the insulation, which may comprise polyethylene, polyvinyl chloride, polytetrafluoroethylene, polypropylene, rubber, or the like, is destructively distilled or decomposed and the copper near the bottom where the temperature is the highest is melted and allowed to collect as a molten bath.

The partially combusted organic vapors are drawn off intermediate the top and bottom of the cupola at a temperature of about 400° to 1200°F via conduit or pipe **14** and conducted to the afterburner **15** where the organic vapors are substantially completely combusted at a temperature ranging from 2000° to 2300° or 2400°F., the resulting effluent gas being then passed through settling box **16** and up through scrubber **17** where the effluent gas is washed with an alkaline solution, e.g., an ammoniacal solution, to remove any acid in the gas, for example, HCl formed from the destructive distillation of polyvinyl chloride. The alkaline wash solution may be lime water, dilute sodium hydroxide or carbonate, or the like.

The effluent gas is removed from the scrubber through the top thereof via pipe **18**, passed through at least one packed tower **19** to complete the gas scrubbing and neutralize the saturated effluent gas and then fed via pipe **20** to electrostatic precipitator **21** where colloidal solids are removed and the relatively clean gas then vented to the atmosphere through the stack. The molten copper and slag are tapped at intervals from the bottom of the cupola into ladles **22**, the slag being poured off as shown at **23**, the blister copper being cast into cakes such as shown at **24**.

FIGURE 62: PROCESS SEQUENCE FOR INSULATION DISPOSAL BY INCINERATION
AND COPPER RECOVERY FROM INSULATED WIRE SCRAP

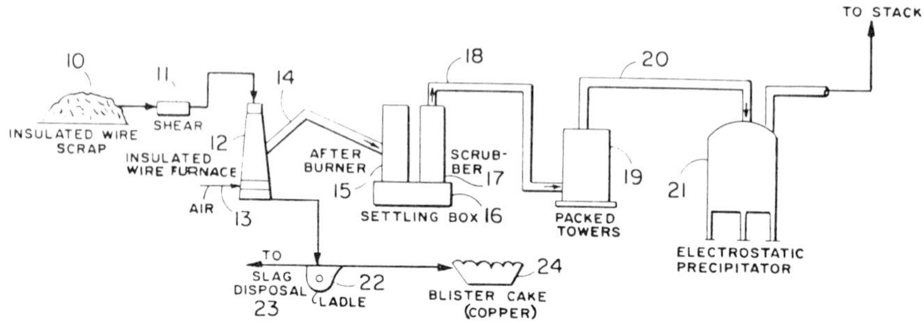

Source: U.S. Patent 3,719,471

A process developed by J.E. Perry (2) relates to the recovery of metal values in scrap materials in which the metal values are associated with nonmetallic materials, such as insulation on scrap wire.

Briefly, this method consists of heating the contaminated scrap, in the absence of oxygen, to a temperature at which all of the volatile components are distilled off, using a continuous feed and removal of residue. At the temperatures involved, some volatile components are formed by destructive distillation. The volatile components, liquid and gas fractions, may be collected for reuse or they may be ignited and burned with a smokeless, odorless flame.

A process developed by H.L. Wentworth and K.H. Seelandt (3) provides a continuous fluid bed furnace system which can be used to remove insulation from wire or other material.

A process developed by Z.J. Przewalski (4) utilizes an incinerating system for recovering copper wire following the burning of the insulation therefor as combustible waste while simultaneously insuring against wire overheating. The process includes the steps of: first, burning the insulated wire in an ignition chamber; second, preheating the secondary chamber; and third, exploiting the fuel generating in the ignition chamber for maintaining the temperature of the secondary chamber while passing the smoke generated in the ignition chamber through the secondary chamber preliminary to passage to the atmosphere.

PRINTED CIRCUIT BOARD SCRAP

Incineration

An apparatus developed by N.D. Hazzard and W.M. Anderson (5) provides a conversion system for waste having organic and inorganic components such as printed circuit boards wherein the organic components are pyrolyzed in an essentially oxygen-free atmosphere so as to preclude the oxidation of the inorganic constituents. An afterburner provides the heat necessary for burning pyrolyzed gases and the carbonaceous particle matter produced in the pyrolyzing chamber. The pyrolyzing chamber is maintained at a suitable temperature for properly pyrolyzing waste material by tempering hot gases exhausting from the afterburner with cool ambient atmosphere, and then directing the mixed gases back over the pyrolyzing chamber.

An apparatus developed by R.F. Stockman (6) can be used for the recovery of metallic components from metal bearing scrap such as foil-backed paper, printed circuit boards and plastic coated wire.

Incinerators for the combustion of organic components in scrap material containing a metallic base have long been utilized as a principle means for reducing the metal to a basic form whereby it may be used again. The salvaging of metallic components from a scrap presents problems not common for the complete incineration of organic waste because the high temperatures necessary for the elimination of the organic components melt, excessively oxidize, vaporize or otherwise damage the metal constituents of the scrap.

This device is one in which waste material having a significant content of metal scrap together with a quantity of organic waste may be incinerated by a controlled process that burns the organic waste without melting the metal, producing excess metallic vapors, or producing undesirable metallic oxides.

Figure 63 shows such an apparatus. In the drawings the incinerator comprises a housing enclosing a chamber **12** having a loading opening **14**, an ash clean door **15** and a gas outlet **16** whereby gas produced therein may be exhausted to an afterburner chamber **18**. The chamber **18** contains an afterburner **22** having a supply for fuel **24** and a supply for air **26** that is provided by a fan **28**. The suction port of fan **28** is connected to passageway **30** containing air that is first circulated through a heat exchanger **32** over duct **34** that contains hot gas exhausting from afterburner chamber **18** whereby air for combustion in afterburner **22** is significantly heated before it is supplied thereto.

Inasmuch as controller **52** responding to thermocouple **48** is adapted to control valve **54**, a supply of air to afterburner **22** sufficient to maintain the temperature in chamber **18** at from 1600° to 1800°F is controlled by a proper setting of the controller. Thus, as the temperature in chamber **18** is maintained near its set-point, combustion of organic matter from the afterburner is so complete that the exhaust gas contains chiefly carbon dioxide, water vapor, metallic oxides, a small quantity of oxygen usually in the amount of 8 to 12%, and minute quantities of carbon monoxide.

Inasmuch as the exhaust gas contains sufficient oxygen to pyrolyze all organic matter in chamber **12**, the exhaust gases from chamber **18** are recirculated to make use of the oxygen therein for the pyrolyzation of all raw waste; no other source of air to chamber **12** is provided.

Thus the exhaust gas from afterburner chamber **18** is directed back through duct **34** to the primary combustion chamber **12**. In order that a sufficient amount of exhaust gas including oxygen therein is returned to chamber **12**, a damper **36** directing the recirculation of exhaust gases is actuated by a controller **38** which is responsive to a thermocouple **42** in chamber **12**. The controller is set to maintain the temperature in chamber **12** at from 700° to 1000°F, a temperature well below the melting point of most metals or the temperature at which the metals freely oxidize. Thus a temperature in chamber **12** below the set-point of 700° to 1000°F would call for an increased rate of combustion.

To provide such an increase, damper **36** would be moved toward an open position to allow an increased flow of exhaust gas containing oxygen into chamber **12**. An increase of temperature in chamber **12** would be similarly sensed by thermocouple **42**. Controller **38** responsive to thermocouple **42** would close damper **36** to accordingly reduce the flow of oxygen, the rate of combustion and the temperature in chamber **12** to meet the set-point temperature. Exhaust gas containing oxygen exhausted from the afterburner chamber **18**, not needed for combustion in chamber **12**, would be vented to the atmosphere through a duct **45**.

A thermocouple **48** in afterburner chamber **18** acting through a controller **52** maintains the flow of oxygen to the afterburner **22** at sufficient levels to insure the burning of all combustibles in the exhaust gas being exhausted from chamber **12**. A valve **54** in air supply passageway **30** is therefore moved toward an open or closed position in accordance with the temperature in chamber **18** sensed by thermocouple **48** and controller **52**, usually set at from 1600° to 1800°F.

A thermocouple **50** in afterburner chamber **18** acting through a controller **55** maintains the flow of fuel traversing valve **56** in line **24** to the afterburner **22** in such a manner that the valve **56** gradually opens or closes until its set-point is reached. Thus at a predetermined setting of controller **55**, as for example 1600°F, the valve **56** closes completely so that no supplementary fuel is being added thereto and only gas exhausting from chamber **12** is being burned in the afterburner. However, if the temperature in afterburner chamber **18** should drop below the set-point, valve **56** would be opened to permit supplementary fuel to flow to the afterburner **22**.

In operation, chamber **12** is at first loaded with metal-bearing scrap through a charging door **14**. Air valve **54** is partially opened to permit air from inlet **30** to be supplied to afterburner **22** and afterburner chamber **18**. Inasmuch as there is at first no combustion in chamber **12**, and the gas exhausting therefrom is cool, damper valve **36** in duct **34** is wide open as determined by controller **38** and thermocouple **42** to permit recirculated air from chamber **18** to be supplied by fan **25** to chamber **12**. When air sufficient for combustion is available, the charge is lighted manually or by an automatic means (not shown) and combustion in chamber **12** is thus effected, the temperature in chamber **12** rises, and damper **36** accordingly moves toward its closed position.

FIGURE 63: INCINERATOR FOR PRINTED CIRCUIT BOARD SCRAP FEATURING METAL RECOVERY

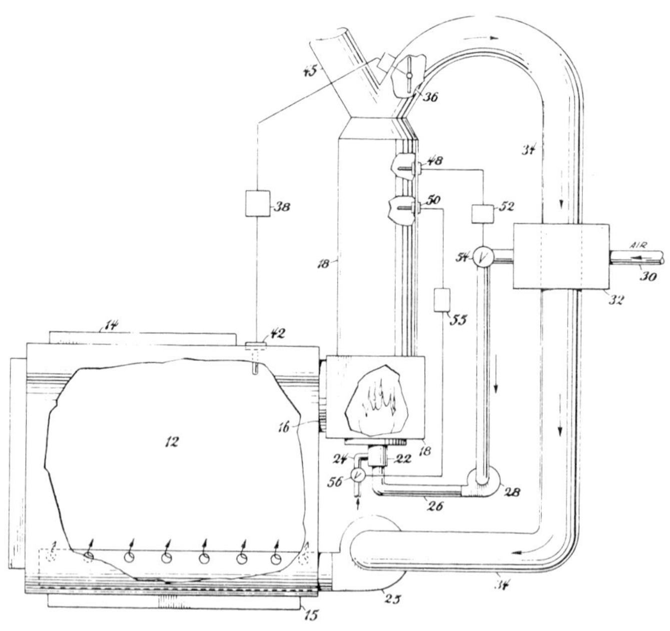

Source: U.S. Patent 3,807,321

As partial combustion or pyrolyzation of the waste material in chamber **12** is effected, exhaust gases given off therefrom pass to chamber **18** whereby they are mixed with oxygen from inlet **26** and burned in afterburner **22**.

When the temperature increases in afterburner chamber **18**, the airflow through valve **54** is also increased by controller **52** in accordance with thermocouple **48** commensurate with the flow capacity of the stack **45** so that excess or cooling air is flowing therethrough. As additional waste material in chamber **12** is subjected to pyrolytic action, more hot gases are given off through exhaust duct **16** to pass to the afterburner chamber **18** where they, along with fuel from source **24** and air being delivered through duct **26**, are subjected to the temperature of combustion so that the temperature continues to climb.

Additional fuel is added to the afterburner **22** until controller **55** reaches a predetermined set-point of approximately 1800°F as determined by thermocouple **50** in chamber **18**. At this point valve **56** moves to a closed position to shut off the quantity of fuel flowing to the afterburner through duct **24**. When the temperature in chamber **18** falls below its set-point, the valve **56** again is moved towards its open position to permit a flow of fuel to the afterburner **22** to compensate for the diminishing flow of pyrolyzed gas from chamber **12** whereby the temperature in the afterburner chamber may be maintained at approximately 1800°F.

Inasmuch as hot exhaust gas in afterburner chamber **18** contains a small amount of oxygen, this exhaust gas alone is recirculated to chamber **12** to provide all the oxygen for pyrolysis of the waste in chamber **12** at a rate determined by thermocouple **42**. Thus at a temperature in chamber **12** below the set-point (approximately 900°F), damper **36** will be moved to open to allow more oxygen to be supplied via duct **34** to chamber **12**. As the temperature therein builds up to 900°F, the damper **36** closes and all excess gas is exhausted to the atmosphere through duct **45**.

Since the metal-bearing waste in chamber **12** is subjected to oxygen in an amount that supports only partial combustion of the waste, the oxidation of the metal is practically eliminated and the temperature is held below that at which the metal vaporizes or melts into a molten mass. However, the temperature within afterburner chamber **18** is at all times maintained sufficiently high to insure the complete incineration of gases given off from chamber **12**.

Various other temperature ranges for the main chamber and the afterburner chamber may be maintained to properly reduce different types of scrap material by adjusting the controllers **38** and **52**.

VARNISH FROM IMPREGNATION AND COATING PROCESSES

Incineration

A device developed by H. Vits (7) was developed for treatment of products containing volatile hydrocarbons from manufacturing operations such as those for producing impregnated hard board, electrical insulation having insulating varnish, aluminum sheets coated with varnish, etc. Such products are dried normally producing, during drying, gaseous hydrocarbons which, for ecological reasons, cannot be discharged directly into the atmosphere but which must be treated. Figure 64 shows such an installation.

A suspension dryer **1** is divided into a first section and a second section disposed in series in the direction of the path **2** for the products to be dried therein, for example, impregnated hard cardboard, electrical insulating material comprising base-material and insulating varnish, varnished aluminum foil (sheets), etc. The dividing of the dryer **1** into a first and a second section is accomplished in such a way so that in the area of the first section, at temperatures of about 130°C, approximately 95% of the light or more volatile hydrocarbons are driven off, while in the second section at temperatures of about 150°C approximately 95%

of the heavy or less volatile hydrocarbons become volatile and are driven off. A mixing of the dryer air from both sections of the dryer (which occurs with regard to the concentration of the hydrocarbons in both parts of the air of the dryer) generally does not occur since a transformation of the dryer air occurs zonally.

On one hand, the dryer must be provided with openings for the path-entry and path-exit for the purpose of the suspended guiding of the products, and on the other hand there should not exit from the opening the dryer-air which is mixed with volatiles. The dryer is maintained under low pressure. Fresh air is continuously supplied to the dryer by means of the suspended jets or nozzles (not shown) which are necessary for the suspended transport of the products. In order to enable the retention of the under-pressure, spent air is continuously removed or drawn from the first and the second section and is directed to an after-burning device **3**.

In addition to that, the heavy hydrocarbons are sprayed as a fuel into the after-burner device **3** by means of an atomizing nozzle **4**. In order to permit after-burning, auxiliary burners **5** are provided through which supplemental fuel which is supplied to the device **3** for after-burning exhausts dryer gases. The portion of the extraneous fuels must be sufficient to ignite and retain the burning of the light and heavy hydrocarbons in the device **3**. At an exotherm of about 400°C of the spent air from the first section, it is necessary to supply enough supplemental fuel in order to heat the exhaust air to an average of about 400°C.

The light hydrocarbons and the heavy hydrocarbons which are contained in the exhaust air of the dryer are then heating the exhaust since they burn at about 800°C. The flue gases which are heated to this temperature level are taken from the device **3** and are directed through an economizer **6** to which they transfer a large part of their heat to a heat-carrier, for example heating oil, which is circulated via a conduit system **7** to the first and the second section of the dryer **1** to thus maintain the dryer-section atmosphere at predetermined drying temperature levels.

FIGURE 64: INCINERATION FOR VARNISH FUMES FROM ELECTRICAL PRODUCTS DRYER

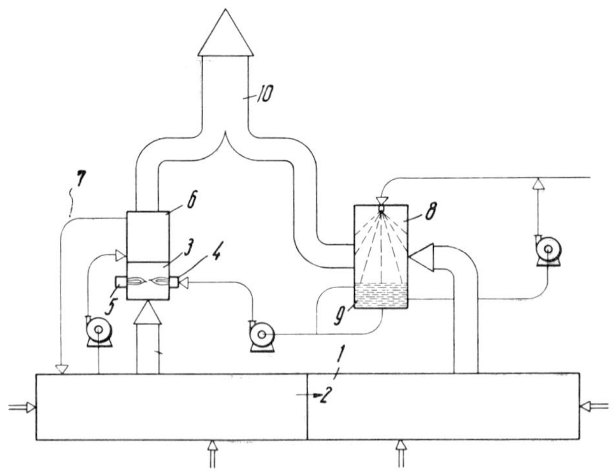

Source: U.S. Patent 3,875,678

The exhaust gases sucked from the second section of the dryer, containing the heavy or less volatile hydrocarbons, are moved through a wash column **8** in which the heavy hydrocarbons are extracted or precipitated by means of water which is sprayed thereinto. From a sump **9** of the wash column **8**, the liquified heavy hydrocarbons, as explained, are drained off in the form of a water slurry and supplied to the after-burner device **3** through nozzle **4** where they are burned. The cleaned humid exhaust gas of the wash column **8** passes into a mutual flue-gas canal **10**, joining the dry flue-gases which exit from the economizer **6**, whereafter they exit as dehumidified dry flue-gas mixture into the open atmosphere.

REFERENCES

(1)   W.D. Jones; U.S. Patent 3,719,471; March 6, 1973; assigned to American Metal Climax, Inc.
(2)   J.E. Perry; U.S. Patent 3,744,779; July 10, 1973; assigned to Horizons Research, Inc.
(3)   H.L. Wentworth and K.N. Seelandt; U.S. Patent 3,841,240; October 15, 1974; assigned to Sola Basic Industries, Inc.
(4)   Z.J. Przewalski; U.S. Patent 3,777,679; December 11, 1973.
(5)   N.D. Hazzard and W.M. Anderson; U.S. Patent 3,780,676; December 25, 1973; assigned to The Air Preheater Co., Inc.
(6)   R.F. Stockman; U.S. Patent 3,807,321; April 30, 1974; assigned to The Air Preheater Co., Inc.
(7)   H. Vits; U.S. Patent 3,875,678; April 8, 1975; assigned to Vits-Maschinenbau GmbH.

FOOD INDUSTRY WASTES

Solids disposal has become a mounting problem for the food industry (1). For those materials that have no economic recovery value, there is a constant cost escalation for disposal. Various methods and associated problems for solids disposal include:

> Incineration: creates air pollution
> Landfill and Land Spraying: limited by available land and high
>   cost of material hauling to suitable locations
> Quick Composting: results in a humus which is a satisfactory
>   soil conditioner; however, suitable markets for the conditioner
>   are not always available
> Ocean Dumping: grinding and dumping from barges beyond a
>   26 mile point from the nearest mainland point are available
>   only for food produced near a coast

Industry sources indicate that the most satisfactory waste disposal process would involve the creation of new by-products from food processing. Until research chemists and engineers are able to create economical by-products, however, it appears that most food processing wastes will have to be treated as garbage and hauled away at considerable expense.

ANIMAL WASTES

Pyrolysis

A process developed by P.R. Kelly (2) is one in which waste materials to be disposed of are fed into a feed hopper, the outlet of which is directed into a rotary vane feeder to automatically control the feed rate and act as an airlock. Shredded waste material is stacked in a pyrolyzing chamber or column through which hot gases, initially containing no oxygen pass upwardly, driving off volatile material.

A condenser reclaims the condensable materials, and the noncondensable gases are recirculated through the combustor and pyrolyzing chamber carbonizing the pyrolyzed waste in the lower reaches of the chamber. After pyrolyzing is completed in the lower portions of the waste material, excess air is admitted into the recirculating hot gas stream. Contact of the oxygen in the excess air with the carbonized material causes substantially immediate combustion. Ash from the combustion zone is then dumped into a discharge chute, or, optionally, removed by traveling scraper bars.

Loading of waste into the described disposal system may be on a batch basis, or, alternatively, by a conveyor as a continuous process, and, in the latter case, a steady state operation is achieved where destructive distillation occurs, in the central regions, while distillation is substantially complete and carbonizing and combustion takes place at the lowermost levels.

Figure 65 shows the application of such an apparatus to the treatment of bovine waste. The excess heat on recovery from the flue gas may be used to drive a turbo alternator, for example, producing electric power, or may also be used for cooking of the cattle feed material, or, if used on a farm, space heating of the barn. The solid residue can be used for land fill, as construction material and has a sufficient salt content such that the recovery thereof may be feasible. As to the liquid condensate, where further refining is required, the recovered heat can be useful in refining the condensate to produce the final by-product. Where refining is not required, the condensate material may be sold to others, as is. Moreover, in the event that the quantities produced are not such as to make utilization of the distilled by-products feasible or desirable, they may be discharged into a sewer, for example.

FIGURE 65: ALTERNATIVES FOR DISPOSAL OF ANIMAL WASTES

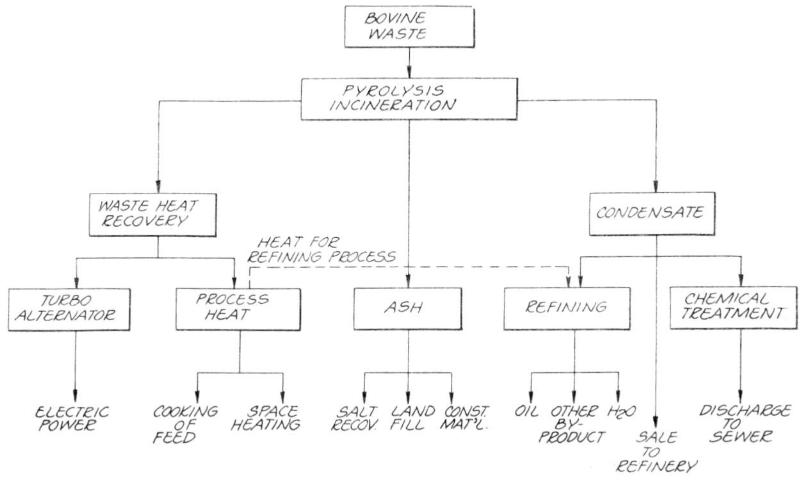

Source: U.S. Patent 3,771,468

A process developed by H.V. Hess et al (3) is one in which animal and dairy wastes are converted to deodorized coke suitable for use as fertilizer or soil conditioner, an aqueous odor-free effluent having low chemical oxygen demand and reduced phosphorus content and an odorless gas suitable for discharge to the atmosphere. The wastes are coked in the liquid phase under pressure to produce wet coke and a malodorous, gas-containing effluent high in chemical oxygen demand. The wet coke is treated with pressurized hot air to remove liquids therefrom and the air stream is mixed with the effluent and the mixture is oxidized, brought into heat exchange relationship with incoming wastes and then discharged.

Figure 66 shows the essential features of this process. The waste stream is pumped up to system pressure in pump 1 through the heat exchanger 2 where it is in heat exchange with the oxidized effluent and through the heater 3 where the cokable materials are coked and from there to a decanter (held at system pressure) 4 where the system separates into a gas phase, a liquid phase (called aqueous coker effluent) and a wet coke phase. The wet coke

FIGURE 66: APPARATUS FOR COKING PUMPABLE ANIMAL AND DAIRY WASTES

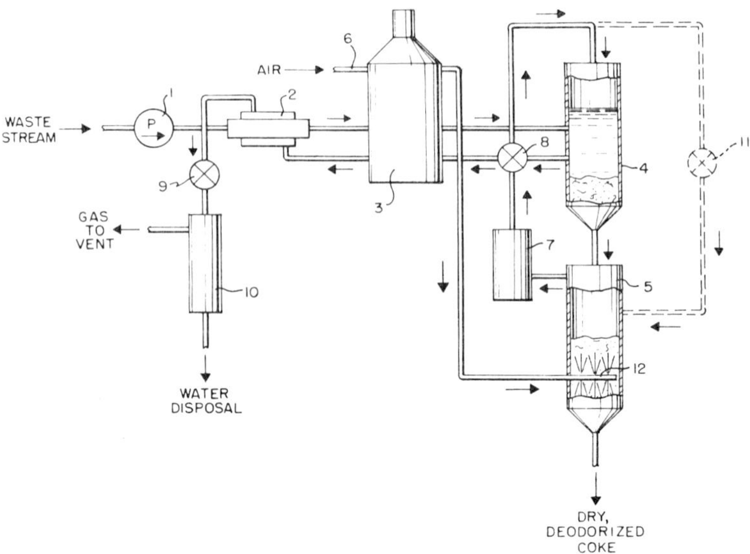

Source: U.S. Patent 3,671,403

phase is drawn down intermittently into vessel **5** which is maintained slightly above atmospheric pressure. Coke is now in the vessel **5** and to dry it and remove malodorous constituents, air is brought in at slightly above atmospheric pressure, preferably 30 to 60 psig, through the top part of heater **3** where it is heated to 250° to 350°F (a temperature sufficient to dry and deodorize the wet coke) and passes through the sparger **12** shown at the bottom of vessel **5**. The temperature range indicated suffices to dry and to deodorize the wet coke but is not sufficiently hot to burn the coke. The flow rate of hot air through the coke is sufficient to expand the coke bed and maintain it in an expanded condition above the sparger. In any case, sufficient air is used so that there will be an excess of oxygen available for burning the COD from the aqueous effluent and gases.

Next, dry, deodorized coke is removed intermittently from the bottom of vessel **5** for use as fertilizer or soil conditioner. The warm, moist and now, smelly, air from vessel **5** is compressed in compressor **7** to system pressure and mixed with aqueous coker effluent from vessel **4** and gas from vessel **4** in mix valve **8** and passes through the lower part of heater **3**. This air-water-gas system is oxidized in heater **3** whereby the chemical oxygen demand is removed from the gas and the aqueous coker effluent while the odor-forming constituents are destroyed. This oxidized stream is passed through heat exchanger **2** in heat exchange with the raw waste feed and then through a pressure reducing valve **9** to a gas-liquid separator **10**. The gases and liquid from separator **10** may be discharged to the atmosphere and ditch respectively.

In another embodiment shown in dotted lines, the gas from vessel **4** can be passed through a pressure reducing valve **11** into the vessel **5** where it is mixed with the gases obtained in drying the coke. The effluent from the coker (produced at conditions of 550°F and 1,150 pounds per square inch gauge has been oxidized with air at 550°F and autogenous pressure to produce an oxidized effluent with a COD of 3,802 mg O_2/ liter and a pH 7.6. In considering the coker effluent alone (which had a COD of 72,076 mg O_2/liter) this represents a COD reduction of 94.7% or basing it back on the COD of the charge it represents a reduction of upwards of 96%. The odor is also substantially removed. This oxidation was carried out in a batch operation. Operation on a continuous basis with excess air leads to essentially complete removal of COD.

The process is also suitable for treatment of other animal wastes such as pig wastes. The effluent from poultry feeding units can likewise be handled. A portion of the water produced from the described operation also can be used to wash down dairy barns, feed lots, and the like and then, in a sense, recycled through the whole system.

FRUIT AND VEGETABLE WASTES

Land Application

This is a common method for the disposal of fruit and vegetable industry wastes as described by H.R. Jones (4). The successful disposal of fruit and vegetable wastes onto land depends on a number of factors. Engineering aspects include infiltrative and percolative capacities of the soil, clogging, quality changes in the soil, and the engineering of soil systems. Other factors are the translocation of the underground water to springs and streams or through faults to water supplies, evaporation, and transpiration through cover crop plant growth.

The effects of disposal of wastes into the soil mantle are as complex as are the wastes. While many operations appear to be successful, the long term effects are less clearly understood. Several considerations are important for successful disposal of food wastes into the soil and they are listed below.

[1] Land Area: sufficient land area must be available to handle the waste during peak operations without overloading. Some provisions should be made for land to accommodate expansion of operations.

[2] Soil: the character of the soil is important to acceptability of land for irrigation.

[3] Slope: some slope is desirable to minimize ponding of water which is followed by destruction of plants, bacterial decomposition, and odor development. Too great a slope may result in excessive runoff.

[4] Rest Interval: the necessary rest interval depends on several factors, including BOD load during spraying, porosity of the soil, and the distribution of the effluent.

[5] Cover Crop: this is essential for the spray irrigation method of soil waste disposal in order to increase absorption and transpiration, and to prevent soil settling and erosion. A dense cover crop protects the soil from physical change, increases transfer of water to soil through the root system, and from the soil to atmosphere by evapo-transpiration.

[6] Waste: wastes vary considerably in characteristics such as pH, BOD, ratios of nutrients, SS (soluble solids) and salt concentrations which all affect the soil mantle. These, ultimately, affect degradation of the wastes.

Spray irrigation consists of spraying screened liquid wasted from vegetable or fruit process operations onto land where it undergoes percolation into the soil and biodegradation. Spray irrigation of sewage wastes was practiced as early as 1860 to 1870, and of cannery waste beginning at least in 1947.

A large variety of cover crops have been successfully used for spray irrigation fields. There is, however, a great difference in the capability of various crops to absorb and transpire water. Cover crops used include planted vegetable crops, grasses, stands of native bush, trees, and orchards. In many instances efforts are made to maintain stands of certain grasses while in other the practice has been to let natural survival determine the stand. A dense growth will transpire 10 to 20 inches of water per season.

Spray irrigation on slopes draining to lagoons appears to be very successful and adaptable to soils which limit lateral movement of underground water; 99% reduction in BOD, total color removal, and 90% reduction in N and P may be expected. Preliminary figures from the *National Canners Association Solid Waste Study* (1970) showed the following percentages of canners and freezers of the listed products disposing of effluents by irrigation in the continental United States.

Product	Percent
Citrus	34
Tomato	13
Corn	44
Potato	21
Peach	11
Apple	30
Snap bean	27
Pea	36
Pear	8
Other fruit	20
Other vegetable	17

It has been estimated there were 2,400 systems for disposing of effluents to the land in 1965 and that 900 of these were used by food processors.

The ditch and furrow system consists of developing shallow ditches in the land through which effluent is directed. At intervals of use the land is permitted to dry, is disked, and refurrowed.

It has been estimated in one operation that 46 inches of effluent were spread over 66 acres in four months operation, refurrowing after 18 inches. About 50% of the effluent was believed to evaporate and the balance to percolate. All accumulated SS were disked into the soil. The estimated soil loading was 0.124 pound of COD per square foot per day, based on 66 acres in use, 100 days operation, and 35,600 pounds daily COD load.

The ridge and furrow system was originally conceived as a system in which the waste overflowed furrows atop ridges on sloped land. Unabsorbed waste was collected in a furrow and directed for further disposal. Ridge and furrow irrigation now consists of confining the wastes to furrows several inches below the ridges. Vegetation may be grown on top of the ridges. A number of installations were made in the 1940 to 1960 period with some success. Generally, this system has been replaced by the spray system.

GREASE VAPORS

Incineration

A process developed by J.L. Newcomer (5) is one in which cooking odors generated by the deep fat frying of potato chips or other food products are eliminated by directing the fumes back to the burner heating the cooking oil for incineration thereof. The flow of the fumes from the cooking process can be controlled to maintain the quantity of fumes no more than the amount of combustion air needed by the burner to insure that all of the fumes are incinerated. Additionally, condensers may be utilized to remove a substantial portion of the moisture carried by the fumes before they are incinerated. Figure 67 shows a suitable form of apparatus for the conduct of this process.

FIGURE 67: INCINERATOR FOR COOKING FUMES

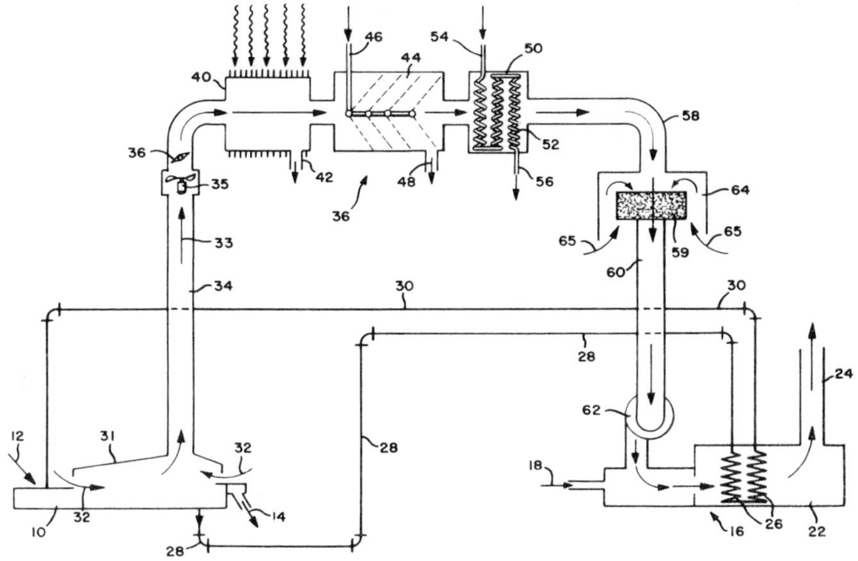

Source: U.S. Patent 3,762,394

As seen in the drawing, the apparatus may include a vat **10** containing a supply of hot
cooking oil and into which the food products, such as raw potato slices, may be deposited
adjacent one end, as indicated by the arrow **12** and removed, after sufficient cooking, at
the opposite end as indicated by the arrow **14**. Generally, the raw potato slices will contain
approximately 85% moisture by weight, and cooking in the hot oil, which is maintained
at approximately 230°F, will reduce the moisture content of the slices to approximately
2% by weight.

In order to maintain the temperature of the cooking oil at the desired level a heat exchanger
16 is provided which may be fired by gas, oil or any suitable fuel. In using a combustible
gas, such as natural gas, for example, such gas may be admitted, as indicated by the arrow
18, to the mixing chamber **20** for mixture with combustion air and then passed into the
combustion chamber **22**, with normal combustion products vented from chamber **22** through
the stack **24**.

The heat exchanger **16** may be of conventional construction, including heat exchanging
coils **26** connected at their intake end to a line **28** and at their outlet end to a line **30**.
As will be apparent from the figure, oil is withdrawn from the cooking vat **10** through the
line **28**, circulated through the coils **26**, where the hot combustion gases pass thereover,
and returned to the cooking vat **10** through the line **30**. In this way and with suitable
control mechanism which may be of conventional construction and is not shown for purposes of simplification, the temperature of the cooking oil in the vat may be maintained
at the desired level. While the furnace is shown, for purposes of illustration, as positioned
remote from the vat, it will be apparent that the two may be positioned adjacent each other
or incorporated in a single unit.

A hood **31** is positioned over the vat **10** and collects the mixture of steam and hot, odor
laden gases which are normally emitted from the surface of the cooking oil during the
cooking process. The hood is open at each end of the vat, usually just enough to permit

the potato slices to be deposited in the vat and the finished product removed at the opposite end. Entrained air, indicated by the arrows **32**, will pass into the hood to mix with the steam and odor laden gases and the resulting mixture, indicated by the arrow **33**, will pass from the hood through the duct **34**. A fan **35** is mounted in duct **34** and together with damper **36** provides control over the amount of entrained air drawn into the system and hence, the amount of gases delivered to the furnace or heat exchanger **16**.

Preferably, before being delivered to heat exchanger **16**, the mixture of gases is passed through a cooling and condensing section **36**. Thus the mixture is conveyed to a condenser **40**, which may conveniently be of the air cooled type, although, as will be explained presently, may also be cooled by flowing water thereover. In any case a portion of the moisture carried by the mixture **33** is removed therefrom and flows from the condenser **40** as indicated by the arrow **42**. From the condenser **40** the mixture is conveyed to the condenser or scrubber **44**, which is serially connected thereto and which may be of the water spray or jet type, falling water film type, or any other type for intimately mixing the steam to be condensed with the condensing water.

Thus, the condensing water may be fed to the condenser **44** as at **46** and the condensing water and condensate removed from the condenser **44** at **48**. From the condenser **44** the mixture, which is now composed of relatively dry but odor laden air substantially free of oil particles, may then be conveyed to a cooler **50** containing coils **52** through which a cooling medium, such as water, may be circulated.

It will be noted that, if desired, the efficient use of water in the system can be improved if it is first introduced at point **54** in the cooler **50** with the discharge from the cooler, as at **56**, being directed to the condenser **44** for introduction at **46** and the mixture of condensate and cooling water removed from the condenser **44** at **48**, sprayed over the condenser **40** so that this condenser then operates as an evaporative cooler type unit to further improve its condensing capacity. Additionally, the water from the cooling and condensing section may be utilized as a source of hot water for the plant in which the cooker operates. Also, the particular order of condensers and cooler may be varied to suit the needs of a particular installation.

From the cooler **50** the dried, cooled, odor laden air is directed by means of a duct **58**, filter **59**, duct and a fan **62** to the heat exchanger **16**, where the odor laden air is incinerated as combustion air for the burner **20**.

It will be seen that the downstream end of duct **58** is enlarged, as at **64**, and extends about filter **59** in spaced relationship thereto. As noted above, the amount of gases delivered to heat exchanger **16** is preferably maintained, by controlling the amount of entrained air passing into the system, at or below the amount of combustion air required by heat exchanger **16**. Therefore, if additional air is needed for combustion it may be drawn into the duct **60** through the filter **59** and enlargement, as indicated by the arrows **65**. Of course, if for some reason an excess of air is delivered to duct **60**, the excess may pass from the system in a flow opposite to that shown by arrows **65**.

In any case, the steam and oil particles carried by the mixture passing up the stack **38** may be removed by the serially interconnected condensers **40** and **44** and the odor laden gas remaining then incinerated by the same burner used to heat the cooking oils for the vat **10**. As a result, the possibly offensive odors emanating from the process are eliminated without the necessity of a separate, specially constructed burner and the expense of separate burner maintenance and fuel consumption.

A device developed by T.E. Hampel (6) is a grease vapor incinerator for disposing of cooking waste in an exhaust system comprising a chamber within the exhaust system. Baffle means is provided within the chamber for intercepting and deflecting the exhaust air and means for heating the baffle means to a temperature sufficient to ignite grease in the system upon contact is provided.

RENDERING OPERATION FUMES

Without controls, rendering plant odors have been reported discernible at distances up to twenty miles from the source. Unfavorable atmospheric conditions serve to magnify the problem so that a serious public nuisance develops, as evidenced by the number of complaints received by air pollution control offices both in the U.S. and abroad.

Incineration

The problem of odor controls for rendering plants has been described by R.M. Bethea et al (7). Data (7) indicate that odors from specified facilities can be held below the probable standard only if incineration is employed. In almost all cases, combinations of incineration and condensation reduced emission concentrations to an acceptable level. The combination is usually more economical as well as more effective because of the 20-fold reduction in gas volume which must be incinerated.

A process developed by H.S. Ashley (8) involves removing atmosphere polluting odors from systems such as rendering operations by introducing an oxidant and in selected instances a fuel into a combustion chamber located in a low pressure region between a cooker and a condenser.

A process developed by C.O. Schmidt (9) is one in which the noncondensable gases from a cooking device are incinerated by the intensely hot high velocity flame of an afterburner which is self-contained and may be associated with an exhaust pipe in which noncondensable gases are present. The flame of the device is produced in an area in which water from the incinerated gases cannot extinguish the flame.

RICE HULLS AND OTHER GRAIN WASTES

Incineration

A device developed by O.R. Gardner (10) is particularly adapted for the burning of light combustible materials such as rice hulls and straw from various grains for example. Various devices and methods are known in the prior art for these purposes. However, there remains a need for improvements to meet various requirements for combustion apparatus and processes. One such general requirement is the provision of efficient and economical means for burning such materials. There are often substantial quantities of such materials which can be processed or disposed of only by burning. In many cases, combustion products from these materials do not provide an economic return. Accordingly, it is desirable that capital and operating costs for the combustion process be minimal.

In the processing of other materials such as rice hulls for example, the ash or combustion product may be usable and thus permit an economic return. However, the need for efficiency and economy remains. In addition for combustible materials of this type, it is necessary to provide for accurate control or regulation of the combustion process in order to provide various desired properties in the combustion products.

An additional requirement is raised by the pollution controls enacted in many states. This consideration has become substantially important. Relatively precise limits have commonly been enacted particularly for the burning of large quantities of materials and these limits may be made even more restrictive in the future. Generally, these pollution controls require quite close control over the combustion apparatus or method.

Particular combustion features offered by this device for meeting these controls include the manner of introducing combustible materials into a combustion region and the degree or rate of combustion which may be uniformly maintained to regulate both the quality and quantity of combustion products, either solids or gases, which are introduced into the atmosphere.

In this apparatus, the combustible material is introduced into the chamber near its base through a feed duct. Air is also introduced. During combustion, the material is dispersed by an agitator and carried upwardly into the combustion zone by air flow or convection currents with the material tending to remain suspended in the combustion zone until combustion is complete to a desired degree. Combustion products, both solid and gaseous, pass upwardly from the combustion zone and out the open top of the chamber.

SMOKEHOUSE EXHAUST FUMES

Incineration

A device developed by S.R. Porwancher (11) is an incinerator for exhausting smoke from a food product smokehouse, including an insulated vertical cylindrical chamber tapered inwardly at the top, a deflector for directing smoke from the smokehouse, in a turbulent flow, into an intermediate combustion region in the chamber, a burner for burning the smoke in the combustion region, and means to introduce cooling air into the base of an exhaust stack extending upwardly from the top of the chamber.

In cooking and smoking meat and other food products, the product to be smoked is suspended within a smokehouse. During a part of the cycle of treatment carried out in the smokehouse, the interior of the smokehouse is flooded with a heavy dense smoke. This smoke is produced by incomplete combustion of wood chips, carried out in an oxygen-starved atmosphere to produce the desired dense form of smoke.

After the smoking operation is completed, it is necessary to remove the smoke from the smokehouse in order to permit completion of the cooking and other processing and to allow for changing of the smokehouse contents. As might be expected, the exhaust from the smokehouse is extremely dirty and presents substantial problems with respect to air pollution control ordinances and similar regulations.

Incinerators have been provided, for use in the smoke exhaust systems of smokehouses, to reduce the undesirable components in the smoke exhaust by burning the smoke. But incinerators employed for this purpose tend to be rather bulky and expensive, particularly because they must handle large volumes of dense smoke within short periods of time in order to permit the smokehouse to be used efficiently. Furthermore, incinerator systems have presented substantial problems due to the fact that the heavy, dense smoke used in the smokehouse tends to foul any exhaust equipment and incinerator apparatus through which it flows. This device provides a compact incinerator that permits the exhaust from a commercial smokehouse to meet antipollution requirements.

REFERENCES

(1) Booz-Allen Applied Research, Inc, *A Study of Hazardous Waste Materials, Hazardous Effects and Disposal Methods,* Report PB 221 466, Springfield, Va., Nat. Tech. Inf. Serv. (July 1973).
(2) P.R. Kelly; U.S. Patent 3,771,468; November 13, 1973.
(3) H.V. Hess and E.L. Cole; U.S. Patent 3,671,403; June 20, 1972; assigned to Texaco, Inc.
(4) H.R. Jones, *Waste Disposal Control in the Fruit and Vegetable Industry,* Park Ridge, N.J., Noyes Data Corp. (1973).
(5) J.L. Newcomer; U.S. Patent 3,762,394; Oct. 2, 1973; assigned to Food Technology, Inc.
(6) T.E. Hampel; U.S. Patent 3,164,445; January 5, 1965; assigned to American Gas Association, Inc.
(7) R.M. Bethea, B.N. Murthy and D.F. Carey; *Environmental Science and Technology* 7, No. 6, 504-510 (June 1973).
(8) H.S. Ashley; U.S. Patent 3,499,722; March 10, 1970.
(9) C.O. Schmidt; U.S. Patent 3,592,614; July 13, 1971; assigned to The Cincinnati Butchers Supply Co.
(10) O.R. Gardner; U.S. Patent 3,744,440; July 10, 1973.
(11) S.R. Porwancher; U.S. Patent 3,511,224; May 12, 1970; assigned to Michigan Oven Co.

GRAPHIC ARTS INDUSTRY WASTES

There are solvent emission problems associated with the various inks and coatings used in the printing of paper and containers. Many plants are small and do not employ air pollution control equipment. Of those that do, the disposal techniques employed were as follows (1):

Equipment	Plants	Process
Incinerators	28	Lithography, gravure
Solvent recovery	2	Gravure
Filters (particulates)	6	Lithography, gravure, letterpress
High stacks (dispersion)	11	Lithography

CAN COATING WASTES

Incineration

Some cans are coated inside after being formed, and some are not coated at all on the inside. Thus, a sheet may be coated lightly or not at all on the nonprinted side if it is to be made into cans.

Figure 68 shows the application of inside can lacquer to a sheet although it can also represent the application of inside coating for bottle caps or reverse side coating for products other than sanitary cans. When a sanitary can (the type of the Food and Drug Administration for food and beverages) is to be produced, soldered seams are employed and the area which will become the can seams is not coated. To accomplish this, a spot coater is used.

Figure 68 is the first stage in the most common three-stage metal decorating operation and also the first stage in the four-stage beverage can coating operation. In this diagram, sheets sized 28" x 35" are fed at a rate of ninety per minute to the roller coater and then are transferred to preheated wickets which enter the oven. The sheets remain in the oven for ten to fifteen minutes, including the time in the cooling zone. The oven temperature is shown at 370°F. After leaving the oven the sheets may be inverted before they are stacked.

The major source of solvent and odor emission is the oven exhaust. A minor source is the coating operation. If there is no local ordinance similar to the Los Angeles Air Pollution Control District's Rule 66, or if exempt solvents are used, a hood will probably not be used at the coating operation. There is very little solvent or odor from the cooling zone exhaust or from the dried sheets coming out of the oven. The solvent content of the oven exhaust can be estimated as 8.9% of the lower explosive limit.

The inside or reverse side lacquer coating, used in this operation contains 30% solids by weight. The solvent used is a mixture of MIBK, xylol and an aliphatic mixture in a weight ratio of 28:36:36. This usually has enough solvent power, but special solvents can be added for vinyl coatings. The dried coating weight is 20 milligrams per four square inches of coated surface.

Extra solvent is added at the coater, especially in spot coating, where the coating roller does not pick up all the material from the feed roller. Thus, a scraper is used on the feed roller and a small amount of solvent is added to the roller surface before it contacts the scraper. Another scraper removes excess coating from the steel roller which holds the sheet against the coating roller (the impression roller), and solvent is added just ahead of this scraper also.

The oven exhaust and hood outlet, at 350°F, are shown entering an incinerator operating at 1400°F. The heat exchanger is being used to preheat the incinerator input to 825°F (about 45% of the difference between 1400° and 350°F). Another heat exchanger could be used to preheat part of the oven input air, if economics indicated that it was advisable. However, this is probably not common in the industry.

Figure 69 is the second step in the most common metal decorating procedure. After the coated sheet is flipped over and stacked, it is coated on the other side with a thin primer or a thicker pigmented primer. This diagram is the same as Figure 68, except that the dried coating thickness is greater. In small metal decorating operations this could be done on the same coating line as that shown in Figure 68, by changing the coating used without changing operating conditions.

As above, the sheets pass through the coater and the oven. The sheets are stacked, but this time they are not turned over. The dried coating thicknesses are shown as 8 mg per 4 sq in for the vinyl sizing and 48 mg/4 sq in for the white primer.

FIGURE 68: INSIDE CAN COATING AND FUME DISPOSAL

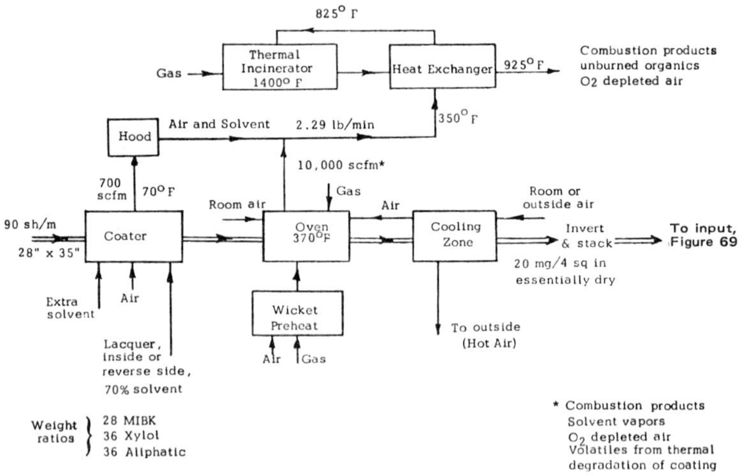

Source: PB 195 770

FIGURE 69: OUTSIDE CAN PRIMER APPLICATION AND FUME DISPOSAL

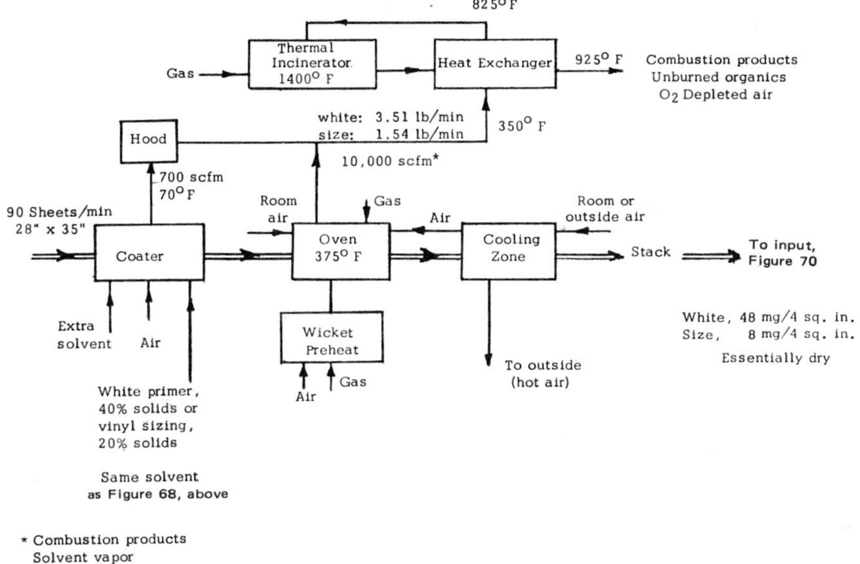

Source: PB 195 770

Figure 70 is the third step in the most common metal decorating procedure. The coated sheet is printed with lithographic ink. Usually there are either one or two offset printing units in series with a varnish coater, which is sometimes called a trailing or backup coater. The varnish is put on over the wet ink. The sheet then passes through the oven, which is operated at a lower temperature than the coater ovens. Some printing lines operate at a lower speed than coating lines; consequently, the rate shown in Figure 70 is sixty sheets per minute.

The source of organic emission is again the heating zone exhaust of the oven, and the roller coater if a hood is used over this area. Ther thermal incinerator and heat exchanger are used, but the input to the heat exchanger is at a slightly lower temperature than in Figures 68 and 69. Water vapor is emitted from the dampener on the printing units, but the inks do not emit significant amounts of solvent. In this process, isopropanol is not used in the lithographic fountain solutions, consequently no alcohol vapor is shown being emitted in any metal decorating flow sheet.

An additional process which is applicable only to sheets which are not going to be made into cans is the waxer. A protective wax coating can be applied to the sheet by melting and spraying on, or by other means. Mixtures of waxes or waxy synthetic resins can be used. Viscosity is controlled mainly by composition and temperature, so no solvent is necessary. Heat input to the waxing machine is intentionally not show since the source of heat presumably does not come into direct contact with the wax. This process is also used for coating cardboard, as in the production of milk cartons. There is no further information about organic emissions from wax coating processes.

FIGURE 70: PRINTING LINE IN METAL DECORATING AND ACCOMPANYING FUMES DISPOSAL EQUIPMENT

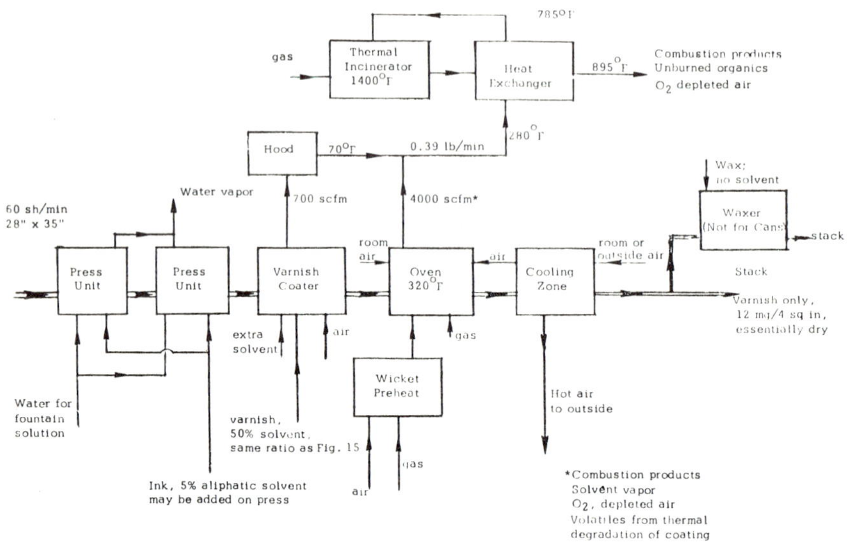

FIGURE 71: INSIDE CAN SPRAY COATING AND FUME DISPOSAL

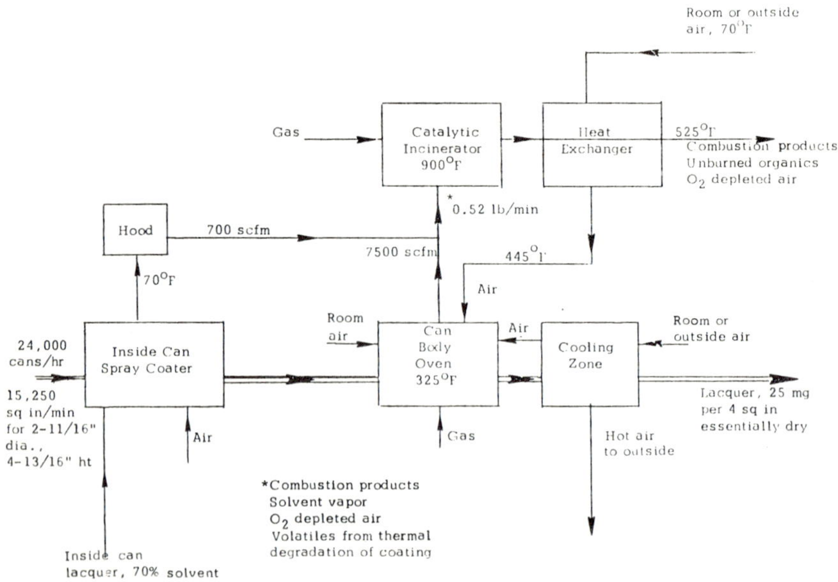

Source: PB 195 770

Figure 71 describes an oven used to dry the inside of can bodies. Beverage cans need a double coating inside to protect against the effects of carbonation. This double coating protects not only by added thickness, but also by eliminating the possibility of pinholes in the coating. Therefore, after the can is formed and a stripe of protective coating is printed over the seam, the cans are sprayed inside with the second coating.

After being spray coated, the cans are fed in an upright position to a steel band which travels through the can body oven. The rate of drying in square feet per minute in the oven depicted is about one-third of the rate at which sheets can be printed on one printing line.

This diagram suggests the possibility that a catalytic unit can be used in this (or any) metal decorating process for removal of solvent fumes. The heat exchanger is used to preheat the oven input air in this diagram, but it could be used to preheat the input to the incinerator instead.

The thickness of the second coating is given as 25 mg/4 sq in, essentially dry. The organic vapors are emitted at only one point, the output of the incinerator. There is very little emission from the oven cooling zone, or from the product. Finally, the heat exchanger is not necessary in the diagram, and would be used only if economics so indicated.

LITHOGRAPH PLATE DRYER EFFLUENT

Incineration

A device developed by R.E. Edwards (2) is a fume incinerator associated with a unit which gives off noxious or other combustible fumes, such as a dryer for lithographed sheets or plates. As an example, such fumes can be aromatic hydrocarbons or generally resinous fumes, such as vinyls.

PRINTING PROCESS WASTES

Incineration

In Figure 72, a 38 inch wide web traveling at 1,000 feet per minute enters a press. When the blanket-to-blanket press configuration is used the paper is fed between two blankets and two different images are printed simultaneously on the two sides of the sheet. After passing through several printing units, the two complete images are printed on both sides and the web enters the dryer. Leaving the dryer, the paper passes over chill rolls and is then ready for folding, cutting or other finishing operations.

It is also possible to feed the paper between a blanket roll and an impression cylinder, printing it on only one side, although this is more usual in sheet-fed operations. Another configuration consists of several blanket rollers and a common impression cylinder. In either case, when the paper is printed only on one side, the dryer exhaust rate may be as much as 30% less than when both sides are printed. A tunnel dryer may be used when a web is printed on one side only and is the only dryer that can be used when both sides of the web are printed simultaneously such as in blanket-to-blanket printing. Conversely the steam drum is not applicable to the drying of blanket-to-blanket printing because of the problem of wet ink contact with the steam drum surface.

Optional additions to the drying process are shown with dotted lines. The smoke and air from the chill roll area is shown entering the dryer exhaust. However, it could probably be added to the dryer recycle as part of the makeup air or simply be exhausted without any correction. The incinerator and two heat exchangers are shown in one possible arrangement. Either or both heat exchangers may be eliminated. If a catalytic incinerator and only one heat exchanger is used, the greater efficiency will be obtained using Exchanger 2,

FIGURE 72: WEB OFFSET PUBLICATION

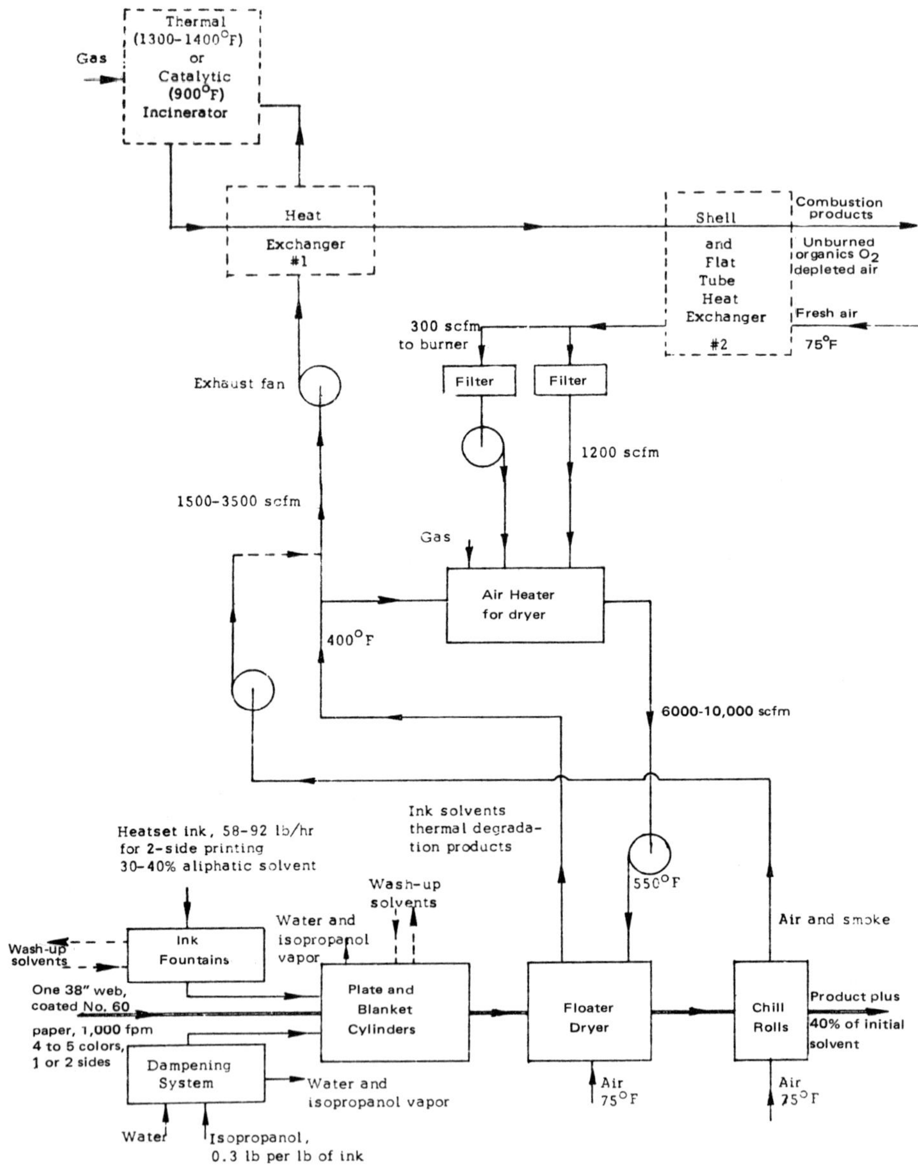

Source: PB 195 770

FIGURE 73: WEB PUBLICATION LETTERPRESS

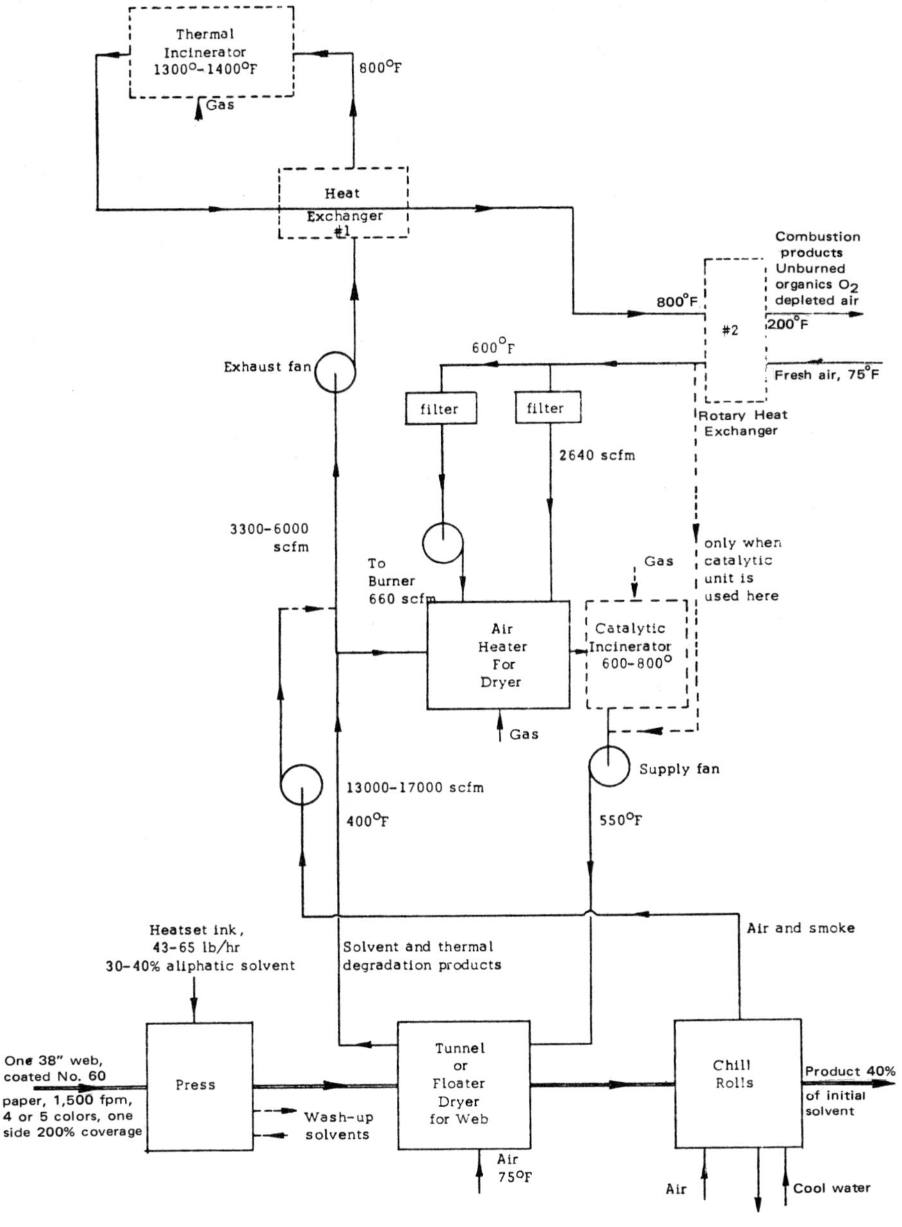

Source: PB 195 770

rather than Exchanger 1, simply because of the greater temperature difference between the two streams.

Figure 73 is shown with an unusually narrow web for this application, so that some comparison with web offset can be made. In this case the increase in web speed from 1,000 to 1,500 feet per minute more than triples the recommended exhaust rate for two-side printing while a one-third cut in this rate is made because only one side is printed. This has resulted in a doubling of the air flow rates relative to Figure 72, Web Offset Publication.

The flow diagram differs from the web offset diagram because no dampening solution is used. However, wash-up solvents are still used on the press. Some air enters the dryer box where the web enters and exits to prevent emission of smoke from these openings. Air also enters the chill roll hood and the burner section of the dryer and may enter at any gas burner although the incinerator sometimes uses only the exhaust air from the dryer. All air and organic emissions are from the dryer unless incineration equipment is used, in which case the only exhaust is from the incinerator.

In a few cases a catalytic incinerator has been installed in the recycle line of the dryer, as shown. This would not necessarily eliminate the need for another incinerator on the dryer exhaust. Two heat exchangers are shown in the incinerator output line. Either Exchanger 1, or both Exchanger 1 and Exchanger 2 may be eliminated. A rotary heat exchanger may be used either in location 1 or 2, or at neither location. If used at location 1, however, there is a possibility that there will be some leakage of fumes to the exhaust side from the input side of the incinerator. In Figure 72 and Figure 73 optional equipment is indicated by dotted lines and letterpress and offset incineration systems are interchangeable.

REFERENCES

(1)   R.R. Gadomski, M.P. David and G.A. Blahut, *Evaluation of Emissions and Control Technologies in the Graphic Arts Industries,* Report PB 195 770, Springfield, Va., Nat. Tech. Information Service (August 1970).
(2)   R.E. Edwards; U.S. Patent 3,251,656; May 17, 1966; assigned to The Roy M. Moffitt Co.

HOSPITAL WASTES

Waste materials generated by hospitals can be divided into several types as described by Booz-Allen Applied Research (1):

> General rubbish
> Food residues
> Pathological wastes
> Radioactive wastes
> Drug residues and solvents
> Disposables—syringes, needles, test tubes, etc.

Disposables have played an increasing part in hospital operations for reasons of infection control and economy.

Accident risk involved in the disposal of needles, syringes, glass and surgical items, as well as personal and/or environmental contamination associated with the handling and disposal of pathological materials, appear to be the unique problems associated with disposal of hospital wastes. The quantities of drug residues, radioactive wastes or other special chemical wastes do not appear to present signal hazards. Although radiopharmaceuticals are being used in increasing quantity, the half-life of the diagnostic agents used is quite short and thus, unused materials and other radioactive wastes decay very quickly to low radiation levels.

The amount of hospital waste has increased from approximately 4 pounds per patient per day, in 1955, to a current figure of approximately 19 pounds per patient per day. The estimated per capita figure for the United States population is approximately 5 pounds per day. One reason for the larger figure estimated for the hospital patient is that there are 2.9 full-time hospital employees per patient, each contributing substantially to the solid waste load of the hospital.

To obtain some estimate of waste material quantities and composition, inquiries were made at a number of general hospitals representing a total of 5,281 beds. At each institution, the hospital administrator or his designate provided available information. A summary of waste quantities estimated for each of the hospitals is presented in the tabulation shown on the following page. In the tabulation it is shown that the estimated average waste load for the hospitals, as a group, is 16.13 pounds per bed per day and the range 5.8 to 45.0 pounds per bed per day.

Hospital	Number of Beds	Estimated Solid Waste (pounds/day)
1	1300	32,000
2	1000	7,200
3	758	7,000
4	600	3,500
5	430	19,000
6	377	3,000
7	350	2,800
8	350	7,650
9	116	2,700
Total	5,281	85,200

Disposal practices employed by some of these hospitals are as follows.

Hospital No. 1: The facility disposes of approximately 32,000 pounds of waste per day using incineration, compacting and sanitary landfill. Approximately 16,000 pounds are incinerated. Pathologic wastes and all microbiological materials are included in that fraction that is incinerated. Radioactive wastes are disposed of by service contract. Unused drugs are returned to the manufacturer or flushed into the municipal sewage system.

Hospital No. 2: The waste is divided into several types: solid wastes including papers, flowers, general trash; food residues; nonburnables such as plastic and rubber materials; pathological wastes, including animal carcasses, autopsy and surgical wastes and microbiological wastes (tubes, cultures, petri dishes); radioactive wastes; syringes and needles; and return or low potency drugs.

Solid wastes (7,200 pounds) are incinerated daily at a municipal incinerator. Food residues (100 pounds per day) are sold to a farmer who, in turn, sterilizes the material and utilizes it as animal fodder. Nonburnables (100 pounds per day) are disposed of at a municipal dump. Hypodermic needles and syringes, included as burnables, are disassembled and broken prior to incineration. Pathological wastes are incinerated before or after autoclaving. Liquid residues are autoclaved, diluted and poured into the drain. Radioactive wastes are disposed of by service contract except for syringes. These are stored in a lead vehicle until a low level of radioactivity is reached at which time they are disposed of by incineration.

Hospital No. 7: Solid wastes (2,800 pounds per day) are disposed of primarily by a contract service. No information as to the final disposition (incineration, landfill) was available. Autopsy remains are either incinerated or buried on the premises. Microbiological wastes are autoclaved prior to incorporation into the general waste collection.

The disposal practices described for these hospitals are typical of those employed by all the hospitals at which inquiries were made. The following conclusions were derived from the information obtained:

The quantity of solid wastes generated by hospitals is increasing annually.

The rate of increase in solid waste generation is largely the result of the broad usage of "use and discard items" in the medical and surgical environment.

The primary methods of solid waste disposal practiced by hospitals are incineration and landfill, or a combination of the two. Pretreatment consists of compacting and in some instances, sterilization.

Although estimates of total solid wastes are available, quantitative data for pathological and other potentially hazardous wastes are limited and unreliable.

Although some information regarding the quantity and composition of solid hospital waste was obtained from several hospitals, minimal information describing the quantities of pathological wastes generated was obtained. One estimate was obtained from a survey of solid waste practices utilized by the United States Air Force (1). A questionnaire survey of solid waste practices was conducted on all Air Force installations and data were made available on 98 major installations.

To obtain an idea of the composition of the pathological wastes generated, the questionnaire requested percentages in the following categories: tissues, plastics, bandages, paper and other. Some items described in the response to the other category included: syringes, kitchen wastes, serum, cardboard, drugs, glass, blood, splints, vials, test tubes, petri dishes, cultures, needles, metal, rubber and cloth. The difficulty in obtaining accurate quantitative composition data is obvious.

Reported pathological waste generated at medical treatment facilities varied from 0 to 22,700 pounds per week. For 77 bases with an in-patient capability, the per capita production of pathological wastes varied from 0.04 to 181.6 pounds per bed per week. There were 21 bases having no beds or no estimate of the amount of pathological waste. The mean for those installations having in-patient capability (77 in all) was determined to be 5.63 pounds per bed per week (median = 1.21 pounds per bed per week).

The majority of hospitals dispose of solid waste by incineration, landfill or a combination of the two processes. Prior to removal to landfill, many institutions utilize compacting to reduce the volume of waste that must be disposed of. Liquid wastes are diluted or neutralized and introduced into the sewage system. Pathological wastes are usually incinerated. However, prior to incineration, microbiological wastes may be autoclaved. Radioactive wastes are given to service contractors or returned to the manufacturer. Unused drug products are flushed into the sewage system, incinerated or returned to the manufacturer. Syringes and needles are broken or crushed prior to incineration or removal to landfill. Several hospitals sterilize food residues and make them available for animal feed.

INCINERATION

A device developed by P. Livengood et al (2) is an incinerator for disposing of biological or pathological wastes and other materials. It is designed for use by veterinarians, hospitals, laboratories, research stations, and similar institutions where biological or pathological waste accumulates and must be disposed of in a manner meeting strict requirements regarding production of odors, smoke and pollution and elimination of pathological organisms.

An apparatus for the incineration of hospital refuse has been described by M. Stloukal et al (3). It comprises an auxiliary chamber having at least three serially arranged mechanical grate platforms, a hopper for the discharge of the refuse above the first of the grate platforms, each of the platforms comprising a plurality of spaced bars. The apparatus includes means for reciprocating the bars of the first platform, means for reciprocally moving alternate bars of a second of the three platforms, the bars between the alternate bars of the second platform being fixed in position. A portion of the bars of the second platform are in vertical alignment with the bars of a third of the platforms. The apparatus further includes means for moving the bars of the third platform in phase opposition with the bars of the second platform. The apparatus also includes a waste heat boiler, the grate platforms being separated from the waste heat boiler, the waste heat boiler being equipped with an independent ignition means, the auxiliary chamber being interconnected by an outlet draught to a combustion chamber of the waste heat boiler.

LANDFILL DISPOSAL

As noted above in the discussion of the survey of Air Force hospital disposal practices conducted by Booz-Allen Applied Research (1), the majority of the installations dispose of these wastes by incineration and landfill. Of the bases reporting, 68.0% use incineration only, 13.4% use landfill only, 16.5% use a combination incineration and landfill, and the remaining 2.1% utilize incineration, landfill and sewage disposal. Approximately 40% of those installations using landfill as the primary disposal technique autoclave the materials prior to disposal.

REFERENCES

(1)   Booz-Allen Applied Research, *A Study of Hazardous Waste Materials, Hazardous Effects and Disposal Methods,* Report PB 221,467, Springfield, Virginia, National Technical Information Service (July 1973).

(2)   P. Livengood and R.L. Christophel; U.S. Patent 3,782,301; January 1, 1974; assigned to Shenandoah Manufacturing Co., Inc.

(3)   M. Stloukal, Z. Syrovatka, M. Dolezel and K. Stripek; U.S. Patent 3,926,130; December 16, 1975; assigned to Prvni Brnenska Strojirna.

INORGANIC CHEMICAL INDUSTRY WASTES

The reader is referred to the earlier chapters on Incineration and Landfills for indications of the applications of those techniques to the disposal of specific inorganic chemicals. More specifically, the reader is referred to the section in the chapter on Incineration entitled Chemicals Which Can Be Disposed of by Incineration (p 56). In the chapter on Landfills, the reader is referred to Table 25 (p 167) for a listing of specific inorganic chemicals which may be disposed of by landfilling.

A discussion of waste production, pollution prevention and ultimate disposal techniques practiced by the inorganic chemical industry has been presented by R.G. Shaver et al (1). A variety of treatment methods and processes are currently being used by manufacturers of inorganic chemicals to control solid, liquid, and thermal wastes including:

    [1] Chemical addition
    [2] Equalization
    [3] Sedimentation
    [4] Filtration
    [5] Reverse osmosis
    [6] Electrodialysis
    [7] Ion exchange
    [8] Multiple effect evaporation
    [9] Deep well injection
  [10] Ocean burial
  [11] Dumping and landfill
  [12] Lagooning/cooling ponds/solar evaporation ponds
  [13] Centrifugation
  [14] Cooling towers

Three typical schemes are indicated in Figure 74 for the treatment of [1] waste containing dissolved and suspended solids, [2] excess thermal energy discharge, and [3] waste containing primarily only dissolved solids respectively. Where wastewater contains appreciable dissolved and suspended solids, a typical treatment process might be 2-3-4-7-11-15 for liquids, and 15-14-12-17 for solids.

In this sequence, the waste flow is equalized, followed by oil removal. Clarification is used for suspended solids removal and the dissolved solids are concentrated and disposed of in deep wells. Effluent distillate is then discharge or reused. Suspended solids slurries are thickened, centrifuged, and lagooned. Alternately, chemical addition could be used for dissolved solids removal if the dissolved ions have a common insoluble salt.

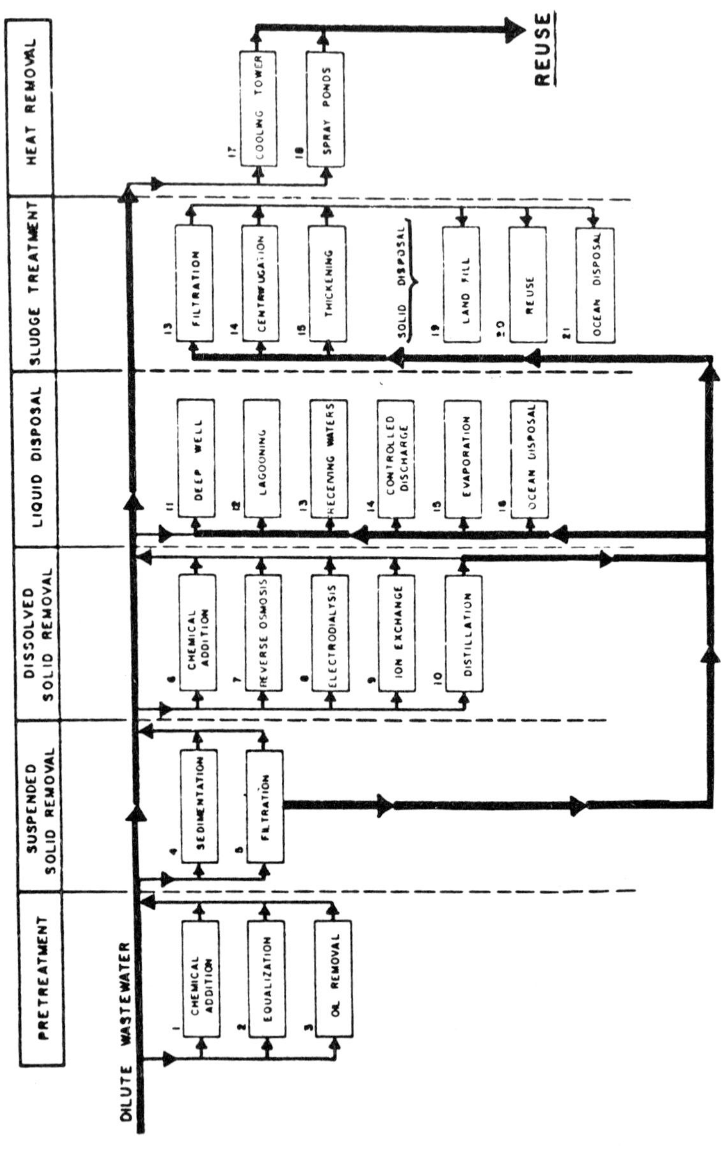

FIGURE 74: WASTEWATER TREATMENT SEQUENCE

Source: PB 221 466

Dissolved solids may also be concentrated by electrodialysis or ion exchange instead of distillation. These may be recovered and/or converted to a marketable product. Where there is excess thermal energy discharge, a treatment sequence might be 17-6-4-13. Here a cooling tower or pond would be used, and the cooled effluent reused or discharged. If the water is recycled, the blowdown from the system may be treated by chemical addition and clarification to remove undesirable components, especially $Cr(VI)$ and zinc added for corrosion control. Suspended matter would then go into a solid disposal sequence. In some cases, such as in the manufacture of H_2SO_4, by-product steam may be used to economic advantage.

Where there is a heavy dissolved solid load but a light suspended solid load, a sequence for acidic effluents would be 2-1-7-11-13. Here neutralization would occur after equalization, followed by reverse osmosis (or distillation, ion exchange or chemical addition), before alternate disposal such as deep well injection or evaporation to dryness.

In general, it is advantageous to keep contaminated and relatively clean effluent streams segregated, since most methods aim at concentrating effluents before discharge, or work better when solutions are more concentrated.

ALKALI AND AMMONIUM FLUORIDES

The procedure recommended (3) for treatment of large quantity, continuous discharge of alkali and ammonium fluorides is continuous reaction with an excess of lime, followed by lagooning, and either recovery or landfill disposal of the separated CaF_2.

ALUMINUM, BARIUM AND CADMIUM FLUORIDES

The Manufacturing Chemists Association recommends packaged lots of soluble or slightly soluble fluorides be slowly added to a large container of water. Then a slight excess of soda ash or slaked lime is stirred into the solution. The slurry formed is allowed to settle for 24 hours. If aluminum fluoride is being treated, the supernatant liquid is decanted or siphoned into another container, and neutralized with dilute hydrochloric acid before being washed into a sewer or stream with large quantities of water. The sludge is placed in a landfill.

If cadmium fluoride is the fluoride being treated, cadmium hydroxide (solubility is 0.0026 g/100 g of water) will be precipitated with the slurry formed upon addition of lime. The mixed calcium fluoride-cadmium hydroxide sludge from treatment of cadmium fluoride should be sent to a landfill of the California Class 1 category.

The supernatant liquid will require treatment via another process such as ion exchange, reverse osmosis, or activated carbon adsorption to reduce the cadmium content of discharge solutions to less than 0.01 mg/l of cadmium. If barium fluoride is being treated, the supernatant liquid may be neutralized with sulfuric acid, instead of hydrochloric acid, to form the insoluble barium sulfate. After removal of the barium sulfate by settling, the effluent will contain about 2 ppm of barium. This effluent may be diluted with additional water to meet the permissible criteria of 1.0 ppm for barium in public water supplies. The barium sulfate sludge may be landfilled.

ALUMINUM FLUORIDE MANUFACTURE

This product is manufactured by a process that generated fluoride wastes. Waste treatment by liming produces calcium fluoride-containing hazardous waste streams for land disposal. One plant uses a unique proprietary process that generates no wastes of this sort. Of the remainder, four plants use on-site disposal by ponding or dumping and one uses a general purpose municipal landfill (1).

ALUMINUM OXIDE

See Tonnage, Insoluble, Nontoxic Products, page 256.

AMMONIA PLANT EFFLUENTS

Incineration

A process developed by H.A. Adams et al (4) is one in which ammonia and organic impurities are removed from an ammonia plant effluent consisting mainly of steam by treating the effluent with an oxidation catalyst which oxidizes the impurities to harmless products of nitrogen, water and carbon dioxide.

The use of a potassium carbonate solution to separate carbon dioxide from a gaseous stream is well known. In addition to the carbon dioxide, the potassium carbonate solution also entraps ammonia and organic impurities. These impurities are stripped along with the carbon dioxide and found in the impure carbon dioxide. The nature of these impurities, of course, is directly related to the process in which the carbon dioxide is generated.

For example, in an ammonia plant where methane is catalytically oxidized in the presence of steam and air to form hydrogen and carbon dioxide, the impurities in the potassium carbonate solution are extraneous low molecular weight organic compounds, such as alcohols and aldehydes. Also found in the solution is ammonia formed from the hydrogen produced in the reforming and the nitrogen from the air. These impurities are stripped from the carbonate solution along with the carbon dioxide. Figure 75 shows such a separation process along with subsequent pollutant disposal step.

FIGURE 75: PROCESS FOR REMOVAL OF AMMONIA FROM AMMONIA PLANT EFFLUENT BY CATALYTIC INCINERATION

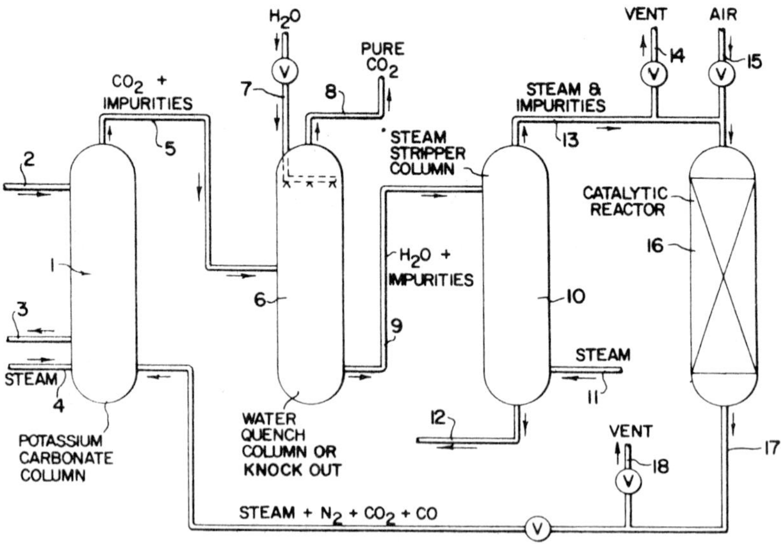

Source: U.S. Patent 3,812,236

In the potassium carbonate column, **1**, potassium carbonate solution having a high concentration of carbon dioxide and impurities enters the column through entrance **2**, the solution flows through the column and exits the column at exit **3**. During the circulation through the column, the potassium carbonate solution is subjected to steam stripping from steam inlet **4**.

This input of heat strips the carbonate solutions of the carbon dioxide, ammonia and organic impurities. These materials along with some of the steam leave the column through conduit **5**. Conduit, **5**, takes the gaseous stream to the water quench column **6**. In this column, water is sprayed from inlet **7**, into the gas stream. This water quench removes ammonia and the other water soluble or condensed impurities, leaving an essentially pure carbon dioxide stream which leaves the water quench column, **6**, through exit, **8**.

When the water quench column **6** is a knock out, **6**, there is no water flow and only the condensibles are recovered. The liquid containing the impurities in the bottoms water quench column **6**, is taken from the column through conduit **9**, to the steam stripper column **10**. In the steam stripper column, steam enters through entrance **11**, to strip the impurities from the contents of the column to give essentially pure water which is taken out of the steam stripper column **10**, through exit **12**. The stripped impurities plus the steam carrier are channeled into conduit **13**.

In the past, this stream in conduit **13**, was either vented to the atmosphere or charged to the sewer as shown by vent **14**, or recycled to the potassium carbonate column, **1**, by a conduit which is not shown in the figure. The venting or disposal of the stripper effluent through vent **14**, is undesirable. This discharge of a combination of steam and impurities is essentially eliminated in the process. When the stream in conduit **13** is recycled to the potassium carbonate column, the concentration of these impurities is increased so that the columns may be overloaded with impurities. This problem is also overcome by the process.

It has been found that the impure gas stream in conduit **13**, containing steam and about 0.01 to 10% by volume of ammonia or organic compounds of up to about 6 carbons or mixtures thereof is conveniently converted to a stream containing essentially no detrimental impurities by contacting the impure stream with a catalyst selected from the group consisting of oxidation catalysts containing the oxides of copper, iron, manganese, bismuth, nickel, cobalt, uranium, molybdenum, vanadium, chromium, tungsten, palladium, platinum, silver, zinc, alkali metals, alkaline earth metals, tin and antimony or mixtures thereof at a temperature of about 200° to 800°C.

The process is carried out by preferably injecting molecular oxygen usually in the form of air through conduit **15**, into the gaseous stream in conduit **13**. This mixture is then fed into the catalytic reactor **16**. The catalytic reactor **16**, contains an oxidation catalyst as just described. The catalytic reactor **16**, is operated under conditions such that ammonia is converted to nitrogen and hydrogen, and the organic impurities are converted to carbon oxides and water. Of course, in the presence of the catalyst and molecular oxygen, any hydrogen formed from the ammonia is almost instantaneously converted to water under the conditions of the reaction.

Thus, the gas stream leaving the catalytic reactor through conduit **17**, contains steam, nitrogen, oxygen and carbon oxides almost all of which are in the form of carbon dioxide. This stream leaving the catalytic reactor has essentially the same temperature as the temperature of the catalytic reaction.

The gas stream in conduit **17**, would not deleteriously affect the environment, and therefore, it may be released through vent, **18**. In preferred practice, however, the heat of the stream in conduit **17**, is preferably used by returning the heat to the potassium carbonate column as heat input to assist in stripping additional carbon dioxide and impurities or as heat input to preheat the gas coming into the catalytic reactor.

A process developed by Y. Kajitani et al (5) involves introducing a substance which generates nitrogen oxide by combustion, such as ammonia, hydrogen cyanide or other nitrogen containing gases or a mixture of these gases, into a high temperature reducing flame produced by the combustion of fuel gas containing hydrogen, carbon monoxide, a gaseous hydrocarbon and mixtures thereof, with a primary air supplied in an amount below the amount of air theoretically required for combustion of the fuel gas whereby the generated nitrogen oxides are reduced to nitrogen in the high temperature reducing flame, and if required supplying a third air to the relatively high temperature portion out of the flame thereby burning any unburned components. Figure 76 shows a suitable form of apparatus for the conduct of the process.

FIGURE 76: INCINERATOR FOR AMMONIA-CONTAINING GASES

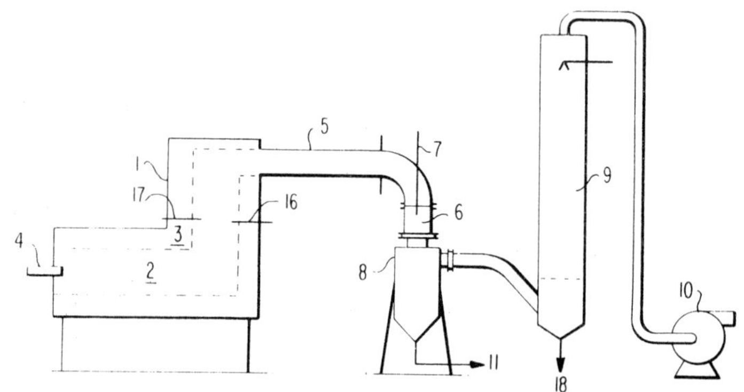

Source: U.S. Patent 3,838,193

The combustion furnace body **1** has a furnace chamber **2** with furnace wall **3** made of heat insulating refractory material and a burner **4** for the combustion of ammonia. Flue **5** is connected to venturi scrubber **6** for cooling the flue gas having a water supply nozzle **7**. The gas and water are separated in chamber **8** with the gas passing to an absorbing tower **9** filled with Raschig rings, and exhausted by blower **10**. Liquids leave via outlets **11** and **18** and port **16** introduces the third air. City gas is supplied via port **12**, the primary air is supplied via inlet **13**. The secondary air is supplied via port **14**. The nitrogen oxide generating substance enters via port **15** and **17** serves as an additional port.

In the operation of this apparatus, city gas is introduced from the gas supply port **12** while primary air is supplied from the primary air inlet **13** in the amount of 80 to 95%, preferably 85 to 90% of the theoretical air amount required for city gas so as to burn it to make a high temperature reducing flame. The substance which normally generates nitrogen oxide upon combustion such as ammonia or a gas containing ammonia is supplied from port **15** so as to be introduced into the high temperature flame. Secondary air is introduced from the port **14**.

Additional air insufficient for complete combustion is supplied from the port **16** as the third air in the relatively high temperature (900° to 950°C) portion out of the flame so that in the interior of the high temperature flame unburned fuel gases remain. The nitrogen oxides NO and NO_2 generated upon combustion of ammonia are reduced to nitrogen and an oxide of the reducing gas by the following reactions: $2NO + 2H_2 \rightarrow N_2 + 2H_2O$ and $2NO_2 + 4H_2 \rightarrow N_2 + 4H_2O$. These product gases are harmless when the reducing gas is hydrogen.

This process may be applied to the substances generating nitrogen oxide by combustion such as not only the aforesaid ammonia or gas containing ammonia, but also, substances such as hydrogen cyanide or other nitrogen containing gases which generate nitrogen oxide upon combustion. Examples of these materials also include liquid and solid ammonium sulfide, ammonium chloride, synthetic high polymeric substances such as nylon, acrylonitrile or carbon dust.

AMMONIA-SODA PLANT EFFLUENTS

Deep Well Disposal

A process developed by A. Bietlot (6) involves disposing of the effluents from the distillers of ammonia-soda plants at the bottom of subterranean cavities of disused salt boreholes, the cavities being filled with a sodium chloride brine to prevent subsidence and an equivalent volume of the brine being recovered. The effluent introduced at the bottom of the borehole comprises a calcium chloride solution which has been concentrated to a density significantly higher than that of the saturated solution of sodium chloride present in the borehole.

AMMONIUM CHLORIDE AND NITRATE

Incineration

Ammonium nitrate and ammonium chloride after dilution with water can be charged into a gas fed incinerator (3). The NO_x and/or HCl formed must be removed by appropriate gas cleaning devices (scrubber for HCl and NO_2, oxidation or reduction for NO). Though this method is possible, it is not in wide use.

Stream Discharge

Both ammonium nitrate and ammonium chloride upon treatment with sodium hydroxide liberate ammonia and form soluble sodium salts, either the chloride or nitrate. The liberated ammonia can be recovered and sold. After dilution, sodium nitrate or sodium chloride can be discharged into a stream or sewer (3).

AMMONIUM PERCHLORATE AND OTHER OXIDIZERS

Chemical Degradation

Disposal of oxidizing agents may be accomplished by dissolving the material, adding the resultant solution to a large volume of a concentrated solution of reducing agent (sodium thiosulfate, sodium bisulfite, or a ferrous salt), and then acidifying the mixture with $3\,M\,H_2SO_4$. When reduction is complete, soda ash is added to make the solution alkaline. (If an ammonium salt is present, ammonia will be liberated and will require recovery). The alkaline liquid is decanted from any sludge produced, neutralized, and diluted before discharge to a sewer or a stream. The sludge (if magnesium or zinc is present) is added to a landfill (3).

Incineration

Though the technique is not in wide use, dilute aqueous solutions of these oxidizer materials can be destroyed by injection into an incinerator supported by a gas flame. Scrubbers are required to remove HCl, NO_x and metal oxide particles from the incinerator vent gases. Properly designed and operated incineration is considered a promising near-future method for the disposal of aqueous wastes containing oxidizing materials but additional research is required before the process could be employed on a large scale (3).

Open Burning

Uncontaminated oxidizers, particularly the perchlorate reject materials, process fines or overruns from grinding or blending operations, are collected and transported to "burn sites" in open-top 30 gallon drums with lids attached. At the burn sites the oxidizers are spread thinly over a layer of excelsior or other flammable dry materials. The flammable materials are ignited by a squib, fired electrically, with the controls at a safe distance. Contaminated materials are left in the containers in which they are collected and the material burned in the container in a similar manner. Though this process is widely used at solid propellant manufacturing sites, it is not considered satisfactory because large quantities of NO_x and HCl are liberated (3).

AMMONIUM PERSULFATE

See Ammonium Perchlorate and Other Oxidizers, page 233.

AMMONIUM SULFATE

Ocean Disposal

The effects of ammonium sulfate mother liquor on brine shrimp and minnows in the laboratory, as well as the diffusion of these wastes during discharge were studied. This waste material is saturated with ammonium sulfate and contains about 10% organic carbon as alcohols, amides, esters, etc.; it is generated in obtaining ammonium sulfate from sulfuric acid in a fertilizer manufacturing process. The work included chemical analyses, biodegradability and toxicity studies, and preliminary diffusion calculations, as well as sampling at sea to determine actual diffusion.

On the basis of the preliminary duffusion and toxicity studies, it was concluded that the wastes would not produce adverse effects on the environment or organisms within the disposal area. A trial disposal operation verified the predictions that diffusion would be sufficiently rapid to minimize the harmful effects to the biota (7).

ANTIMONY FLUORIDES

Manufacturing process effluent streams containing the antimony fluorides can best be handled at the sites where they originate by the continuation of current, acceptable disposal processes (8). Small contaminated or surplus quantities of antimony trifluoride and antimony pentafluoride should either be returned to the manufacturers for reprocessing, or disposed of.

A proposed disposal process is to combine Methods 27d and 11 of the *Manufacturing Chemists Association Laboratory Waste Disposal Manual*. Briefly stated, the disposal concept is as follows: dissolve the antimony fluoride in the minimum quantity of dilute hydrochloric acid. Saturate with hydrogen sulfide. Filter, wash and dry the antimony sulfide precipitate; sell the precipitate to a primary antimony metal producer. Air strip the filtrate of dissolved H_2S, passing the effluent air into a controlled incineration device, equipped with a lime scrubber. React the stripped filtrate with an excess of lime; evaporate to dryness, and dispose of the lime-CaF_2-$CaCl_2$ mixture by land burial.

ANTIMONY SULFATES AND SULFIDES

The materials in this section are insoluble in water and therefore may be conveniently disposed of by landfill. However, antimony compounds are highly toxic and the disposal of dilute wastes containing these constituents by landfill should be in Class 1 sites located over nonwater-bearing sediments or with only unusable ground water underlying them. For particularly antimony sulfide and sulfate wastes, recovery should be considered and preferred over disposal by landfill (3).

ARSENIC TRICHLORIDE

Recycling

Rocky Mountain Research, the only current producer of arsenic trichloride, has indicated its willingness to accept excessive or waste arsenic trichloride for reprocessing, purification or disposal (3). It is fed into their reactor and purified in the same manner as a normal batch. Rocky Mountain Research was not willing to discuss the details of their process and hence an evaluation of the process and the purification technique for $AsCl_3$ were not available.

Rocky Mountain Research's alternate procedure would be to add water to the waste under ventilation control to hydrolyze the arsenic trichloride to arsenic trioxide which is much easier to handle.

It is conceivable that any industry that handles $AsCl_3$ in significant amounts would have special facilities for hydrolyzing the $AsCl_3$ and scrubbing the evolved HCl.

Landfill

Landfilling of arsenic trichloride is not considered as an acceptable disposal option (3) because of: [1] the high degree of toxicity of the compound; [2] the nondegradable nature of the toxic arsenic component of arsenic trichloride, whereby the compound would not be converted to a less toxic or nontoxic form in the soil environment; and [3] the reactivity of arsenic trichloride with water to yield arsenic trioxide and corrosive hydrochloric acid. The corrosive potential of arsenic trichloride also indicates that the land burial of arsenic trichloride wastes containerized in metal drums is not an adequate method of disposal.

ASBESTOS

See Tonnage, Insoluble, Nontoxic Products, page 256.

In view of the newer knowledge of asbestos fiber pollution of water and its possible adverse health effects, it would seem that asbestos is in a somewhat different category than the other commodities in this group. Perhaps a Class 1 Landfill should be specified for asbestos.

BERYLLIUM AND BERYLLIUM COMPOUNDS

Recycling

This is most desirable from the standpoints of users, producers and environmentalists (3). Beryllium and beryllium compounds are difficult to produce and the primary producers repurchase all available material at $10 to $20/lb contained beryllium. This situation is expected to continue indefinitely.

Landfill

Liquid, solid or particulate waste which is too dilute to recycle is buried on private property or in public landfills. Often waste is first burned to produce the insoluble, chemically inert oxides. This is easily and safely done, providing the exhaust gases are scrubbed to remove any particulates. These procedures were verified (3) with KBI Industries, two other large beryllium users, and the County of Los Angeles, California, which operates several landfills which receive liquid coolant wastes containing beryllium. All independently agreed with this analysis. Since there have been no new reported cases of berylliosis in 20 years and demand is expected to remain static for the indefinite future, it may be concluded that such practices are adequate at present and for the foreseeable future (3).

BORIC ACID MANUFACTURE

U.S. Borax & Chemical Corporation has a plant that manufactures boric acid in Wilmington, California. The effluent from the plant contains about 3,000 ppm of boron and has been discharged directly into the ocean for many years. Although U.S. Borax has indicated plans to discontinue this practice, it is claimed that marine life around the discharge point has not been affected.

Boric acid manufacture in only one location uses a process that generates wastes that go to land disposal.(1). This waste is arsenic sulfide from product purification and wastewater treatment by sulfide precipitation. The waste is sent to secured municipal landfill.

BROMIC ACID

Bromic acid, like bromine, is recovered from a dilute solution or waste stream by passing an aqueous solution over iron turnings to produce the so-called ferrosoferric bromide. This is decomposed by sodium carbonate, the excess carbon dioxide boiled off and the sodium bromide crystallized and sold.

Small packaged lots of bromic acid may be decomposed by the addition of reducing materials such as sodium thiosulfate, bisulfites or ferrous salts. This method is recommended by the Manufacturing Chemists Association, but procedures for recovery of the bromides produced are not provided. Instead, when reduction is complete, the treatment solution is neutralized with soda ash, and the solution is washed down the drain with a large excess of water. This process is satisfactory for small quantities of bromic acid, but the process is not recommended for larger quantities because valuable bromides will be lost (3).

BROMINE

Bromine is a valuable commodity and is seldom disposed of as a waste. If disposal of contaminated bromine or bromine solution is required, the bromine or bromine solution is returned to the manufacturer for recovery. The manufacturer will in most cases buy the bromine (3). If bromine vapor is in a process waste stream, it can be condensed easily. If the bromine is in an aqueous waste stream that is too dilute for shipment, concentration is required.

Concentration is accomplished in the same manner used to recover bromine from seawater, i.e., chlorine is used to oxidize any bromide to bromine and the solution is air stripped of the bromine, which is subsequently trapped in an ice cooled condenser. The impure bromine can be used or returned to the manufacturer.

An alternate recovery process is to pass an aqueous waste stream containing bromine over iron turnings to produce so-called ferrosoferric bromide. This is decomposed by sodium carbonate, the excess carbon dioxide boiled off and the sodium bromide crystallized and sold.

CALCIUM CARBIDE

For disposal of the calcium carbide waste, the material is slowly added to a large container of water. The acetylene gas liberated is burned off with a pilot flame. The remaining residue is lime and can be sent to landfill.

CALCIUM HYPOCHLORITE

See Ammonium Perchlorate and Other Oxidizers, page 233.

CALCIUM PHOSPHATE

See Tonnage, Insoluble, Nontoxic Products, page 256.

CAUSTIC-CHLORINE PRODUCTION WASTES

Diaphragm Cell Process Wastes

Three types of hazardous wastes can be generated by this process that may go to land disposal: asbestos, lead-contaminated treatment sludges, and chlorinated hydrocarbons. The latter, when not incinerated or disposed of in adjacent organic chemicals manufacture, are generally landfilled in drums or sent to disposal contractors (1). The asbestos and lead-contaminated solid wastes and sludges are generally sent to landfill or pond disposal on-site.

The treatment generally consists of a greater or lesser amount of segregation and dewatering prior to disposal. Of the 38 plants using this process known to be land disposing hazardous wastes, 21 use on-site landfills, 11 use on-site pond storage, and 11 send wastes to contractors having general purpose landfills.

Five plants incinerate their chlorinated hydrocarbons and 3 plants sewer them. 25 plants have either negligible chlorinated hydrocarbons or do not separate them from the products. In any event, the amounts of chlorinated hydrocarbons generated from these industries are small, and are the product of diaphragm cell circuits where treated graphite is used.

The lead-contaminated wastes are ponded on-site in 11 plants, land-filled on-site in 3 plants and off-site in 2 plants. An estimated 22 plants generate no lead-contaminated wastes for land disposal.

Mercury Cell Process Wastes

Mercury-contaminated wastes and chlorinated hydrocarbons are the hazardous wastes generated by this process (1). Of 28 plants having land-destined hazardous wastes from this process, one plant retorts all mercury-containing sludges, 8 others retort the mercury-rich wastes only, and 4 are storing the high mercury sludges in drums until decisions are made on final disposal.

One plant disposes of all sludges in a brine cell off-site, and one plant sends mercury-rich sludges to contractors for recovery. This latter disposal is used by others occasionally. 9 plants are now using on-site pond storage of sludges and 7 use on-site landfill. 4 plants send wastes to contractors for secured landfilling. Several plants employ combinations of these treatment/disposal technologies. Chlorinated hydrocarbon wastes from these plants are either not separated from product streams, separated and used elsewhere, or drummed for land storage.

A process developed by A.T. McCord and L.E. Wagner (9) involves disposing of wastes containing metallic mercury by treating such wastes with sulfuric acid and then neutralizing the treated wastes with a lime slurry to convert the metallic mercury into an insoluble form of mercury under neutral and alkaline conditions.

One of the major problems encountered in the disposition of mercury wastes, particularly those mercury waste sludges or muds generated from mercury cathode electrolytic cells, is the safe disposition of the metallic mercury contained therein. The problem resides in the vapor pressure of metallic mercury which is 0.00277 mm at ambient temperatures.

Investigations have demonstrated that if metallic mercury is placed in an enclosed flask, the atmosphere in the flask will eventually become saturated with mercury vapor to the extent of about 24 parts per million. Therefore, if mercury sludges or muds contain metallic mercury, it follows that the atmosphere above such muds can contain dangerous quantities of mercury.

Attempts have been made to destroy the metallic mercury in these wastes or muds by converting the metallic mercury into compounds. One common approach is to treat the mercury muds with sulfide to produce mercury sulfide, which is considered insoluble and is buried along with the muds as landfill. However, it has been found that any alkalinity will cause some dissolution of the mercury and the liquid portion obtained by redissolution can obtain several parts per million of soluble mercury which can leach out and ultimately find its way to our streams and other natural waters for possible consumption by marine life and directly or indirectly by human beings.

In accordance with the process, the waste muds containing metallic mercury are treated in a manner producing an insoluble form of mercury which remains insoluble in the muds under neutral or alkaline conditions, thereby rendering the waste muds harmless or more desirable as a landfill. To this end, the waste muds generated in the mercury cell process, and which contain approximately 35% $CaCO_3$ along with metallic mercury, are mixed with dilute sulfuric acid.

In lieu of dilute sulfuric acid, the muds can be blended with a pickle liquor, which is generated as a waste material in the pickling of steel and which contains sulfuric acid. The usual spent pickle liquor contains a mixture of about 3 to 10% sulfuric acid (H_2SO_4) and about 5 to 19% ferrous sulfate ($FeSO_4$). Using spent pickle liquor solves the problem of disposing of this waste material along with the disposal of waste mercury muds.

Mixing the diluted sulfuric acid or pickle liquor with the waste muds converts the calcium carbonate to calcium sulfate according to the following equation:

$$CaCO_3 + H_2SO_4 \longrightarrow CaSO_4 + CO_2 + H_2O$$

The sodium chlorate and perchlorate in the muds react with the acid to produce chloric and perchloric acid, respectively, according to the following:

$$2NaClO_3 + H_2SO_4 \longrightarrow 2HClO_3 + Na_2SO_4$$
$$2NaClO_4 + H_2SO_4 \longrightarrow 2HClO_4 + Na_2SO_4$$

Any finely divided mercury is oxidized to mercury chloride or mercuric oxide as follows:

$$Hg + 2HClO_3 \longrightarrow HgCl_2 + H_2O + O_2$$
$$Hg + HClO_3 \longrightarrow HgO + HCl + O_2$$

Thus, all the metallic mercury is destroyed by conversion into some mercury compound.

The resulting muds are further diluted with an equal or greater amount by weight of spent pickle liquor. The ratio of spent pickle liquor to the muds can range from 1:1 to 20:1 by weight, so long as there is sufficient sulfuric acid in the pickle liquor to react with the minor amount of mercury present in the muds. All of the calcium carbonate is converted to calcium sulfate and carbon dioxide is released. The above mercury compounds, and others that may be present in the waste muds, are converted to basic sulfates as illustrated for example, by the following equations:

$$3HgO + H_2SO_4 \longrightarrow HgSO_4 \cdot 2HgO + H_2O$$
$$HgCl_2 + H_2SO_4 \longrightarrow HgSO_4 + 2HCl$$
$$HgSO_4 + 2HgO \longrightarrow HgSO_4 \cdot 2HgO$$

The resulting slurry is then neutralized to about 7.5 pH with a thick lime slurry. The percentage of lime or calcium oxide in the slurry can range from about 5 to 28%, and preferably 15 to 25%. Calcium oxide concentrations above and below this preferred range are practical, but not desirable. The calcium carbonate present in the lime slurry reacts with the mercuric sulfate to produce insoluble basic mercury carbonate according to the following equation:

$$HgSO_4 \cdot 2HgO + CaCO_3 \longrightarrow HgCO_3 \cdot 2HgO + CaSO_4$$

Thus, any metallic mercury present in the waste sludges is converted to an insoluble form of mercury, namely basic mercuric carbonate, which remains insoluble under neutral or alkaline conditions and, as such, forms a highly desirable landfill material. Large concentrations of chlorides, such as sodium chloride or calcium chloride, do not solubilize the insoluble basic mercury carbonate.

CHLORATE-PHOSPHORUS MIXTURES

These are used in fireworks (cap and torpedo) manufacture.

Burning Pit Detonation

Torpedoes and caps in large quantities are detonated by burning in pits by applying heat from a fire. The devices are placed on top of a flammable substance such as straw and the flammable substance is ignited by means of a squib. Other explosives must be kept behind a barricade with overhead protection during destruction operations and located at a distance that assures safety.

Personnel should be similarly shielded. This destruction process is not satisfactory because individual components may not detonate and will constitute a personnel hazard during clean-up and because NO_x, metal oxides, P_4O_{10}, SO_2 and hydrogen chloride will be liberated during the destruction process on an uncontrolled basis (10).

Incineration

The approved method for the destruction of detonators and primers may be used for torpedoes containing chlorate-phosphorus mixtures (10). In this method the components are fed to the combustion chamber of a specially designed detonator destruction furnace by means of a channel chute and a special conveying device. Scrubbers are employed to remove toxic fumes and dusts such as NO_x, P_4O_{10}, HCl, SO_2 and metal oxides.

Potassium chlorate mixtures, toy torpedoes and toy caps which have been properly marked to permit due precautions in handling can be destroyed safely by burning in municipal or plant incinerator systems equipped with scrubbing and mist removal systems.

The incinerators should be fed by automatic conveyor devices, and the scrubber effluent should be neutralized and processed in a standard secondary treatment sewage system before discharge. This technique is recommended for disposal of the mixtures and devices.

CHLOROSULFONIC ACID

The method of disposal of packaged lots of chlorosulfonic acid recommended by the Manufacturing Chemists Association is satisfactory if the effluent from the disposal process is within the limits for pH, chloride ion and sulfate ions.

In this procedure, chlorosulfonic acid is decomposed by pouring small quantities from behind a shield onto a dry layer of sodium bicarbonate. After mixing thoroughly, the sodium bicarbonate, while being stirred, is sprayed with 6 M ammonium hydroxide. Then the sodium bicarbonate is covered with a layer of crushed ice; and while continuing stirring, the sodium bicarbonate mixture is again sprayed with 6 M ammonium hydroxide. When evolution of ammonium chloride (which must be trapped and disposed of) has partially subsided, the sodium bicarbonate solution is neutralized with hydrochloric acid. Following dilution to the concentration permitted for effluents, the treated solution is discharged into a stream or storm sewer (3).

CHROME PIGMENTS MANUFACTURING WASTES

The plants and plant complexes manufacturing chrome pigments, iron blues and cadmium colors generate land-disposed hazardous wastes containing lead, chromium, zinc, cyanide and cadmium hazardous constituents (1). Of the eight plants land-disposing such wastes, one disposed on-site by ponding and one by landfill after treatment, which consists principally of neutralization and precipitation. Six other plants use off-site disposal, one to municipal landfill, four to secured private landfill, and one to a land dump. Two remaining chrome pigments plants do not land dispose, one claiming to recover all wastes and the other discharging.

CHROMIUM SALTS

Soda ash or other alkaline chemicals are added to trivalent chrome or other heavy metal solutions to adjust the pH to about 9.5; thereby precipitating the insoluble heavy metal hydroxide gels. Aluminum sulfate (alum) or other suitable flocculating agents are also often used to aid in precipitation, settling of the gels and clarification of the effluent. Slaked lime must also be added to precipitate insoluble calcium fluoride, a reaction which requires at least 24 hours. The effluent is neutralized before sewering and the resultant sludge is landfilled (10).

The procedure can be carried out in either batch or continuous processes. Complete equipment systems are available for performing the precipitation on an automated or semi-automated basis. The biggest drawback of this procedure is the removal and disposal of the highly hydrated metal hydroxide sludges. It is not unusual for these sludges to contain upwards of 75 to 80% water by volume.

Precipitation and settling is normally a slow procedure and with a high volume of effluent it is normally necessary to have settling ponds or lagoons in which to allow the slow coagulation process to function. The clear effluent can be removed and the residual sludge dried. No feasible economic process has yet been developed for the reclamation of valuable metals or other credits from this type of sludge since it often contains other organic and inorganic materials which hinder effective purification and recovery (10).

Three ultimate disposal options for these sludges are: [1] landfill, [2] incineration, and [3] ocean disposal. Landfill is believed to be, by far, the most prevalent final disposal procedure but the problem of potential contamination of surrounding land or water tables due to the leaching of poisonous materials from the sludge must be considered. It is therefore recommended that California Class 1 type landfill be utilized for these wastes. Incineration reduces the sludge to an ash residue which can be more easily handled for ultimate disposal but at the same time could also lead to the formation of poisonous soluble metal oxides. Incineration of course requires additional equipment and a combustible fuel source as well as equipment for air pollution control.

Ocean disposal is currently being practiced to a considerable degree by municipal waste treatment plants located near the coast where the primary and secondary sludges can be either pumped or barged out to submarine disposal areas. This disposal approach has been the subject of much recent debate and it is likely that it will not long remain one of the significant sludge disposal options.

Chromite-contaminated land-destined hazardous wastes are generated by the manufacture of these products in three locations (1). One plant disposes of the chrome ore gangue off-site to a contractor for land and road construction purposes. One uses on-site land dumping and another uses approved private landfill.

COPPER HYDROXIDE AND SULFIDE SLUDGES

Three ultimate disposal options for these sludges are: [1] incineration, [2] landfill, and [3] ocean disposal.

Incineration

Incineration reduces the sludge to an ash residue which can be more easily handled for ultimate disposal but at the same time could also lead to the formation of poisonous, soluble metal oxides. Incineration also requires additional equipment and a combustible fuel source as well as equipment for air pollution control.

Landfill

Landfill is believed to be, by far, the most prevalent final disposal procedure but the problem of potential contamination of surrounding land or water tables, due to the leaching of poisonous materials from the sludge, must be considered.

Ocean Disposal

Ocean disposal is currently being practiced to a considerable degree by municipal waste treatment plants located near the coast where the primary and secondary sludges can be either pumped or barged out to submarine disposal areas. This disposal approach has been the subject of much recent debate and it is likely that it will not long remain one of the significant sludge disposal options.

CYANIDES

The reader is referred to the section which follows in this chapter on Hydrogen Cyanide and also the the section on Cyanides in the subsequent chapter on Metal Finishing Industry Wastes.

HYDRAZINE

Incineration

The U.S. Air Force has trailer-mounted incinerators capable of incinerating up to 6 gpm of hydrazine in a variety of mixtures with water (from 100% hydrazine to 100% water). The effluent from the units is limited to 0.03 pound per minute NO_x when incinerating hydrazine. These units are acceptable for disposing of large quantities of hydrazine (3).

Open Pit Burning

Hydrazine poured into an open lined pit is burned to nitrogen and water. The transfer of the hydrazine and the ignition must be accomplished by a remote means. For drum quantities of hydrazine this method is generally acceptable although since excessive NO_x might be generated another option would be preferred (3).

HYDRAZOIC ACID

Hydrazoic acid as a waste will not be encountered in large quantities and a satisfactory method for its destruction exists (10). An excess of nitrous acid decomposes hydrazoic acid in accordance with the following equation: $HN_3 + HNO_2 \rightarrow N_2 + N_2O + H_2O$. This reaction is accomplished by diluting the hydrazoic acid with 500 times its weight of water, slowly adding 12 times its weight of a 25% solution of sodium nitrite, agitating, and then slowly adding 14 times its weight of a 36% solution of nitric acid or glacial acetic acid.

A red color produced on addition of ferric chloride solution indicates azide to be present; more time or additional reagents are required to effect complete destruction of azide. After pH adjustment to 6.0 to 9.0, the reacted solution is diluted to a nitrite and nitrate concentration of less than 450 ppm, a typical maximum allowable concentration in an effluent being discharged to a storm sewer or stream.

HYDROFLUORIC ACID MANUFACTURING WASTES

Hydrofluoric acid manufacture generates fluoride-containing hazardous waste streams that are land-disposed (1). The principal fluoride constituent is calcium fluoride, which results from waste treatment by lime neutralization. Of the twelve plants generating such wastes, six are known to use on-site ponding disposal, two to use contractor general purpose landfills and one to send to a contractor who uses it for roadfill. The remainder is unknown.

HYDROGEN CYANIDE

Chemical Degradation

Hydrogen cyanide in aqueous streams is easily destroyed by first adjusting the pH to the 10 to 11 range and adding chlorine or hypochlorite which converts the cyanide to cyanate. Then the pH is adjusted to the 8 to 9 range and further oxidizing agent converts the cyanate to carbon dioxide and ammonia. Other methods though not often used are available. In an alternate equally acceptable method, hydrogen cyanide is removed from aqueous solution by passing air through the solution. The venting of this air containing hydrogen cyanide is not recommended, but instead the contaminated air must be passed through an alkaline scrubber (10).

Incineration

The destruction of hydrogen cyanide is usually accomplished by incineration which produces only carbon dioxide and nitrogen along with possibly some NO_x that can be stripped from the incinerator effluent with an alkaline scrubber, or reduced over a catalyst (10).

LEAD

Because lead is insoluble in water and not attacked by most inorganic acids or bases, lead in small quantities is often disposed of in landfills (10). Though this disposal method is satisfactory under most conditions, it is not recommended because lead is lost to future use. In addition, under certain conditions, lead is solubilized as either the chloride or the sulfate, which are both slightly soluble in water.

LEAD OXIDES

Since the lead oxides are insoluble in water, and also are very stable under ordinary temperatures, disposal of lead oxide wastes by means of landfill is acceptable. Mixed wastes of lead oxides such as glass and ceramic products often do not contain any water-soluble toxic components. Hence, the risk of contaminating the ground and surface waters by these waste materials is very minimal (10).

LEAD SALTS

Disposal of soluble lead compounds such as the acetate, nitrate and nitrite by landfill technique is generally unacceptable (10). Even the slightly soluble lead chlorite should not be disposed of by landfill. However, lead carbonate is practically insoluble in water and, in fact, occurs naturally as the mineral cerussite. Similarly, lead sulfide is also insoluble in water and occurs naturally as the mineral galena. Therefore, disposal of these two lead compounds by the landfill technique in California Class 1 type disposal sites might be considered adequate.

MAGNESIUM CHLORATE

See Ammonium Perchlorate and Other Oxidizers, page 233.

MAGNESIUM OXIDE

See Tonnage, Insoluble Nontoxic Products, page 256.

MANGANESE SALTS

Lagooning

All Mn-containing waste from the production of $MnSO_4$ as a by-product of hydroquinone is moved to a lagoon, where the manganese is precipitated with lime as $Mn(OH)_2$. The overflow from the lagoon contains less than one ppm Mn, which is below the limit set by the State of Illinois. This does, however, exceed the drinking water standard so the water should not be used as a potable source. There is no air emission except for an occasional, accidental release of quinone (10).

Landfill

The use for landfill of the insoluble residues from the production of manganese sulfate and manganese chloride by acid leaching is satisfactory (10). These residues consist mostly of silicone dioxide, with some highly-insoluble manganese oxides included. There is no air emission or leaching into any groundwater.

NICKEL ANTIMONIDE, ARSENIDE AND SELENIDE

Landfill

Some landfills are chartered to accept these compounds for ultimate encapsulation and landfill. They would most likely be combined with other compliant materials in a steel drum, and the steel drum would be given a concrete collar before burial. Landfilling is adequate in California Class 1 type facilities (10).

Recycle

The amounts of concentrated waste nickel antimonide, arsenide, or selenide which could arise are expected to be very small. For this reason, careful packaging and shipment back to the vendor or manufacturer by prior arrangement is an easy and safe means of disposal. The manufacturer can either reprocess the material, if economically feasible, or could dispose of it in his own facilities. Return to the manufacturer or vendor should be considered before any other disposal option is considered (10).

NICKEL SULFATE, CHLORIDE, NITRATE

Nickel-containing hazardous wastes for land disposal in relatively small amounts are generated by the manufacture of nickel sulfate (1). They result from the treatment of waste waters by raising the pH to precipitate metallic salts. Two plants in this segment directly sewer their wastes. One plant uses on-site landfill for the precipitates and another uses contractor landfill (general purpose). Perhaps the major source of nickel salts, however, is metal plating effluents.

Nearly all the small and medium size plating shops, and even some of the large ones, discharge their metal laden aqueous effluents to municipal sewers. Some plants have facilities to precipitate and clarify before discharge, but they are estimated to be about 15% of the total shops. In many areas, especially large metropolitan areas, this practice is allowed to continue as long as the toxicity of the nickel dishcarges has little or no effect on the continuous operation of any of the biological waste treatment processes of the sewage treatment plant.

Many of the metal cations, nickel included, precipitate with sulfides in the domestic sewer lines while in transit to the sewage treatment plant. Thus, a great proportion of them are removed in the primary treatment process and most of the remainder are bound up in the sludge resulting from the secondary treatment processes.

The method is considered adequate provided that: [1] the sewage treatment authorities can effectively regulate the discharge of toxic metal compounds into the sewer system at a rate which will not disturb the efficiency of the sewage treatment processes; and [2] the operation and design of the sewage treatment plant is such that effective treatment is accomplished and the effluent from the treatment plant complies with the local purity regulations for discharge to open waters. The secondary sludge must be landfilled such that the provisional limits are not exceeded or they must be placed in a California Class 1 type landfill (10).

Landfill

When caustic or an alkaline stream in the plant is combined with an acid plating shop effluent, the heavy metal hydroxides precipitate. These precipitates are normally gelatinous sludges with very high water content (70 to 80%). These sludges normally contain a wide mixture of materials and treatment after precipitation includes passage through a sand bed or similar clarifier. The sludge is removed for landfilling and the effluent is discharged to a municipal sewer or open waterway if permitted by the regulating authorities (10).

NITROGEN OXIDES

Incineration

Burning of nitrous oxides can produce the undesirable oxides of nitrogen, NO and NO_2, and therefore uncontrolled incineration is not recommended for the disposal of nitrous oxides. Nitrous oxide and other nitrogen oxides can be incinerated in a unit designed and operated to produce elemental nitrogen and water vapor, however. An apparatus developed by T.D. Felder, Jr. (11) is a burner as shown in Figure 77.

The device is composed of inlet conduits **1** and **2**, a stack or base cylinder **3**, and a flared conical discharge cap **4**. Off-gas, or any gas containing nitrogen oxides in concentrations that are noxious, is fed into conduit **2**; a reducing gas such as natural gas is fed into conduit **1**; these gases enter the cylindrical portion **3** of the stack via ports **5** and **6** and as

FIGURE 77: APPARATUS FOR CONVERTING NITROGEN OXIDES TO INNOCUOUS
GASES

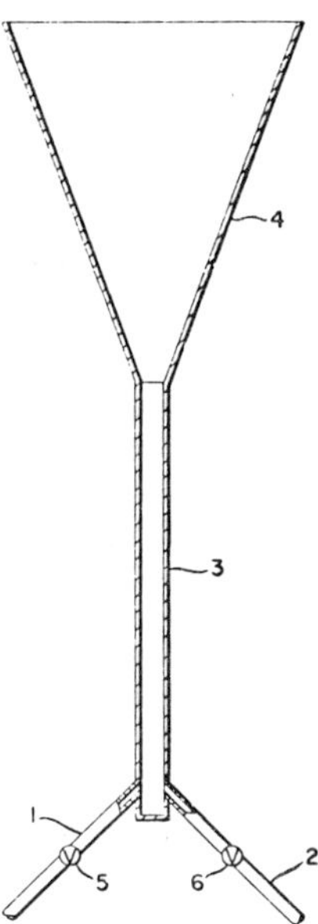

Source: U.S. Patent 3,232,713

the resulting mixture flows into the conical (inverted) discharge cap **4** the mixture is ignited and the nitrogen oxides present converted to nitrogen and water.

A device developed by P. Kandell et al (12) involves the total abatement of nitrogen oxides
in tail gas streams from nitric acid plants. Total abatement is conventionally carried out
by mixing preheated tail-gas with a fuel and passing the mixture over a catalyst. For
methane fuel or natural gas the preheat temperature is approximately 900°F whereas for
other fuel gas, such as hydrogen, the temperature may be somewhat lower. In passing
over the catalyst, the residual oxygen is removed and some reduction of the nitrogen oxides
occurs.

The outlet temperature of the catalyst bed is limited to about 1450°F maximum because
higher temperatures cause damage and deactivation to the catalyst. Because of the limitation in temperature rise across the bed, the gas is generally cooled to approximately the
same temperature as the inlet gas, mixed with additional fuel gas, and passed over a second
catalyst bed where most of the remaining nitrogen oxides are reduced to N_2.

Interstage cooling is generally accomplished by a waste-heat boiler or other heat exchanger,
the outlet gas from the second catalyst bed is then cooled and sent to a hot gas expansion
turbine which recovers power and helps to drive the nitric acid plant process air compressor.
The amount by which the gas is cooled is determined by the metallurgy of the hot-gas expansion turbine and is generally limited to about 1000°F for most commercial units. The
cooling is similarly performed in a waste-heat boiler or other heat exchanger. Exit gas from
the expansion turbine goes to a vent stack.

The process represents an improvement on the above described tail-gas treatment. In this
process the hot gas between catalyst stages is cooled with a direct water quench. Also, the
hot gas after the second catalyst stage is similarly cooled with a direct water quench. This

method of operation accomplishes the following: [1] the quantity of hot gas to the expansion turbine is increased by evaporation of quench water, thus leading to greater power recovery on the expansion turbine, [2] the exit gas from the expansion turbine is diluted with water vapor thus further reducing the concentration of residual nitrogen oxides in the stack gas discharged to the atmosphere, [3] the necessity for a waste-heat boiler or other heat exchanger is eliminated between catalyst stages and again after the second stage, thus reducing the plant cost, and [4] the reduction of NO to N_2, which is the more difficult and controlling reaction, is improved as follows (assuming methane fuel gas):

$$5CH_4 + 8NO + 2H_2O \longrightarrow 5CO_2 + 8NH_3$$
$$2NH_3 + 3NO \longrightarrow \tfrac{5}{2}N_2 + 3H_2O$$

In the first of these two reactions, the H_2O is provided by the water vapor arising from the direct water quench, and an increase in the amount of this reactant drives the reaction in the forward direction.

The apparatus used in the following example consisted of two conventional combustors each 2½ feet deep, modified to accept water spray nozzles at the outlet. Eight water nozzles spaced around the internal circumference of the shell are used for good dispersion or alternatively sprays can be installed in the outlet piping. The nozzle heads are adjusted to give a fairly fine spray. The two combustors are operated in series. Such an apparatus is shown in Figure 78.

FIGURE 78: APPARATUS FOR REMOVING NITROGEN OXIDES FROM GASES

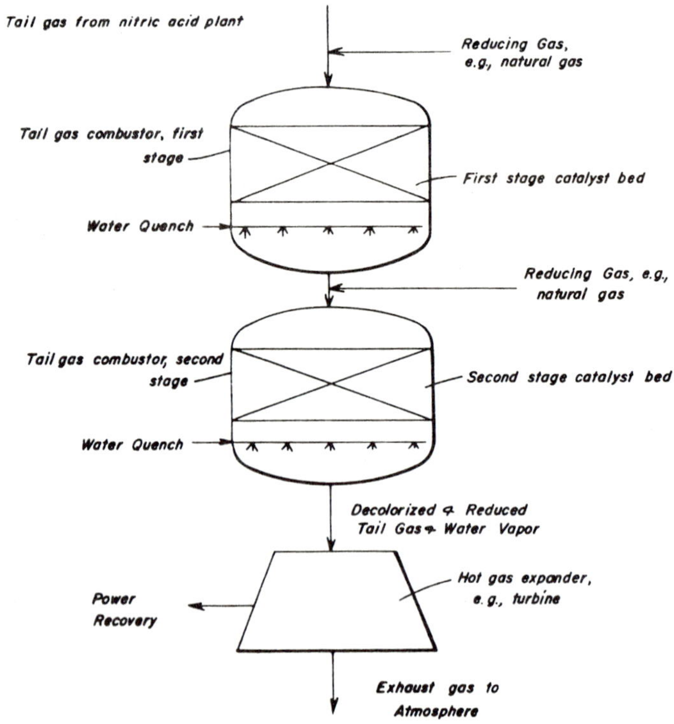

Source: U.S. Patent 3,567,367

PHOSPHATE SLIMES

Landfill

A process developed by H.P. Ledden et al (13) comprises just before deposition in a landfill area admixing: (a) waste slimes from phosphate ore processing operations, which consist essentially of an aqueous suspension of solid particles at least 99% by weight of which
on a dry basis have a particle size smaller than 105 microns (–150 mesh), the suspension
having a solids content of from 5 to 25% by weight at the time of admixing; and (b) per
part by weight on a dry basis of the slimes, from 0.8 to 5 parts by weight on a dry basis
of waste tailings, i.e., the sands from an ore processing operation, which comprise solid
particles at least 95% by weight on a dry basis of which have a particle size in the range
of 105 to 1,000 microns (–16+150 mesh).

About 0.05 to 0.5 pound of a polyacrylamide which is up to 40% hydrolyzed and has a
molecular weight of at least 1,000,000 per ton of slime solids is added to the sands. The
water released from the above mixture for a period of at least one month (i.e., until the
solids compact sufficiently to possess load bearing characteristics) is then separated off.

A process developed by C.C. Cook et al (14) is one for effective phosphate slime disposed
in landfills. Figure 79 is a flow diagram of the process.

FIGURE 79: PROCESS FOR LANDFILL DISPOSAL OF PHOSPHATE SLIMES

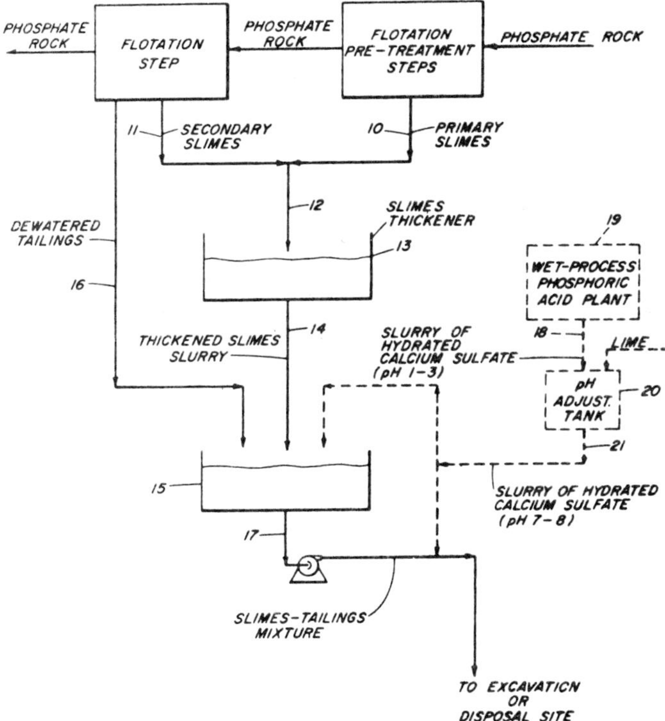

Source: U.S. Patent 3,718,003

The waste primary slimes **10** from the flotation pretreatment process steps in a phosphate ore processing plant and the waste secondary slimes **11** from the flotation step itself are admixed to form a dilute aqueous suspension of slime solids **12** which typically contains 1 to 2% solids. Usually about 90% of the total slimes produced are primary slimes with the remaining 10% being secondary slimes.

All or a portion of the primary or secondary slimes may be used to produce stream **12**. It is, of course, possible to use only the primary or only the secondary slimes to make up stream **12** if desired. Slimes stream **12** is fed to thickener **13** where it is typically thickened for 20 to 50 hours, and preferably 22 to 26 hours, to produce a solids content in the slimes of about 3 to 6%.

The thickener underflow **14** containing 3 to 6% solids and typically 3 to 4% solids, is then fed to tank **15** whereupon it is admixed with waste tailings **16** from the flotation step. Tailings **16** are dewatered and typically contain about 20% water. The thickened slimes stream **14** is used to repulp tailings **16** in tank **15** to produce an aqueous slurry of a mixture of slimes and tailings **17** which is then pumped out of tank **15** to the waste disposal excavation or other suitable land reclamation site.

The respective quantities of streams **14** and **16** fed to tank **15** will, of course, depend on several factors such as, for example, the ratio of tailings to slimes solids desired in the landfill composition and provision of sufficient water to render stream **17** pumpable to the excavation site. The required proportions of tailings to slimes in the landfill compositions have been discussed hereinabove. Generally, stream **17** will contain a solids content of from about 15 to 45% and preferably 40 to 45%.

If it is desired to add hydrated calcium sulfate to the landfill composition, an aqueous slurry of hydrated calcium sulfate **18** from the wet process phosphoric acid plant **19** or from the calcium sulfate ponds of such a plant is fed to a suitable vessel **20** where the acid pH of stream **18** is rendered alkaline by treatment with lime. The alkaline calcium sulfate slurry **21** is thereupon added to tank **15** where, along with the slimes stream **14**, it serves to repulp tailing **16**. Optionally stream **21** can be injected directly into the pipeline which conducts the slimes-tailings mixture **17** to the excavation site where, due to the continuous churning and mixing which occurs in the stream flowing in the pipeline, the calcium sulfate admixes with the slimes-tailings mixture. The quantity of stream **21** required will, of course, depend on how much hydrated calcium sulfate is desired in the land fill composition. Stream **21** can contain a wide range of solids content, usually up to about 20%.

The slimes-tailings mixture **17** is then pumped to an excavation which is typically a ditch a half to one mile or longer in length, 40 to 60 feet deep, 30 to 50 feet wide at the base and from 100 to 150 to 200 feet wide at the top. Mixture **17** is introduced at one end of the ditch. Once mixture **17** is deposited in the ditch, a substantial portion of the slimes tend to physically separate from the tailings and flow away from the tailings.

These slimes roll on ahead of the tailings accumulating ahead of the tailings where they are then given the opportunity to concentrate to the required solids content of 10 to 25%. Almost as soon as mixture **17** is deposited in the ditch, water is rapidly released from the mixture and proceeds by gravity down the ditch until it eventually reaches a dam at one end of the ditch. This dam will typically have an overflow weir which directs the overflow from the accumulated water into the fresh water return canal of the phosphate rock processing plant.

After those discharged slimes which have accumulated ahead of the tailings have become concentrated to the required 10 to 25% solids level, the discharge pipe for mixture **17** is advanced to discharge mixture **17** onto these concentrated slimes. As the mixture is discharged, most of the slimes discharged again separate from the tailings in the mixture and flow on ahead of the tailings. The discharged tailings, however, settle by gravity into the concentrated slimes and mix with these slimes to form a fertile landfill composition of acceptable bearing strength.

In effect, these concentrated slimes remix with the slimes-tailings mixture as the deposited mixture advances along the ditch. The discharged slimes-tailings mixture is gradually advanced along the length of the ditch until the ditch is substantially filled with the landfill composition. Where desired, low check dams can be placed at intervals across the ditch to impede slimes flow and assist in concentrating the slimes solids to the 10 to 25% level where the tailings will mix with the slimes instead of merely displacing them or capping them.

It should also be noted that despite the fact that substantial portions of the slimes flow away from the tailings when the slimes-tailings mixture is deposited in the ditch, the tailings will retain as much as 1 to 3% by weight on a dry basis of the slimes even at the surface of the deposited tailings. This clay provides nutrients for plant growth and also acts as a sponge to retain the moisture necessary for plant growth and erosion inhibition.

In another embodiment, landfill is achieved by draining the water from a slimes disposal area and pumping a slimes-tailings mixture, prepared as previously described to one end of the disposal area. The advancing mixture has a rolling motion which increases mixing with the slimes and brings about the release of substantial amounts of water ultimately providing a firm fertile soil.

In yet another embodiment, slimes from a previously employed disposal site, which have settled to between 10 and 25% solids, preferably 15 to 25% solids, are pumped to a mixer such as a blunger, pug mill, or the like, where from 60 to 99%, and preferably 60 to 70%, by weight of tailings are mixed with the slimes whereupon the resultant mixture is deposited in an excavation from which released water can drain. The deposited mixture is firm and fertile.

PHOSPHORUS

Incineration

Phosphorus in phossy water is oxidized by spraying into an incinerator equipped with an alkaline scrubber or it is oxidized by blowing through a flame followed by an alkaline scrubber. Rat and vermin poisons which contain white phosphorus and other small lots of wet phosphorus wastes, if properly segregated to permit safe handling, can be burned in municipal incinerators equipped with automatic conveying equipment and scrubbing and mist removal devices. The fly ash produced in such incinerators will generally neutralize the H_3PO_4 made (10).

Open Pit Burning

Rat and vermin poisons and other solid or pasty residues containing phosphorus should be burned in an isolated location, the residue neutralized and diluted with water to a phosphate concentration less than 50 milligrams per liter (10).

PHOSPHORUS CHLORIDES

The major manufacturers of phosphorus pentachloride and phosphorus oxychloride admit that they have never had the problem of disposing of large quantities of concentrated waste material (10). The major sources of concentrated wastes are spills in loading material left in emptied containers and heels left in tank cars or tank trucks.

Chemical Degradation

All three phosphorus compounds decompose readily in water. Therefore, flushing with water is an extremely effective and economical method of disposal. Under controlled conditions, utilizing large amounts of water, phosphorus trichloride, pentachloride and oxychloride are reacted to phosphoric and hydrochloric acids which are easily neutralized. Monsanto in Sauget, Illinois occasionally uses a degradation technique to scrap phosphorus trichloride and phosphorus oxychloride. The plant practice is to slowly add phosphoric acid to the material and then very slowly neutralize the mass with a sodium carbonate solution. The neutralized mass is then sewered to the waste treatment plant. This proves to be an adequate disposal technique for larger quantities of waste material (10).

Recycle

Waste material is easily recycled to the plant production system. Drums, tank cars and tank trucks can be drained or vacuumed to holding tanks, returned to the production process and purified. FMC Inc., Monsanto and Hooker Chemical Co., said all material returned to them is handled in this manner (10).

PHOSPHORUS PENTASULFIDE

Chemical Degradation

Several manufacturers of phosphorus pentasulfide recommend dousing small quantities of the chemical with copious amounts of water, while taking precautions to keep hydrogen sulfide generation at a low level. Large volumes of water keep the decomposition temperature low, thereby reducing the H_2S emission rate. The aqueous decomposition product (phosphoric acid) can then be treated by standard industrial wastewater treatment techniques (e.g., neutralization) (10).

Standard neutralization techniques and dilution of the resultant stream are the simplest and cheapest method of disposal. The danger is thereby limited to the increase in phosphate concentration which could possibly lead to an increased rate of water eutrophication.

Incineration

Phosphorus pentasulfide can be incinerated to yield phosphorus pentoxide (P_2O_5) and sulfur dioxide (SO_2). Large quantities of waste material should be disposed of by incineration only where the combustion equipment is capable of removing the combustion products from the exhaust gas stream (10).

Landfill

Phosphorus pentasulfide can be buried in a sanitary landfill disposal site. This essentially is a slow hydrolysis process which must be carefully done to minimize the evolution of hazardous hydrogen sulfide. Very small quantities may be adequately handled in this manner (10).

SELENIUM

The selenium containing solid wastes from the manufacture and reconditioning of Xerox drums are currently disposed of by burial in a segregated section of a landfill site for industrial wastes. Waste sludges containing selenium and toxic heavy metals from the manufacture of paint and pigments are also disposed of in approved sanitary landfill sites. It is felt that sanitary landfill is an acceptable and adequate method for the disposal of selenium wastes, provided that the sites used are located over nonwater-bearing sediments or have only unusable groundwater underlying them and are completely protected from flooding, surface runoff or drainage, so that all waste materials and internal drainage are restricted to the site. In other words, any sanitary landfill sites approved for the disposal of selenium wastes must meet the criteria for a Class 1 disposal site in California (10).

SODIUM AZIDE

Chemical Degradation

Fairmont Chemical's method for disposal of sodium azide is to react it with sulfuric acid solution and sodium nitrate in a hard rubber or teflon vessel (10). Nitrogen dioxide is generated by this reaction and the gas is run through a scrubber before it is released to the atmosphere. It is conceivable that any industry that handles sodium azide in significant amounts would probably have facilities for reacting the sodium azide and treating the evolved nitrogen dioxide gas.

Deep Well Disposal

A process developed by F.C. Dehn (15) is one in which the disposal of salt solutions contaminated with azide is accomplished by depositing the solution in a subterranean cavity, maintaining the solution in the subterranean cavity until it is essentially free of azide, and pumping the essentially azide-free solution to the surface.

The subterranean cavity should be deep within the earth so that the solution will not be mixed to any substantial degree with surface groundwaters. Most subterranean cavities used for the practice of this process are at least 300 feet from the surface of the earth. More often, they are at least 500 feet deep. It is preferred that they be at least 1,000 or even 1,500 feet below the surface.

Many wells especially suitable for the practice of this process have been or are presently used for the solution mining of salts. Examples of such salts are sodium chloride, potassium chloride, and sodium carbonate. Wells used for the recovery of sulfur by the Frasch process are also satisfactory. These wells may be abandoned for production purposes. Alternatively, the well may be located in an active field where one or more other wells are currently producing. In either case, the by-product salt from the production of hydrazoic acid may be the same or different from the salt produced from the field.

In a preferred embodiment of this process, the salt is the same as the principal salt comprising the subterranean deposit. Because hydrazoic acid is most often prepared from sodium azide, it is convenient to use hydrochloric acid as the feed acid and to deposit solutions of the by-product sodium chloride in a sodium chloride well. The azide contaminant often decomposes with the passage of a short time, usually a matter of a few weeks or months. Rarely does azide persist longer than about 4 weeks. Upon reduction of the azide concentration to any insignificant values, e.g., 30 parts per billion parts of by-product salt, the well may be placed into production retrieving most of the salt deposited. Figure 80 is a diagram showing the operating sequence of the process.

As shown there, equimolar portions of sodium azide and sodium hydroxide are fed through line **1** to mix tank **2** where they are contacted with hydrochloric acid introduced through line **4**. A gas stream containing mostly nitrogen, a small percentage (usually about 3%) of hydrazoic acid, some water vapor, and some hydrogen chloride is introduced through line **6** and sparger **7** and bubbled through the mixture in mix tank **2** to dilute any hydrazoic acid formed and sweep it from the mix tank through line **8**. Because hydrazoic acid is explosive in concentrated amounts, dilution by an inert gas such as nitrogen, helium, argon, neon, xenon, krypton, air, methane, ethane, water vapor, CF_4, CF_3Cl, or mixtures thereof is desirable to reduce the possibility of an explosion.

The gas in line **8** advantageously contains about 5 to 8% hydrazoic acid. Greater or lesser proportions may be used if desired. Liquid from mix tank **2** having a pH of about 5 to 6 is forwarded to mix tank **10** through line **12** by pump **14**. Hydrochloric acid is introduced through line **16** to lower the pH of the reaction mixture to about 2 to 3. Nitrogen containing small amounts of hydrazoic acid and water vapor is introduced through line **18** and sparger **20** to sweep hydrazoic acid from mix tank **10** through line **6** and into the mixture contained in mix tank **2**.

FIGURE 80: PROCESS FOR DISPOSING OF AZIDE-CONTAMINATED SALT SOLU-
TIONS USING DEEP WELLS

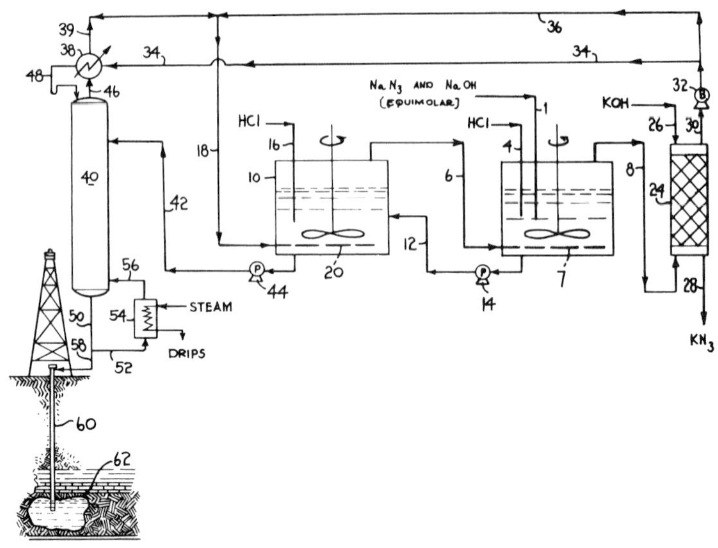

Source: U.S. Patent 3,768,865

A gas stream including nitrogen, hydrazoic acid, and water vapor is passed through line **8**
to scrubber **24** where it is contacted with aqueous potassium hydroxide introduced through
line **26**. The scrubber may be of any of many conventional designs such as, for example,
bubble plate column, sieve tray column, packed column, spray column, disc and donut
column, or liquid filled column. Flow may be either countercurrent or parallel; a packed
column utilizing countercurrent flow is shown. Potassium azide solution removed from
scrubber **24** through line **28** may be used directly as an aqueous solution or it may be
forwarded for further processing such as crystallization, drying and packaging.

Nitrogen, some water vapor, and a small amount (on the order of about 1%) hydrazoic
acid are removed from scrubber **24** through line **30** where they are forwarded by blower
32 to lines **34** and **36**. Most of the gas leaving the scrubber **24** passes through line **36** to
line **18** and then into mix tank **10**. A minor portion of the gas leaving scrubber **24** passes
through line **34**, through condenser **38** of stripping column **40**, through line **39**, and onto
line **18** where it combines with the gas flow from line **36** for passage through the liquid in
mix tank **10**.

It may be seen that nitrogen is recycled through mix tanks **2** and **10** and through scrubber
24, thereby containing most of the hydrazoic acid within the system. Make-up nitrogen
may be added at any convenient location in the nitrogen system. Should there by any
build-up of inert gases in the nitrogen system, the excess may be vented, preferably through
a basic solution, to neutralize any hydrazoic acid present. Liquid from mix tank **10** usually contains some dissolved hydrazoic acid which is often advantageously recovered.

Stripping column **40** is suitable for this purpose. The stripping column itself is of conventional design and may be a bubble cap column, a sieve tray column, a packed column, or
some similar device. Liquid from mix tank **10** is passed through line **42** by pump **44** and
introduced to stripping column **40** at or near the top thereof. Vapor taken overhead is
passed through line **46** to condenser **38**. Condenser **38** is preferably run at total reflux.

Reflux may be returned to stripping column through line **48**. Nitrogen from line **34** is passed through condenser **38** where it entrains noncondensibles and hydrazoic acid and passes through lines **39** and **18** to mix tank **10**. Bottoms, comprising principally sodium chloride solution, are removed from the stripping column through line **50** and split into two streams. One of these streams passes through line **52** to reboiler **54** where it is heated and returned to stripping column **40** through line **56**. The other stream passes through line **58** to a salt (NaCl) well where it is passed through shaft **60** and deposited in subterranean cavity **62**.

Incineration

The complete and controlled oxidation of sodium azide in air or oxygen with adequate scrubbing and ash disposal is possible if the sodium azide is mixed with other combustible wastes. Since the combustion process produces nitrogen and sodium oxides, an adequate gas clean-up system must be installed to alleviate the air pollution problem (10).

SODIUM CHLORATE

Chemical Degradation

The most acceptable method for disposing of both solid and aqueous sodium chlorate waste is to reduce it with a reducing agent in a plastic vessel. If the waste is present in the solid form, it is first dissolved and then reduced in the vessel. The most economical and readily available reducing agent is iron filings. Pickle liquor from the iron and steel industry is also a good reducing agent. The sodium chlorate is reduced to sodium chloride in the reduction process and the resulting iron precipitate can be removed with a lime slurry, soda ash, or sodium hydroxide followed by a lagooning operation (10).

Incineration

A small amount of sodium chlorate waste can be incinerated if it is mixed with a large quantity of other combustible waste (10). An adequate gas clean-up to remove the gaseous combustion products must be installed to alleviate the air pollution problems.

SODIUM HYPOCHLORITE

See Ammonium Perchlorate and Other Oxidizers, page 233.

SODIUM METAL

Primary Na producers either drum their waste and dump it at sea with perforation in the drums to assure complete reaction and destruction (Ethyl Corporation, Baton Rouge, Louisiana) or burn it mixed with kerosene in a closed chemical reactor (DuPont Corp., Niagara Falls, New York) (10).

Incineration

The use of a reactor allows the burning to be controlled by varying the feed rates of sodium and air to prevent a violent conflagration and also allows the exhaust gases to be scrubbed with water to remove the sodium oxide and carbonate.

Ocean Disposal

The 30,000 pounds per year of Na waste produced by Atomics International is packed in barrels on site. The California Salvage Company picks up this material under contract, fits the barrels with concrete collars and dumps them at sea off Santa Catalina Island. The cost of this service is $2,200 on an annual basis. This method of disposal is currently considered satisfactory to all parties concerned, including the California Water Quality Control Board and the California Department of Fish and Game (10).

A study of the disposal of sodium sludge off the Texas coast was performed (7). The sludge consisted of approximately 75% metallic sodium and 24% metallic calcium, and was a waste product from a petrochemical plant. The barreled sodium sludge was punctured and dropped overboard; when seawater made contact with the sodium, the drums exploded. Field observation and sampling indicated that the explosions and elevated pH (resulting from the interaction of the sodium with seawater) had no significant effects on the Sargassum and zooplankton communities as compared to those in the control areas. An absence of floating dead fish may have been due to the barrenness of the area.

Corps of Engineers records indicate that on two occasions unpierced barrels loaded with sodium sludge have been retrieved from Gulf waters by fisherman. In these cases, it is almost certain that the disposal operators did not dispose of the barrels in the designated disposal area.

SODIUM PERCHLORATE

See Ammonium Perchlorate and Other Oxidizers, page 233.

SODIUM SILICOFLUORIDE MANUFACTURING WASTES

The manufacture of this product generates fluoride-containing land-destined wastes (1). Of the three plants generating such wastes, one is known to deep well the fluoride waste and one uses on-site pond disposal. The other is unknown.

SULFURIC ACID

Sulfuric acid is used in large quantities by the steel mills to pickle (remove surface rust and mill scale from) steel stock (rod, bar, sheet, plate, and the like) prior to fabrication or other manufacturing processes. The acid (pickle liquor) is usually removed from further process use when between one-half and two-thirds of the free acid has been converted to ferrous sulfate by reaction with the steel and iron oxides. The spent acid sent to disposal contains up to 7% free H_2SO_4 and up to 30% $FeSO_4$.

In some steel plants, hydrochloric acid is used for all or part of the pickling, and the spent acid wastes contain free hydrochloric acid and ferrous chloride instead of, or in addition to, the sulfuric acid and ferrous sulfate resulting from the use of sulfuric acid for pickling.

Sulfuric acid is also used in large quantities by titanium pigment plants, which digest the ore (whose principle impurity is iron) with H_2SO_4. The process wastes are in the form of a liquor which contains 7 to 9% free H_2SO_4, 8 to 10% $FeSO_4$, and a mud slurry with 15 to 20% inert solids.

Incineration

Waste sulfuric acids from many organic chemical and petroleum refining processes contain organic impurities which are burned out by flames in fired vessels which also concentrate the sulfuric acid for reuse.

Ocean Disposal

Spent acid wastes dumped at sea comprise 58% of all industrial wastes so disposed of, or about 2,700,000 tons.

THALLIUM COMPOUNDS

Small quantities may easily be disposed of by sealing in polyethylene bags and burying in Class 1 landfill, i.e., landfills with no access to groundwater.

TITANIUM DIOXIDE PROCESS WASTES

Titanium dioxide may be made by the older sulfuric acid route or by the newer chloride route. Wastes are generated by the chloride process by solubilizing of ore impurities and subsequent treatments (1). Of the nine plants now generating such wastes, one deep wells, one disposes at sea by barging, two send the wastes to contractors for off-site disposal, and four dispose on land by on-site ponding. The remaining plant is now discharging these wastes. Prior treatment in some cases consists of neutralization.

Ocean Disposal

Studies of the dispersion and environmental effects of acid-iron water discharged from barges in the New York Bight disposal area show that wastes consisting of ferrous sulfate and spent sulfuric acid in a freshwater solution, are the by-products of a titanium dioxide paint pigment manufacturing process using the sulfuric acid route (7).

The studies showed that the toxic effects of these wastes are minimal. Laboratory toxicity tests to determine if the acid-iron wastes had any effect on the growth of green plankton algae showed concentrations of wastes which severely limited growth are found only close to the discharge barge and are diluted rapidly at sea. Zooplankton from samples immediately behind the barge were temporarily immobilized, but recovered rapidly on dilution with un-polluted water at a short distance astern of the disposal barge.

Additional findings showed that the accumulation of iron hydroxide, resulting from the chemical reaction of the wastes with seawater, was about twice as great in 1957 as that observed in 1948, and 50% more than that observed in 1950. This was less than might have been expected, however, from the increase in the discharge rate that had occurred over the same period. No evidence was found to indicate that the general turbidity of the waters caused by the iron precipitate has increased outside the general disposal area. Since establishment of the acid-iron disposal operation in 1948, a concentration of blue-fish has developed on the outer boundaries of the disposal area, thus creating a highly popular sport fishery that did not previously exist.

TITANIUM TETRACHLORIDE PROCESS WASTES

Incineration

A process developed by S.F. Brzozowski (16) is one in which waste gas from the chlorination of titaniferous ores is treated with steam before being vented and burned.

Ocean Disposal

As noted in *Chemical Week* (17), the chloride process generates 1.2 pounds of solid wastes for each pound of TiO_2 produced. This requires 150 barge loads per year with a capacity of one million gallons per barge from the DuPont Edgemon, Delaware plant alone.

Recovery of ferric chloride is one partial answer to EPA criticisms of ocean disposal. However, miscellaneous metal chlorides make up 25% of the waste stream. Conversion of all waste chloride to recoverable HCl and metal oxides is under study. A new DuPont plant in DeLisle, Miss. is scheduled for startup in 1977 and will use deep well disposal to avoid ocean disposal problems.

TONNAGE, INSOLUBLE, NONTOXIC PRODUCTS

Materials placed in this category include: aluminum oxide, asbestos, calcium phosphate, magnesium oxide, vanadium pentoxide, and zinc oxide. For the waste disposal of the materials in this category, the landfill method is recommended because the waste materials are all insoluble in water and nontoxic in nature. In fact, most of them originally come from the land as naturally-occurring minerals. Should they be washed into any water source, by rainfall, for instance, they can be easily removed from water by filtration. If a trace concentration of the material remains in the water supply, it would not constitute a hazard. Also, they are in general, chemically stable and will not degrade to yield air or water pollutants. The landfill method therefore offers a convenient and economic way to dispose of the waste material (3).

VANADIUM PENTOXIDE

See Tonnage, Insoluble, Nontoxic Products, above.

ZINC CHLORATE

See Ammonium Perchlorate and Other Oxidizers, page 233.

ZINC OXIDE

See Tonnage, Insoluble Nontoxic Products, above.

REFERENCES

(1) R.G. Shaver, L.C. Parker, E.F. Rissman, K.M. Slimak and R.C. Smith, *Assessment of Industrial Hazardous Wastes Practices: Inorganic Chemical Industry,* Report PB 244 832, Springfield, Va, Nat. Tech. Information Service (March 1975).

(2) Booz-Allen Applied Research, Inc., *A Study of Hazardous Waste Materials, Hazardous Effects and Disposal Methods,* Report PB 221 466, Springfield, Virginia, Nat. Tech. Information Service (July 1973).

(3) R.S. Ottinger, J.L. Blumenthal, D.F. Dal Porto, G.I. Gruber, M.J. Santy and C.C. Shih; *Recommended Methods of Reduction, Neutralization, Recovery or Disposal of Hazardous Waste,* Vol. XII, Report PB 224 591, Springfield, Va., Nat. Tech. Information Service (August 1973).

(4) H.A. Adams, K. Kunchal and J.L. Callahan; U.S. Patent 3,812,236; May 21, 1974; assigned to The Standard Oil Company (Ohio).

(5) Y. Kajitani, H. Ito and S. Mitsuda; U.S. Patent 3,838,193; September 24, 1974; assigned to Kawasaki Jukogyo KK.

(6) A. Bietlot; U.S. Patent 3,914,945; October 28, 1975; assigned to Solvay & Cie.

(7) A.W. Reed, *Ocean Waste Disposal Practices,* Park Ridge, N.J., Noyes Data Corp. (1975).

(8) R.S. Ottinger, J.L. Blumenthal, D.F. Dal Porto, G.I. Gruber, M.J. Santy and C.C. Shih, *Recommended Methods of Reduction, Neutralization, Recovery or Disposal of Hazardous Waste,* Vol. VIII, Report PB 224 587, Springfield, Va., Nat. Tech. Information Service (Aug. 1973).

(9) A.T. McCord and L.E. Wagner; U.S. Patent 3,804,751; April 16, 1974; assigned to Chem-Trol Pollution Services, Inc.

(10) R.S. Ottinger, J.L. Blumenthal, D.F. Dal Porto, G.I. Gruber, M.J. Santy and C.C. Shih, *Recommended Methods of Reduction, Neutralization, Recovery or Disposal of Hazardous Waste,* Vol. XIII, Report PB 224 592, Springfield, Va., Nat. Tech, Information Service (Aug. 1973).

(11) T.D. Felder, Jr.; U.S. Patent 3,232,713; February 1, 1966; assigned to E.I. du Pont de Nemours & Company.

(12) P. Kandell and G. Nemes; U.S. Patent 3,567,367; March 2, 1971; assigned to W.R. Grace & Co.

(13) H.P. Ledden and K. Muller; U.S. Patent 3,707,523; Dec. 26, 1972; assigned to American Cyanamid Company.

(14) C.C. Cook and E.M. Haynsworth; U.S. Patent 3,718,003; Feb. 27, 1973; assigned to American Cyanamid Co.

(15) F.C. Dehn; U.S. Patent 3,767,865; October 30, 1973; assigned to PPG Industries, Inc.

(16) S.F. Brzozowski; U.S. Patent 3,615,163; October 26, 1971; assigned to PPG Industries, Inc.

(17) *Chemical Week,* pp. 26, 29 (January 1, 1975).

IRON AND STEEL INDUSTRY WASTES

COKE OVEN EFFLUENTS

Incineration

A process developed by F. Breitbach et al (1) involves decomposing the ammonia of impure ammonia-containing vapor clouds, particularly ammonia-containing vapor clouds emanating from coke oven plants and gas works and contaminated by hydrogen sulfide, cyano compounds like HCN, benzene naphthalene and the like. The vapor clouds are first heated to the decomposition temperature of the ammonia, whereafter the hot clouds are passed at this temperature through a decomposition zone to cause decomposition of the ammonia into hydrogen and nitrogen. The combustible matter in the decomposed gases is then burned in an intermediately installed combustion zone by supplying combustion air. The combustible matter including the hydrogen may be burned at a place remote from the decomposition plant or may be added to other gases.

Figure 81 shows a suitable form of apparatus for the conduct of this process. Referring to Figure 81, **1** indicates schematically a reactor to be heated to the required decomposition temperature. For this purpose the reactor is heated by the combustion of a gaseous or liquid fuel which is supplied to a burner **3** through line **2**. The combustion air necessary to accomplish combustion of the fuel flowing through this line is supplied to the burner through line **4**. The amount of combustion air is adjusted so that a reducing atmosphere is produced in the reactor. The ammonia vapor clouds to be decomposed can be supplied to the reactor in different ways.

According to one embodiment, these ammonia clouds may be supplied through the line **5** so that they will directly enter the combustion chamber **8** of the reactor. According to a second possibility, the ammonia clouds may be supplied jointly with the combustion air through the lines **5**, **6** and **4**. Further, it is also possible to supply the ammonia clouds jointly with the fuel through the lines **5**, **7** and **2**.

In all these instances, the ammonia clouds are introduced into the chamber **8** of the reactor, which constitutes the heating zone, and are heated to the reaction or decomposition temperature. In the embodiment shown in the figure, there thus occurs actual mixing of the heating medium and the ammonia clouds. After the ammonia clouds have been heated to the necessary decomposition temperature, they are then, jointly with the off-gases from the combusted heating gases, passed through the decomposition zone **9** for complete decomposition of the ammonia. The decomposition zone may be filled with filling bodies or with a suitable nickel catalyst.

FIGURE 81: COKE OVEN WASTE GAS DISPOSAL UNIT FEATURING AMMONIA
DECOMPOSITION FOLLOWED BY COMBUSTION

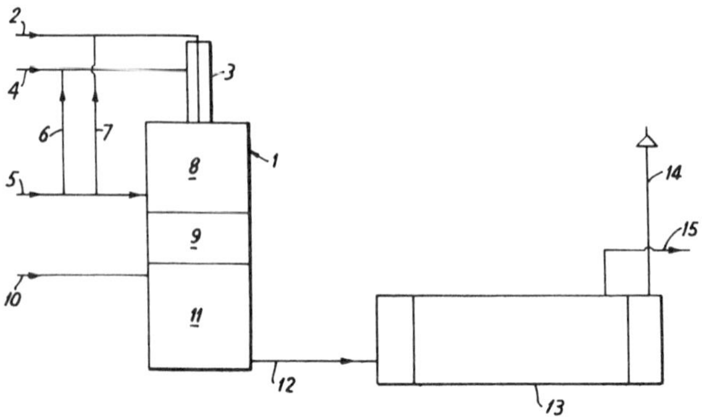

Source: U.S. Patent 3,661,507

The admixing of a part or the whole of the vapor clouds to the heating gas or to the combustion air before the burner makes it possible that during the combustion with a deficiency of air no carbon black is formed. By supplying secondary air which, through line **10**, enters a combustion zone, combustion of the combustible matter formed in the decomposition zone may be effected close to the decomposition zone. This means that any combustible matter which is formed in the decomposition of the ammonia is completely burned up in the combustion zone **11** due to the introduction of the secondary air through line **10**.

The hot off-gases flow subsequently through line **12** to reach the waste heat boiler **13**. The off-gases are discharged to the atmosphere through the chimney **14**. The steam which is produced in the waste heat boiler flows through line **15** to any suitable point of utilization, that is, it may be supplied to the steam network of the plant. The hot off-gases of the after-combustion zone may, of course, also be utilized or cooled in any other suitable manner.

The burnable hydrogen-containing gas may, however, be burned at a place remote from the plant. In other words, the burning may be delayed until the gas has been conveyed to a suitable place of utilization where the gases are burned in situ, if desired, after admixture with other gases. In this case the gas, which leaves the reactor with about 1000°C, is also conducted into a waste heat vessel for using its content as waste heat. In the following example, the terms kg/d and m^3/d signify kilograms per day, and cubic meters per day, respectively.

Example: The coking of 1,000 tons of dry coal results in the formation of 2,600 kg/d of ammonia. This ammonia may, for example, be contained in 220 m^3/d of wash water and condensate. The ammonia is expelled from this liquor in known manner so that the ammonia clouds formed contain about 250 kg of ammonia per long ton and in addition to the other customarily accompanying substances 750 kg of steam per ton.

These ammonia clouds are supplied to the reactor **1** through line **5**, as shown in the figure. 260 Nm^3/hr (atmospheric pressure) of coke oven gas are supplied through line **2** while 1,100 Nm^3/hr of air are introduced through line **4**. The coke gas amount is adjusted in such a manner that the heat resulting from its combustion is utilized for the decomposition of the ammonia into nitrogen and hydrogen and that the gas mixture which exits from the decomposition zone **9** has a temperature of about 1000°C.

This hot mixture enters into the combustion zone **11** into which are supplied about 750 Nm³/hr air through line **10**. After the combustion of the combustible gases which have been formed, the combustion gases leave the after-combustion zone **11** through line **12** and enter the waste heat boiler **13**. 40,000 kg/d of steam (40 atm) are obtained which are discharged from the waste heat boiler through line **15**. The off-gases which leave through the chimney **14** do not contain any significant amount of nitrogen oxides.

If, by contrast, ammonia clouds are supplied to an ordinary prior art combustion furnace, where the clouds are burned up in a single stage with the same coking gas amount and in the presence of 1,850 Nm³/hr of air, so that nitrogen and water are formed from the ammonia, an off-gas leaves the chimney which contains up to 4.8 grams of nitrogen oxides per Nm³.

A process developed by H. Grulich et al (2) involves removing and burning ammonia vapors and the like which are stripped from a waste liquor such as ammonia water which may be produced during the purification of distillation gases obtained from coke gases. The apparatus used comprises a stripper column means for stripping ammonia vapors from a waste liquor having an inlet for communicating with a waste liquor supply and having an outlet in the top portion for stripped ammonia vapors; and a burner member having a plurality of feed ducts.

Each of the feed duct members is separated from the other feed duct members and arranged to form a cone of flame having a hollow interior region. The furnace has the burner located therein at the upper portion thereof and has an exit for exhaust gases. The system includes a closed system heat transfer means operatively connected between the furnace and the stripper column for transferring heat for use in the stripper column.

Figure 82 shows a suitable form of apparatus for the conduct of this process. Referring to the figure, **1** denotes a pipe furnace having a burner **2** located therein. Ammonia vapors **4** are supplied to the burner through pipe **3**. The ammonia vapors flow through distributor nozzle **5** into a region **6** of a truncated hollow cone of flame **7**. The cone is formed by means of an annular outlet opening **8** of a nozzle **9**. A coke gas **12** which is ejected from the nozzle is provided thereto through a pipe **10** and a duct **11**. The air **13** which is required for combustion is supplied to the nozzle from outside of the system to flame cone **7**, through a pipe **14** and a duct **15** that surrounds at least outlet-opening **8** of the nozzle.

FIGURE 82: COKE OVEN WASTE GAS DISPOSAL UNIT FEATURING HEAT RECOVERY

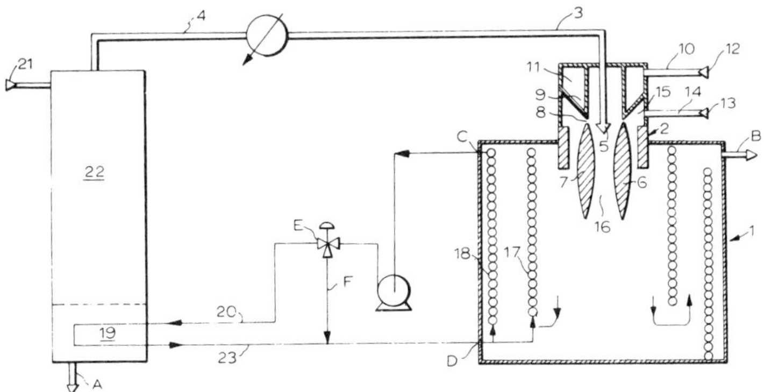

Source: U.S. Patent 3,915,655

The flame cone **7** extends downwardly toward the bottom of the oven **1**, and one of the means of combustion **4**, **12**, **13** or a plurality thereof is blown under pressure into the burner **2**. Naturally, the gases of combustion also have the same direction initially. The direction of the gases of combustion is reversed substantially in the area of the bottom of the oven, in particular through the flue draft, as shown by the arrows in the drawing. Hot gaseous combustion products **16** are cooled on pipe coils **17** and a hot oil **18** flowing therein is thus heated.

Owing to the arrangement of the flame cone, wherein the air required for combustion is supplied from outside, a region of excess oxygen exists in the outer regions of the flame while a deficiency in oxygen is encountered as one moves toward the central axis of the cone of flame. Consequently, ammonia vapors **4** introduced into the oxygen deficient central portion of the cone flame burn at a relatively low temperature in a reducing atmosphere, and the amounts of nitric oxide produced during the combustion are kept at a low level.

Stripper column **22** and evaporator **19** are combined in the drawings for the sake of simplicity. Steam produced in the evaporator is introduced into the base of the stripper column and ammonia water (ammoniacal liquid) is introduced at inlet **21** and flows downwardly from the top portion of the stripper column. The free ammonia is stripped from the ammonia water by means of the steam flowing upwardly. The stripped water collected below in the stripper column is fed into the evaporator and is employed for generating the steam which is required for the stripping. Any excess stripped water is removed at **A** as wastewater.

The heat which is generated during the combustion of the ammonia vapors **4** which is supplied from the stripper column is employed for heating a heat exchange medium **18** such as oil. The oil enters the furnace at inlet **D**, flows upwardly through coil pipes **17** and exits the furnace at outlet **C**. The temperature of the hot oil at outlet **C** is kept approximately constant at 300°C. The hot oil is transferred to the evaporator **19** through pipe **20** in order to generate the steam which is employed in the stripping process.

The amount of hot oil introduced into the evaporator is regulated according to the amount of desired steam. The remaining oil in the closed oil circulating system is caused to bypass the evaporator and is mixed with the oil leaving the evaporator through a pipe **23**. The temperature of this mixed oil is equal to the inlet temperature of the oven, which depends on the amount of steam to be generated. The potential danger that the gases of combustion **16** might cool to below the dew point of the sulfuric acid present therein, owing to the low inlet temperature of the hot oil **18**, which would promote corrosion in furnace **1**, is obviated by the burner of the process, particularly in connection with the closed cycle system of the hot oil.

A variable amount of coke gas **12** is added as a function of the inlet temperature of the oil and at the substantially constant outlet temperature of the oil, to the substantially constant amount of ammonia vapors to be burned. In this manner the temperature of the gaseous products of combustion **16** at outlet **B** of the furnace is maintained at above 350°C and, consequently, the temperature in the furnace certainly does not drop below the dew point of the sulfuric acid. The operation of the device is as follows. Concentrated ammonia solution is introduced into the stripper column **22**, at inlet **21** and free ammonia is stripped therefrom by upwardly flowing steam which is produced by evaporator **19**.

Some of the stripped water is fed into the evaporator and is employed for generating steam and the remaining water is removed at **A**. The resulting ammonia vapors **4** are introduced into the hollow region **6** of the flame cone **7** by pipe **3**. The temperature of the ammonia vapors may be as low as 85° to 90°C. The amount of coke gas **12** and air **13** which are supplied to the flame is automatically varied to warm the hot oils up to 300°C as they leave the furnace at exit **C**. The hot oils are then transferred to evaporator **19** via pipe **20**. The amount of hot oil required at the evaporator depends on the percentage of ammonia contained in the ammonia water and consequently on the quantity of steam which is required to strip the free ammonia. Valve **E** may be manually set to appropriately proportion the flow of oil between pipe **20** and pipe **F**. If, for example, the quantity of oil flowing through pipe **20** is increased, the oil mixture in pipe **23** as the oil returns to the furnace **1** at **D** has a lower temperature than before. Consequently the coke gas and air supplied to the furnace is automatically increased to maintain the temperature at exit **C** at 300°C.

As shown in the figure the gases generated by the combustion first flow downwardly toward the bottom of the pipe and then flow upwardly through the coils **17** and then through outlet **B**. The temperature of the outlet gases never falls below 350°C as a result of maintaining the outlet **C** temperature of the circulating oil at 300°C. As a result of the relatively low temperature of combustion, the rate of combustion is at a slower, more advantageous rate than is achieved with the prior art devices. Consequently there is provided process and apparatus of simple construction, which is economical to construct and to operate and which does not contribute to the pollution problem.

CUPOLA FURNACE EXHAUSTS

It is common in cupola practice to discharge a jet of air into the upper portion of the cupola for the purpose of oxidizing oxidizable matter passing to the stack but there is no control and no provision for reaching with the jet all or even a considerable part of the oxidizable components passing toward the stack. As a result the practical effect of the jet in cupola practice is slight and it does not prevent or minimize emission of unoxidized matter into the atmosphere.

Incineration

A device developed by J.M. Phillips (3) has as its purpose the cleanup of smoke and noxious fumes, particularly from iron and steel plants. This device is also capable of completely oxidizing oxidizable components of products of combustion emanating from the generally coking chamber of a continuous coke oven.

The apparatus comprises surfaces heated to a temperature sufficient to substantially completely oxidize such components, means for conducting efflux containing oxidizable components into contact with such surfaces in an oxidizing atmosphere and means controlling the movement of the efflux to maintain it in contact with such surface for a time sufficient for substantially complete oxidation of such components.

Desirably the mixing zone is generally circular and means are provided for admitting the gaseous material to be admixed with the efflux generally tangentially into the zone to cause the efflux to swirl turbulently in an oxidizing atmosphere to promote homogeneity and uniformity of temperature thereof.

A device developed by R.J. Frundl et al (4) is an afterburner system and arrangement for cupola furnaces wherein a series of burners are arranged near the charging opening of the cupola stack and ignite and combust the carbon monoxide gas emitted during furnace operation as it leaves the furnace bed. The afterburning raises the stack gas temperature thereby permitting complete combustion of high molecular weight organic compositions and minimizing the amount of free air pulled in through the charging opening.

Another device developed by R.J. Frundl et al (5) is an in-bed burner system and arrangement for stack furnaces wherein a series of burners are mounted on the stack below the charge door opening to direct a jet of burning gases into the stack below the top surface of the charge bed. Supplementary air is supplied to the stack from a position adjacent the burners to support relatively complete combustion of the products of combustion rising in the stack. Oxygen can augment or replace the supplementary air. The in-bed burners and supplementary air effect burnout of carbon monoxide in the stack exhaust gases and maintain a relatively high temperature in the top portions of the charged bed to thermally crack and burn out organic materials present in the charge.

FOUNDRY WASTES

Landfill Disposal

The landfill disposal of solid foundry wastes has been outlined by Senske (6). Such solid wastes include:

> waste sand from the foundry sand reclamation process
>
> sludge from a settling basin which is the product of particulate collection systems for air pollution control in the plant.

REFERENCES

(1)   F. Breitbach and G. Choulat; U.S. Patent 3,661,507; May 9, 1972; assigned to Firma Carl Still, Germany.

(2)   H. Grulich, E. Hackler and M. Galow; U.S. Patent 3,915,655; October 28, 1975; assigned to Didier-Kellogg Industrieanlagenbau GmbH, Germany.

(3)   J.M. Phillips; U.S. Patent 3,215,501; November 2, 1965; assigned to Salem-Brosius, Inc.

(4)   R.J. Frundl, V.W. Hanson, R.M. Jamison and O.M. Arnold; U.S. Patent 3,545,918; December 8, 1970; assigned to Ajem Laboratories, Inc.

(5)   R.J. Frundl, V.W. Hanson, R.M. Jamison and O.M. Arnold; U.S. Patent 3,666,248; May 30, 1972; assigned to Ajem Laboratories, Inc.

(6)   M.L. Senske, *Solid Waste Management Practices in a Foundry,* Washington, D.C., Environmental Health Service, Bureau of Solid Waste Management (1970).

METAL FINISHING INDUSTRY WASTES

CADMIUM WASTES

Landfill Disposal

Cadmium and cadmium compounds are considered as waste stream constituents requiring centralized treatment for the following reasons: [1] the extremely high degree of toxicity of all cadmium compounds; [2] the nondegradable nature of the toxic cadmium component of all cadmium compounds; [3] cadmium compounds are present in sizable quantities as wastes from electroplating, paint manufacture, battery manufacture, and the metals industry; [4] a significant portion of cadmium wastes is contributed by small plating shops where treatment is either technically or economically infeasible; and [5] the cadmium hydroxide waste sludges resulting from the treatment of cadmium wastes should be disposed of in California Class 1 type landfills only and are not being adequately handled at present (1).

The only adequate method for the disposal of concentrated cadmium wastes is coagulation with lime, then sedimentation followed by sand filtration. The effluent from this process would probably have to be treated further (for example, adsorption with activated-carbon or ion exchange) to reduce the cadmium concentration to a level in compliance with the U.S. Public Health Service recommendation for public drinking water (0.01 mg/l). The cadmium hydroxide sludge produced in this process can be dried and placed in an approved chemical landfill area of the California Class 1 type. Cadmium hydroxide is not very soluble (0.00026 g/100 cc), so contamination of water supplies from the landfill operation should not be a problem. If the cadmium hydroxide sludge is relatively pure, it can be dissolved in sulfuric acid and the cadmium metal recovered by zinc dust precipitation.

CHROME WASTES

According to R.S. Ottinger et al (1), the chromates and dichromates are also considered as waste stream constituents for centralized treatment for the following reasons: [1] the high degree of toxicity of the compounds; [2] hexavalent chromium oxides cannot normally be reduced to the nontoxic elemental chromium form; [3] the chromate and dichromate wastes appear in sizable quantities; [4] a substantial portion of the chromate and dichromate wastes are contributed by small plating shops that cannot afford waste treatment equipment; and [5] the disposal of waste chromium hydroxide sludges can be adequately handled at a centralized facility and chromium and heavy metal values present can be more readily recovered if such a need arises. For the disposal of concentrated hexavalent chromate wastes at National Disposal Sites, it is recommended that a reductive-precipitative facility be constructed. This scheme is by far the most applicable approach for treating chromate concentrates. The sub-process of electro-reduction might have merit to replace the chemical reduction step of the operation and save on necessary reducing chemicals. The technique is efficient and adequate for large-scale removal of chromates. The effluent can have 1 to 2 ppm levels of Cr^{+3} or lower.

Landfill Disposal

The permanent containerized storage or proper landfilling of chromium waste sludges or their incinerated residues are recommended for ultimate disposition. The landfill sites used must be located over nonwater-bearing sediments or have only unusable groundwater underlying them and must also be completely protected from flooding, surface runoff or drainage, so that all waste materials and internal drainage are restricted to the site. The economics of transportation to the landfill or storage site must be considered in the decision to transport the hydrated sludge or to incinerate it prior to shipment. Such studies will be necessary in site selection and process recommendation for sites in various parts of the nation. The very same facility is usable for many of the toxic metal compounds which would also be likely to need disposal from time to time including cadmium, mercury, lead, copper, and arsenic.

CYANIDES

Chemical Destruction

The chemical destruction of cyanides in liquid waste streams has received a great deal of attention and numerous methods have been proposed. Some of the more attractive methods are briefly discussed in the following paragraphs. For dilute cyanide wastes, chlorination under alkaline conditions is generally favored; too much heat is generated if the technique is used on concentrated wastes and undesirable and dangerous side-reactions will take place unless the operation is carried out very slowly, or the waste is diluted. When considerable amounts of concentrated wastes must be treated, e.g., from cyanide heat-treating, the wastes must be diluted or another method for cyanide destruction such as electrolytic oxidation must be used.

Chlorination: Oxidation of cyanides by the hypochlorite ion (which may be furnished by either chlorine or sodium or calcium hypochlorite) proceeds in three stages.

$$[1] \quad CN^- + H^+ + OCl^- \longrightarrow CNCl + OH^-$$

$$[2] \quad CNCl + 2OH^- \longrightarrow CNO^- + Cl^- + H_2O$$

$$[3] \quad 2CNO^- + 3OCl^- + H_2O \longrightarrow 2CO_2 + N_2 + 3Cl^- + 2OH^-$$

Reaction [1] is very fast; reaction [2] is very slow below pH 9 unless excess hypochlorite is present; at pH 10 or higher it is rather rapid and oxidation to the cyanate stage is complete in 5 minutes or less, provided no nickel ion is present. If nickel is present, reaction [2] is not completed in less than 30 minutes, and then only if 20% excess reagent is used. Reaction [3] is very slow above pH 9 requiring at least an hour, and many hours if the pH is 11 or more. The best practice is to adjust the pH to 8.5 and allow one hour reaction time. About 10% excess hypochlorite should be used, or destruction of the cyanides will be incomplete.

The heavy metals present will be precipitated as hydroxides or carbonates in total chlorination treatment. An exception is copper which will not be precipitated in wastes containing copper and rochelle salts unless sufficient calcium is also present as chloride or hydroxide, so that the tartrate will be precipitated as the calcium salt, thus allowing precipitation of copper. It has been shown that iron cyanide complexes are not destroyed by chlorination and that a cyanide residual reappears on long standing (100 hours).

Total chlorination (i.e., to N_2 and CO_2) does not lend itself readily to continuous treatment processes because current control methods are not adequate. If a packaged cyanide treatment system is employed, continuous operation for oxidation to cyanate (1,000 times less toxic than cyanide) is possible. A second system is required for oxidation of cyanates, with appropriate holding time due to the slow reaction.

Kastone Process: DuPont has introduced a process which appeals primarily to small plant operators using cyanide baths to plate zinc or cadmium. This process oxidizes cyanides to cyanates and simultaneously precipitates zinc or cadmium complexes by simple filtration. The Kastone Process uses a proprietary peroxygen formulation that contains 41% hydrogen peroxide with trace amounts of stabilizers. The cyanates, though 1,000 times less toxic than cyanides, cannot be discharged into most natural streams. Therefore, this process has only limited application (2).

Electrolytic Oxidation: Automatic electrolytic oxidation units are marketed (2) for complete decomposition of cyanide ion content in waste streams. Some difficulties have been reported with this unit for dilute solutions, but this problem has been circumvented by using a semiconductive bed in the cell. The bed serves as an intermediate electrode that provides in effect more than a million anode and cathode sites per cubic foot.

Radiation Decomposition: A process has been developed for destroying cyanides by gamma radiation which serves to rupture the C≡N triple bond and converts the cyanide ion into harmless by-products (2). This method is not in commercial use at present.

Conversion of Cyanides to Ferrocyanide by Ferrous Sulfate: The formulation of less toxic cyanide complexes such as ferro and ferricyanides has been used as a method for disposing of cyanide wastewaters. This process involves the use of iron salts to form complex compounds with the free cyanide in the wastes. These cyanide complexes are precipitated and removed as sludge. The major advantage of this treatment method is that it is relatively inexpensive where waste ferrous sulfate is available. However, considerable quantities of sludge are formed, and the treated solutions are strongly colored. There is also evidence that ferrocyanides may decompose to free cyanide in the presence of sunlight. The regeneration of cyanide can then contaminate the receiving stream.

Reaction with an Aldehyde: A process has been developed (2) for the removal of cyanide from a waste stream by reaction wth an excess of an aldehyde according to the following:

$$KCN + CH_2O \text{ (aqueous)} + H_2O \longrightarrow HOCH_2CN + KOH$$

$$CNCH_2OH + KOH + H_2O \longrightarrow HOCH_2COOK + NH_3$$

It is claimed that nearly all cyanides, even stable complexes, are destroyed in this manner. However, though the reaction products are not toxic, there is the problem of disposal of the organic compounds formed.

Ozonation: Ozonation is reported to be more economical and easier to control than chlorination. Ozonation, however, oxidizes cyanides only to the cyanate in accordance with the reaction given below,

$$3CN^- + O_3 \longrightarrow 3CNO^-$$

and the oxidation of the cyanate is too slow to be practical, according to Ottinger et al, (2). A process has been developed by R.L. Garrison et al (3), however, in which the destruction of cyanides in an aqueous cyanide solution, particularly an aqueous solution of cyanides complexed with iron is carried out by contacting an aqueous solution thereof with an ozone-containing gas, while simultaneously irradiating the aqueous cyanide solution with ultraviolet light. The method is preferably carried out while maintaining the pH of the aqueous cyanide solution within the range of pH 5 to pH 9. Increased reactivity can also be achieved by heating the aqueous cyanide solution.

The method is preferably carried out by contacting the aqueous cyanide solution and an ozone-containing gas in a plurality of separate contact zones, countercurrently or with parallel flow, with the irradiation of the aqueous cyanide solution with ultraviolet light being carried out in at least one of the separate contact zones, preferably at least the final contact zone.

Acidification: Waste acid solutions have been used to acidify cyanide waste solution. Air is then passed through the solution and the liberated hydrogen cyanide is discharged up a high stack or is passed through a burner. This method is not recommended because of the danger involved.

In summary, the oxidation of cyanides in alkaline solution by chlorine or hypochlorites is an acceptable method for destroying cyanide. However, because cyanide wastes are generated by a large number of metal treating operators, some of whom are small, in some cases proper treatment is an economic burden and not always complete. It has, therefore, been recommended (2) that centralized facilities have the capability for treatment of cyanide wastes by the use of the chlorination techniques.

Lagooning

Though the methods discussed above are available for treating cyanide wastes, lagooning of cyanide wastes from small and medium sized metal processing and plating operations is widely practiced. For example, in the Los Angeles area commercial waste disposal companies collect cyanide wastes and truck these wastes to a large abandoned rock quarry. If precious metals, such as silver, are present, these are first removed by salvage companies.

The use of lagoons for cyanide wastes cannot be recommended (2) because the cyanides may some day leak into underground water supplies and the wastes, if acidified, will liberate hydrogen cyanide.

HYDROCARBON SOLVENTS

Incineration

A device developed by P.H. Stibbe (4) is an apparatus for incinerating gaseous waste material, for example, the combustible portions of the effluents from various industrial equipment such as wire enameling ovens, metal coating ovens, paint ovens, web dryers or the like. The apparatus includes a diffusor for the combustible chamber and which acts to create turbulence and flow disturbance to the gaseous material as it proceeds through the combustion chamber, thereby promoting effluent mixing for complete combustion.

A device developed by K.H. Hemsath et al (5) is one for incinerating gases which contain varying quantities of combustibles therein. Means are provided for controlling combustion air added to the flue gases when they contain combustible matter and means are provided for sensing the temperature of the combusting waste gases. When a preset temperature is exceeded, additional air is added to reduce the temperature of the gas, but when the temperature falls below the preset point, the burner firing rate is increased and the added air is cut off to insure proper incineration and achieve fuel efficiency.

PAINT DRYING FUMES

Incineration

A scheme developed by L.C. Hardison (6) is a multistage system for treating a discharge air stream which contains particulates as well as volatiles that are obnoxious or combustible so as to permit recirculation of at least a major portion of the stream in a substantially particle free, purified state. It is also an object to provide a heat resistant continuous belt means for use in filtering out and burning up the particulates from the discharge stream so as to provide a convenient disposal method for such material, as well as provide in addition thereto both catalytic oxidation and or adsorbent bed means for effecting the removal of volatile and combustible components from, respectively, the vent stream and the recirculating air stream.

Figure 83 shows the overall scheme in some detail. There is indicated a spray booth zone **1** which is generally an elongated type of room designed to accommodate a continuous conveyor means, such as **2**, for carrying material that is to be painted as it moves continuously through the zone. The figure indicates auto bodies **3** being carried by the conveyor means **2** and such car bodies being subjected to enameling or lacquering from operators stationed in the spray booth. For simplicity, the drawing shows merely half of the spray booth zone and the accompanying treating stages in combination therewith. Actually the total unit may be considered as being symmetrical about a vertical center line extending through the middle of the auto body.

The floor of the spray booth is provided with a continuous grill or grid means **4** which permits the down flow of a heating and/or ventilating air stream through the booth whereby entrained volatiles and particulates will be carried into a lower particulate removing zone **5** that is housing a continuously moving mesh type filter belt **6**. The latter passes over a plurality of spaced crown rollers **7** within zone **5**, a solvent removal zone **8** and a burn-off zone **9**. The latter is equipped with a plurality of burners **10** being supplied by a fuel-air mixture from line **11** having control valve **12**.

FIGURE 83: SYSTEM FOR ADSORBING, DESORBING AND CATALYTICALLY COMBUSTING PAINT SPRAY BOOTH FUMES

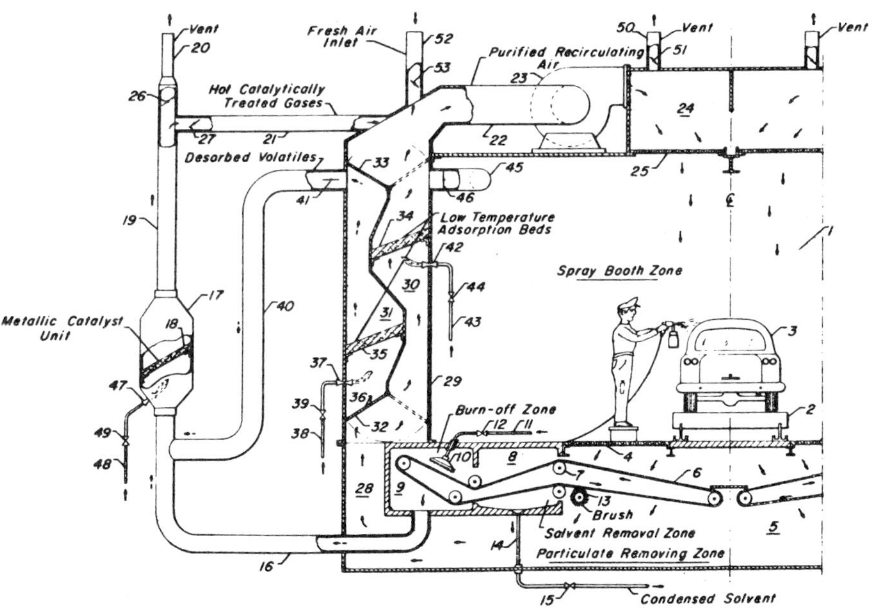

Source: U.S. Patent 3,395,972

Preferably, as indicated hereinbefore, the burners are of the infrared type suitable for providing controlled high energy heating to a confined relatively small area of the surface mesh type filter belt within the burn-off zone. The impinging temperature should of course be adequate to effect the complete burn-off of the deposited material on the belt surface and permit the belt to be returned to the filtering or particulate removing zone in a cleaned state. A brush means **13** may be positioned at the belt outlet section of the burn-off zone or at the solvent removal zone such that oxidized ash-like particles will be completely removed from the openings in the mesh of belt **6** prior to its reuse for filtering the air stream descending through grid **4**.

The temperature from the burner means **10** in zone **9** will of course provide some transfer of heat into the adjacent zone **8**, which first receives the filter belt from zone **5**. Thus, there will be some heating and removal of the volatile components entrained with the deposited paint particulates in zone **8**. At the same time, any condensation of volatile solvent materials that may be collected in zone **8** can be removed from the floor by way of line **14** having control valve **15**.

However, in a preferred operation, the burner means shall be operated in a rich manner to minimize the presence of oxygen and to preclude any upstream burning of paint or other deposited particulates on the moving belt.

In this figure, the gases from the burn-off zone are passed by vent means **16** through an oxidizing catalyst zone **17** having a catalyst bed **18**. The latter may comprise a permeable unit containing subdivided particles of an active oxidation catalyst or a mat-like unit of crimped alloy ribbon that is coated with a noble metal, and particularly a platinum group metal, whereby there will be oxidation and conversion of combustible materials in the gas stream to provide harmless, odorless, oxidation products which are primarily carbon dioxide and water. For example, one desirable form of catalytic fume incinerating means may comprise the all metal catalyst unit such as described in U.S. Patent 2,658,742.

The treated oxidized gas stream from the catalyst oxidation zone may be discharged into the atmosphere by way of duct **19** and stack **20** or in part recirculated and reused within the spray booth zone by passage through duct **21** and **22**, fan means **23** and air distributing plenum **24** which in turn releases the gases downwardly through grill means **25** into the spray booth zone. Suitable valve means **26** and **27** can be adjusted to accommodate the desired gas flow through the recirculation fan and to the spray booth zone or to the atmosphere by way of stack **20**.

The air stream from zone **5** is shown to flow by way of passageway **28** into an adsorption section **29** which in turn accommodates two separate adsorption zones **30** and **31**. The air flow to one or the other of the zones is controlled by a lower damper means **32** and an upper damper **33**. As shown for one case, air flow with entrained volatile components carries by way of zone **30** into the transfer duct **22** after first passing through the adsorption bed **34**. When the bed becomes heavily saturated with volatile materials then it is subjected to desorption while air flow is routed (by means of the movement of damper means **32** and **33** into the dashed lined positions) such that adsorption is effected by bed **35** in zone **31**.

As previously noted, the beds **34** and **35** for completing the purification of the air stream, may utilize activated carbon particles as the adsorption media since such material is particularly adapted to provide high adsorption activity for the volatile components being encountered from paint and lacquer spraying operations.

During the desorption cycle, as indicated in the drawing for zone **31**, there may be heated air provided by a bleed opening **36** in damper **32** and a burner means **37** for fuel being supplied by line **38** and control valve **39**. A low velocity hot gas stream will thus pass through the bed **35** at a desorption temperature and at a rate sufficient to substantially and completely effect the removal of adsorbed volatiles. The desorption stream is carried by way of duct **40** and control valve **41** into duct **16** at a point upstream from the catalyst zone **17**. Thus, the desorbed volatiles can be catalytically oxidized and removed from the air stream to permit discharge into the atmosphere or be recirculated into the air stream by way of duct **21** with control valve **27**.

In the alternative operation, where the adsorption bed **34** is undergoing desorption, then the main air flow is through bed **35** and a desorbing heated air stream is passed upwardly through bed **34** by means of burner **42** which receives fuel by way of line **43** and control valve **44**. The desorbed volatiles carry downstream from bed **34** through duct **45** which, although not shown on the drawing, may be made connective with duct **40** to in turn pass the volatiles into contact with the catalyst unit **18** in zone **17**.

During this cycle, the dampers **32** and **33** will be in the dash line positions and valve **46** at the inlet of duct **45** opened to accommodate the desorption stream and the entrained desorbed volatiles. Generally, the desorbing gas stream used during the reactivation cycle for each of the adsorption beds **34** and **35** will be at a temperature of the order of say 500°F or more, so as to provide an effective removal of all of the adsorbed components. The streams passing by way of lines **40** and **45** into the oxidation zone **17** may generally be at a temperature sufficient to maintain catalytic oxidation within bed **18**, particularly where the latter utilizes an active catalyst coating.

However, where additional heat may be required to sustain complete incineration of the combustible entrained components, then a burner **47**, being supplied fuel by line **48** and control valve **49**, will assist in adding heat to the catalyst oxidation zone and insure complete conversion of the combustible materials as they pass through the bed.

In an operation where the oxidized gases leaving the zone **17** are maintained in the system, by recirculation through ducts **21** and **22** into fan **23**, there may be provision for a continuous slip-stream removal of a portion of the combustion gases in order to preclude any buildup of carbon dioxide and, at the same time, provision to continuously introduce a small portion of fresh air into the system. Venting may be effected from above the plenum by duct **50** with adjustable valve **51** and fresh air introduced by duct **52** with adjustable valve **53** at a point connective with duct **22** just upstream from the fan **23**.

In a normal operation, the air stream through the spray booth zone will be maintained at a relatively constant temperature of 75°F, or any other desired room temperature. The filtering and adsorption steps will have no effect on the air stream inasmuch as it is continuously circulating through only the moving filter belt **6** and the adsorption beds **34** and **35** which in turn will operate at room temperature in their respective contact zones.

As indicated briefly hereinbefore, in an operation where there are substantially no obnoxious volatiles in the filtered stream, then it may be returned directly to the processing section. In other words, the absorption beds may be by-passed or eliminated from the system. The combustion products stream of oxidation zone **17** which can be passed by way of duct **21** into the recirculating air system will be at a high temperature of 1000°F or more and thus be of advantage for some heating in cool seasons. During the summer months, such stream may be discharged to the atmosphere to prevent any heat buildup in the system.

An apparatus developed by B.P. Shiller (7) is a fume combustion apparatus incorporating a regenerative type of heat exchanger mounted for rotation on an axis passing through the combustion chamber and a means for dividing the heat exchanger into a plurality of gas flow passages, is combined with a combustion chamber in such a way that fuel requirements for maintaining combustion of fumes in the combustion zone are reduced to minimum levels.

A device developed by D.M. Wilkinson (8) is a paint drying system in which the vapors of the volatile solvents are burned to eliminate pollution of the atmosphere and the heat generated by such combustion is employed to preheat the incoming fresh air thereby reducing the fuel required to dry the paint.

A system developed by H.A. Price et al (9) is a waste gas incineration system for preventing the discharge into the atmosphere of oxidizable waste particles in exhaust gases of ovens. Exhaust gases are introduced into a combustion chamber of an incinerator where they are mixed with a combustible gas and ignited. The gases are retained in the combustion chamber sufficiently long to assure substantially complete incineration of all waste particles. An impeller withdraws the waste gases from the chamber. Ports are provided for mixing the gases with ambient air to reduce the temperature of the air and gas mixture. The air-gas mixture is transported to locations where heat or thermal energy is required.

Figure 84 shows the essential features of such a system. An incineration unit **84** is fluidly connected with the discharge side of an oven **86**, such as a paint bake oven for example, by suitable conduits **88**. A conventional duct heater **90** is connected with the oven and circulates heated air therethrough. If required, ambient air may enter the oven at **92**.

Exhaust gases discharged by the oven contain evaporated paint solvents, generally a plurality of hydrocarbons. These waste particles are incinerated in the incineration unit **84** and the hot gas is discharged at the outlet **94** of the unit. There it is mixed with cooled air at **96**, introduced into a blower **98** and discharged into conduits **100**. A properly sized T-section **102** diverts a portion of the gas stream discharged by the blower into a recirculation line **104**. The cooled gas stream entering the recirculation line may have a temperature of about 400°F and is reintroduced into the bake oven **86**, there heating the atmosphere in the oven and reducing the amount of heat that must be supplied by the duct heater **90**.

The remainder of the gas stream is transported to a dry-off oven **106** through a pipe line **108** and upon leaving the oven it is discharged into the atmosphere through a stack **110**. In painting operations it is particularly advantageous to use gas discharged by the incinerator for removing moisture from articles which are subsequently to be painted. The low oxygen content of the gas reduces surface oxidation of the articles. The dry-off oven can, of course, be replaced with any other oven, heat exchanger (not shown), etc. where heat and thermal energy is required.

**FIGURE 84:   INCINERATOR SYSTEM FOR PAINT DRYING FUMES FEATURING
HEAT RECOVERY**

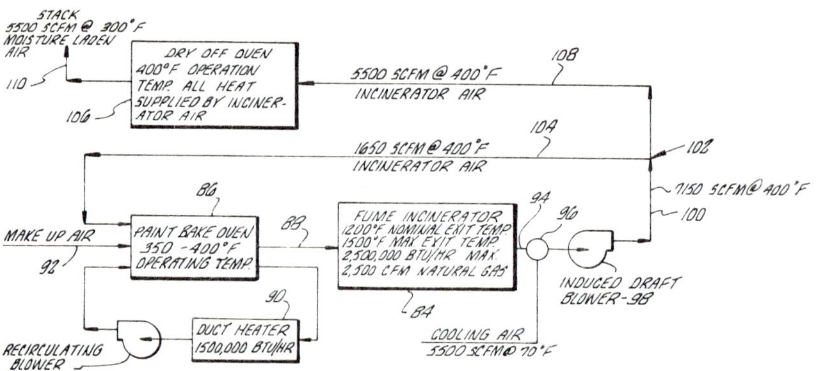

Source: U.S. Patent 3,472,498

A device developed by E.C. Betz (10) is one in which a liquid heat transfer medium, such as Dowtherm, is used in a closed and pressured heat exchanger section of a heat recovery system for integration with catalytic oxidation of an oven discharge stream in order to provide a form of "heat-sink" and a more uniform heat release to a circulating gaseous stream. The system is of particular advantage for use with a discharge stream having a cyclical release (or varying quantity) of oxidizable volatile materials.

A device developed by W.A. Phillips (11) is an incinerator for a drying oven and the like to remove combustible fumes such as solvent vapors from the exhaust from such ovens. The incinerator is formed from a narrow, elongated, heat conducting, metal incinerator conduit having at least one U or reverse bend to increase turbulance within the conduit and to increase heat exchange of hot gases with the exterior of the conduit. The incinerator conduit is formed of a plurality of elongated arcuate sections which are secured together at the edges. Heat conducting fins extend through the conduit between the interior and the exterior thereof at the edges of the arcuate sections to facilitate heat transfer between the interior and exterior of the incinerator conduit.

The fume-laden gases to be incinerated are channelled along the exterior surface of the incinerator conduit in contact with the heat exchange fins by a larger preheat conduit surrounding the incinerator conduit. A fan draws the solvent-laden gases through the larger preheat conduit and forces the preheated gases into the incinerator conduit at a higher pressure than that of the preheat conduit. A similar device has been developed by C.B. Gentry (12).

A process developed by C.R. Wilt, Jr. et al (13) involves determining concentration of evaporated solvent in gases emitted from a drying oven and delivered to an incinerator, by simple measurement of the temperature of gases constituting products of combustion exhausted from the incinerator. By maintaining the fuel input to the incinerator constant and also maintaining the temperature of the gases entering the incinerator constant, the variation of the temperature of the products of combustion emitted from the incinerator with respect to a standard value reflects the variation in the concentration of evaporated solvents supplied to the incinerator.

A valve means, responsively controlled according to the temperature of the products of combustion emitted from the incinerator, functions to regulate the admission of a diluent, such as air, to the oven, thereby regulating the concentration of evaporated solvent to a substantially constant and safe value and effecting regulation of a uniform temperature of gas products of combustion at the outlet of the incinerator.

A process developed by M.K. Carthew (14) involves heat recovery from incinerators used to clean the effluents from industrial furnaces and ovens used in metal and like surface finishing plants in the painting or anticorrosive treatment of articles.

REFERENCES

(1)   R.S. Ottinger, J.L. Blumenthal, D.F. Dal Porto, G.I. Gruber, M.J. Santy and C.C. Shih, *Recommended Methods of Reduction, Neutralization, Recovery or Disposal of Hazardous Waste,* Vol. VI, Report PB 224 585 Springfield, Va., Nat Tech Information Service (August 1973).

(2)   R.S. Ottinger, J.L. Blumenthal, D.F. Dal Porto, G.I. Gruber, M.J. Santy and C.C. Shih, *Recommended Methods of Reduction, Neutralization, Recovery or Disposal of Hazardous Wastes,* Vol. V, Report PB 224 584, Springfield, Va., Nat Tech Information Service (August 1973).

(3)   R.L. Garrison, H.W. Prengle, Jr. and C.E. Mauk; U.S. Patent 3,920,547; Nov. 18, 1975; assigned to Houston Research Inc.

(4)   P.H. Stibbe; U.S. Patent 3,736,103; May 29, 1973; assigned to TEC Systems, Inc.

(5)   K.H. Hemsath and A.C. Thekdi; U.S. Patent 3,838,974; October 1, 1974; assigned to Midland-Ross Corp.

(6)   L.C. Hardison; U.S. Patent 3,395,972; August 6, 1968; assigned to Universal Oil Products Co.

(7) B.P. Shiller; U.S. Patent 3,404,965; October 8, 1968.

(8) D.M. Wilkinson; U.S. Patent 3,437,321; April 8, 1969; assigned to B & K Machinery International, Ltd.

(9) H.A. Price and D.A. Price; U.S. Patent 3,472,498; October 14, 1969; assigned to Gas Processors, Inc.

(10) E.C. Betz; U.S. Patent 3,486,841; December 30, 1969; assigned to Universal Oil Products Co.

(11) W.A. Phillips; U.S. Patent 3,670,668; June 20, 1972; assigned to Granco Equipment, Inc.

(12) C.B. Gentry; U.S. Patent 3,706,445; December 19, 1972; assigned to Granco Equipment, Inc.

(13) C.R. Wilt, Jr. and F.L. Schauermann; U.S. Patent 3,868,779; March 4, 1975; assigned to Salem Corp.

(14) M.K. Carthew; U.S. Patent 3,917,444; November 4, 1975; assigned to Carrier Drysys, Ltd.

NUCLEAR INDUSTRY WASTES

COMBUSTIBLE RADIOACTIVE WASTES

Incineration

A device developed by J.L. Tarbox et al (1) is a portable incinerator for efficient and economical concentration and disposal of combustible radioactive waste comprising an ash pit;
three tiers or sections having steel panels in each tier, the panels being bolted together and
the tiers nested within each other and located over the ash pit; primary burners located in
the lower tier; secondary burners located in the center tier; dampers located in the lower
and middle tiers; and a spark arrester.

The advent of nuclear activities has necessitated incineration as a means of disposing of
combustible radioactive waste. To appreciate the need for a portable incinerator, one need
only consider the event of a large-scale radioactive contamination resulting from an accidental release of radioactivity as through a weapons accident or from enemy action which
would result in many radioactive contaminated combustible structures, such as buildings
and materials, which could be effectively disposed of, to remove the radioactivity hazard,
through means of portable incinerators erected at the contaminated site in a few man-hours.

Figure 85 shows such an incinerator. This portable combustible radioactive waste incinerator comprises three tiers or sections, **1, 2,** and **3,** nesting within and held within each other
by the tier's own weight. Each tier comprises 12 panels, character **12** in the drawing, fabricated from standard four by eight-foot steel plate, $\frac{3}{16}$-inch thick, the panels being bolted
together and having the geometrical configuration of a twelve-sided truncated cone. When
assembled, the three tiers have a base diameter of about ten feet, a top diameter of about
three feet, and a total height of about thirteen feet. Mounted on top of the three tiers is
a three-foot high cylindrically shaped screen structure of 12-gauge locomotive netting, 11-
gauge woven wire 4 x 4 openings per inch, **11,** which serves as a spark arrester, during the
burning operation.

Three oil-fired burners, **4,** are located in the primary tier, **3,** at a height of two feet above
the ground on 120° centers and fired toward the center of the pit, **8,** at an angle of about
30° downward, relative to the horizontal plane. Two oil-fired burners, **5,** are mounted in
secondary tier, **2,** at a height of five and two-thirds feet above the ground, on 180° centers,
and offset from each other. The purpose of such an offset arrangement is to induce tangential flow and produce more efficient combustion of the gases. Dampers, **6,** are provided
in all panels of the lower two sections to provide both excess air for combustion and cooling air for the walls of the incinerator. Structure **10** is merely a mounting structure for

instrumentation used in testing the efficiency of operation, and structure **10** is not necessary to the operation, but could be utilized to test efficiency of operation when desired. Door, **7**, is provided to place the combustible radioactive waste charge within the incinerator to be burned. When assembled, the three tiers are located over a four-foot-deep ash pit, **8**, lined with a steel basin, **14**, the pit having a top diameter of about nine feet and a base diameter of about six feet. The pit is merely a hole dug in the ground at the contaminated site.

As can readily be seen, when the burning operation is completed the incinerator can be collapsed into the pit, and the pit closed with earth due to economical cost of the incinerator component parts and materials. The pit is provided with a grate, **9**, fabricated from railraod rail and a six-inch-diameter pipe, **13**, for externally supplied forced-air draft. The oil-fired burners **4** and **5** are conventional commercially available units such as utilized in home heating systems. Each of the burners has a continuous ignition, not shown in the drawing, and an oil recycle system, not shown in the drawing. A common motor, not shown in the drawing, on each burner drives both the air blower section, not shown in the drawing, and the oil pump, not shown in the drawing, and the ignition system is electrically independent of the blower-pump section.

FIGURE 85: FREED INCINERATOR FOR COMBUSTIBLE RADIOACTIVE WASTES

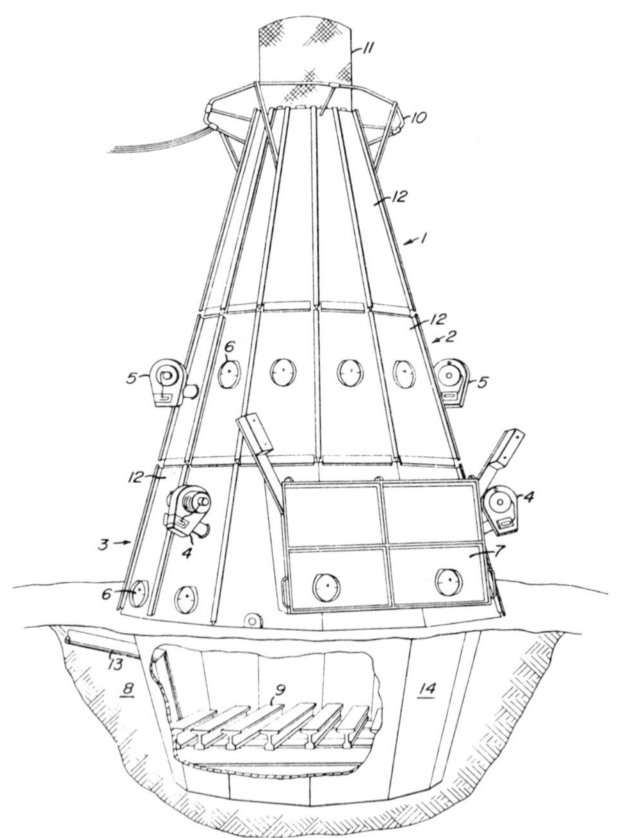

Source: U.S. Patent 3,452,690

With the ignition off and the blower-pump section operating, fuel is recycled from a fuel storage tank, not shown in the drawing, through the oil pump and discharges through a solenoid valve, not shown in the drawing, back to the fuel storage tank. This recycle system was necessary due to the type of oil pump utilized, a pump which utilized the fuel as a pump lubricant. Also, with the ignition off, the burner units can be used as a source of overfire combustion air. In operation, primary burners, **4**, are used in the firing mode until the charge of combustible radioactive waste begins to burn freely by itself.

When free burning of the charge begins, the primary ignitions are switched off and primary burners, **4**, utilized for overfire air supply only. The two secondary burners, **5**, are fed continuously for the duration of the burning operation and serve as after burners to minimize discharge of any unburned radioactive material that might rise from the primary combustion section **3**. As mentioned above, combustion air and surface cooling was provided by damper means, **6**. Underfire air was supplied through six-inch ducting, **13**, connected to a centrifugal blower, not shown in the drawing, having a capacity of 570 cfm.

A device developed by J.L. Tarbox (2) is an improvement for a nonportable incinerator for radioactive wastes. Included in such incinerators there must be a means of air dilution prior to discharge of the combustion gases to cyclones, electrostatic precipitators and filters. The air dilution means in the prior art incinerators comprised nothing more than a simple register, similar to heat registers, for bleeding in make-up air for cooling purposes. The make-up air feed must be regulated to obtain the proper air-flue gas mix, and such regulation was difficult to satisfactorily obtain with the simple register means.

Also, a turbulent flow has been found to be the only efficient method to satisfactorily lower the flue gas temperature from about 2000° to 600°F, the lower temperature being required due to the utlization of mild steel in air dilution means and following component fabrication. A turbulent stream could not be obtained with the prior art register means and bad leakage around the register means were problems encountered with the prior art incinerators.

The device simply incorporates an efficient gas mixer, or fluid turbulator, to introduce dilution air to the flue gas to lower gas temperatures without permitting leakage of radioactive materials to the atmosphere.

A process developed by A. Taeymans (3) is one in which the fumes and particle waste produced by a furnace are passed through a fluidized bed of granules wherein they are further incinerated and they then pass through a filter of granules where the particles are removed; when the filter becomes contaminated, the pressure drop caused thereby is sensed and contaminated granules are ejected therefrom into the fluidized bed; granules in the fluidized bed are simultaneously transferred back to the filter.

Figure 86 shows a suitable form of apparatus for the conduct of such a process. The drawing shows a furnace **1** with external gas heater **1a**. The furnace incinerates the material in it at 900°C and operates at a pressure slightly below the atmospheric pressure. The furnace comprises an air inlet **2**, a thermocouple **3**, a device for supplying burnable material comprising a supply duct **4** and a supply container **5**, an ash outlet **6**, and a fume outlet **7**. The fumes containing unburnt particles, soot, and ash dust then pass through the fluidized bed **10**. Bed **10** is comprised of a column filled up to 40 cm in height with sand. The grains of sand have a diameter between 0.60 and 0.15 mm.

The temperature of the bed **10** is raised to 900°C by external gas heater **10a**. The bed, at its lower part, has a grid for supplying and distributing the gases. Grid **11** is comprised of a perforated plate, the orifices of which are less than 0.10 mm in diameter. The upper part of bed **10** is comprised of an expansion chamber **12**, the pressure in which is adjustable by means of the pump **19** to cause the fumes and particles to pass through filter **30**. The bed **10** is also provided with an inlet and outlet **13** for the filling material, an air inlet **14**, and a thermocouple **15**. The bed has a transfer mechanism **16** in it to transfer the sand from the bed through transfer mechanism inlet **17** to enclosure **20**. The fumes pass through the fluidized bed where the combustion of the unburnt particles takes place, and leave by the expansion chamber **12**. According to an alternate embodiment, the fumes from the oven **1** may be directly conveyed to the expansion chamber **12**, without passing through a fluidized bed. This altered embodiment may, for instance, be considered if the furnace is of an improved type, wherein the combustion ordinarily takes place at very high temperatures. The expansion chamber **12** is connected to the stack **45** through a panel bed filter **30**. This filter comprises, on the gas inlet side, inclined metallic shutters or louvers **31**, and, on the gas outlet side, a perforated plate **32**. The filter is filled with sand of the same characteristics as the sand used in the fluidized bed.

The filter also comprises a thermocouple **34**, a sand removal duct **33** in its lower part, and, in its upper part, a supply tank **35**, connected to the enclosure **20** by the duct **21**. The outlet side **32** of the filter is connected to a device **36** that is adapted to supply compressed air and that starts operation as soon as the pressure drop across the filter passes beyond a predetermined level. The apparatus also comprises manometers **41, 42**, and **43**, respectively indicating the pressure on the furnace **1**, the pressure drop across the fluidized bed **10** and the pressure drop across the filter **30**.

FIGURE 86: INCINERATOR FOR RADIOACTIVE WASTES HAVING FILTER FOR COMBUSTION PRODUCTS

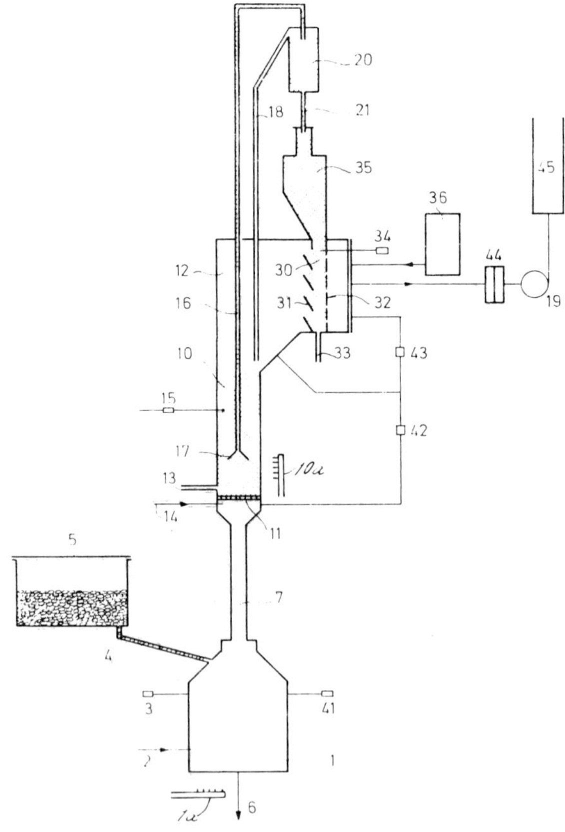

Source: U.S. Patent 3,847,094

A control filter **44** allows inspection of the ash particles carried along by the fumes after their passage through the filter **30** and before their evacuation through the stack **45**. The transfer mechanism is comprised of inlet bell **17** and a transport tube **16** that operates as an elevator. Owing to the greater pressure in the bottom part of bed **10** from the fumes entering the bed, the gas-sand mixture passes through the tube **16** into the enclosure **20** wherein gas and sand are separated by expansion and gravity. The gas returns to the fluidized bed by means of overflow pipe **18**. The position of the bell **17** may be adjusted in height, so that it may open into the upper or the lower part of the fluidized bed as a particular effective operation requires.

If necessary, the incineration gases are burnt once more in the fluidized bed **10**. From bed **10** the gases pass through the filter **30** and are evacuated by the stack **45**. When the pressure drop across the filter **30** passes beyond a certain level, for instance 10 mm mercury, manometer **43** senses this and is connected with compressed air mechanism **36** to immediately start it operating and driving out a quantity of sand grains through the metallic louvers or shutters **31** into chamber **12** of bed **10**.

Mechanism **36** continues to operate until the pressure drop across filter **30** decreases to a desired level indicating a sufficient quantity of dirty granules and fume particles have been expelled from the filter. The expelled sand grains and particles fall in the fluidized bed where the sand granules are regenerated by combustion. The supply tank **35** automatically refills emptying filter **30**. Tank **35** is filled by means of the duct **21**, the enclosure **20** and the transfer mechanism.

The advantage of the process is that it enables continuous incineration of residue without requiring regular replacement and regeneration of the filtering system, because the proposed filter system is automatically regenerated. The filling material for the fluidized bed may be adapted to also absorb noxious materials contained in the incineration gases.

An apparatus developed by W. Hempelmann (4) is a furnace which has air inlet conduits and a flue gas outlet conduit and air heaters as well as blowers connected to the air inlets for forcing hot air into the furnace. The apparatus further has a feeding device connected to the charging end of the furnace for introducing liquid or solid radioactive wastes thereinto and a device which communicates with the discharge end of the furnace for removing solid reaction products from the furnace.

In the flue gas conduit there is connected a plurality of flue gas filters each containing filter candles, a flue gas chamber and a mechanism for removing ashes from the flue gas chamber. The apparatus also includes a mixer section connected with the outlet of each flue gas filter and having a mechanism for mixing cool air with the flue gas filtered by the flue gas filters. Gas blowers connected to the output of the mixer section draw the gas from the apparatus.

COMMERCIAL RADIOACTIVE ISOTOPE WASTES

Carbon 14, cobalt 60, iridium 192 and radium 226 are representative of radioisotopes of commercial interest which are generally produced, distributed and used by the private sector of the economy.

Plutonium, americium and curium are artifically produced radioisotopes. Plutonium 238 is used extensively for radioisotopic power devices for space electric power, for radioisotopic heaters, and as an energy source for cardiac pacemakers, heart pumps and small undersea propulsion systems. Plutonium 239 is used in nuclear weapons and in experimental reactors. Curium and americium are used in various national laboratories for research purposes and to produce heavier elements.

Land Burial

Land burial of carbon 14, cobalt 60, and iridium 192 wastes, in small concentrations, at approved sites that are acceptable from a geologic standpoint, is an acceptable means of disposal. Their concentration should not be in excess of 10^4 times the maximum permissible concentration for the general population (5). All wastes to be disposed of should be in a solid form and encapsulated in a suitable container. The burial site design, geology, and hydrology should be in conformance with the criteria used in selecting and licensing the present commercial burial sites. This method of disposal is not considered satisfactory for the disposal of radium 226 because radium 226 has such an extremely long half-life and a high radiotoxicity.

Disposal of plutonium, americium and curium, in small concentrations, by direct burial in unlined trenches is not considered a satisfactory means of disposal. These types of wastes are presently disposed of in this manner at approved sites. This method of disposal does not provide for any type of secondary containment barrier to prevent their release by leaching with the local ground water. Since these materials have an extremely long half-life and a high radiotoxicity, this method of disposal is not satisfactory.

Long-Term Storage

The storage of soldified radium 226 wastes in engineered storage facilities offers the best intermediate method for storage of these wastes. The technology for these facilities has been developed and the wastes will be under surveillance and control and can be retrieved, should this be required. The wastes will be stored in stainless-steel-lined concrete vaults (5).

Near-surface storage of radionuclides such as plutonium, americium and curium in stainless steel tanks encased in concrete and buried underground is not considered as a satisfactory means of disposal. Aqueous solutions of spent fuel reprocessing wastes have been stored in this manner over the past 25 years. These tanks are considered as an interim storage technique due to a general lack of confidence in their long-term integrity.

Since present regulations require the solidification of all reactor wastes within five years following reprocessing, the near-surface storage of these wastes in steel tanks should only be considered as an interim storage technique and not as a permanent storage or disposal technique.

The storage of solidified plutonium, americium and curium wastes in engineered surface facilities offers the best immediate method for storage of these wastes. The necessary technology for these facilities has been developed. The wastes will be under constant surveillance and control and can be retrieved, should this be required. The wastes should be solidified and encapsulated in a suitable container (steel). Of the four high-level solidification processes developed for reactor-produced wastes, spray or phosphate glass solidification offers the better solidified waste characteristics than the pot calcination or fluidized bed calcination processes. The wastes will be stored in stainless-steel-lined concrete vaults which will be either air or water-cooled.

Salt Deposit Disposal

This method offers the best potential for the disposal of radium 226 wastes since bedded salt deposits are completely free of circulating ground waters. This method of disposal has been under study by the Oak Ridge National Laboratory since 1957, and in November of 1970 a committee of the National Academy of Sciences recommended that the use of bedded salt for the disposal of radioactive waste is satisfactory (5). Recent questions concerning the adequacy of this method have resulted in the need for further development work before it can be accepted as an ultimate method of disposal. The critical problem is the selection of a site that meets the necessary design and geological criteria. Plutonium, americium and curium wastes can also be handled expeditiously using this technique. The wastes must be solidified and disposed of in the solid form encapsulated in a suitable container. The solidified wastes are buried in rooms carved in the salt deposits approximately 1,000 feet below the ground. The salt is a good heat transmitter, provides about the same radioactive shielding as concrete, and can heal its own fractures by plastic flow.

REACTOR OFF-GASES

Nuclear chain fission reactions and the reactors in which they take place are well known. A typical reactor includes a chain reacting assembly or core made up of nuclear fuel elements. The fuel material is generally encased in corrosion-resistant heat-conductive containers or cladding. The reactor core, made up of a plurality of these fuel elements or rods in spaced relationship plus control rods or blades, in-core instrumentation, etc., is enclosed in a container or core shroud through which the reactor coolant flows. As the coolant passes between the spaced fuel rods, it is heated by thermal energy released in the fuel during the fission reaction. The heated coolant then leaves the reactor, and the heat energy is used to perform useful work, such as by driving a turbine-generator set to produce electrical power. The now-cooled coolant is purified by removing any particulate material and/or noncondensable gases from the coolant and the coolant is recycled back to the reactor.

In boiling-water-type reactors, the coolant is water which is partially evaporated in the core. The resulting steam is separated from the water within the reactor vessel and is directed to a load, such as a turbine. After passing through the turbine, the steam is condensed. The condensate is demineralized and treated to remove any particulate material, such as corrosion products, and the water is returned to the reactor.

Under commercial practice, the noncondensable gases which were mixed in the steam are removed in the condenser, held for a suitable period to permit any shortlived fission and activation product gases which may be present to decay to safe levels, then the gases are vented to the atmosphere through a stack. These reactor off-gases primarily consist of air which has leaked into the system through various flanges and fittings, and hydrogen and oxygen produced by radiolytic decomposition of water in the reactor core. In addition, small amounts of radioactive noble gases, such as xenon and krypton are present in the off-gas. In a typical large reactor system, the total off-gas is about 200 cubic feet per minute, of which about 90% is a radiolytically formed stoichiometric mixture of oxygen and hydrogen, about 10% is water vapor and air in-leakage.

Also, there are a few cubic centimeters per minute of fission product gases such as xenon and krypton. It is generally necessary to retain the off-gas for about 30 minutes to permit the xenon and krypton isotopes and the short-lived activation products to decay to safe levels before venting the gases to the atmosphere. This necessary delay requires that large volumes of gases containing potentially explosive mixtures of hydrogen and oxygen must be stored for this period before venting. Standard practice has been to provide a large volume piping system designed to withstand explosions to hold this gas for the required period.

For example, such a hold-up pipe may be a few feet in diameter and extend for several hundred feet. This hold-up system is expensive because of its size and the necessity of an explosion-proof design. Thus, it would be highly desirable to eliminate the hazards and expense resulting from the need to retain large quantities of explosive gases for appreciable periods.

It has been proposed that this volume be reduced by recombining the stoichiometric hydrogen-oxygen content of the off-gas to form water, which would be condensable. This would both greatly reduce the quantity of gas being stored and remove the danger of an explosion. Attempts have been made to use catalytic recombiners in this manner. However, these recombiners have not been entirely successful since they require steam dilution of the hydrogen to a concentration of about 4 percent by volume, superheating of the steam diluted mixture to prevent catalyst poisoning and periodic catalyst replacement because of the accumulative poisoning by water and/or organic materials. Thus, these catalytic recombiners have not been entirely successful in this application which requires the continuous treatment of large volumes of gases. Recombining of the hydrogen and oxygen in the reactor off-gases is further made difficult by the fact that the gas flow may be continuous or intermittent, may vary somewhat in composition and may have variable velocity through the system.

Incineration

A system developed by W.A. Hartman, Jr. et al (6) is a system for burning flammable gas in the off-gas from a nuclear reactor power plant. The system includes three porous plugs in series in the off-gas line. The first plug serves as an upstream flame arrestor in case of burner failure. The second porous plug serves as a burner with a flame ignition means adjacent the downstream face of the plug. The third porous plug serves as a downstream flame arrestor and condenser for the water vapor formed in the flame reaction. Each of the porous plugs includes internal cooling means to remove heat. Appropriate temperature sensing means and control means are provided to insure safe, reliable operation. Figure 87 shows such a system.

FIGURE 87: NUCLEAR REACTOR OFF-GAS BURNER SYSTEM

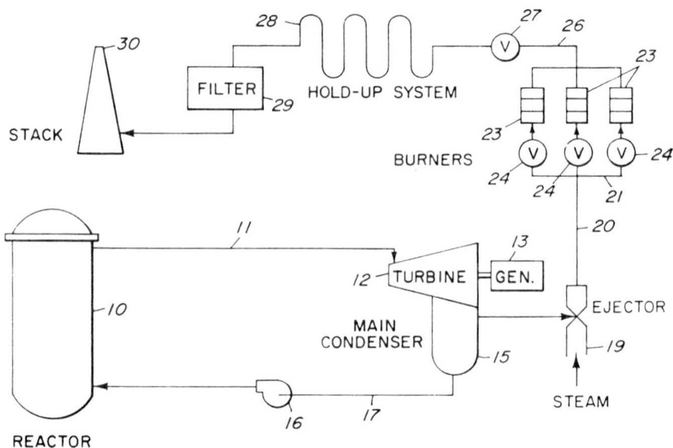

Source: U.S. Patent 3,598,699

There is seen a general schematic representation of a nuclear power generating system including provision for disposing of off-gas from the system. In such a system, steam generated in the reactor 10 passes through a line 11 to a turbine 12 which drives an electrical generator 13. Spent steam from turbine 12 is condensed in main condenser 15. The resulting condensate is pumped back to reactor 10 by pump 16 in line 17.

Noncondensable gases in main condenser 15 are extracted by a steam driven ejector 19. These gases pass through line 20 to mainifold 21. Three off-gas burner systems 23 are arranged in parallel. Isolation valves 24 are provided so that each burner 23 can be isolated from the system. Burners 23 are preferably sized so that the system can be operated with one burner set closed off for replacement or repair. The off-gases, greatly reduced in volume, leave the burner array and pass through line 26 and shut-off valve 27 to a hold-up system 28. Hold-up system 28 contains sufficient volume so that incoming off-gas has a sufficient residence time in the system that short-lived radioactive isotopes decay. From the hold-up system the gases pass through an absolute filter 29 and then are vented to the atmosphere through a stack **30**. Hold-up system **28** can be much smaller than presently used systems which must provide temporary storage volume for the large quantities of oxygen and hydrogen in the off-gas leaving the main condenser. Also, hold-up system **28** need not be of explosion-proof construction since the explosive stoichiometric mixture of oxygen and hydrogen has been eliminated in the burner system.

A very similar system has been described by W.A. Hartmann, Jr. et al (7).

A system developed by G.E. Moore and L.H. Tomlinson (8) utilizes a cooled porous plug device in combination with a continuous ignition source for burning off explosive mixtures of contaminated radiolytic hydrogen-oxygen gas flow discharged from the steam turbine cycle of a boiling water nuclear power reactor. Optional use of a second cooled porous plug is shown, the second porous plug being located downstream of the burner to function as a heat exchanger to cool the combustion products for controlled condensation of the water vapor. Figure 88 shows such a system.

FIGURE 88: BURNER FOR OFF-GAS FROM BOILING WATER REACTOR POWER PLANT

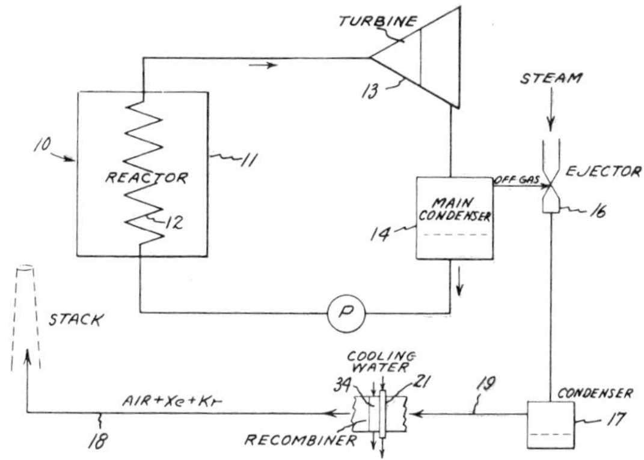

Source: U.S. Patent 3,660,041

Water passing through reactor case **11** in conduit **12** of a boiling water nuclear power plant **10** is converted to steam. Thereafter, the steam so generated is passed through steam turbine **13**. After the low-pressure stage of turbine **13** has been traversed, the steam (now at low pressure) passes through main condenser **14** at which point the gas content thereof is removed via the steam jet air ejector **16**. Ejector **16** pumps the off-gas up to atmospheric pressure, or slightly above atmospheric pressure. The off-gas then passes to condenser **17**, where the pumping steam used in operation of ejector **16** is condensed. From this point the off-gas is an explosive mixture.

The dry composition of this off-gas would typically be 60% H_2; 30% O_2; 10% air plus about 10^{-6}% radioactive gaseous isotopes. The volume flow of this gas varies in proportion to the reactor output, so that when the power load is low, much less off-gas is produced than at full load, although air in-leakage in the low-pressure stages of the turbine tends to be constant. Typically, the volume flow of off-gas for an 1,100 Mw electrical (Mwe) boiling water reactor may be as high as **300** standard cubic feet/minute having the proportions set forth above. Thus, by the practice of this process, whereby the hydrogen and oxygen can be safely recombined to form water, about 90% of the off-gas flow volume can be eliminated enabling a great reduction in the size of the off-gas holding system, pipe **18**, and equally important, eliminating the expense of explosion-proof design for the off-gas system. The off-gas flow from condenser **17** is passed through pipe **19** to the cooled porous plug burner **21**, where the explosive hydrogen/oxygen mixture is burned in a stable flame as rapidly as it enters the ignition region and, at the same time, functions as a flame arrester reliably preventing the propagation of flame upstream into the explosive mixture in pipe **19**. The porous plug burner **21** is made of sintered metal, preferably copper particles in the size range from about 1 to 200 microns.

A very similar system is described by G.E. Moore et al (9).

SPENT FUEL PROCESSING WASTES

The problem of disposal of nuclear wastes from fuel reprocessing for nuclear fission reactors has been reviewed by A.S. Kubo and D.J. Rose (10). A diagram of the waste disposal options presented by these authors is shown in Figure 89.

FIGURE 89: THE TAXONOMY OF NUCLEAR WASTE DISPOSAL OPTIONS

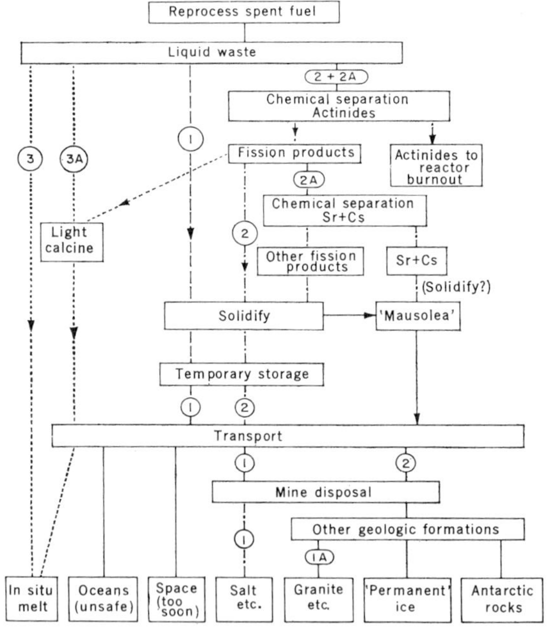

Source: Science 182, 1205-1211, December 21, 1973

The primary difficulty with radioactive wastes, particularly fission product waste, is that they cannot be disposed of by ordinary dilution methods (11). In order to bring solutions to accepted AEC activity tolerances, the dilution is so great that the method is unfeasible. Retainment until sufficient radioactive decay has taken place to allow dilution seems to be the only solution. In the normal distribution of elements from the fission process, this means a contaminant of approximately 500 to 600 years' life. As the volume of these wastes increases, it is apparent that the problem of properly containing these wastes will become enormous if not already so. For example, one reactor is estimated to consume 13,500 kg of fuel per year which will yield 3,370 kg fission product and, after reclaiming the unused fissionable elements, the waste products will amount to about 16,000,000 gallons per year. Prior art methods for the disposal of such wastes which have been suggested include the injection of radioactive waste into depleted oil reservoirs or other subterranean porous formations, underground caverns, abandoned mines, salt domes, aquifers, clays having ion exchange ability, and into cribs and the like; storage in steel-lined concrete tanks; and injection into the sea. None have been entirely successful. Specifically when such materials are injected into subterranean formations, there is the ever-present danger of subsurface movement or migration of the radioactively hot waste, a potential health and economic hazard.

In summary, any real solution to the problem of disposal of radioactive wastes must necessarily cope satisfactorily with extremely large volumes of substances having ionic compositions of relatively high concentration and often containing substantial amounts of solids, high-intensity radiation, high levels of heat energy, and long retention times.

Fixation of Wastes

Fuel reprocessing plants have, in addition to some upgrading of low-level radwaste treatment systems, a much more complex problem to solve, namely, that of converting the stored, high-level fission product and actinide liquid waste, which are the result of reprocessing, to an acceptably immobile solid waste form which will permit its ultimate disposal (12).

In this regard, the U.S. AEC has for some time at its Brookhaven, Oak Ridge and Hanford facilities been developing techniques for concentrating and solidifying the high-level radioactive wastes from reprocessing into an acceptable form for long-term storage. Recently the work on the performance of three waste solidification techniques was completed at Battelle Northwest in Richland, Washington. These techniques are phosphate glass solidification, spray solidification, and pot solidification.

The phosphate glass process, originally tested by Brookhaven National Laboratory, consists of an evaporative denitration of the high-level waste after the addition of phosphoric acid, sodium hydroxide, and ferric nitrate. The bottoms discharge of the denitrator is melted and collected in a pot for final storage. The overheads from denitration are condensed and re-evaporated with the bottom, recycled to the denitrator feed tank, and the overheads fractionated for nitric acid recovery. The aqueous effluent was decontaminated by a factor of 10^7 to 10^8 for the nonvolatiles, and by 10^5 to 10^6 for the radio-ruthenium. Storage of the phosphate glass product was adequate; there was no increase in pressure after the pot cap was welded in place.

The spray solidification process utilized a spray calciner to solidify the incoming high-level waste liquor followed by a fusing of the solids directly in the storage container using a sodium borosilicate frit as a melt additive. The off-gas from the calciner is condensed and re-evaporated, and the overheads are fractionated for HNO_3 acid recovery. The bottoms of the evaporator are recycled to the calciner feed tank. The aqueous effluent was decontaminated by a factor of 10^9 for nonvolatiles and 10^5 or 10^6 for radio-ruthenium. The waste containers were capped and welded and showed no pressure buildup.

The pot solidification process, originally tested at Oak Ridge National Laboratory, is a direct solidification in the storage container used for ultimate disposal. The use of auxiliary evaporator, fractionator, and scrubber equipment provided a 10^9 decontamination factor for the liquid effluent and greater than 10^3 for radio-ruthenium. The sealed containers have undergone tests at the Solids Storage Engineering Test Facility at Hanford.

One of these basic techniques will undoubtedly be utilized in order to comply with the recent AEC regulation making it mandatory to convert stored high-level waste liquids to a form acceptable for ultimate disposal.

A process for calcination of radioactive waste solutions has also been described by B.M. Johnson, Jr. et al (13) and is shown in Figure 90.

FIGURE 90: PROCESS FLOW SHEET FOR THE CALCINATION OF RADIOACTIVE
WASTE SOLUTIONS

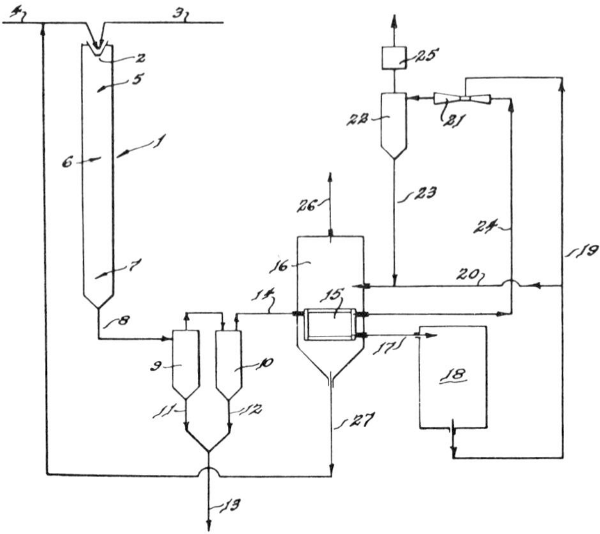

Source: U.S. Patent 3,008,904

Referring to the drawing, numeral **1** designates a dehydration and calcination tower, a calciner equipped at its top with a fluid-spray nozzle **2**. An inlet pipe **3** for steam and an inlet pipe **4** for the waste solution to be treated lead into the spray nozzle. The tower has
a dehydration zone **5**, a calcination zone **6**, and a cooling zone **7**. To create these zones,
independent heating means (not shown) arranged around zones **5** and **6** and cooling means
(not shown) arranged around zone **7** are operated. At the bottom of the tower there is an
outlet pipe **8**, leading to a cyclone **9**, which is connected in series with a glass-bag filter **10**.
Discharge pipes **11** and **12** of the cyclone and the glass filter, respectively, lead to a common pipe **13** through which the solids deposited at the bottoms of the cyclone **9** and the
glass-bag filter **10** are withdrawn and charged to a packaging unit (not shown).

At the top of filter **10** there is arranged a pipe **14** which is connected with a heat exchanger
and condenser **15** located near the bottom of an evaporator **16**. An outlet pipe **17** for
the condensate is arranged in the bottom of heat exchanger **15**; it leads to a holdup tank
18. Tubes **19** and **20** connect the holdup tank **18** with the evaporator **16** and tube **19**
connects it also with a scrubber **21**. The scrubber is linked to a separator **22**, and the latter
is connected with the evaporator **16** through pipes **23** and **20**. Pipe **24** connects the condenser and scrubber **21**.

A filter **25** is arranged above and connected with the separator **22**. The top of the evaporator **16** is equipped with an exhaust line **26** which leads to a vacuum condenser (not shown)
where the steam and any other vapors escaping the evaporator are condensed. A discharge
line **27** connects the bottom of the evaporator with inlet pipe **4**.

The waste solution is introduced in the form of a spray at the top of the calciner where it
passes through zones **5**, **6** and **7** for dehydration, calcination and cooling, respectively. The
powder suspended in the gas leaving the calciner at its bottom through pipe **8** is separated
roughly from the gas in cyclone **9**, the solid being discharged through pipe **11**. The gas
which still contains some fine powder particles is withdrawn at the top of the cyclone and

introduced into the top of glass-bag filter **10** where further gas-solid separation is accomplished. The gaseous component is then subjected to a cooling and condensation step by heat exchange in condenser **15**; noncondensed fractions leave the evaporator at **26**. The cooled condensed liquid is withdrawn through pipe **17** and passed into holdup tank **18**. Part of the condensate accumulated in tank **18**, preferably about 90% of it, is cycled as cooling medium into evaporator **16**, while the rest is introduced into scrubber **21** and utilized for washing the gas. The scrubbed gas leaving the separator is filtered in **25**, while the scrubbing liquor is recycled, via pipes **23** and **20**, into evaporator **16**. Liquid condensed in evaporator **16** is withdrawn at **27** and recycled into the feed line **4** and back into the calciner. Gas noncondensable in the heat exchanger **15** leaves through pipe **24** where it passes through scrubber **21**, etc.

While phosphate and borate anions are the preferred ions for the purpose of converting the calcined waste solution to a glassy water-insoluble material, silicate and mixtures of any of the three can also be used to advantage. The disadvantage of silicate is that a higher processing temperature is needed for the material to melt.

A process developed by M.L. Spector (14) relates to the decontamination of the volatile effluent in the calcination of radioactive waste materials. Troublesome features of the calcination include the volatilization of cesium and ruthenium which are abundant fission products in the waste mixture and entrainment of solids. Both of these elements are volatile in the upper temperature range at which calcination occurs. The volatile form of ruthenium is the tetroxide state. While normally ruthenium tetroxide is unstable at high temperatures, decomposing to insoluble ruthenium dioxide and oxygen, the oxides of nitrogen in the radioactive mixture serve to maintain the higher oxidation state of the ruthenium and thus the volatilization of this radioactive component accompanies the calcination. For this reason, prior practice in calcining radioactive wastes has been a difficult problem requiring expensive equipment operated by remote control and extensive shielding to protect the area in which the calcination is performed.

In the process, as shown in Figure 91, the vapors and any entrained solids are passed out of the calcination zone into a cartridge containing a fusible solid mixture on which the cesium oxide and/or ruthenium oxide are sorbed. The cartridge removed from operation, which contains sorbed radioactive volatile materials can be separately sealed and fired, can be added to the mixture in the fixation zone before firing or as a further alternative, the contents of the cartridge can be introduced to the fixation zone along with the radioactive solids from the calcination zone for ultimate fixation therein.

As shown in the figure, the radioactive waste material comprising a mixture of nitrates of aluminum, sodium, mercury, ruthenium, cesium and nitric acid are passed into calcination zone **2** from feed line **4**. In the calcination zone, the feed is heated from 100°C up to about 500°C. The radioactive mixture remains in the calcination zone for a period of from about 10 minutes to a period of one hour or longer depending upon the size of the charge. During the heating operation, the solid nitrates are converted to their corresponding oxides; the free water and oxides of nitrogen derived from nitric acid in the waste mixture are vaporized at lower temperatures and pass out of zone **2**.

Ruthenium tetroxide and a minor portion of the cesium oxide which are volatilized at temperatures between about 300° and 500°C, are then passed to the upper portion of the calcinator and out of zone **2**. The solid materials in the lower portion of zone **2** can be aerated, if desired, by an upwardly flowing gas such as steam, entering zone **2** under grid **7**. Alternatively, mixing or agitation in zone **2** can be achieved by carrying out the calcination in a ball mill or similar apparatus. The decomposed solid mixture containing aluminum oxide, sodium oxide, cesium oxide, strontium oxide and small amounts of entrained ruthenium oxides is passed downwardly through conduit **10** into fixation zone **12** which contains a solid mixture of nonradioactive metal oxide and a nonradioactive elemental metal.

In zone **12** the solid radioactive material is mixed with the nonradioactive solids and the resulting mixture is chemically reacted by heating the mixture to its ignition temperature by means of removable furnace **14**, which encloses zone **12** when in use. For best results, it is recommended that the radioactive solids be distributed below the upper level of the bed of nonradioactive material. This can be accomplished by hollowing out an area of the bed or by providing a delivery conduit **10** slidably mounted to extend below the bed level during delivery of material. One modification which is beneficial as a means of mixing, entails alternately feeding radioactive solids and nonradioactive solids into the fixation zone; another embodiment involves continuous mixture of the feeds.

FIGURE 91: APPARATUS FOR THE FIXATION OF RADIOACTIVE WASTES IN FUSED CARTRIDGES

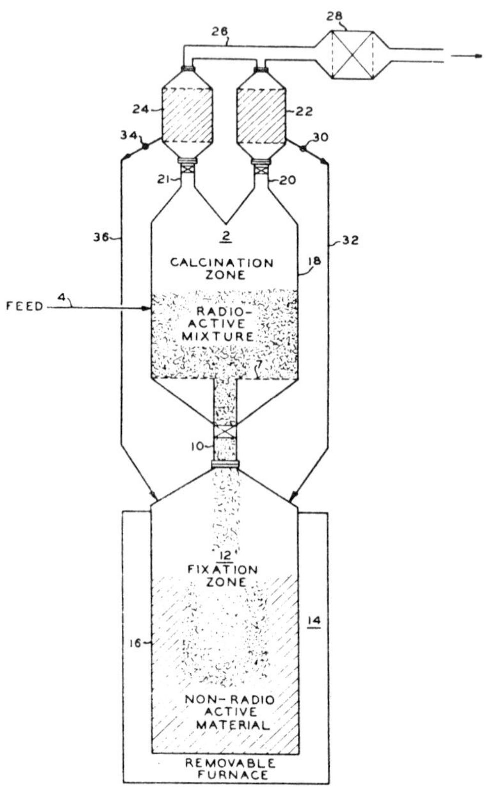

Source: U.S. Patent 3,153,566

The fixation is generally conducted at a temperature between about 500° and 1200°C although, after ignition, the temperature generated by the system may rise as high as 2800°C or higher. In certain instances, when the particle size of the solids is extremely small, for example, less than 400 mesh, the ignition temperature may be lowered. During the reaction of the radioactive materials with the nonradioactive materials, the feed of radioactive materials through conduit **10** is discontinued, thus avoiding the deposition of radioactive materials on top of the fused fixation product. After fixation, the radioactive and nonradioactive metals are chemically bonded through oxygen atoms in a fused crystalline mass which has extremely low water leachability. At this point, removable furnace **14** is withdrawn and container **16**, containing the fused mass can be sealed and removed for ultimate deposition in a safe storage area. For ease in removal of the fixation container, it is recommended that the calcination container **18** be connected to fixation container **16** by a threaded conduit **10**. Any vaporous materials entering the lower portion of calcination zone **2** pass upwardly and are withdrawn from the top of the calcination zone together with vaporous and solid entrained effluent from zone **2** through valved conduits **20** and/or **21** connected to filter cartridges **22** and **24** respectively.

For ease of operation it is suggested that one of the filter cartridges, or if more than two are employed, one of a series or one series of filter cartridges be used until the desired extent of saturation is reached at which point the cartirdge or cartridges in use, or materials contained therein, are removed by closing the respective valve and a fresh cartridge or series of cartridges or sorbent material put on stream by opening the respective valve or valves. In the drawing, alternate operation of cartridges **22** and **24** is preferred.

The vapors from the calcination zone pass upwardly into the cartridge in operation and cesium oxide and ruthenium tetroxide are sorbed by the solid mixture contained therein. The deposition of ruthenium oxide is observed as a bluish-black film which collects in the lower portion of the cartridge. Before the black film reaches the top of the cartridge, the delivery valve is closed and the second cartridge (or cartridges) containing fresh sorbent material is put into use. The spent cartridges or spent material in the cartridge is disposed of and replaced by a fresh cartridge or fresh sorbent material for use after the second cartridge or series of cartridges become spent.

In the calcination zone, as the temperature is increased from 100° to 500°C, the first vapors leaving zone **2** are water and oxides of nitrogen, which pass through the filter cartridges. Some entrained solid material may also pass upwardly in zone **2** and enter cartridge **22** and/or **24** where they are entrapped along with ruthenium oxide and cesium oxide which vaporizes at higher temperatures. The ruthenium oxide passing upwardly in calcination zone **2** is in the tetroxide state, being oxidized from ruthenium dioxide by the oxides of nitrogen. However, upon deposition the ruthenium tetroxide reverts to the dioxide state which is responsive for the black deposit which appears on the solid mixture of elemental metal and metal oxide. In the dioxide state the ruthenium is no longer volatile but is retained in the solid sorption medium.

The gases passing through cartridges **22** and/or **24** are removed by conduit **26** and are passed to cartridge **28** which may contain the same solid mixture as cartridges **22** and **24** or may be any ordinary filter. The gases leaving cartridge **28** are then vented to the atmosphere. When the desired level of deposition of radioactive components has been observed in cartridge **22**, valve **20** is closed and valve **30** opened and the contaminated solid mixture in cartridge **22** is downwardly withdrawn through detachable line **32** and introduced into the upper portion of fixation zone **12** and valve **30** is closed. Fresh sorbent solids are then added to cartridge **22**. This same operation is repeated when, after cartridge **24** has approached the desired saturation limit, it is withdrawn from operation, except that valve **21** is closed and valve **34** open to permit withdrawal of contaminated solid through detachable line **36** for delivery to the upper portion of zone **12**.

It is to be understood, however, that in place of withdrawing the contaminating solids from cartridges **22** and **24**, the entire cartridge can be withdrawn and disposed of by the chemical fixation reaction carried out in zone **12** and a fresh cartridge substituted in its place. The following example serves to illustrate the process.

Example: A simulated waste mixture having the following composition was prepared: 1.7 molar aluminum nitrate, 1 molar nitric acid and 0.1 molar sodium nitrate. To this mixture of about 20 grams was added, 0.1 gram of ruthenium dioxide. Of this mixture, 0.63 gram was calcined at a temperature of from 100° to 475°C and a yellowish vapor containing ruthenium tetroxide and oxides of nitrogen was volatilized. All of the vapors were passed from the calcination zone into a cartridge containing 4 parts by weight ferric oxide and 1 part by weight silicon. The cartridge was maintained at a temperature of about 400°C. The black deposit of RuO_2 began to collect and continued to collect on the solids in the cartridge until vapors ceased to form in the calcination zone. No trace of ruthenium oxide was found leaving the cartridge either during or after the use of the cartridge. The RuO_2-containing cartridge was then sealed and fired to red heat at which temperature the reaction between silicon, ferric oxide and ruthenium oxide is initiated (between about 500° and 700°C). The solid fused mass was then removed from the cartridge and since ruthenium dioxide is not soluble in water, the leachability of the cartridge is substantially zero. No black desposit of ruthenium oxide was noted on the walls of the cartridge thus insuring complete chemical incorporation of ruthenium in the fused mass.

A process developed by J.D. Murphy et al (15) is one in which radioactive waste is collected as a slurry in aqueous media in a metering tank located within the nuclear facilities. Collection of waste is continued from time to time until a sufficient quantity of material to make up a full shipment to a burial ground has been collected. The slurry is then cast in shipping containers for shipment to a burial ground or the like by metering through a mixer into which fixing materials are simultaneously metered at a rate to yield the desired proportions of materials.

Land Burial

Land burial of cesium 134 and cesium 134 in low-level wastes, at approved sites that are acceptable from a geologic and hydrologic standpoint, is an acceptable means of disposal. The cesium 134 and cesium 137 concentrations should not be in excess of 10^4 times their maximum permissible concentrations for the general population (4). All cesium wastes to be disposed of should be in a solid form and encapsulated in a suitable container. Liquid wastes should be solidified, preferably using asphalt.

The burial trenches should be designed not to intercept the groundwater table and constructed with a bottom drain and sump for water monitoring. The trenches should be covered with either asphalt or vegetation to limit infiltration of water. The burial site design, geology, and hydrology should be in conformance with the criteria used in selecting and licensing the commercial burial sites. Since land burial sites meeting the criteria are operating on a commercial scale currently, they are considered the most satisfactory method of disposing of dilute concentrations.

Since land burial has successfully been practiced on a commercial scale, it should also be considered as the most satisfactory method of disposing of all dilute concentrations of strontium 90 wastes. Since land burial has successfully been practiced on a commercial scale and both zirconium 95 and niobium 95 have short half-lives, it should be considered as the most satisfactory method of disposing of all dilute concentrations of these wastes.

Land burial of iodine, krypton, and xenon wastes, in small concentrations, at approved sites that are acceptable from a geologic and hydrologic standpoint, is an acceptable means of disposal (5). Their concentrations should not be in excess of 10^4 times their maximum permissible concentrations for the general population. The burial trenches should be designed not to intercept the groundwater table and constructed with a bottom drain and sump for water monitoring. The trenches should be covered with either asphalt or vegetation to limit infiltration of water. The burial site design, geology, and hydrology should be in conformance with the criteria used in selecting and licensing the commercial burial sites. Since land burial has successfully been practiced on a commercial scale, it should be considered as the most satisfactory method of disposing of low concentrations of these wastes.

Land burial of Ru 106, Ce 144 and Pm 147 wastes, in small concentrations, at approved sites that are acceptable from a geologic and hydrologic standpoint, is an acceptable means of disposal. Their concentrations should not be in excess of 10^4 times the maximum permissible concentration for the general population. All wastes to be disposed of should be in a solid form and encapsulated in a suitable container. Liquid wastes should be solidified, preferably using asphalt, in accordance with the methods described in the radioactive waste solidification report. The burial trenches should be designed not to intercept the groundwater table and constructed with a bottom drain and sump for water monitoring. The trenches should be covered with either asphalt or vegetation to limit infiltration of water. The burial site design, geology, and hydrology should be in conformance with the criteria used in selecting and licensing the commercial burial sites. Since land burial is successfully practiced on a commercial scale and since these isotopes have moderate half-lives, it should be considered as the most satisfactory method of disposing of all dilute concentrations of these wastes.

Long-Term Storage

Near-surface storage (using carbon steel and stainless steel tanks encased in concrete and buried underground) of high-level, reactor-produced aqueous solutions of cesium and other fission product salts such as strontium salts is not considered as a satisfactory means of storage. Aqueous solutions of high-level wastes from reactor fuels have been stored in this manner over the past 25 years. The tanks range in size from 0.33 to 1.3 million gallons and are equipped with devices for measuring temperatures, liquid levels, and leaks. These tanks are considered as an interim storage technique due to a general lack of confidence in their long-term integrity. Therefore, the near-surface storage of aqueous solutions of cesium and other fission product salts in steel tanks should only be considered as a near-term storage technique and not as a permanent storage or disposal technique (5).

The storage of high-level solidified cesium 134 and cesium 137 wastes and other waste salts such as strontium 90 salts in engineered storage facilities offers the best method for intermediate storage of these concentrated wastes. The advantages of this method are that the wastes will be under surveillance and control and that they can be retrieved, should this be required. The high-level wastes from fuel processing should be solidified and packaged in a suitable container (steel). Of the four high-level solidification processes developed, spray or phosphate glass solidification processes offer better solidified waste characteristics than the pot calcination or fluidized bed calcination processes. The wastes will be stored in stainless steel vaults encased in concrete and will be either air or water-cooled.

The disposal of high-level, aqueous solutions of cesium 134 and cesium 137 wastes, along with other radioactive wastes such as low-level strontium 90 wastes or high levels Zr 95 or Nb 95 wastes in vaults excavated in crystalline rock over 1,500 feet beneath the ground is currently being evaluated by E.I. du Pont de Nemours, at their Savannah River Plant near Aiken, South Carolina (5). The wastes would be stored in six tunnels and once in the tunnels the wastes will seep into the surrounding rock.

Located above the crystalline rock is the Tuscaloosa formation, a good source of fresh water, which is separated by a layer of clay that would act as a barrier to the leakage of radioactive wastes. An advisory committee appointed by the National Academy of Sciences recommended abandonment of the project. In May 1972, another National Academy of Sciences panel concluded that bedrock storage provides a reasonable prospect for long-term safe storage, but precise information is needed to decide if and where underground storage vaults should be built. Another method has also been proposed for disposing of liquid wastes by in situ incorporation in molten silicate rock.

However, both of these methods are unproven since sufficient engineering data or exploration has not been completed to verify their suitability. Due to the long half-lives of certain of the elements in the mixed fission products and actinides, it is doubtful that data could ever be accumulated to prove that the geological characteristics of the site over the next few hundred years are acceptable to ensure the absolute safety of such a disposal method.

The storage of high concentrations of krypton, xenon and iodine wastes in engineered storage facilities offers the best immediate method for storage of these wastes. The technology for these facilities has been developed. The wastes will be under surveillance and control and can be retrieved, should this be required. The wastes will be stored in stainless-steel-lined concrete vaults which will be either air or water-cooled.

Near-surface storage of aqueous Ru 106, Ce 144 and Pm 147 wastes in stainless steel tanks encased in concrete and buried underground is not considered as a satisfactory means of disposal. Aqueous wastes have been stored in this manner over the past 25 years. The tanks range in size from 0.33 to 1.3 million gallons and are generally equipped with devices for measuring temperatures, liquid levels, and leaks. These tanks are considered as an interm storage technique due to a general lack of confidence in their long-term integrity. These same comments apply to Zr 95 and Nb 95.

The storage of high-level solidified Ru 106, Ce 144, and Pm 147 wastes in engineered surface facilities offers the best immediate method for storage of these wastes. The technology for these facilities has been developed. The wastes will be under surveillance and control and can be retrieved, should this be required. The aqueous wastes should be solidified and packaged in a suitable container (steel). Of the high-level solidification processes developed, spray and phosphate glass solidification offer the best solidified waste characteristics. The wastes will be stored in stainless-steel-lined concrete vaults which will be either air or water-cooled.

The periodic replacement of waste containers probably will not be required since their activity is reduced by a factor of 1,000 in 10 years for Ru 106 wastes, in 21 years for Ce 144 wastes, and in 26 years for Pm 147 wastes. The replacement of the waste containers might be required if other long-lived isotopes are also present. Similarly the periodic replacement of the waste containers probably will not be required in the case of Zr 95 and Nb 95 since the activity of zirconium 95 is reduced by a factor of 100 in 432 days and niobium 95 activity is reduced by a factor of 100 in 232 days. Again, however, the replacement of the containers may be required if other long-lived isotopes are present.

A method developed by A.L. Backstrom (16) utilizes a rock-cavity reservoir for storing radioactive residues comprising: [1] a container disposed within an underground rock-cavity; [2] resilient means connecting top and bottom walls of container to the walls of the rock-cavity such that the container is spaced from the walls of the rock-cavity; and [3] a liquid having a higher density than water disposed within the space between the rock-cavity walls and the container, the liquid having a sealing action on the walls of the rock-cavity and the container.

Salt Deposit Disposal

This method offers the greatest potential for the disposal of high-level wastes from fuel reprocessing since bedded salt deposits are completely free of circulating groundwaters. This method of disposal has been under study by the Oak Ridge National Laboratory since 1957, and in November of 1970 a committee of the National Academy of Sciences recommended that the use of bedded salt for the disposal of radioactive wastes is satisfactory. Recent questions concerning the adequacy of this method have resulted in the need for further development work before it can be accepted as a method of ultimate disposal.

The cesium or strontium wastes must be solidified and disposed of in the solid form, encapsulated in a suitable container. The solidified wastes are buried in rooms carved in the salt deposits approximately 1,000 feet below the ground. The salt is a good heat transmitter, provides about the same radioactive shielding as concrete, and can heal its own fractures by plastic flow. The critical problem is the selection of a site that meets the necessary design and geological criteria (5). This technique is also applicable to zirconium 95 and niobium 95 disposal.

However, as pointed out by P.M. Boffey (17) problems continued to be encountered in attempts to build storage caverns in bedded salt layers deep underground. The search in New Mexico was launched after previous efforts to build the facility in an abandoned salt mine near Lyons, Kansas, were abruptly terminated because of unanticipated problems. A survey for possible sites conducted by the U.S. Geological Survey resulted in attention being focused next on the salt beds of southeastern New Mexico. At the site, however, brine containing hydrogen sulfide and methane was encountered. The presence of the brine was disturbing for two reasons. One was that the gases could pose a safety hazard for workers building the facility or operating it. The second disturbing aspect is that the presence of the brine solution may indicate that fluids have been migrating underground, thereby threatening the integrity of the site.

The search for a site to handle the radioactive wastes generated by the commercial nuclear power industry continues. More geophysical surveying and more test drilling will hopefully yield such a site.

A process developed by G.H. Billue (18) involves the storage of surplus products from chain reaction piles in spaces formed by leaching underground strata in which there are alternately present salt beds and layers of shale. As shown in Figure 92, the reservoir is produced by first drilling a bore hole through earth in which there are present alternate layers of soluble salt and insoluble shale strata. A casing **17** is set in the upper portion of the bore hole and is cemented to the surface of the earth. Water is then introduced into the bore hole to remove portions of the salt layers and to thus form reservoir chambers **10**, **11** and **12**. The chambers **10**, **11** and **12** correspond with the regions that were initially occupied by the salt, and are vertically connected by undisturbed portions **13** and **14** of the original bore. The undisturbed portions correspond to layers of shale rock which of course are not attacked by the water used to dissolve the salt.

The radioactive solution is deposited in the bottom storage chamber **12** until it reaches the level **15**, which is well below the top of the chamber **12**. In addition to only partly filling the storage reservoir with the radioactive solution, another security feature involves mixing sodium hydroxide or other alkaline material with the radioactive solution before its introduction into the storage hole. The sodium hydroxide or other alkaline solution neutralizes any nitric acid contained in the radioactive solution, thereby preventing attack of the solution on metal fittings.

The neutralized solution consisting of the reaction product of the radioactive solution and the sodium hydroxide or other alkaline reagent is transferred from the tank cars into the underground reservoir by means of a tube **16** which projects through the chambers into the bottom cavity **12**. In the event the tube **16** becomes contaminated, a wash-down with water is employed to remove the contamination. Successive wash-downs with small quantities of water may be employed, with draining intervals between the wash-downs being provided in order to allow any material adhering to the tube to reach the bottom of the reservoir.

Acids and ethylenediaminetetraacetic acid sodium salt or other suitable substances may be used in the final cleaning of the pipes used in conveying the solution in accordance with established decontamination proactices. This ensures that the pipes are safe for personnel above the ground. After all piping has been installed, all other outlets from the storage reservoir are properly plugged.

Another feature involves a refluxing principle which affords even greater security against the upward loss of any radioactive material. The radioactive solution present in the bottom of the reservoir will, at least initially, evolve steam due to the thermal effects resulting from the radioactive decay that inevitably takes place. The radioactive substances themselves, however, are not volatile but small amounts may be picked up and become entrained with the steam. The steam or water vapor that is constantly distilling upwardly is contacted by condensing vapor from the cooler upper portions of the reservoir, and this brings about a constant refluxing action which tends to prevent any radioactive substance that may be picked up or entrained with the distilling vapors from ever reaching the surface of the bore hole. These are shown as droplets **22** which form on the underneath sides of the shale ledges and then drop back into the radioactive mass.

There is shown an auxiliary bore hole **18** which extends from the surface of the earth to the upper surface of the uppermost chamber **10**. A filter **19** of positive filter material is placed in the bottom of the bore hole **18** just above the chamber **10**, and a condenser **20** is located at the upper end of the bore **18**. A vent **40** is provided at the condenser inlet, and a trap **41** and drain **42** are located at the outlet of the condenser **20**. A tube **21** extends from the outlet of the condenser **20** into the reservoir so as to complete a closed loop refluxing system. The evolved steam flows from the upper chamber **10** through the filter **19**, where it is purified, up through the auxiliary bore hole **18**, and through the condenser **20** where it is condensed into liquid. Purified condensate may be dropped back into the remaining mass via the tube **21** for cooling. The inlet to the condenser **20** is also connected to a vacuum pump **43**, to allow the cavity to be operated at subatmospheric pressures, if desirable, to control the condensate temperatures within the cavity.

FIGURE 92: SCHEME FOR UNDERGROUND STORAGE AND DISPOSAL OF RADIOACTIVE PRODUCTS

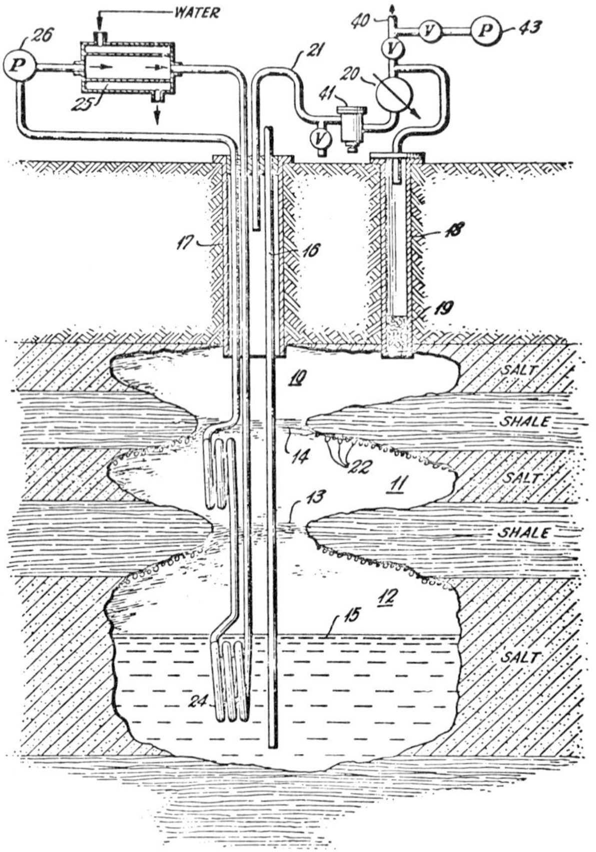

Source: U.S. Patent 3,236,053

The underground storage reservoirs may also be provided with cooling means, preferably in the form of a heat transfer system such as is commonly used in shipboard nuclear reactors. The cooling system comprises cooling coils **24**, preferably of stainless steel piping, located inside the reservoir and connected to a cooler **25** and a coolant circulating pump

26 located above the surface of the earth to form a closed coolant circulation loop. Liquid sodium may be employed as the coolant because it does not attack stainless steel or carbon steel and also because it remains in a liquid state over a wide temperature range (i.e., from about 200° to 1600°F). In the event it is desired to lower the freezing temperature of the coolant to extend its useful range, an alloy of liquid sodium and potassium in the ratio of about 25% of sodium and 75% potassium, which flows freely at room temperature, may be employed.

A radioactive waste disposal process has been described by M.D. Nelson, Jr. (19) and is shown in Figure 93. A cavern 13 is formed in salt bed 11 and disposed a sufficient distance below the earth's surface 12 that the emissions from the radioactive wastes stored in the cavern 13 are substantially adsorbed. Thus, no emissions escape to the earth's surface to become a hazard to public health. The cavern 13 may be formed by any suitable method. For example, the steps of the method are as follows: the cavern 13 is formed by providing a passageway 14 from the earth's surface 12 into the salt bed 11 and then forming the cavern 13 by conventional water leaching processes.

FIGURE 93: RADIOACTIVE WASTE DISPOSAL PROCESS INVOLVING PUMPING
WASTES INTO AN UNDERGROUND CAVERN

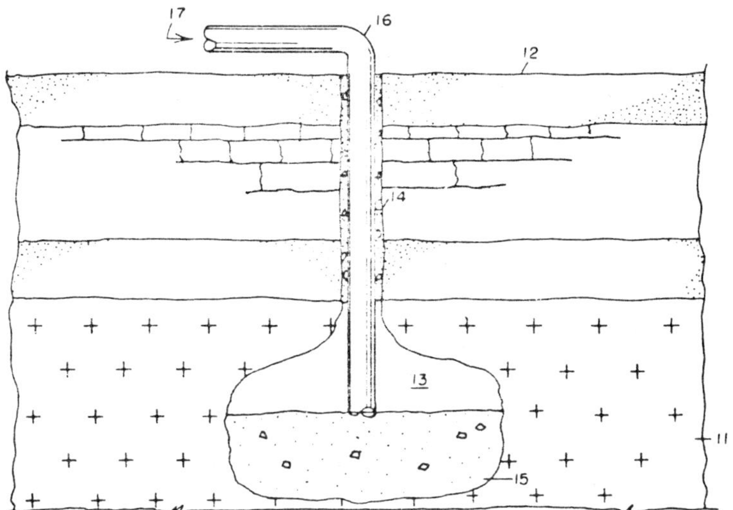

Source: U.S. Patent 3,262,274

As the next step, after the cavern 13 is completed, means for transporting a prepared containment composition 15 into the cavern 13 are provided by inserting a conveying means, such as fluid conduit 16, into the passageway 14. Fluid conduit 16 preferably extends from the earth's surface 12 through passageway 14 and into the cavern 13. It is preferred to cement, or otherwise seal, the conduit 16 into the passageway 14. Fluid conduit 16 may also be used in the leaching process for forming cavern 13, if desired. In another step, the prepared containment composition 15 is conveyed through conduit 16 into the cavern 13 by any suitable means, such as gravity flow, in the direction of arrow 17. After the composition 15 is in place, the final step of this method is to maintain the composition in the cavern 13 until it sets to a substantially rigid and nonflowing mass. This mass is the cement-like radioactive material. The method produces a system that provides for the absolute containment of radioactive wastes. If desired, a quantity of cement, without any radioactive wastes, may be passed through the conduit **16** to cleanse it of radioactive materials and to assist in maintaining the containment composition **15** in the cavern **13**.

This method is of advantage in that the high heat conductivity of salt dissipates the heat created by the decay of high level radioactive nuclides. Further, solid radioactive wastes of large-sized aggregates which can pass through conduit **16** can be readily stored. Of course, the normal precautions regarding the handling of radioactive wastes should be observed when practicing the process.

Deep Well Disposal

The direct disposal of aqueous, low-level radioactive wastes into shale formations has been investigated by Oak Ridge National Laboratory (5). The method consists of mixing the aqueous wastes with cement and pumping the resulting slurry down a well out into a nearly horizontal fracture in a thick shale formation. Additional work is required to demonstrate that this method of disposal is satisfactory. Since strontium 90 is leached from cement and no additional encapsulation is provided and since it has long half-life this method of disposal is not recommended for strontium 90. Since zirconium 95 and niobium 95 are leached from cement and no additional encapsulation is provided, this method of disposal requires additional work to determine its suitability for those materials. Figure 94 is a diagram showing a technique developed by J.J. Reynolds et al (20).

FIGURE 94: WELL FOR UNDERGROUND DISPOSAL OF RADIOACTIVE LIQUIDS OR SLURRIES

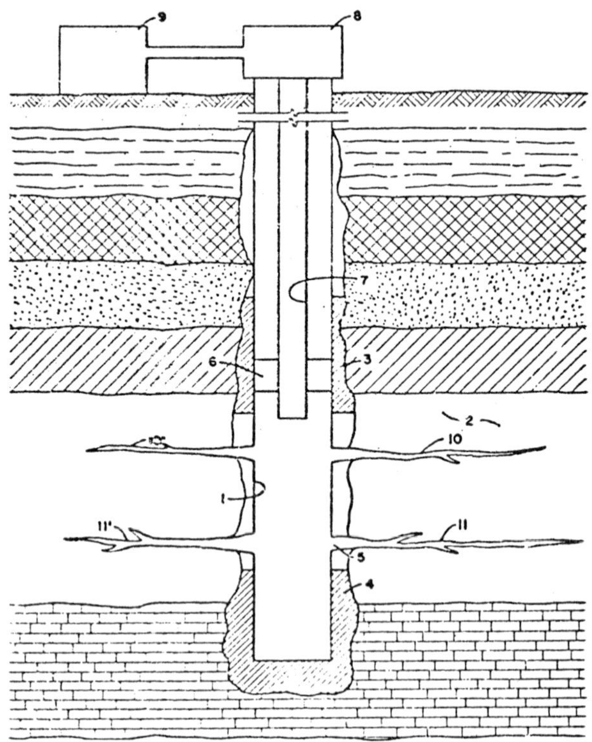

Source: U.S. Patent 3,108,439

This is a method of subterranean disposal of radioactive waste which comprises providing
a wellbore having a casing therein which penetrates an impermeable rock formation, sealing
between the casing and the impermeable rock formation to provide a sealed zone and fracturing through the zone with a fracturing fluid to provide at least one fracture which extends into the formation at an angle with respect to the axis of the wellbore. The fracture
is confined within the impermeable rock formation and at a depth sufficient to provide
shielding at the surface from the most energetic fraction of radioactive waste to be injected
thereinto. The radioactive waste is injected into the aforementioned fracture, and a sealing
material injected into the impermeable rock formation to seal the radioactive waste therein.

As shown in the figure, the casing **1** penetrates the nominally nonporous formation **2** in
which the radioactive waste is to be stored. This formation is properly isolated by seals **3**
and **4** to prevent dissipation into the surrounding formation. The lower portion of the casing **1**, located in stratum **2**, is perforated as at **5**. Within casing **1** is a packer **6** and tube **7**.
Attached at the upper portion of casing **1** and tube **7** is suitable fracturing apparatus **8** and
suitable waste reservoir **9**. Fracture **10** or its counterpart **11** are the artificial reservoirs into
which the radioactive waste is injected through tube **7** from reservoir **9** by means of apparatus **8**.

A process developed by D.A. Shock (21) is another deep well disposal process. It involves
injecting a stream of solidifiable material free from radioactive waste into a fracture formed
in an impermeable rock formation, thereafter introducing radioactive waste to the stream,
subsequently discontinuing the introduction of radioactive waste and continuing the injection of solidifiable material free from radioactive waste to seal the radioactive waste in the
impermeable rock formation. The apparatus used is shown in Figure 95.

**FIGURE 95: PROCESS FOR FIXING ATOMIC WASTE AND DISPOSAL OF PRODUCT
THEREOF**

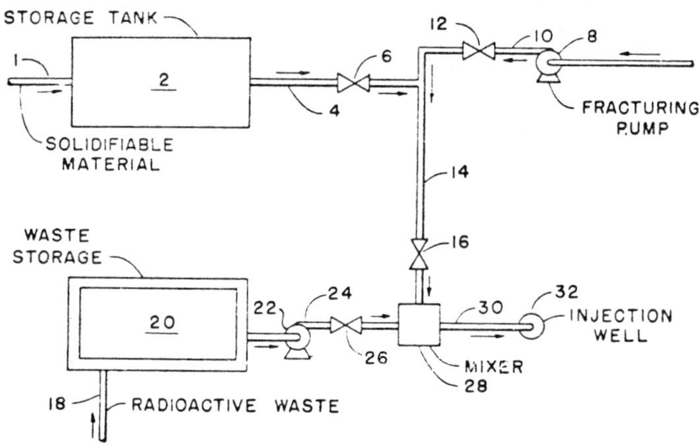

Source: U.S. Patent 3,274,784

The pump **8** is fracturing pump which is connected to an injection well **32** through conduits **10**, **14** and **30**, containing valves **12** and **16** and a mixer **28**. Tank **2**, which is adapted
to contain a mobile solidifiable material, is connected to conduit **14** through conduit **4** and
valve **6**. Shielded storage **20**, adapted to contain radioactive waste, connects with mixer
28 through pump **22**, conduit **24** and valve **26**. In the preparation for disposal of radio-

active waste, injection well **32** is first provided in an impermeable formation. A suitable fracture or fractures are established in the injection well, utilizing fracturing pump **8** and a suitable fracturing fluid, the fracturing being carried out in the ordinary and conventional manner well known in the art. Any of the usual fracturing fluids can, of course, be employed in this operation.

When the underground storage is ready for use, the disposal process is initiated by commencing flow of solidifiable material (e.g., Portland cement) from tank **2** through conduits **4**, **14** and **30** and mixer **28** into the injection well **32**. After the desired amount of solidifiable material has been introduced to the injection well, the addition thereto of radioactive waste from storage **20** is commenced through pump **22** and conduit **24**. The radioactive waste is admixed with the solidifiable material in mixer **28**, and the admixture thereafter flows into injection well **32**.

After the desired amount of radioactive waste has been added, the flow of this material is terminated; and the flow of solidifiable material in the injection well is thereafter continued for a period of time to provide sealing of the radioactive waste in the impermeable formation. This process can be repeated at various levels and utilizing various fractures in the formation. As necessary, additional solidifiable material and radioactive waste can be introduced to the storage vessels through conduits **1** and **18**, respectively.

A process developed by K.A. Slagle et al (22) is one in which radioactive waste materials, are added to a hydraulic cement slurry comprising attapulgite and illite-type clays, the cement slurry is pumped into an earth formation under fracturing pressure and allowed to set in the fractures created.

Figure 96 shows graphically the utilization of this process. In the drawing, radioactive waste fluids or materials are stored in suitable storage tanks **10**. Underground lines connect the storage tanks to a source of supply (not shown) and to the mixing equipment **11**. A pump house **12** has pumps located therein and suitably connected to the underground lines for pumping the waste material to the mixing equipment in the mixer cell.

Several dry solids storage bins **13** are conveniently located around the mixer cell for introducing cement, clays and other additives into the waste fluid for blending therewith. The mixer cell contains suitable blending and mixing equipment **11**, such as a dry solid blender and a jet mixer. A water supply source **14** is also located nearby for providing water when the liquid content of the waste is less than that desired for preparing a suitable injection slurry or for such other uses as may be desired, and is connected to the mixing equipment or apparatus **11** by an underground line.

Additional underground lines connect the mixing equipment **11** to the wellhead fracturing manifold or Christmas tree **15**. A high-pressure fracturing pump **16** is installed adjacent the mixer cell for pumping the slurry of the process from the mixing equipment **11** into the wellhead **15** and down the cased wellbore **17** into the earth formation **18**. A waste pit **19** is located near the mixer cell to catch any overflow or leakage from the wellhead **15** or the mixing apparatus **11**. Suitable drainage is provided for enabling any excess or overflow slurry to run into the waste pit **19**. The waste pit is connected to both the mixing apparatus **11** and the storage tank **10** by underground lines so that fluids collected therein may be disposed of when convenient.

Also, an emergency trench or pit **20** is located near the waste pit **19** for collecting any overflow fluids therefrom. Sufficient drainage is provided for this. Suitable shielding such as concrete, earth or other materials for protection from radioactive materials is placed around the various plant components. The injection well **17** is cased from the wellhead **15** to its lower end **17a** to prevent fluids from escaping into areas of the formation in which no slurry is injected. The drawing of the earth formation is partially cut away so that the distribution of the slurry **21** into the earth formation **18** may be readily seen. One or more cased observation wells **22** are located on the perimeter of the plant site for viewing the distribution of the slurry **21** in the earth formation **18**.

FIGURE 96: SCHEMATIC VIEW OF NUCLEAR WASTE DISPOSAL WELL INSTALLATION

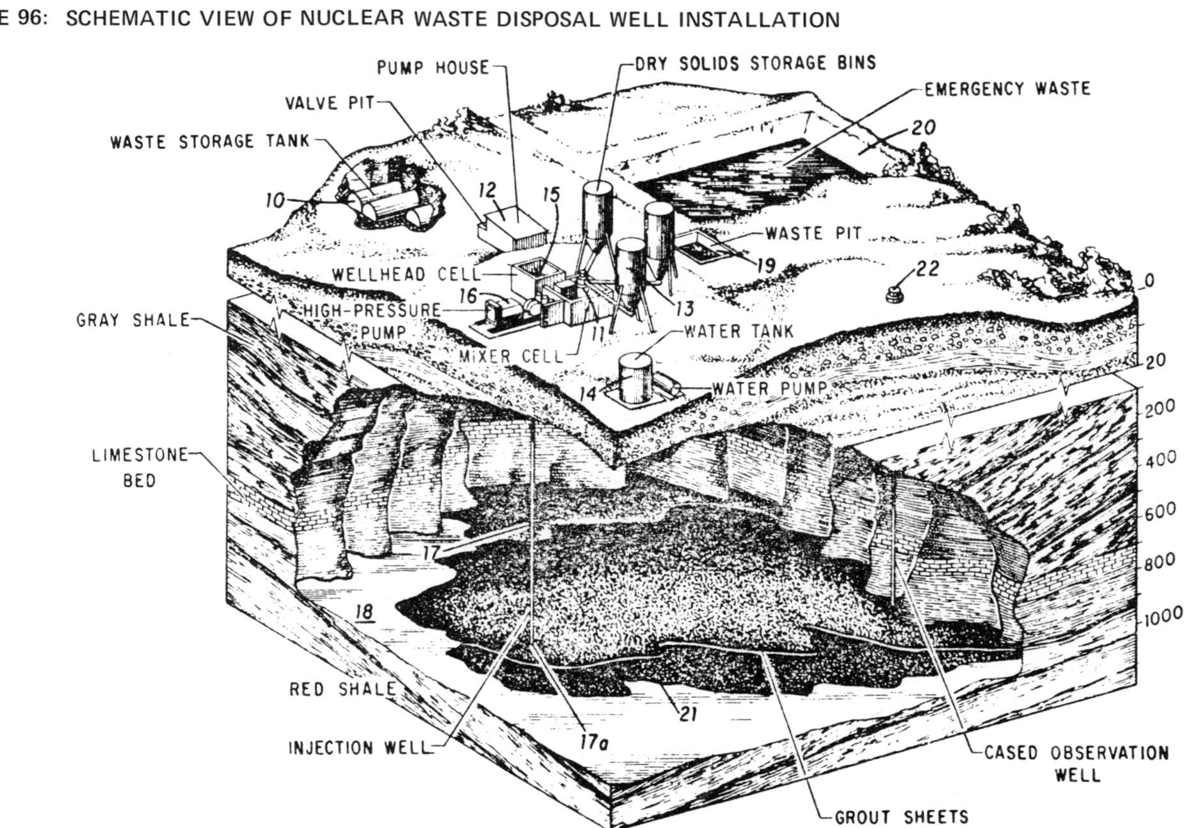

Source: U.S. Patent 3,379,013

A process developed by J.M. Stogner (23) involves disposing of high-level, solid, radioactive waste material by delivering such material in a continuous, water-phase cement to a subterranean formation. The solid waste is used as a propping agent to hold formation fractures open as a slurry containing the solid, radioactive waste material is pumped thereinto.

A technique developed by J.J. Cohen et al (24) is one in which a nuclear explosive is emplaced in an impermeable silicate rock formation and detonated to form a nuclear chimney partially filled with rubble. Thereafter, vapor output and fluid input conduits, e.g., wellbores are drilled or are otherwise provided to establish communication, respectively, with the top of the chimney and to intermediate peripheral sections of the chimney. In the first phase of operation of the process, radioactive wastes in a fluidic form, i.e., as an aqueous solution or as a slurry of particulate material in water are introduced through the input conduit into the nuclear chimney, i.e., into the rubble zone therein. The radioactivity level of the wastes are regulated to assure that the radioactive heat output is sufficient to heat chimney contents and the introduced material above the vaporization point of the water medium.

The vapor, ie., steam produced thereby which may contain radioactive waste particles, noncondensible radioactive materials, radiolytic products and vaporized solution components is withdrawn upwardly through the rubble zone in which at least some of the particulate portions, spray particles and other components are filtered therefrom and refluxed back into lower portions of the chimney. After passage through the rubble zone the steam is withdrawn to the surface, particulate solid and gaseous contaminants are separated and the steam is condensed for recycle or disposal in a fashion appropriate to the composition thereof. Regulation of the vapor discharge pressure and of the rate of introduction of coolant and diluted radioactive wastes will generally control operating temperatures in the chimney.

The radioactive wastes may usually be introduced over an extended period of time as they are produced in the normal course of operation of a nuclear fuel reprocessing plant so as to avoid the necessity for a large temporary surface high-level radioactive waste storage facility. A relatively large quantity of high level waste may be introduced initially, to establish a desired operating temperature level. Thereafter, high-level waste, low-level wastes and recycle condensed coolant water, which may contain radioactive material separated in the surface facility are introduced in amounts regulated to assist in maintaining the chimney contents at the desired operating temperature level. Introduction of the indicated materials is continued until a predetermined or selected quantity of concentrated radioactive waste residue is accumulated in void spaces of the rubble zone in the chimney.

Eventually, in a second phase of operation, when the ultimate capacity of the nuclear chimney is achieved or it is otherwise desired to terminate operation thereof, introduction of the various wastes and coolant water is terminated. Radioactive heat output then gradually increases the temperature of the chimney contents vaporizing residual water and other volatiles if present which vaporized material also may be withdrawn as described above. The quantity of radioactive waste introduced into the chimney should be sufficient to compensate for the decrease in radioactive heat output with passage of time and to assure that the radioactive heat output exceeds that lost by conduction to the formation so as to be sufficient to increase the temperature sufficiently to melt the rubble to eventually form a molten mixture with the radioactive waste components.

The quantity of radioactive wastes may be in excess of that required to melt the chimney contents and sufficient to melt rock adjacent to the chimney walls. As the melt volume increases, its surface area increases and heat loss by conduction increases so that the melting process is self-limited at the point where such loss balances the radioactive heat output. When vapor output from the molten mixture has substantially ceased, to instigate a third phase of operation of the process, the input and output conduits may be sealed. Once the heat balance limit described above is achieved, a gradual cooling of the molten chimney contents occurs since the radioactive heat output decreases as the radioactive species therein decay. Eventually, in the fourth phase of operation, the molten mass solidifies as cooling progresses so that the residual radioactive materials are entrapped in an insoluble solidified silicate rock matrix.

TRITIUM

Most tritium release is in the form of tritiated water in liquid waste streams at reactor or spent fuel processing sites and only a very small percent is released to the air.

Land Burial

Land burial of tritium wastes, in small concentrations, at approved sites that are acceptable from a geologic and hydrologic standpoint, is an acceptable means of disposal. Its concentrations should not be in excess of 10^4 times its maximum permissible concentrations for the general population. The burial trenches should be designed not to intercept the groundwater table and constructed with a bottom drain and sump for water monitoring. The trenches should be covered with either asphalt or vegetation to limit infiltration of water. The burial site design, geology, and hydrology should be in conformance with the criteria used in selecting and licensing the commercial burial sites. Since land burial has successfully been practiced on a commercial scale, it should be considered as the most satisfactory method of disposing of low concentrations of these wastes (5).

Long-Term Storage

The storage of high concentrations of these wastes in engineered surface facilities offers the best intermediate method for storage of these wastes. The technology for these facilities has been developed and the wastes will be under surveillance and control and can be retrieved, should this be required. The wastes will be stored in stainless-steel-lined concrete vaults which will be either air or water-cooled (5).

Salt Deposit Disposal

This method offers the best potential for the disposal of this radionuclide. This method of disposal has been under study by the Oak Ridge National Laboratory since 1957, and in November 1970 a committe of the National Academy of Sciences recommended that the use of bedded salt for the disposal of radioactive wastes is satisfactory. Recent questions concerning the adequacy of this method have resulted in the need for further development work before it can be accepted as an ultimate method of disposal. The wastes are buried in rooms carved in the salt deposits approximately 1,000 feet below the ground. The salt is a good heat transmitter, provides about the same radioactive shielding as concrete, and can heal its own fractures by plastic flow. The critical problem is the selection of a site that meets the necessary design and geological criteria (5).

REFERENCES

(1) J.L. Tarbox and D.G. Lachapelle; U.S. Patent 3,452,690; July 1, 1969; assigned to Secretary of the Army.
(2) J.L. Tarbox; U.S. Patent 3,464,375; September 2, 1969; assigned to Secretary of the Army.
(3) A. Taeymans, G. Dumont and W. Balleux; U.S. Patent 3,847,094; November 12, 1974; assigned to Belgonucleaire.
(4) W. Hempelmann; U.S. Patent 3,922,974; December 2, 1975; assigned to Gesellschaft fur Kernforschung mbH.
(5) R.S. Ottinger, J.L. Blumenthal, D.F. Dal Porto, G.I. Gruber, M.J. Santy and C.C. Shih, *Recommended Methods of Reduction, Neutralization, Recovery of Disposal of Hazardous Waste,* Vol IX, *Radioactive Materials,* Report PB 224 588, Springfield, Va., Nat Tech Information Service (August 1973).
(6) W.A. Hartman, Jr. and M. Siegler; U.S. Patent 3,598,699; August 10, 1971; assigned to General Electric Company.
(7) W.A. Hartman, Jr. and M. Siegler; U.S. Patent 3,679,372; July 25, 1972; assigned to General Electric Company.
(8) G.E. Moore and L.H. Tomlinson; U.S. Patent 3,660,041; May 2, 1972; assigned to General Electric Company.
(9) G.E. Moore and L.H. Tomlinson; U.S. Patent 3,740,313; June 19, 1973; assigned to General Electric Company.

(10) A.S. Kubo and D.J. Rose, *Science* 182, 1205-1211 (December 21, 1973).

(11) M. Sittig, *Water Pollution Control and Solid Wastes Disposal,* Park Ridge, N.J., Noyes Deveopment Corporation (1969).

(12) Booz-Allen Applied Research, Inc., *A Study of Hazardous Waste Materials, Hazardous Effects and Disposal Methods,* Vol III, Report PB 221 467, Springfield, Va., Nat Tech Information Service (July 1973).

(13) B.M. Johnson, Jr. and G.B. Barton; U.S. Patent 3,008,904; November 14, 1961; assigned to U.S. Atomic Energy Commission.

(14) M.L. Spector; U.S. Patent 3,153,566; October 10, 1964; assigned to Pullman, Inc.

(15) J.D. Murphy, J. Pirro, Jr., M. Lawrence and S.F. Wisla; U.S. Patent 3,883,441; May 13, 1975; assigned to Atcor, Inc.

(16) A.L. Backstrom; U.S. Patent 3,925,992; December 16, 1975; assigned to Svenska Entreprenad AB Sentab.

(17) P.M. Boffey, *Science* 182, 361 (October 24, 1975).

(18) G.H. Billue; U.S. Patent 3,236,053; February 22, 1966.

(19) M.D. Nelson, Jr.; U.S. Patent 3,262,274; July 26, 1966; assigned to Mobil Oil Corporation.

(20) J.J. Reynolds and D.A. Shock; U.S. Patent 3,108,439; October 29, 1963; assigned to Continental Oil Company.

(21) D.A. Shock; U.S. Patent 3,274,784; September 27, 1966; assigned to Continental Oil Company.

(22) K.A. Slagle and T. Tamura; U.S. Patent 3,379,013; April 23, 1968; assigned to Halliburton Company.

(23) J.M. Stogner; U.S. Patent 3,513,100; May 19, 1970; assigned to Halliburton Company.

(24) J.J. Cohen and A.E. Lewis; U.S. Patent 3,706,630; December 19, 1972; assigned to U.S. Atomic Energy Commission.

ORDNANCE INDUSTRY WASTES

EXPLOSIVES MANUFACTURING WASTES

The disposal of wastes from explosives manufacture has been reviewed by I. Forsten (1) of Picatinny Arsenal, Dover, N.J.

CHEMICAL WARFARE AGENT WASTES

Controlled Incineration

When chloroacetophenone and mixtures of chloroacetophenone are packaged with easily combustible materials, the following process recommended by the Manufacturing Chemists Association, appears to be satisfactory (2). The chloroacetophenone-containing waste is dissolved in an organic solvent (usually benzene or an alcohol) and sprayed into an incinerator equipped with an afterburner and alkaline scrubber.

In April of 1969, the Army prepared a plan for development of a transportable system capable of disposing of any of the chemical munitions now stockpiled by the Department of Defense (2). Because of the presence of a large variety of chemical munitions and because of the availability of engineering personnel and shop facilities, the South Area of Tooele Army Depot was selected for test and operation of the Transportable Disposal System (TDS). The system design is based on the removal of the chemical warfare agent from the various munitions and the separate incineration of the agent, casings and explosives and propellants.

The TDS consists of twelve major subsystems: [1] Explosive Containment Cubicle (ECC) for the removal of explosives from explosive-loaded munitions; [2] Projectile Demilitarization Facility (PDF) for the removal of agents from projectiles and mortar ammunition and the decontamination of the empty hardware; [3] Bulk Item Facility (BIF) for the removal of agent from nonexplosive containing bombs and ton cylinders and the decontamination of the empty hardware; [4] Deactivation Furnace (DF) for burning propellant, explosives, and empty rocket and mine bodies; [5] Deactivation Furnace Scrubbers System (DFSS) for pollution control of the effluent from the DF; [6] Metals Parts Furnace (MPF) for the thermal decontamination of empty metal parts that previously contained agents or which might have been contaminated by agent; [7] Air Pollution Control System (APCS) for scrubbing the gaseous effluent of the MPF; [8] Agent Incinerator-Scrubber System (AISS) which incinerates agent and reduces pollutant to acceptable low levels; [9] Dunnage Incinerator (DI) with scrubber used for nontoxic combustibles; [10] Sludge Removal and

Treatment System (SRTS) used to remove explosive materials from spent decontaminating solutions and evaporate the water; [11] Control Module (CM) which monitors and controls the above system; and [12] Personnel Support Complex (PSC) providing change rooms, locker rooms, toilet and shower facilities, lunchroom and laundry.

The Agent Incinerator/Scrubber System is of particular importance to the disposal procedure for chemical warfare agent wastes. All agents are drained from the munitions and fed at a constant, controlled rate to the agent incinerator. The agent feed is continuously supplemented by a fuel oil flame. All incinerator controls are fail-safe and the agent feed is equipped with a fast-acting cut-off valve in the event of loss of flame.

The combustion gases leaving the incinerator are water-quenched before being processed through cross-flow packed scrubbers in series to remove pollutants. The scrubbing media employed in the cross-flow scrubbers are calcium hydroxide and sodium hydroxide for GB and nitric acid, sodium hydroxide and calcium hydroxide for VX incineration products. The decontaminating solutions used in the PDF and BIF are 10% sodium carbonate solutions. Sodium carbonate is also the scrubbing medium in the other furnace systems. All solutions are dried except the water-fly ash mixture from the Dunnage Incinerator and the salts are stored. The TDS represents a very adequate disposal method which might serve as a model for pesticide and other disposal systems.

It may also be used for the disposal of munitions or tanks containing lewisite in concentrated form. The combustion products are carbon dioxide, water, HCl and arsenic trioxide. The arsenic trioxide removed by alkaline scrubbing should be converted to the insoluble magnesium salt and placed in controlled storage.

Nitrogen mustard may also be handled using this system. The combustion products of the nitrogen mustards are carbon dioxide, water, HCl and nitrogen oxides. The nitrogen oxides require scrubbing or reduction to nitrogen and oxygen before the combustion gases are released to the atmosphere.

This technique can also be applied to munitions or tanks containing sulfur mustard in concentrated form. The sulfur mustard may be dissolved in gasoline and the gasoline solution incinerated. The combustion products will be carbon dioxide, water, sulfur oxides, and hydrogen chloride which are removed by alkaline scrubbing.

Chemical Destruction

Chloroacetophenone reacts with sodium sulfide in alcoholic or alcohol-water solutions to form bis-(acetylphenyl)-thioether (melting point 74°C) which shows no physiological effect. This process is attractive because it not only converts the chloroacetophenone into a compound that has no physiological effect, but most explosive nitro compounds, when present, are decomposed to nonexplosive compounds. This process results in the liberation of hydrogen sulfide, which must be collected by an alkaline scrubber. The alcohol solution containing the decomposed compounds may be sprayed into an incinerator equipped with an alkali scrubber to remove any hydrogen chloride, nitric oxide and sulfur oxides formed (2).

By boiling chloroacetophenone in an alcohol-water solution with sodium thiosulfate, the sodium salt of acetylphenyl thiosulfonic acid is obtained.

$$C_6H_5COCH_2 \cdot S_2O_3Na$$

The thiosulfonate solution must then be sprayed into an incinerator equipped with an alkali scrubber. This process converts the chloroacetophenone into a compound that shows no physiological effect, but incineration of the thiosulfonate is required. This option does not, in most cases, offer any advantages over incineration or sodium sulfide treatment.

Water containing wastes from the manufacture, handling or disposal of organophosphorus agents can most easily be treated using caustic soda to accelerate the hydrolysis reactions. The products of these reactions for GB and VX solutions have a much lower toxicity than the agents themselves.

GB undergoes its first reaction at the P-F bond and then at the P-O single bond. Normal reaction products are sodium fluoride and the sodium salt of methylisopropylphosphoric acid. Specific information for VX is not available although similar compounds are known to have much slower rates of hydrolysis. The thio-choline type compounds are first attacked at the P-S bond. Quantitative conversion is possible only by the use of strong alkalies in moderately concentrated solutions. Heating of the hydrolysis mixture increases the rate at which the reactions occur. These techniques are adequate for dilute systems if agent concentration is monitored and the salt products dried and stored.

Under the auspices of Project Eagle the Army has developed a chemical detoxification procedure for the demilitarization of weapon systems containing GB. The reactions are those described earlier for the dilute wastes. 5% excess sodium hydroxide in an 18% solution is reacted with GB to form sodium salts of low toxicity which are then recovered by evaporation and stored in plastic-lined drums. VX may be detoxified with chlorine in acidic aqueous media to give O-ethyl phosphoric acid, and diisopropylthurine hydrochloride and hydrochloric acid. This mixture is subsequently neutralized with caustic and spray-dried to give the sodium salts. These salts may also be stored or used as raw material for other processes. This method is considered adequate for VX.

In dilute aqueous solution chlorine converts lewisite into arsenic trioxide and dichloroethene. Dichloroethene requires further treatment as a chlorinated organic compound. Arsenic trioxide is slightly soluble and hence should not be placed in a landfill as such. If arsenic trioxide is reacted with a suspension of magnseium hydroxide, insoluble magnesium arsenite is formed. The magnesium arsenite thus precipitated and the excess magnesium hydroxide can be stored under controlled conditions, or the slurry can be evaporated and stored in a permanent disposal area. Aqueous chlorine or sodium hypochlorite treatment, followed by reaction with magnesium hydroxide, is satisfactory for decontamination or disposal of lewisite even though an arsenite is produced that must be in indefinite long-term storage (2).

Both cis- and trans- isomers of lewisite are completely decomposed in sodium hydroxide solutions at temperatures above 40°C. The reaction proceeds as follows:

$$Cl-CH=CH-AsCl_2 + 6NaOH + 2H_2O \longrightarrow Na_3AsO_3 + 3NaCl + CH\equiv CH\uparrow + 5H_2O$$

The sodium arsenite produced is soluble and requires treatment with magnesium hydroxide, as noted above, to form insoluble magnesium arsenite before disposal in a controlled storage facility (lined lagoon, tank, abandoned quarries or mines, etc.).

The nitrogen mustards, when acidulated, react with calcium hypochlorite in solution to yield compounds with a much lower toxicity including aldehydes, chloramines and chlorates. The reactions are violent with dry or highly concentrated hypochlorite. Care must be taken to insure that sufficient agitation and time are allowed for the reaction due to the phase separation problem.

Sulfur mustard added to 10% calcium hypochlorite solution is decomposed according to the following equation:

$$S(CH_2CH_2Cl)_2 + 7Ca(OCl)_2 + 2Ca(OH)_2 \longrightarrow CaSO_4 + 8CaCl_2 + 4CO_2 + 6H_2O$$

Thus 7.0 lb of $Ca(OCl)_2$, which contains 0.7 lb of CaO equivalent to 0.93 lb of $Ca(OH)_2$, are theoretically required to detoxify 1.0 lb of sulfur mustard. Some mustard sulfone may be formed in accordance with the following reaction:

$$2CaCl(OCl) + (ClCH_2CH_2)_2S \longrightarrow 2CaCl_2 + (ClCH_2CH_2)_2SO_2$$

Mustard sulfone produces severe burns if left on the skin. In practice, therefore, 1.2 times the required amount of Ca(OCl)$_2$ is used. The decontamination is conducted in a closed system where all air leaving the system is scrubbed through Ca(OCl)$_2$ solution and filtered. The salts formed upon detoxification are placed in a landfill after evaporation of the water. This treatment process is used to treat sulfur mustard that is excess, contaminated or loaded in surplus munitions (2).

Ocean Disposal

In the fall of 1968, the Department of the Army determined that the disposal of certain chemical munitions was necessary to remove excess and unserviceable material from our national deterrent stockpile. The proposed disposal plan (Operation CHASE) was reviewed by the National Academy of Sciences which recommended that the particular disposal should proceed as planned with certain modifications and which further recommended that ocean disposal not be used for later chemical munitions disposals.

The munitions in question were sealed in concrete in blocks which were subsequently covered with steel plates which were welded. Because of their appearance, size and shape they were called and referred to as "coffins." The coffins were shipped to a port facility, loaded on a hulk and towed out to sea. The hulk was sunk in approximately 16,000 ft of water and the area has been monitored since the disposal without any evidence of damage to the ocean ecology near the disposal site. In view of the National Academy of Science recommendation it is concluded that ocean disposal is not adequate (2).

PYROTECHNIC WASTES

Controlled Incineration

A process developed by R.D. Altekruse (3) is a process for combustion of pyrotechnic waste which involves controlled burning of the waste in an incinerator with scrubbing of the smoke and fumes produced. Figure 97 shows a suitable form of apparatus for the conduct of this process.

FIGURE 97: BURNER FOR DESTRUCTION OF PYROTECHNIC WASTE

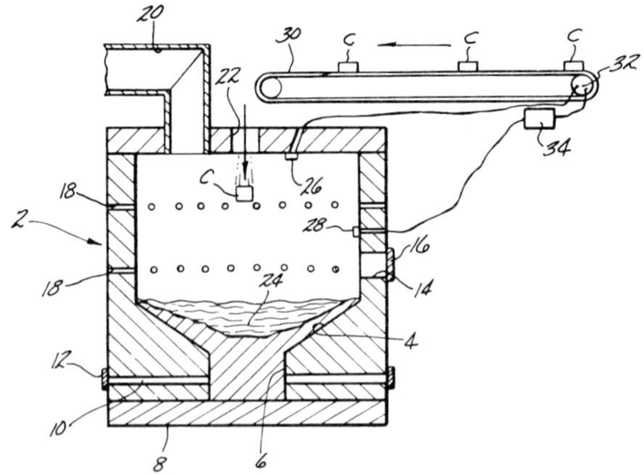

Source: U.S. Patent 3,903,814

Referring to the drawing, there is shown an incinerator 2 made of refractory brick and having a cubical or cylindrical shape. The bottom wall 4 of the incinerator is sloped down-wardly and inwardly at an angle of about 45° to a centrally located sump 6. A foundation 8 closes off the bottom of the sump 6. The bottom wall 4 of the incinerator is preferably constructed so as to be blast-proof. A plurality of draft ducts 10 extend into the sump 6 and may be closed with doors 12. An access opening 14 which may be closed with a door 16 is disposed in the side wall adjacent the bottom wall for charging the incinerator with fuel and for cleaning out the incinerator. A plurality of air admission ports 18 are formed in the side wall of the incinerator. A flue 20 of internally insulated sheet steel is disposed in the top wall of the incinerator and is equipped with a blower and fume scrubber (not shown).

A charging port 22 is disposed in the center of the incinerator top wall and is preferably just large enough to pass a charge of material to be incinerated. A charge of fuel in the form of coke 24 is disposed in the bottom of the incinerator and preferably covers the bottom wall 4 and fills the sump 6. An infrared radiation cell 26 is focused on the coke bed and an ultraviolet cell 28 is focused on the center of the incinerator chamber. A con-veyor 30 is disposed above the incinerator and is powered by an electric motor 32. The infrared cell 26 is connected to the motor 32 so that the conveyor 30 cannot be operated unless the coke bed 24 is incandescent. The ultraviolet cell 28 is connected to a timer 34 which resets itself to zero each time a combustion flash is detected by the cell 28, and the timer 34 is in turn connected to the electric motor 32. In this manner the conveyor will be shut off by the timer if no ignition flash is detected by the ultraviolet cell 28.

As will be readily seen, charges C of pyrotechnic waste material of predetermined weight are placed on the conveyor 30 at spaced intervals and transported to the incinerator charging port 22 where they fall from the conveyor into the incinerator. As the charge C falls through the incinerator it approaches the incandescent coke bed and is heated thereby. When the temperature around the descending charge reaches its ignition point, the charge ignites and burns in a ball of flame. The size of the incinerator is selected so as to insure that the side and top walls of the incinerator are not touched by the fire ball and are thus not prematurely broken down during incineration of the pyrotechnic material. The bottom wall of the incinerator is insulated against the fire ball temperature by the incandescent coke. The smoke and fumes produced by burning the pyrotechnic material are drawn out through the stack 20 and are scrubbed therein in a conventional manner so that the incin-erator does not emit harmful material into the environment.

HIGH EXPLOSIVE AND PROPELLANT WASTES

Open Burning

Nitrocellulose for destruction is usually placed in fiberboard drums, metal drums or cans lined with a heavy-gauge leakproof liner and the nitrocellulose covered with water (2). The nitrocellulose is then transported to an approved burning ground. The NC wastes are placed on a noncombustible pad such as asbestos, the water allowed to drain off, and the NC wastes are then covered with a combustible material such as fuel oil. The flam-mable material is ignited by firing a black powder squib or other device placed in the wastes. Although the products of combustion contain considerable NO_x, better methods for disposal are not currently in wide use.

As regards gelatinized nitrocellulose or plastisol nitrocellulose (PNC) as it is often known, the only process widely used for disposal of surplus, scrap or obsolete PNC is burning. The methods used for burning are generally in compliance with the procedures outlined in military safety manuals.

The current procedure employed for disposal of the majority of ammonium picrate manu-facturing wastes is open burning in a safe area. This practice is unacceptable, however. The current procedure employed for the disposal of the bulk of PETN and TNT manu-facturing wastes is open burning in a safe area. This practice is not acceptable. The majority of picric acid wastes from manufacturing operations are now disposed of by open burning in a safe area. This practice is not satisfactory because of the nitrogen oxides emitted.

Controlled Incineration

Nitrocellulose wastes and combustible materials contaminated with nitrocellulose are readily amenable to disposal via controlled incineration. Two types of equipment are currently in prototype use for the controlled combustion of explosive wastes. The first disposal system is similar to conveyor-fed devices used as municipal incinerators, operates on an induced draft, and is equipped with afterburner and scrubbing systems to abate NO_x and particulate emissions. The second type of equipment, similarly equipped with scrubbing systems for the abatement of NO_x and particulate emissions, is a rotary kiln incinerator, which is fed with a slurry of nitrocellulose in water, in a 1 to 3 ratio. The rotary kiln incinerator has secondary fuel oil or natural gas burners. This method of disposal is preferred (2). The above statements regarding nitrocellulose are equally applicable to plastisol nitrocellulose (PNC).

Solutions of PETN and TNT in acetone can be incinerated readily. The explosives are dissolved in at least eight times their weight of technical acetone. The solution is often burned in a shallow container; this type of burning is not recommended. It is recommended that destruction be carried out in an injection-type incinerator equipped with an afterburner and a caustic soda solution scrubber. The technique is acceptable for small to moderate quantities of explosives, and for decontaminating equipment (2).

The Chemical Agent Munition Disposal System (formerly Transportable Disposal System) under development by the U.S. Army Materiel Command includes a deactivation furnace which is particularly suited to the disposal of small charges of high explosive (such as nitrocellulose, PNC, ammonium picrate) as well as detonators and primers. High explosives up to about 7 lb in weight per charge, produced by disassembly of scrap munitions, are fed via an automated conveyor to an explosion-resistant steel rotary kiln, countercurrent to an oil or gas flame.

The rotary kiln is equipped with steel screw flights to isolate the explosive charges from each other. The explosive charge end of the kiln is at about 500°F gas temperature; the kiln is about 25 ft in length, and the fired end opposite the explosive feed end is maintained at a gas temperature of about 1200°F. Combustion product gas exits through a cyclone. In practice, the exit gases should go through an afterburner, to complete oxidation of CO prior to the cyclone, and then be scrubbed in a packed tower with caustic soda or soda ash solution recirculated as scrubbing medium. Bleed-off alkaline solution, after neutralization, would exit to sewer. The metal components of the primers and detonators are recovered as scrap.

A process developed by S.J. Yosim et al (4) is one in which unwanted explosives and propellants are destroyed in a safe nonpolluting manner by being contacted with a hot molten salt containing as essential reactive component alkali metal carbonate or hydroxide or mixtures thereof. Some of the resulting decomposition products that would ordinarily be released as atmospheric pollutants react with or are retained in the melt. Any combustible matter present in the melt may be further reacted by use of oxidative molten salts alone or in combination with oxygen added to the melt.

Formed gaseous products may be treated in a second reaction zone to complete oxidation of any combustible matter present before discharge of the gases to the atmosphere. The reactive component of the treated melt may be regenerated, or the melt may be disposed of, preferably by reaction with lime in an aqueous or molten medium to form a water-insoluble calcium salt residue. Particularly effective and preferred reactive molten salts are the NaOH-KOH and Li_2CO_3-Na_2CO_3-K_2CO_3 eutectic mixtures.

A process developed by J.W. Bolejack, Jr. et al (5) is one in which waste propellant or explosives in particulate form are mixed with water forming an aqueous suspension. The aqueous suspension is burned in a rotary incinerator under conditions which permit sequential evaporation of water from the suspension, drying of the particulate propellant or explosive, and then ignition of the propellant or explosive. The combustion gases are scrubbed with water prior to passing into the atmosphere. Figure 98 is a block flow diagram of this process.

FIGURE 98: SEQUENCE OF OPERATIONS IN WASTE PROPELLANT AND EXPLO-SIVE INCINERATION

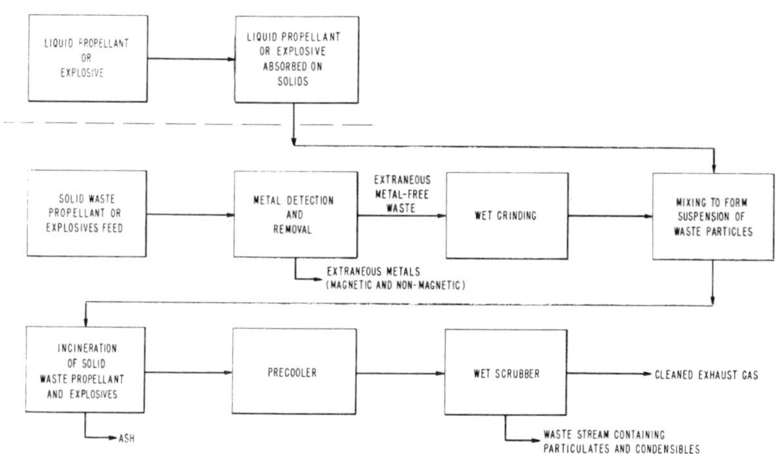

Source: U.S. Patent 3,848,548

It is seen that either solid or liquid waste propellants or explosives can be incinerated in accordance with this process by forming a suspension of solid or liquid propellant or explosive, in solid particulate form in water, and incinerating the suspension. The following is one specific example of the conduct of the present process.

Example: A waste single-base propellant charge is passed through a metal detection and removal device wherein extraneous metals are removed from the propellant charge. The propellant is then passed at a controlled rate into a flying blade grinder where it is mixed with water and ground into small particles. The particle size of the ground propellant is determined by Tylers Screen Analysis. The propellant particles have an average length or maximum dimension of about 0.04 inch.

The water-wet particles are transferred to a mixing vessel wherein additional water is added. The weight ratio of water to particles in the mixing vessel is 2.8:1. The particles are slurried in the mixing vessel to form a suspension of the propellant particles in water. The propellant suspension is introduced into an incinerator operating at a temperature of 1600°F and in which the air flow rate through said incinerator is 1,248 cfm. The incinerator consists of a high-alumina refractory brick-lined rotating chamber, 6 feet long and having an inside diameter of 5 feet. The chamber rotates at a speed of 1.1 revolutions per minute.

The feed line in the rotary incinerator is cooled to maintain the liquid phase of the suspension in the liquid state. The suspension of waste propellant in water is fed at one end of the rotary kiln near the rotating walls. The suspension immediately starts to flow and very rapidly forms a thin layer of suspension on the moving walls of the incinerator. The water phase of the suspension is rapidly evaporated and the propellant particles are sequentially dried and ignited. The product gases from the combustion are exhausted through a second incinerator operating at a temperature of about 1600°F. The residence time of the exhaust gases in said second incinerator is about 0.3 second.

The second incinerator, which is sometimes referred to as an afterburner, is a high-alumina refractory-lined chamber, seven feet in length and having an inside diameter of 2¾ feet. The hot exhaust gases are drawn through a precooler where the temperature of the gases is reduced to about 250°F and the gases are then drawn through a wet scrubber in which the gases are in intimate contact with water. In the wet scrubber, water-soluble gas contained in the exhaust gases and some particulate matter are removed from the exhaust gas stream. The resulting gases are subsequently exhausted to the atmosphere.

A process developed by C.D. Kalfadelis et al (6) involves incineration of nitrogenous waste materials such as TNT in a manner which minimizes NO_x pollution. The material is burned with a fuel and less than a stoichiometric quantity of air in the presence of a catalyst in a fluid bed. Secondary air is added to the gaseous products, and the resulting gas mixture is burned to yield a stack gas which has minimal amounts of NO_x, carbon monoxide and hydrocarbons. Nickel or a compound thereof is preferred as the catalyst.

Chemical Destruction

Small quantities of nitrocellulose may be decomposed by adding it with agitation to five times its weight of a 10% solution of sodium hydroxide that has been heated to 70°C. Agitation is continued for at least 15 minutes after all the nitrocellulose has been added. The products of this decomposition process require additional treatment. After pH adjustment and dilution, the cellulose can be handled by a sewage treatment plant (2).

Ammonium picrate in aqueous waste streams or excess ammonium picrate may be decomposed by reaction with a considerable excess of sodium sulfide solution. The proportions employed are, by weight, one part explosive to 30 parts of 14% (by weight) hydrated sodium sulfide solution. The H_2S and NH_3 liberated must be scrubbed from the vent air, or burned in an appropriate scrubber-equipped incinerator. The solution from the disposal process must be neutralized, stripped of H_2S, and chlorinated to oxidize the remaining phenolics. The H_2S stripped from the solution should be burned in an appropriate, scrubber-equipped incinerator. As an alternative to chlorination, the phenolics may be removed by adsorption on activated carbon. This technique is acceptable and is employed where the quantities of explosive treated are small or where explosive contaminated equipment must be decontaminated.

Picric acid in aqueous waste streams or excess picric acid is decomposed by dissolving the material in 25 times its weight of water containing one part sodium hydroxide and 21 parts hydrated sodium sulfide. The hydrogen sulfide and ammonia liberated must be absorbed or scrubbed from the vent air. The solution from the disposal process should be neutralized, and the phenolic material remaining oxidized by chlorine or removed by adsorption on carbon. This disposal technique is considered satisfactory where the quantity of picric acid is too low to make recovery economically attractive, or when small quantities of the material are contaminated.

Small quantities of PETN can be dissolved in acetone, and decomposed by reaction with a concentrated aqueous solution of sodium sulfide. The technique employed is to add a hot (80°C) 33% solution of $Na_2S \cdot 9H_2O$ to an 11% solution of PETN in acetone at such rate that the acetone does not boil. Seven parts by weight of sulfide solution are used per part PETN. Stirring is continued for 30 minutes after mixing is completed. The reaction products should be burned in a spray injection-type incinerator equipped with a caustic soda solution scrubber. The technique is acceptable for small quantities of manufacturing waste, and for decontamination of equipment.

TNT may be decomposed by adding it slowly, while stirring, to 30 times its weight of a solution prepared by dissolving one part of sodium sulfide ($Na_2S \cdot 9H_2O$) in six parts of water. The reaction products should be incinerated in an incinerator equipped with caustic soda solution scrubbers. The technique is acceptable for small quantities of TNT.

INITIATING OR PRIMER EXPLOSIVE WASTES

This category includes such materials as:

- copper chlorotetrazole
- gold fulminate
- silver styphnate
- silver tetrazene
- diazodinitrophenol (DDNP)
- dipentaerythritol hexanitrate (DPEHN)
- lead-2,4-dinitroresorcinate
- lead styphnate

Lead dinitroresorcinate, for example, is a primary explosive and detonating agent that is so sensitive to electrical discharge, impact and friction that it undergoes detonation when subjected to very mild mechanical or electrical shock by percussion or electrical discharge. Because of their extreme sensitivity, and the high degree of explosive hazard, wastes containing lead dinitroresorcinate should be handled only by an experienced explosives and ordnance disposal team. Normal procedures are to handle lead dinitroresorcinate as a slurry in water to minimize hazards.

The sensitivity of lead 2,4-dinitroresorcinate to shock and friction, as with most initiating agents, requires that all scrap and waste from preparation and purification be maintained wet for destruction. Lead 2,4-dinitroresorcinate is packaged wet for storage and shipment. Packaging is accomplished by placing approximately 25 lb, wet with 20% water or a 50:50 water-alcohol mixture for low-temperature storage, in a duck or rubberized cloth bag covered with a cap of the same material. The bag is then tied securely.

Not more than six such bags are placed in a large bag of the same material. The large bag is tied and placed in the center of a watertight metal or wooden barrel, drum or keg lined with a heavy close-fitting jute bag. The large bag containing lead 2,4-dinitroresorcinate is surrounded with well-packed sawdust that has been saturated with water or water-ethanol mixture. The bag forming a liner is sewn closed before closing the barrel, drum or keg. Not more than 150 lb of initiating explosive is permitted in a single container. It is shipped wet under the Department of Transportation (DOT) regulations for an Explosive, Class A.

Controlled Incineration

The most promising technique under development for disposal of initiating explosives is the controlled combustion process which employs a rotary kiln incinerator equipped with appropriate scrubbing devices. This technique can probably be used with copper chlorotetrazole, gold fulminate, silver styphnate and silver tetrazene. The explosive is fed to the incinerator as a slurry in water, at a weight ratio, explosive to water, of 1:3. The scrubber effluent would require treatment for recovery of particulate copper, gold or silver compounds formed as a combustion product. Additional copper, gold or silver could be recovered from the incinerator. This process is recommended for disposal of the four compounds being discussed (2).

The controlled slurry incineration technique currently being developed under Army cognizance is also recommended for disposal of waste DDNP, as well as for waste lead dinitroresorcinate and lead styphnate.

Small arms cartridges containing DPEHN in the primer should be disassembled from the projectile and deactivated by burning or detonation in a specially designed furnace similar to the Deactivation Furnace under development for the Chemical Agent Munition Disposal System (formerly Transportable Disposal System) by the U.S. Army Material Command.

The disassembled cartridges would be fed via an automated conveyor to an explosion-resistant steel rotary kiln, countercurrent to an oil or gas flame. The rotary kiln is equipped with steel screw flights to isolate the explosive charges from each other. The explosive charge end of the kiln is at about 500°F gas temperature; the kiln is about 25 ft in length, and the fired end opposite the explosive feed end is maintained at a gas temperature of about 1200°F. Combustion product gas exits through a cyclone. In practice, the exit gases should go through an afterburner, to complete oxidation of CO prior to the cyclone, and then be scrubbed in a packed tower with caustic soda or soda ash solution recirculated as scrubbing medium. Bleed-off alkaline solution, after neutralization, would exit to sewer. The metal components of the cartridges would be recovered as scrap.

Open Pit Burning

Detonators and primers discarded as manufacturing wastes may be detonated in burning pits by applying heat from a fire or by electrical ignition. The components are placed on top of flammable substances such as straw and the flammable substance is ignited by means of a squib. Other explosives must be kept behind a barricade with overhead protection during destruction operations and located at a distance that assures safety. Personnel should be similarly shielded. This destruction process is not entirely satisfactory because individual components may not detonate and will constitute a personnel hazard during cleanup. Some NO_x and toxic metal particles or compounds will also be liberated during the destruction process on an uncontrolled basis (2).

Detonation

The Ordnance Safety Manual recommends detonation as the best method for disposal of all initiating explosives (2). In the use of this procedure, bags containing the wet explosive are transported to the demolition area. Several bags are removed from the container and carried to the destruction pit, placed in intimate contact with each other, and a blasting cap placed between the bags to initiate the explosives. All remaining explosives should be kept behind a barricade with overhead protection during destruction operations. In the use of this method, the metal components of the initiators present cannot be recovered, but will be lost to the soil in the demolition area. This method is not environmentally acceptable and should not be used unless the hazards of transportation and handling for disposal via controlled incineration outweigh the ecological impact of detonation.

When more than a few pounds of diazodinitrophenol are to be destroyed, detonation is the usual current disposal procedure. Wet diazodinitrophenol is transported to the disposal area. Then several bags are removed from the transporting container, carried to the destruction pit, placed in intimate contact with each other, and blasting caps are placed between the bags to initiate the diazodinitrophenol. Remaining explosives must be kept behind a barricade with overhead protection during the destruction operations and located at a distance that assures safety. Personnel must be behind a similar barricade.

This destruction process which liberates NO_x is only satisfactory for small quantities of diazodinitrophenol. If large quantities are to be destroyed, the NO_x liberated can cause an environmental problem. Detonation is not recommended as a disposal procedure for DDNP unless safety considerations rule out transportation and handling for disposal via controlled incineration.

The comments above are essentially applicable to the ultimate disposal of DPEHN as well. Also the above text is applicable to lead azide with the additional proviso that the destruction pit by necessity must be in a remote location that should assure limited hazard to the public from contamination of the atmosphere by lead. The destruction pit must also be in alkaline soil or soil treated with lime to avoid leaching of lead into drainage water.

These same precautions are applicable to lead-2,4-dinitroresorcinate.

Chemical Destruction

Those primers and detonators which are charged with explosive materials which can be decomposed by acids may be chemically "killed" by immersion in an acid bath of sufficient strength to destroy the seals. This method permits recovery of the metallic components as scrap, but is limited in application because the items must be segregated by explosive mixture prior to treatment.

Small quantities of diazodinitrophenol such as residues left on parts can be decomposed to an unknown but nonexplosive substance by adding the water-wet material to 100 times its weight of 10% sodium hydroxide. Nitrogen gas is evolved. The remaining residue, though not identified in the publications referenced, requires isolation and development of a method for destruction such as neutralization and solvent extraction, followed by incineration.

DPEHN as well as PETN is rapidly decomposed in a boiling solution of ferrous chloride. The products of this reaction are dipentaerythritol and ferric chloride, which are at present being diluted, neutralized and discharged into streams. The method is satisfactory for use if the iron hydroxide and organic matter are removed via the secondary sludge sanitary waste treatment technique.

Lead azide may be destroyed chemically by mixing it with at least five times its weight of a 10% sodium hydroxide solution. The mixture is allowed to stand 16 hours and the supernatant solution containing sodium azide is decanted. The sodium azide solution is disposed of by draining into the ground. The lead is precipitated and can be recovered. This method is extensively used for waste streams from lead azide manufacture. This procedure is not recommended unless the effluent is treated by either of the two disposal techniques listed as acceptable for sodium azide: reaction with H_2SO_4 and sodium nitrate or incineration (2).

Lead azide may also be converted to potassium azide by reacting the explosive with a 10% solution of potassium bichromate until no more lead chromate is precipitated. This method is not satisfactory for other than laboratory use with very small (milligram) quantities of lead azide.

Lead azide wetted with 500 times its weight of water may be destroyed by adding to the lead azide 12 times its weight of 25% sodium nitrite, stirring, and then adding 14 times its weight of 36% nitric acid or glacial acetic acid. Aliquots of this solution are tested for complete azide destruction by testing with a ferric chloride solution; a red color indicates the presence of azide.

After complete destruction of the azide, the pH should be adjusted to 6.0 to 9.0 and the solution diluted to a nitrite and nitrate concentration of less than 45 ppm, the typical maximum allowable concentration in an effluent being discharged to a storm sewer, lake or stream. Lead will be precipitated when the reaction solution is neutralized to the pH shown above. This is a satisfactory method that permits recovery of the lead, if the NO_x vapors liberated are destroyed by controlled incineration, or removed by scrubbing.

Finally, lead azide may be decomposed by reaction with 50 times its weight of 15% ceric ammonium nitrate. The azide is oxidized to form nitrogen and the lead is subsequently precipitated as lead sulfate. This method, though satisfactory, requires the use of a more expensive chemical than is used in the other processes described (2).

Small quantities of lead styphnate left on manufacturing equipment or in filter or wash solution are usually decomposed chemically. Lead styphnate is decomposed by first reacting it with at least 40 times its weight of a 20% sodium hydroxide solution or 100 times its weight of a 20% ammonium acetate solution. Then a 10% solution of sodium dichromate is added until the weight of sodium dichromate equals the weight of lead styphnate.

The lead is thereby converted to insoluble basic lead chromate which is separated and disposed of in a landfill. The trinitroresorcinol formed is washed into the industrial waste drain along with any excess sodium dichromate. This process is not recommended.

Techniques under investigation include disposal of waste lead styphnate by reacting the material with sodium carbonate solution to yield insoluble basic lead carbonate, and an alkaline solution of trinitroresorcinol (TNR). The basic lead carbonate is separated, washed and recycled, and the trinitroresorcinol recovered for reuse in the manufacture of lead styphnate. The process is in the early development stage, and will be satisfactory if the TNR is recovered so that outfall effluents contain less than 0.5 ppm.

A current desensitization process is to react the lead styphnate with sodium hydroxide, sodium carbonate and aluminum, using live steam as a heat source. The reaction products are nonexplosive and include sodium aluminate, insoluble (basic) lead carbonate, and the disodium salt of aminoresorcinol. With the exception of the lead carbonate, which is separated, the materials are discharged as a solution to the industrial waste system. The technique produces effective desensitization of the styphnate, but requires secondary treatment to destroy the aminoresorcinol and to precipitate the aluminum.

Electrolytic Destruction

Because about 300,000 lb of special purpose lead azide are stored in the United States, the Burlington AEC plant has developed in the laboratory a new electrolytic process. This process converts the lead azide to metallic lead and nitrogen. The lead azide is placed into solution by treatment with sodium hydroxide in an electrolytic cell with a lead cathode and a stainless steel anode. This method is recommended as most satisfactory of the options reviewed here (2).

The reactions at the cathode and anode are as follows:

Cathode

$$Pb^{++} + 2e^- \longrightarrow Pb^\circ$$

$$4H_2O + 4e^- \longrightarrow 2H_2 + 4OH^-$$

Anode

$$2N_3^- \longrightarrow 3N_2 + 2e^-$$

$$2H_2O \longrightarrow O_2 + 4H^+ + 4e^-$$

OBSOLETE MUNITIONS WASTES

The disposal of unserviceable ordnance is a problem common to all branches of the armed services (7). As a matter of historical development, however, the largest quantities of large-scale ordnance have been dealt with by the Navy, on behalf of all three services. Earlier forms of ordnance were relatively easy to desensitize, disassemble and dispose of. In the more recent past, efforts to prevent use of captured ordnance by enemy forces have resulted in design of "tamper-proof" munitions. Unfortunately, this design makes disposal of unserviceable ordnance by "demilitarization" both difficult and dangerous, and the armed forces established the practice of dumping at sea as the safest and most effective method of disposal.

Efforts to resolve the safety problem in other forms of disposal have continued, but the amount of material to be dealt with is enormous and the results of accidents are extremely serious. No deep water dumping is now permitted, and current demilitarization facilities are not capable of disposing of the quantities of material involved, so the ordnance is currently being accumulated.

To exemplify the magnitude of the problem, current estimates of the backlog of ordnance requiring disposal range between 80,000 and 120,000 tons, including:

Explosive projectiles	Firing devices
Small arms ammunition	Pyrotechnics
Fuses	Ejection cartridges
Detonators	Rockets
Primers	Bombs
Grenades	Depth charges
Solid propellants	Rocket Motors

These items range from small handgun cartridges to 16" shells. Individual items weigh up to several tons in some cases. Numerically, as examples, some 100,000,000 rounds of ammunition below 20mm are included; no numerical estimates of the number of other items have been located.

Demilitarization

Time-consuming expensive and dangerous demilitarization operations could, it is estimated, result in disposal of roughly 50,000 tons of the total backlog, allowing reclamation of 7,000 tons of HBX, 6,000 tons of TNT and 15,000 tons of smokeless powder. The procedure varies as a function of the specific type of ordnance being treated. Small rounds (up to 20mm) are separated (projectile form cartridge case), the powder is burned and the projectile and cartridge case "popped" in a retort furnace ("popping" = detonation of primers by heat). The metal is recovered for scrap, and it is possible to design the system to recover and reprocess the powder. Larger units are first defused; the explosive charges are removed by washing, steaming or drilling. The explosive is then either recovered or burned, and the metal is recovered for scrap (7).

The hazards of the demilitarization operations are significant. A single MK-51 underwater mine contains 3,200 pounds of TNT, for example. Successful demilitarization of these mines has resulted in reclamation of over 2,400 tons of TNT in the immediate past. The problems of successful demilitarization are apparent to those who have experience with use of ordnance devices and explosives and are compounded by the fact that large quantities of high explosive devices are handled simultaneously. Many explosives which are easily burned in the open without detonation are extremely subject to detonation when encased in metal sheaths. At least one incident has occurred in the fairly recent past in a 20-mm demilitarization line where a sudden accumulation of some 20,000 rounds detonated in a retort furnace. The details of the incident are not clear, but the entire processing unit was completely destroyed, leaving a sizeable crater.

Destructive Disposal

Two other methods of disposal are currently in use which involve intentional discharge or detonation of ordnance. Smaller items are simply dropped into contained fires through tubes; in the fire they explode and the metal is recovered for scrap. A second program involves burial of large amounts of ordnance in pits; these charges are primed and tamped with 10 to 16 feet of earth and detonated. Roughly 40 pits are detonated daily, disposing of some 40 tons per day. The entire series is detonated over a 200-second interval.

As of September 1971 one such group was in operation on a five-day week basis. At this rate, this group would require eight to twelve years to work off the current inventory; alternatively, ten groups might dispose of the inventory in one year. Thereafter, however, as much as 20,000 tons of material would be handled annually, requiring two groups full time to merely detonate the normal accumulation. In addition to the problems of manpower, and cost and environmental unpleasantness at the least, the long duration safety and land use problems are unclear (7).

At least one incinerator for destruction of some 500 tons of miscellaneous fuses is under construction at the Earle Naval Facility. This unit is designed to meet both safety and pollution regulations, but is as yet a prototype design. The performance of this unit from both points of view is still undefined.

The method approved by the Armed Forces for the disposal of detonators, primers, blasting caps and disassembled small arms ammunition containing DDNP and other primary initiating explosives is by burning or detonation in a specially designed detonator furnace. In this furnace the components are fed to the combustion chamber by means of a channel chute and a special conveying device. The detonator furnace should be equipped with an afterburner to abate NO_x, and cyclones and scrubbing towers for the removal of metallic dust and fumes. The bleed-off from the recirculating scrubbing solution should be treated to prevent lead and copper pollution (2).

Deep Water Dumping

The practice of deep water dumping, as mentioned above, has been discontinued. This method of disposal accounted for some 100,000 tons of unserviceable ordnance over the period 1964-1971 in the Maritime Administration Hulk Numbered Deep Water Dumps system alone (7). According to this system, an obsolete vessel (merchantman) was stripped free of all but fixed elements of the structure and the fuel tanks cleaned thoroughly. The ordnance was then stowed to give maximum density (more buoyant items packed in 55-gallon drums filled with concrete), the hulk was towed at least ten miles from shore and scuttled in at least 500 fathoms depth.

The dumping areas were selected to reduce the probability of fish kills in the event of detonations, which were sometimes intentional. Such intentional detonations were performed at the request of the international scientific community, occasionally unplanned. In most cases, the hulk bottomed without detonations, and in no case did detonation occur prior to scuttling. A dump of this type might range up to 8,500 tons at a time, but was generally in the broad vicinity of 5,000 tons.

Small-scale deep water dumping practices were commonly conducted for many years as ordnance disposal methods. Such operations involve jettisoning up to 250 tons of material at a time in sites meeting the ten-mile, 500 fathoms criteria mentioned above. The actual dumping is performed over a short period of time, rather than all at once. No detonation has ever been experienced in operations of this type. Of 13 sites selected for such dumps in 1971, only one lay less than 20 miles from shore (12,000 feet deep). All liquid propellants, industrial chemicals and chemical agents were excluded from this type of disposal.

Other Proposals

Proposals have been made regarding design of mobile demilitarization lines to reduce the need for overland transportation of munitions. The question of safety is not attacked in this particular solution to part of the problem, but further developments in both demilitarization technology and new forms of ordnance construction are expected to alleviate the danger in the future. Detonation in AEC caverns, in conjunction with AEC tests, and in abandoned mine shafts has been proposed also. This reduces the processing danger somewhat but requires substantially more overland transportation and handling. Biodegradation and more effective forms of chemical treatment are two possible routes to safer demilitarization, but we still are very far from achievement, so far as has been discovered.

REFERENCES

(1) I. Forsten, *Environmental Science and Technology*, 7, No. 9, 806-810 (September 1973).

(2) R.S. Ottinger, J.L. Blumenthal, D.F. Dal Porto, G.I. Gruber, M.J. Santy and C.C. Shih, *Recommended Methods of Reduction, Neutralization, Recovery or Disposal of Hazardous Waste, Vol. VII: Propellants, Explosives and Chemical Warfare Materiel,* Report PB 224 586, Springfield, Va., Nat Tech Information Service (August 1973).

(3) R.D. Altekruse; U.S. Patent 3,903,814; September 9, 1975; assigned to Olin Corp.

(4) S.J. Yosim, L.F. Grantham and D.A. Huber; U.S. Patent 3,778,320; December 11, 1973; assigned to Rockwell International Corp.

(5) J.W. Bolejack, Jr., T.K. Daniel and D.E. Rolison; U.S. Patent 3,848,548; November 19, 1974; assigned to Hercules, Inc.

(6) C.D. Kalfadelis and A. Skopp; U.S. Patent 3,916,805; November 4, 1975; assigned to Exxon Research and Engineering Co.

(7) Booz-Allen Applied Research, Inc., *A Study of Hazardous Waste Materials, Hazardous Effects and Disposal Methods,* Vol. II, Report PB 221 466, Springfield, Va., Nat Tech Information Service (July 1973).

ORGANIC CHEMICAL INDUSTRY WASTES

Table 31 shows the forms of preliminary treatment and of ultimate disposal for a variety of organic wastes, inorganic wastes and plant product wastes from the petrochemical industry as taken from a report by Booz-Allen Applied Research (56). In this table the various treatments are represented by the letters (A) through (U).

Physical Treatment
 (A) = Sedimentation
 (B) = Filtration
 (C) = Flotation
 (D) = Separators (API)
 (E) = Stripping
 (F) = Adsorption and Extraction
 (G) = Evaporation
 (H) = Submerged Combustion
Chemical Treatment
 (I) = pH Adjustment
 (J) = Chemical Oxidation
 (K) = Coagulation and Chemical Precipitation
Biological Treatment
 (L) = Biological Filters
 (M) = Activated Sludge
 (N) = Lagoons
Ultimate Disposal
 (O) = Controlled Dilution to Streams and Bays
 (P) = At Sea
 (Q) = On Land Surfaces
 (R) = Dumping or Burial
 (S) = Deep Wells
 (T) = Incineration
 (U) = Salvage

The processes being used are indicated by a reference number or an X. The reference numbers refer to the bibliography at the end of this chapter.

The reader is also referred back to the listing in the earlier chapter of the volume on Incineration under Chemicals Which can be Disposed of by Incineration which tabulates a number of organic chemicals which can be disposed of by incineration.

TABLE 31: ALTERNATIVE TECHNIQUES FOR THE DISPOSAL OF ORGANIC CHEMICALS

Wastes from Petrochemical Operations	Physical Treatment								Chemical Treatment			Biological Treatment			Ultimate Disposal						
	(A)	(B)	(C)	(D)	(E)	(F)	(G)	(H)	(I)	(J)	(K)	(L)	(M)	(N)	(O)	(P)	(Q)	(R)	(S)	(T)	(U)
Organic:																					
Acetaldehyde				(1)	(2)							(3) (4) (5)	(6)	(7)	(1)						
Acetic acid												(4) (8)	(3) (6) (9)	(7) (10)	X					(11)	
Acetone				(1)	(4) (12)							(4) (8)	(3) (6) (13)							(11)	
Acetylene derivatives				(1)											(1)						
Acrolein						X				X		X	X							X	
Alcohol, general						(14)						(4)					(14)			(14)	
Alcohols, high boiling												(15)	(13)		(1)				(14)		(1)
Allyl alcohol										X		(15) (16)							X		
Allyl chloride																				X	
Aromatics, C₆				(1)																	X
Benzene				(1)								(17)		(1)	(1)					(11)	
Butadiene				(1) (18)								X		X						X	
Carbonyl products															X				(19)		
Carbon tetrachloride																					X
Chlorinated hydrocarbons												(5)				(20)	(14)		(19)		(22)
Chloroform																				(21)	X
Cresols					(23)					(23)					X					(19)	X
Cuprous ammonium acetate											(1)				(1)						X
Cyanides				(24)	(23)					(25) (26)		(26)			(27)			(28)		(29)	
Detergents				(1)					(1)												
Dichloropropane																				(1)	
Diethyl ether				(1)											X					X	
Diethyl formamide															X						

(continued)

TABLE 31: (continued)

Wastes from Petrochemical Operations		Physical Treatment								Chemical Treatment			Biological Treatment			Ultimate Disposal						
	(A)	(B)	(C)	(D)	(E)	(F)	(G)	(H)	(I)	(J)	(K)	(L)	(M)	(N)	(O)	(P)	(Q)	(R)	(S)	(T)	(U)
Organic:																				X	X
Distillate, aromatic				(1)																	
Divinylacetylene																				X	
Dodecyl															X					X	
Dowtherm					(4)							(15)	(6)		(1)				X		
Ethyl alcohol					(12)							(17)									
Ethyl benzene				(1)																X	
Ethyl chloride															(1)				X		
Ethyl ether				(1)																X	
Ethylene diacetate																				X	
Ethylene dichloride															X						
Ethylene oxide																					
Ethylidene chloride												(30)								X	
Epichlorohydrin																				X	(1)
Flux oil				(1)															(11)		
Formaldehyde						(21) (31) (32)						(30) (33)	(3) (21) (32)	(3) (7) (30) (33)							
Formic acid												(34)	(3)						(11)		(1)
Gasolines				(1)									(3)		X		X	X	X	X	
Glycerol													(3)						(11)		
Glycols													(13)								
Hydrazine									X						X					X	(18)
Hydrogen cyanide													(13)		(1)						
Isobutyl alcohol													(6)		(1)				X		
Isobutyl and isopropyl alcohol													(13)								
Isopropyl ether					X										X					X	(1)
Kerosene				(1)								(14)	(1)		X		(14)		(14)	(14)	
Ketones						(14)							(14)	(14)							

(continued)

TABLE 31: (continued)

Wastes from Petrochemical Operations	(A)	(B)	(C)	(D)	(E)	(F)	(G)	(H)	(I)	(J)	(K)	(L)	(M)	(N)	(O)	(P)	(Q)	(R)	(S)	(T)	(U)
				Physical Treatment					Chemical Treatment			Biological Treatment			Ultimate Disposal						
Organic:																					
Mercaptans					(36)	X				(36)										X	X
Methanol												(6) (37)	(3) (6) (32)	(7)						(11)	
Methyl acetylene															X						
Methyl chloride															X				(38)	X	X
Methyl formate																				X	
Methylene chloride																					
Monoethanolamine												(15) (17)			(1)						(1)
Naphthas				(1)																	(1)
Naphthene						X														X	
Octyl alcohol															(1)						X
Octyl aldehyde															(1)						X
Picolines					(1)										(1)						
Phenolics					(39) (40)	(25) (31) (37) (39) (41) (42)			(40)	(23)		(43) (44)	(43) (44)	(44)	(1)				X	X	X
Polybutene				X											X					X	X
Polychloroethane																				X	X
Polyethylenes, solid																				X	
Polyisobutylene																			(19)		X
Polymers				X																X	X
Propadiene																				X	X
Propylene dichloride																				X	
Resins				(33)								(33)		(33)						X	X
Rubber				(1)		X													(19)	X	X
Soaps						X									X						X
Sulfonates, petroleum			(1)						(1)						(1)						(1)

(continued)

TABEE 31: (continued)

Wastes from Petrochemical Operations	Physical Treatment								Chemical Treatment			Biological Treatment			Ultimate Disposal						
	(A)	(B)	(C)	(D)	(E)	(F)	(G)	(H)	(I)	(J)	(K)	(L)	(M)	(N)	(O)	(P)	(Q)	(R)	(S)	(T)	(U)
Organic:																					
Tars	(37)			(1)																(19)(38)(45)	(1)
Terpenes												(1)(46)		(1)							
Terpenes, sulfurized												(1)		(1)							
Toluene				(1)										(1)							(47)
Urea				(1)																X	(1)
Vinyl acetate															X						
Zylene				(1)																X	(1)
Inorganic:																					
Aluminum chloride															(1)			(1)		(1)	(22)
Ammonia					(24)				X			X	(44)	(1)	(1)						X
Ammonium nitrate												X	X	(1)	(1)						
Barium sludge														(1)	(1)						
Carbon bisulfide																				X	X
Carbon black		X																		X	X
Chlorine					X									X	X						
Chromates										(25)					X						
Chromic oxide catalyst										(25)											
Chromium oxide															X			X			X
Copper and compounds		X													(1)						
Copper chloride															X			X			X
Ferrous dichloride															X			X			
Hydrochloric acid									(1)(48)						(1)				(14)		
Hydrogen sulfide					(36)				X	(36)										X	X
Lime		X							X						X		X	X			
Nitric acid									(35)(49)(22)						X						
Phosphoric acid catalyst															X		X	X			(48)

(continued)

TABLE 31: (continued)

Wastes from Petrochemical Operations	Physical Treatment								Chemical Treatment			Biological Treatment			Ultimate Disposal						
	(A)	(B)	(C)	(D)	(E)	(F)	(G)	(H)	(I)	(J)	(K)	(L)	(M)	(N)	(O)	(P)	(Q)	(R)	(S)	(T)	(U)
Inorganic:																					
Potassium permanganate															X						
Sodium chloride															(1)				(14) (38) (50)		
Sodium hydroxide									X						X						
Soot	X	X	X																	(29)	
Sulfur dioxide						X			X						(1)						
Sulfuric acid									(1) (49) (51)						(1)				(14)	X	(48)
Sulfates															X				(14)		
Sodium sulfide	(1)								(1)						X			X			X
Waste caustic sodas										(52)					(1) (52)				(52)	(52)	(52)

Summary of Treatment Methods for Petrochemical Wastes Classified by Plant Product

Plant Product	Physical Treatment								Chemical Treatment			Biological Treatment			Ultimate Disposal						
	(A)	(B)	(C)	(D)	(E)	(F)	(G)	(H)	(I)	(J)	(K)	(L)	(M)	(N)	(O)	(P)	(Q)	(R)	(S)	(T)	(U)
General chemicals	(53)				(39)	(39)			(53)	(29) (53)				(53)							
Nylon								(50)				(50)									
Nylon chemical intermediates	(45)	(45)			(45)		(35) (45)			(45)		(45)		(35) (45)	(35) (45)		(45)		(35) (45)	(35) (45)	(45)
Organic chemicals				(54)			(54)	(54)													(11)
Oxygenated hydrocarbons													(3)								
Photochemicals		(55)									(55)	(1)									
Powders												(30)									
Resins						(25) (31)						(33)		(33)							
Rocket fuels				(35)					(35)			(35)									
Rubber, textiles and plastics													(3)								
Synthetic rubber				(18)											(18)					(18)	

Source: PB 221 467

ACETONE

Incineration

A process developed by P.S. Sharpe (57) involves the thermal oxidation of gaseous, liquid and solid wastes by passing air spirally around the exterior length of a cylindrical oxidation vessel countercurrent to the direction of flow of combustion gases within the chamber and toward the feed end of the oxidation chamber to preheat said air. It then involves introducing the preheated air in rotating motion through tangential air ports toward the periphery from the center of the feed end of the oxidation chamber so that the air moves in a spiral and rotating motion to the exhaust end of the oxidation chamber. The fuel which may be liquid or gaseous or mixed and may include high energy waste for thermal oxidation is introduced through a port or ports in the central portion of the feed end of the oxidation chamber.

The fuel is ignited near the introduction ports and if desired, low energy gas, liquid or solid waste is introduced through a port or ports in the central portion of the feed end of the oxidation chamber. The fuel, air and waste pass through the oxidation chamber in an intimate mixing, spiral, rotating motion at low pressure to effect high-efficiency oxidation of the waste in the central portion of the oxidation chamber. The vessel walls of the oxidation chamber are cooled by excess air moving spirally in the peripheral portion of the oxidation chamber. The products of oxidation and excess air from the oxidation chamber pass through an open exhaust end. The oxidation chamber in combination with a water scrubber and fail-safe exhausting means provides a chemical waste disposal apparatus resulting in minimal pollution. Figure 99 shows such an apparatus.

FIGURE 99: THERMAL OXIDATION APPARATUS SUITABLE FOR DISPOSAL OF ACETONE-RICH WASTES

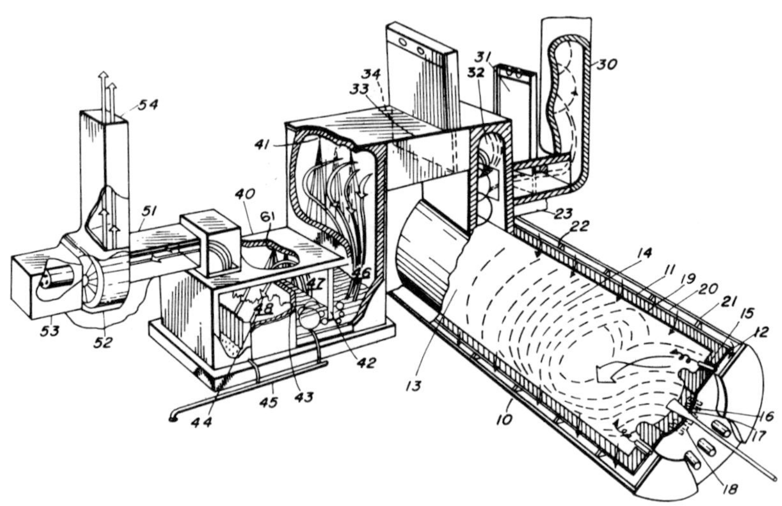

Source: U.S. Patent 3,892,190

The thermal oxidizer shown as **10** comprises cylindrical oxidation vessel wall **11**, having closed feed end **12** and open exhaust end **13** defining oxidation chamber **14**. The closed end has air ports **15** tangential to the axis of chamber **14** and toward the periphery from the center of feed end **12**. The number of air ports toward the periphery of the feed end is not critical, but must be adequate to furnish sufficient air for oxidation and combustion with an excess of air for cooling the oxidation chamber walls. Liquid port **16** is located in the central portion of feed end **12** and has nozzle **17** which provides for injection of liquid fuels and wastes and waste solid containing slurries in a manner which provides conical distribution as the waste progresses away from the nozzle through combustion chamber **14**. The nozzle represents a single nozzle, but multiple nozzles are also suitable. Surrounding central liquid port **16** and spaced about the central portion of the feed end are a series of gas ports **18** providing injection of gaseous fuel or waste which may be in the same general tangential direction as air from tangential ports **15** or may be parallel to the longitudinal axis of the oxidation chamber. The number of gas ports is not critical but must be adequate to furnish sufficient gaseous fuel for efficient combustion and oxidation when gaseous fuel is utilized.

Cylindrical oxidation vessel wall **11** is shown comprised of fire-brick lining **19** and outside the fire-brick may be a circumferential support of noninsulating material **20**. Jacket **21** surrounds the cylindrical oxidation vessel and is spaced therefrom. Baffles **22** are positioned in the space between the jacket and the vessel wall so that when air is forced into the space between the jacket and vessel walls at entrance **23** the air is spirally directed along the exterior of the oxidation vessel toward feed end **12**, as shown by the arrows, preheating the air and cooling the exterior of the oxidation vessel. Air is supplied to the entrance by a supply blower, not shown. Air is supplied to the entrance at suitable pressures which may be in the range of about 2 to 6 inches of water.

The air having a rotating motion imparted to it by the spiral path through the space between the vessel wall and jacket passes through air ports **15** with a rotating motion in addition to the tangential direction imparted by the tangential arrangement of the ports with respect to the longitudinal axis of the oxidation chamber. When liquid fuel is used supplemental air may be provided through gas ports **18** to provide sufficient air for combustion. Thus, the air may enter the oxidation chamber at pressures in the range of about 1 to 4 inches of water and in tangential streams with rotation of the air within each stream. The motion provides intimate mixing and energy exchange between the air, waste and fuel, thereby affecting efficient oxidation of the waste.

Fuel is provided in the gaseous form to gas ports **18** or in the liquid form to liquid ports **16** (from a fuel source not shown) in sufficient quantity to insure complete combustion within chamber **10**. Liquid and gaseous fuels are suitable for the process and apparatus of this process. Any fuel having sufficiently high energy to effect substantially complete oxidation of the waste material is suitable for use in the apparatus and may be supplied to the introduction ports by any fuel supply means known to the art. It is preferred to use natural gas as the fuel although other natural or synthetic fuels are suitable including a wide range of fuel oils, heavy or light, kerosene and the like.

The waste to be oxidized may be supplied to the gaseous or liquid introduction ports, as suitable, by any appropriate means. The waste may be in the form of gas, liquid or solids, and may be mixtures of any such forms. The waste may be of sufficiently high energy to support combustion and thus promote oxidation, or it may be of low energy which requires the transfer of energy from fuel to promote oxidation. When liquid or gaseous waste which supports combustion is used, supplementary fuel may not be necessary. The ratio of waste feed to fuel feed is governed by the required energy transfers necessary for oxidation of waste material. When liquid wastes are used, it is desired that they be atomized in introduction to the oxidation chamber. Suitable atomizing nozzles are well known in the art.

When solid wastes are used, it is preferred that such solid material be finely divided and suspended in liquid slurry for introduction into the oxidation chamber. Such slurry formation of solids is well known and the liquid vehicle may be any suitable liquid such as one which will support combustion such as water. The process has been found to efficiently burn and oxidize solid wastes in slurry form wherein the slurry is 90 weight percent water. The amount of air introduced into the oxidation chamber is in excess of the air necessary for oxidation and combustion so that the periphery of the chamber will continuously be filled with circulating air to cool the inner surface of the refractory material. The air ports are arranged in a tangential fashion so that the air stream will make multiple revolutions in its spiral passage through the combustion chamber. The tangential air ports must also be so arranged as to direct the air into intimate mixing relationship with the incoming fuel and waste chemical. The high volume, low pressure operation of the oxidation chamber and the introduction of the air, fuel and waste chemical causes intimate mixing, spiral rotating motion within the oxidation chamber to effect high efficiency oxidation of the waste chemical and combustion of the fuel in the central portion of the oxidation chamber while maintaining excess air moving spirally in the peripheral portion of the oxidation chamber. When the waste chemical is a combustible liquid or gas, the waste chemical may be mixed with fuel and such mixture introduced with the fuel.

The air ports, gas ports, and liquid ports each may contain means for introduction of the materials into the oxidation chamber in the manner specified above. The specific design of each of these ports or injection nozzles may be any suitable design for such purposes as is known in the art. The open end of cylindrical oxidation chamber 13 is in communication with exhaust means generally shown as 32 for emission of the effluent of the oxidation chamber to the atmosphere. It is desirable to have an expansion chamber located between open end 13 and the final stack open to the atmosphere. It is desirable to pass the products of combustion from the exhaust of the combustion chamber into the expansion volume reducing the velocity of flow of the gases and causing particulate matter to settle to the bottom of the expansion chamber prior to passage into the final stack to the atmosphere.

It is apparent that any pollution control device for removing undesired materials from the stack effluent which operates under low pressure conditions with high volume may be positioned between the exhaust from the combustion chamber and the stack to the atmosphere. The exhaust gas from the oxidation chamber passes through conduit 33 into liquid scrubber 40. The conduit may contain any suitable energy recovery system to utilize the heat in the exhaust gases. The gas containing particulate matter and/or noxious gases enters the liquid scrubber in section 46 and the spray nozzles 41 remove the larger particulate matter to the bottom of section 46. The gas moves downwardly in this section as the spray from the nozzles increases its velocity and strikes the fluid in 46 at a relatively higher velocity.

The greater portion of the remaining particulate matter is removed in the liquid by change of direction and relatively high velocity in passing beneath partition 42 into section 47. The gas rises through the liquid in section 47 at a relatively slower rate permitting solution or desired reaction of the noxious gases in the liquid and the treated gas stream then passes out of the liquid and upward countercurrent to the direction of liquid spray from nozzles 61 in section 47. The treated gas passes through the opening above the partition 43 into section 48 wherein the gas is passed through any desired demister and/or a packed column indicated as 44 through conduit 51 and forced by blower 52 driven by motor 53 up the clean effluent stack 54 to the atmosphere.

In order to provide an apparatus of the greatest versatility, that is which is applicable to the oxidation of a wide variety of chemical wastes, effluent stack 30 may be provided to the open atmosphere having shutter 31 open and shutter 34 closed for passage of effluent gases directly from the oxidation chamber to atmosphere. As pointed out above, the oxidation of some chemicals results directly in gases which may be passed to the atmosphere without further treatment. For use in such applications, the gases are advantageously directed through stack 30. For use in oxidations resulting in gases and particulate matter which require further treatment such as the liquid scrubber 40, shutter 31 is closed and shutter 34 is opened directing the gaseous exhaust from the oxidation chamber through liquid scrubber 40.

Stack 30 is also useful as a fail-safe system permitting uninterrupted operation of the oxidation chamber even in the event of a power failure or over-temperature conditions. The fail-safe system then diverts the flow of gases from the scrubbing systems directly to stack

30 by closing shutter **34** and opening shutter **31** to permit quick venting. The oxidation chamber, scrubber and effluent stacks may be constructed from conventionally available materials, usual construction being steel casing lined with fire brick refractory materials where necessary. The following example illustrates the application of this apparatus to the incineration of acetone-rich wastes.

Example: The thermal oxidation apparatus as shown without any scrubber and passing the exhaust from the oxidation chamber directed to the atmosphere was operated utilizing a high Btu liquid waste as fuel to oxidize a low Btu liquid waste which would not support combustion. The high Btu liquid waste used as fuel had the following analysis:

Chemical	Volume Percent
Acetone	36.8
Methyl alcohol	9.2
Methyl isobutyl ketone	4.6
Xylene	41.4
Water	8.0

The low Btu waste had the following analysis:

Chemical	Volume Percent
Acetone	15
Methyl alcohol	5
Methyl isobutyl ketone	1
Water	77
Residue	2

The liquids were fed to the oxidation chamber through a five-nozzle cluster in the center of the feed end of the chamber, two nozzles being used for the high Btu liquid and three nozzles for the low Btu liquid. The high Btu liquid waste was delivered to the oxidation chamber at 240 gallons per hour and the low Btu liquid waste was delivered to the oxidation chamber at 800 gallons per hour. The total heat input rate was 40×10^6 Btu per hour. The undesired liquid wastes were completely oxidized in the oxidation chamber according to the following reactions:

Acetone
$$C_3H_6O + 4O_2 \longrightarrow 3CO_2 + 3H_2O$$

Methyl alcohol
$$CH_3OH + \tfrac{3}{2}O_2 \longrightarrow CO_2 + 2H_2O$$

Methyl isobutyl ketone
$$C_6H_{12}O + 17\tfrac{1}{2}O_2 \longrightarrow 6CO_2 + 6H_2O$$

Xylene
$$C_8H_{10} + 21\tfrac{1}{2}O_2 \longrightarrow 8CO_2 + 5H_2O$$

Residue (assuming benzene)
$$C_6H_6 + 15\tfrac{1}{2}O_2 \longrightarrow 6CO_2 + 3H_2O$$

The stack gas was measured and analyzed generally following federal EPA test procedure (*Federal Register,* Vol 36, No. 247 Part II, Dec. 23, 1971) showing the following properties:

Temperature, average	$1335°F$
Velocity, average	36 ft/sec
Volume (dry basis)	5.80×10^5 standard cubic feet per hour
Moisture	17.4%
Particle concentration (dry basis)	0.0335 grains per standard cubic foot (corrected to 12% CO_2, 0.0556)

(continued)

Isokinetic ratio	106%
CO_2	7.23 volume %
O_2	12.87 volume %
N_2	80.9 volume %
SO_2	0.116 lb/hr
SO_3	<0.05 lb/hr
NO_x	4.35 lb/hr
HCl	0.117 lb/hr
Unburnt hydrocarbons	45.6 ppm
Ash:	
High Btu waste	24 ppm
Low Btu waste	28 ppm
Particle size	<1 micron

ACROLEIN

Due to its extreme chemical reactivity, toxicity, flammability, and the processing difficulties associated with those characteristics, it is anticipated that disposal systems to handle both dilute and concentrated acrolein wastes will be required at sites located near manufacturers and users. The dilute acrolein wastes that will require treatment include spent cleaning solutions for acrolein containers and any other on-site generated wastewater containing acrolein. The concentrated acrolein wastes that will require treatment include any surplus, contaminated, or degraded material (58). The processes recommended for the treatment of dilute acrolein wastes are:

Process	Order of Preference	Remarks
Biological	First choice	In principal, this method appears adequate when followed by proper disposal of the concentrated effluent.
Submerged combustion	Second choice	Method currently used; effectiveness under study; requires neutralization prior to lagooning.

The processes for treatment of concentrated acrolein wastes are:

Process	Order of Preference	Remarks
Recycle	First choice	Major producer has indicated willingness to accept concentrated acrolein waste.
Incineration	Second choice	Demonstrated technology; applicable to most organic wastes.

ACRYLONITRILE PROCESS WASTEWATERS

Deep Well Disposal

A process developed by W. O. Fitzgibbons et al (59) involves the deep well disposal of a wastewater stream resulting from a catalytic ammoxidation process for the production of acrylonitrile comprising adding minor amounts of acrolein or a mixture of acrolein and ammonium sulfate to the wastewater stream prior to pumping the stream into the well.

The mechanism whereby the addition of acrolein and ammonium sulfate to the wet acetonitrile bottoms stream eliminates plugging of the deep disposal well is not known. However, plugging is believed to be caused by suspended particles in the waste stream in the particle size range of about 0.45 to 5.0 μ which cause a decrease in the permeability of the rock core and a corresponding increase in the injection pressure at the well. It is preferred, and plugging of the well is virtually eliminated, when the wet acetonitrile bottoms stream is filtered through sand after it is treated with the acrolein or the acrolein-ammonium sulfate mixture. The process for acrylonitrile manufacture resulting in the production of the waste acetonitrile bottoms stream is shown in Figure 100.

FIGURE 100: DEEP WELL DISPOSAL PROCESS FOR ACRYLONITRILE PROCESS WASTEWATER

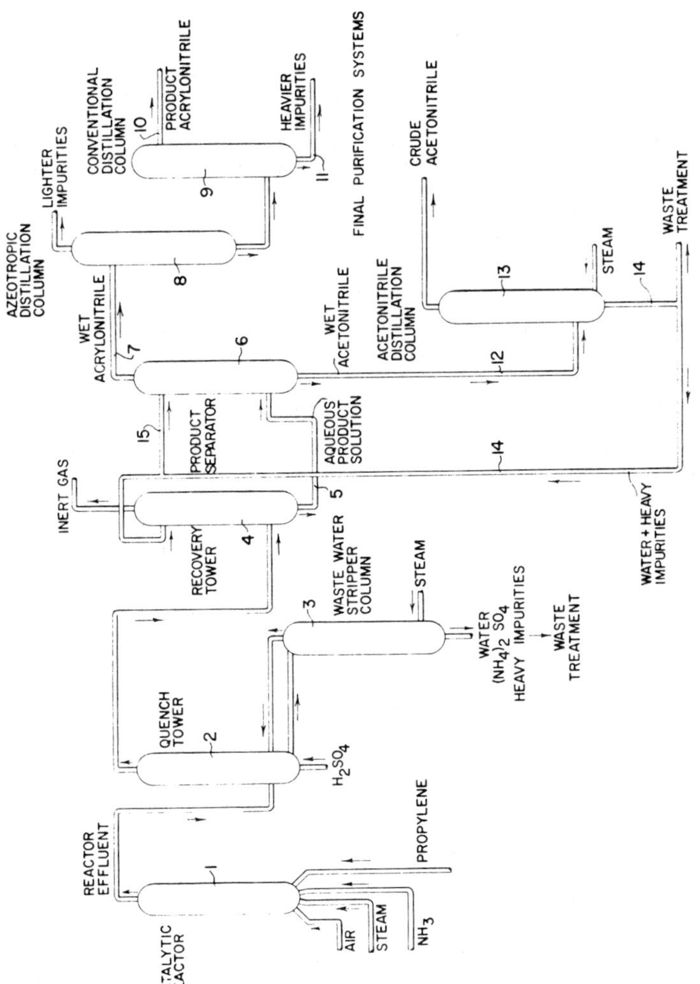

Source: U.S. Patent 3,734,943

In this process, approximately stoichiometric quantities of propylene, ammonia and oxygen (as air) are introduced via spargers and are reacted at elevated temperatures and essentially atmospheric pressures in catalytic reactor **1**. The effluent from the reactor is introduced into a quench tower **2** where the unreacted ammonia in the reactor effluent is neutralized with sulfuric acid and the resulting ammonium sulfate, water and heavy impurities are separated from the stream in the wastewater stripper column **3**. The overhead from the quench tower is taken to a recovery tower **4** where it is scrubbed with water in order to remove soluble organic products as an aqueous solution. The aqueous solution of organic products is removed from the product scrubber by means of line **5**.

This solution is then taken to a product separator **6** into which is also added an aqueous stream containing minor amounts of heavy impurities by means of line **15**. This column provides a wet acrylonitrile overhead **7** and a wet acetonitrile bottom steam **12**. The wet acrylonitrile overhead is dried and purified by azeotropic distillation in column **8**, and conventional distillation in column **9**. The overhead **10** from distillation column **9** consists of product acrylonitrile, and the residue removed through line **11** comprises a mixture of small amounts of by-products and heavier impurities. The wet acetonitrile is transferred from the product separator **6** via line **12** to the acetonitrile distillation column **13**. The bottom stream **14** from the acetonitrile distillation column is divided and a portion of the stream is recycled to the product separator **6** and the remainder is treated and disposed of by deep well disposal.

The overhead acetonitrile from the acetonitrile distillation column **13** is further concentrated and dried. The wet acetonitrile bottoms stream **14**, obtained from the acetonitrile distillation column **13** contains various by-products and heavy organic material resulting from the ammoxidation reaction. These products have been characterized as polynitriles, partially hydrolyzed polynitriles and cyanoethylated products of ammonia which in turn react with other constituents to form polymers, and polymers of hydrogen cyanide, unsaturated aldehydes, ketones, cyanohydrins, and the like. Chemical analysis of a typical wet acetonitrile bottom stream indicates the following composition:

	Percent Concentration
Water	98.2
Ammonium sulfate	0.02
Heavy organic material	1.74

In the treatment of the aqueous stream of wet acetonitrile bottoms a synergistic effect on flow rate of the wastewater into the disposal well is observed for a mixture of acrolein and ammonium sulfate. However, satisfactory flow rates can also be maintained by treating the wastewater stream with small amounts of acrolein per se. Conveniently some waste streams from the acrylonitrile process, as for example, the bottom stream from the wastewater column **3** contain sufficient amounts of acrolein and ammonium sulfate, which are present as a reaction product of acrolein and ammonium sulfate, so that when this stream is mixed with the waste stream from the acetonitrile distillation column, no plugging is observed in the well as evidenced by the lack of increase in injection pressure. Ammonium sulfate by itself however has little effect on the injectability when added to the wet acetonitrile bottoms waste stream.

The elimination of plugging with acrolein and mixtures of acrolein and ammonium sulfate is quite specific for these materials, since analogous aldehydes such as propionaldehyde and methacrolein have little or no effect on reducing plugging. The concentration of acrolein and ammonium sulfate required to minimize or eliminate plugging and to obtain satisfactory flow rates in the disposal well are within the range of from about 0.05 up to about 2.0% by weight of acrolein, and from about 0 to 15% by weight of ammonium sulfate, based on the weight of the wastewater stream. Preferred concentrations are in the range of from about 0.1 to 1.0% by weight of acrolein and from about 0.25 to 10% by weight of ammonium sulfate based on the weight of the wastewater stream, while optimum results are obtained with equal weight concentrations of approximately 0.5 weight percent of each component.

The bottom stream from the wastewater column **3** as indicated contains sufficient amounts of by-product acrolein and ammonium sulfate so that it can be advantageously combined with the waste stream from the wet acetonitrile distillation column to eliminate plugging. The composition of a typical wastewater column stream is shown as follows:

	Percent Concentration
Water	85.5
Ammonium sulfate	8.1
Acrolein-$(NH_4)_2SO_4$ reaction product	0.4
Heavy organic material	6.1
Acrylonitrile	0.02
Acetonitrile	0.01
Maleonitrile	0.20
Fumaronitrile	0.08

Satisfactory flow rates are obtained on combining these streams in a ratio of about 0.2 to 1.8 volumes of the wastewater column stream per volume of the bottom stream from the acetonitrile distillation column. However the ratio is not necessarily limited to this range and may vary in accordance with the concentration of acrolein and ammonium sulfate occurring in the wastewater column effluent.

Incineration

A process developed by H.R. Sheely (60) involves disposing of waste materials from a plant for manufacturing unsaturated aliphatic nitriles or aromatic nitriles whereby wastewater, unreacted ammonia and by-products such as HCN and acetonitrile are not condensed but remain with the absorber off-gas for ultimate disposal by incineration. The method employs a hydrocarbon solvent to adiabatically quench the reactor effluent and, after removal of polymer by-products, the partially quenched effluent is passed to a hot absorber column where the nitrile product but no ammonia and only some of the HCN are absorbed by the hydrocarbon solvent. The nitrile-solvent mixture is distilled to separately recover the solvent and nitrile product.

The solvent is recycled. The hot absorber off-gases are cooled to recover water and then incinerated, with ammonia, HCN, acetonitrile and some vaporized solvent furnishing the necessary fuel values. In a preferred embodiment, solvent in the absorber overhead vapors is recovered by scrubbing with a high boiling oil. Figure 101 is a flow diagram showing the application of this process.

As shown there, propylene, ammonia and oxygen in the form of air are reacted in reactor **2** and the reaction effluent is passed to quench column **6**. In the quench column the reaction effluent is subjected to a hot adiabatic quench in which evaporation of water and a hydrocarbon solvent provides direct cooling sufficient to condense any polymer by-products present in the effluent. Any such condensed polymer by-products are washed out of the bottom of the quench column via line **11** in a stream of unvaporized solvent. The overhead from the column is passed to an absorption column **14A** via line **12** and condenser **16**. Some of the water and hydrocarbon solvent in the overhead are condensed in the condenser and recycled to the quench column via line **18**. Additional hydrocarbon solvent is fed to the quench column via line **46**.

The partially quenched overhead from the quench column is contacted in absorption column **14A** with the hydrocarbon solvent as the absorption medium. Absorption column **14A** is the same as absorption column **14** except that it includes a scrubbing section **60** at its top end in which the absorber overhead vapors are scrubbed with a high boiling oil to reduce process oil losses. Absorption column **14A** comprises a plurality of trays or plates of suitable design for providing countercurrent contact and includes a large liquid collector tray **62** demarcating the bottom of its scrubbing section. The hydrocarbon solvent, e.g., a kerosene boiling at about 350°F, is introduced to the absorption column **14A** at a selected point close to but below its scrubbing section via line **49**.

FIGURE 101: ACRYLONITRILE PROCESS WITH OFF-GAS INCINERATION

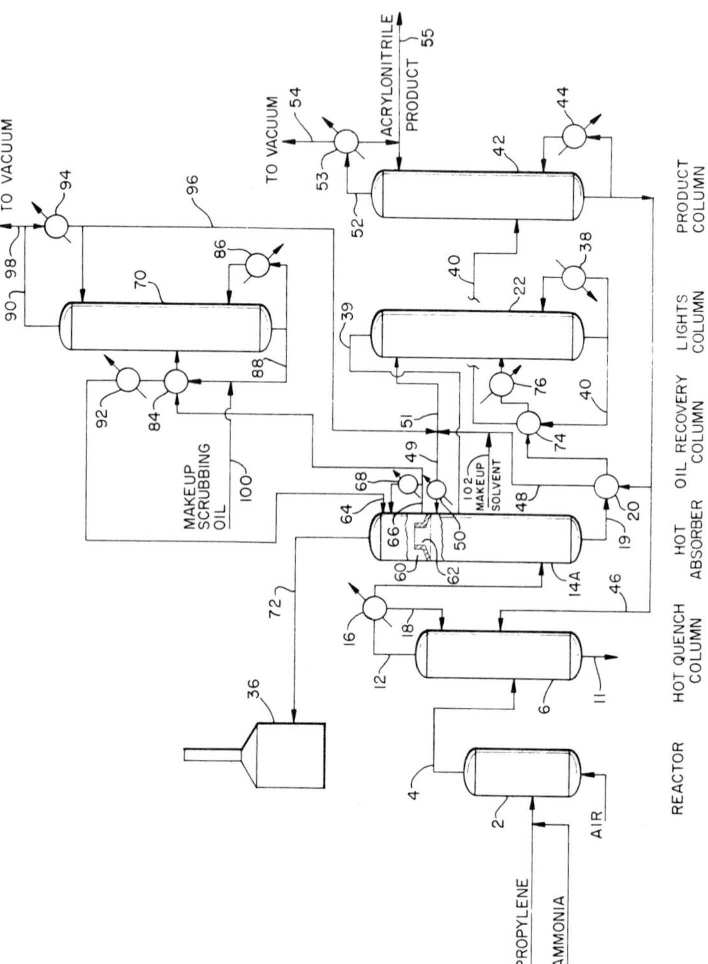

Source: U.S. Patent 3,895,050

The high-boiling scrubbing oil, e.g., a hydrocarbon fraction boiling at about 550°F is introduced to the upper end of scrubbing section **60** via a line **64**. The same oil and hydrocarbon solvent scrubbed out of the absorber overhead vapors are removed from collecting tray **62** via a line **66**. Part of this recovered oil-solvent mixture may be cooled in a water-cooled heat exchanger **68** and returned to the scrubbing section **60**. The remainder is passed to an oil recovery column **70**. The scrubbed absorber overhead vapors are passed directly to incinerator **36** via a line **72**. Substantially all of the acrylonitrile in the overhead from quench column **6** is absorbed by the hydrocarbon solvent in absorber column **14A** and is recovered as bottoms via line **19**. This bottoms stream, which is substantially free of the high boiling scrubbing oil, passes to lights column **22** via heat exchanger **20** and two additional heat exchangers **74** and **76**.

It recovers heat in exchanger **20** from the bottoms of product column **42**, picks up additional heat in exchanger **74** from the bottoms of lights column **22**, and then is raised to a still higher temperature by exchange of heat with steam supplied to exchanger **76**. In lights column **22**, the bottoms product from the absorber is separated by distillation so that HCN, acetonitrile, propionitrile and the like are recovered as overhead and the acrylonitrile and solvent are recovered as bottoms. The lights column overhead is recycled via line **39** to the absorber, while the lights column bottoms is passed via a line **40** and heat exchanger **74** to the product distillation column **42** where substantially pure acrylonitrile is recovered as overhead product and the solvent is recovered as a bottoms product.

The latter is recycled to quench column **6** via line **46** and also to absorber column **14A** via lines **48** and **49** and heat exchangers **20** and **50**. In heat exchanger **50** the solvent is cooled by exchange of heat with cooling water. Some of the recovered solvent is also refluxed to lights column **22** via lines **48** and **51** to permit recovery of lights by extraction distillation. The acrylonitrile product recovered as overhead from column **42** is condensed in condenser **53** and then part of it is recycled to column **42** while the remainder is passed to product storage means (not shown) via line **55**. Any inert gases present in the overhead are withdrawn via line **54** which is connected to a source of vacuum. These inert gases may be vented directly to the atmosphere or may be passed to incinerator **36** for dilution of absorber off-gases.

The mixture of scrubbing oil and solvent withdrawn from the absorber via line **66** is introduced to oil recovery column **70** via a heat exchanger **84** where it is heated by recovery of heat from the bottoms fraction of recovery column **70**. The latter preferably is operated at a pressure less than atmospheric so as to reduce the amount of heat required to be supplied to its reboiler **86** which is heated by steam or hot oil. In column **70** the mixture of scrubbing oil and solvent are separated by fractional distillation into a bottoms product and an overhead product recovered via lines **88** and **90** respectively. The scrubbing oil bottoms product is passed through heat exchanger **84** and then through a water-cooled heat exchanger **92** back to the absorber via line **64**, while the solvent overhead is condensed in a condenser **94** and then returned to the absorber and lights columns via a line **96** and lines **49** and **51**.

The reduced pressure in column **70** is established by connecting the overhead-carrying side of condenser **94** to a source of a vacuum via a line **98**. Makeup scrubbing oil is supplied to the system via a line **100** that connects to line **88**. Makeup solvent is supplied to the system via a line **102** that connects to line **48**. It is believed to be obvious that the purpose of using the scrubbing oil to recover solvent from the absorber overhead vapors is to minimize loss of solvent to the incinerator. Accordingly, a variety of oils (aromatic or paraffinic) may be used for scrubbing the hydrocarbon solvent. The primary requirement is that the scrubbing oil must have a higher boiling point than the solvent. Additionally it must (a) not react (under the operating conditions to which it is exposed) with the acrylonitrile and the other components of the reactor effluent, (b) not react with the solvent, and (c) be thermally stable.

The difference in the boiling points of the solvent and scrubbing oil should be such as to permit ready separation by distillation. Preferably the scrubbing oil should boil at a temperature at least about 40°F higher than the solvent and the latter should have a boiling point that is higher than that of the nitrile product (e.g., acrylonitrile) to permit separation therefrom by distillation. However, the boiling point of the solvent must not be so high in relation to that of the nitrile product as to require a large heat input for reboiling purposes in the lights and product columns. Preferably the solvent has a boiling point at least about 40°F higher than that of the nitrile product and at least about 100°F less than that of the scrubbing oil. The choice of scrubbing oil and solvent as far as their boiling points are concerned is also influenced by how much oil and solvent is to be allowed to go off in the overhead from the absorber to provide fuel values for the incinerator.

The higher their boiling points, the less scrubbing oil and solvent appears in the absorber overhead. By way of example, a suitable scrubbing oil is a recycle gas oil boiling at about 550°F and a suitable solvent is a kerosene fraction boiling at about 350°F. The following is a specific example of the operation of this process.

Example: Propylene, ammonia and air are reacted in reactor **2** which includes a catalyst bed that is fluidized by the incoming air stream. The catalyst may consist of the combined oxides of antimony and uranium prepared as described in U.S. Patent 3,427,343. The propylene, air and ammonia are introduced at rates providing a molar ratio of oxygen to propylene of about 1.5 to 1 and a molar ratio of ammonia to olefin of about 1 to 1. The reaction effluent has a composition approximately as set forth in Table 32 and is cooled to about 450°F before entering quench column **6** which is operated at a pressure of about 9.0 psig. Water and hydrocarbon solvent (consisting of a kerosene oil having a boiling point of 350°F) are continually recovered from the quench column and returned to the quench column via line **18**.

Additional hydrocarbon solvent is recycled to the quench column from product column **42** via line **46**. The water-solvent mixture returned via line **18** has a temperature of about 160°F; the hydrocarbon solvent delivered by line **46** is at a temperature of about 250° to 290°F. As a consequence of contact with the reaction effluent, the water is vaporized and thereby effects cooling of the reaction effluent to an extent sufficient to condense any polymers present in the effluent. Sufficient water and hydrocarbon solvent are delivered to the quench column to cool the reaction effluent to about 180°F and to wash out of the bottom of the column any condensed polymers and any catalyst fines carried over from the reactor.

The partially condensed reaction effluent recovered from the top of the quench column is cooled to about 160°F in quench cooler **16** before entering the absorber column **14A** which is operated at an average temperature of about 165°F and an average pressure of about 5 psig. A scrubbing oil having a boiling point of about 540°F is delivered to the top end of the scrubbing section of the absorber column via lines **64** at a temperature of about 160° to 165°F. Simultaneously solvent is fed to the absorber via line **49** at a temperature of about 160° to 165°F. In the lower section of absorber column **14A** the partially condensed reaction effluent undergoes countercurrent contact with the solvent, with the result that the acrylonitrile is absorbed by solvent and recovered therewith as a bottoms product.

Also absorbed by the solvent are any reaction by-products remaining in the reaction effluent recovered from the quench column that boil at a higher temperature than acrylonitrile. The remaining unabsorbed vapors of the reaction effluent plus solvent vapor pass up into the scrubbing section **60** of column **14A** where all but a small amount of solvent is scrubbed out by the high-boiling scrubbing oil. The scrubbed gases pass out of the absorber column while a solvent-scrubbing oil mixture is removed from tray **62** via line **66**. Substantially all of the ammonia, HCN and acetonitrile and most of the propionitrile present in the reaction effluent recovered from the quench tower appear in the off-gases that pass out of the top of the absorber. The absorber bottoms product is heated to about 350°F in passing through heat exchangers **20, 74** and **76** to the lights column **22**. The latter column is operated at an average pressure of about 6.0 psig with a bottom temperature of about 350°F and a top temperature of about 190°F. Solvent for reflux is introduced to the lights column via lines **48** and **51** at a temperature of about 190°F. The product column is operated at an average pressure of about 6 psia. With a bottom temperature of about 300°F and a top temperature of about 120°F. The overhead from product column **42** is recovered via line **52** and the bottoms fraction is recycled via lines **46, 48, 49** and **51** to quench column **6**, absorber **14A** and lights column **22**. The solvent-scrubbing oil mixture recovered from absorber **14A** is heated to about 395°F as it passes through heat exchanger **84** to oil recovery column **70**. The latter is operated at a pressure of about 6 psia with a bottom temperature of about 400°F and an upper temperature of about 270°F.

The bottoms product from column **70** is cooled to about 160°F as it passes through exchangers **84** and **92** back to the absorber via line **64**, while, the overhead from the same column is cooled in condenser **94** and returned to columns **14A** and **22** via line **96**. During the run makeup hydrocarbon solvent and scrubbing oil are introduced via lines **102** and **100** respectively and off-gases from absorber **14A** are burned in incinerator **36**. The overhead from product column **42** and the off-gases from decanter **36** have compositions approximately as set forth in Table 32 (all values are on the basis of mols).

TABLE 32

Component	Reactor Effluent	Quench Column Bottoms	Off-Gas to Incinerator	Product Column Overhead
Noncondensible gases	240	0	240	0
Water	830	0	830	0
NH_3	50	0	50	0
HCN	50	0	50	0
Acetonitrile	5	0	5	0
Acrylonitrile	150	0	1	149
Other C_3 compounds*	1	0	1	0
Polymer by-products	Trace	Trace	0	0
Solvent	0	1.5	0.05	0
Scrubbing oil	0	0	1.0	0

*The other C_3 compounds are acetone, acrolein, and propionitrile.

Thus all of the NH_3, HCN, acetonitrile, water and noncondensible gases present in the reaction effluent are removed and passed to the incinerator. Additionally, most of the propionitrile is passed to the incinerator and the remainder appears as impurity in the product column overhead.

ALCOHOLS

Incineration

Bulk quantities of contaminated alcohols that can not be reprocessed or released by controlled dilution can best be disposed of by incineration. Since this will probably be an unusual occurrence, this is best done at municipal or industrial incineration sites (62).

A process developed by J.L. Callahan et al (61) is one in which environmentally dangerous organic pollutants, such as hydrocarbons, alcohols, ethers, aldehydes, ketones, esters, acids, amines and the like are readily disposed of in an ecologically-safe manner by entraining the organic pollutants in a stream containing at least about 90% by volume of steam and passing the stream over an oxidation catalyst at a temperature of 250° to 700°C. The organic pollutants are conveniently converted to harmless nitrogen, water and carbon dioxide and valuable high temperature steam is obtained. Figure 102 shows a suitable form of apparatus for the conduct of this process.

FIGURE 102: CATALYTIC COMBUSTOR FOR THE DISPOSAL OF ORGANIC POLLUTANTS

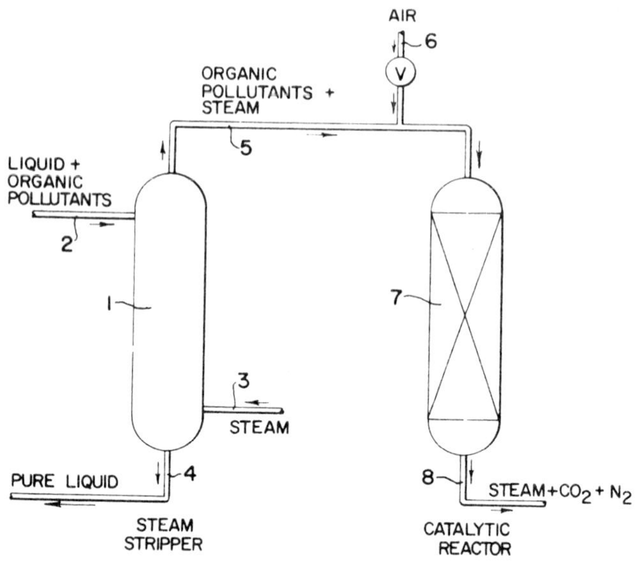

Source: U.S. Patent 3,804,756

The disposal apparatus essentially consists of a steam stripper **1** and a catalytic reactor **7**. A liquid containing organic pollutants is fed through conduit **2** to the steam stripper **1**. Steam is passed through conduit **3** into the liquid in the steam stripper. The passing of steam through the liquid in the steam stripper separates the organic pollutants from the liquid. The resulting purified liquid is removed through conduit **4**. The organic pollutants are entrained in steam, and a stream of organic pollutants in steam leaves the steam stripper through conduit **5**. Air or molecular oxygen is optionally injected into the contents of conduit **5** through conduit **6**.

The resulting mixture in conduit **5** is then passed into the catalytic reactor **7**. This catalytic reactor contains a temperature of about 250° to about 700°C. In the catalytic reactor the organic pollutants are at least partially converted to the harmless products of carbon dioxide, nitrogen and steam. The effluent from the catalytic reactor contains steam and products as removed through conduit **8**.

The following is a specific example of the operation of this apparatus as applied to the removal of methanol and ammonia from a waste stream.

Example: An aqueous stream containing very small amounts of ammonia and methanol is charged to a stripper. In the stripper steam is passed through the liquid containing the pollutants to strip the pollutants from the liquid. In this operation, a gaseous stream consisting of steam containing 2,500 ppm of methanol and 8,450 ppm of ammonia was obtained. A sample of this gas stream was recovered for reaction. A catalytic reactor consisting of a metal tube with an inlet for reactants and an outlet for products was constructed. Into this reactor was charged 205 grams of a catalyst containing 10% copper oxide on alumina. The catalytic reactor was maintained at 900°F and the pressure was 20 psig. The gas stream from the CO_2 recovery system was mixed with 4% of air and passed through the reactor at an hourly space velocity of 2,000. The effluent of the reactor was analyzed, and it was observed that no methanol or other organic impurities were found and only 8 ppm of NH_3 were unreacted. No nitrogen oxides were detected. In the same manner as shown by the example above, other organic pollutants, such as hexane, butanol, diethyl ether, phenol, formaldehyde, acetaldehyde, acetone, valeric acid, ethyl acetate and trimethylamine, are converted to harmless products by passing a mixture of steam containing 1 volume percent of these pollutants over a copper oxide catalyst at a temperature of 600°C.

ALDEHYDES AND KETONES

Incineration

Bulk quantities of contaminated aliphatic aldehydes and ketones that cannot be reprocessed or released by controlled dilution can best be disposed of by incineration. This is the method of choice where adequate biological treatment facilities are not available (62).

ALIPHATIC AMINES

Incineration

Badly contaminated amines that cannot be reclaimed, as well as by-products that cannot be sold and unusable end-products, can be disposed of by incineration. A scrubber and/or catalytic or thermal unit may be necessary to reduce NO_x emissions to required levels (62).

Landfills

Deposit or landfill of amine-containing waste presents a serious long-term environmental hazard to man, animals and fish. These materials are toxic and to some extent water-soluble. Biodegradation of polymers or polyamines will further increase mobility by molecular weight reduction. Thus, landfill disposal of these wastes represents an excessive, long-term threat to underground water supplies and must be considered much less acceptable than incineration. Landfill disposal is acceptable only when landfill sites meet California Class 1 specifications (62).

ALLYL CHLORIDE

Incineration

Most of the allyl chloride produced by Shell Chemical Company is for captive use and they have not experienced the need to dispose of concentrated waste. It was recommended that concentrated waste be incinerated with proper means of removing hydrogen chloride from the effluent stream. This is easily done through the use of aqueous or caustic scrubbers. High temperature incineration (1800°F minimum, 2 seconds minimum) of allyl chloride with hydrogen chloride removal from the effluent gas is an acceptable means of disposing of this material.

Most production of allyl chloride at the Shell Chemical Company plant in Houston is used to produce allyl alcohol. Dilute organic allyl alcohol waste is incinerated. If dilute organic waste containing allyl chloride is generated, it is recommended that it be incinerated as discussed in the preceding paragraph on concentrated allyl chloride.

Every effort is made to prevent allyl chloride from entering sewage because of its high flammability. Spills are allowed to evaporate when practical, or a chemical fog is used to suppress fumes and the material is collected with sawdust or absorbent clay. The sawdust and clay are combusted in incinerators equipped with particulate removal equipment and caustic or wet scrubbers for hydrogen chloride removal. The clay may then be reused as an adsorbent or buried in an approved landfill. If aqueous allyl chloride waste is generated, it is recommended that it be concentrated and incinerated with hydrogen chloride removal from the effluent gas or hydrolyzed to allyl alcohol and treated with acclimated activated sludge. It was reported that as much as 100 ppm allyl alcohol in water will not upset a well acclimated activated sludge (62).

AMYL ALCOHOL PLANT WASTES

Deep Well Disposal

A process developed by D.L. Miller (63) is a process for the disposal of a liquid waste, which comprises mixing said waste with water derived from an underground water-yielding stratum, the proportion of water to waste being sufficiently high to permit any chemical reaction induced by the presence of said water to go to substantial completion, thereafter clarifying the liquid mixture of water and waste, and flowing the clarified liquid mixture thus obtained into a water-yielding underground stratum in which the water has approximately the same chemical composition as the water initially mixed with said waste.

As described in the earlier section of this book on the deep well disposal technique, this process was specifically applied at Riverview, Michigan to the receipt of wastes from amyl chloride, amyl alcohol and amyl amine manufacture.

AROMATIC NITROGEN COMPOUNDS

Incineration

In the event that it becomes necessary to dispose of a significant quantity of organic waste and purification/recyling is impractical, incineration of the materials is the recommended method of disposal. The material must be incinerated under controlled conditions whereby oxides of nitrogen are removed from the effluent gas by scrubber, catalytic and/or thermal devices (62).

BENZENESULFONIC ACID

Chemical Degradation

In those cases where the total quantity of benzenesulfonic acid cannot be handled by the local wastewater treatment plant, benzenesulfonic acid can easily be degraded biologically and chemically on site by conventional wastewater treatment methods. Depending on whether or not the treatment is partial or complete, the wastewater can be discharged either into the sewer or the storm drain (62).

Incineration

Badly contaminated bulk material that cannot be reclaimed, can be disposed of by incineration. This method of disposal will require the use of an SO_2 scrubber on the incineration unit.

Landfill

Disposal of benzenesulfonic acid can be accomplished by mixing it with a slurry of lime to form unsoluble calcium benzenesulfonate. The resulting solid has a small but significant solubility, making it readily mobile in underground water. Thus, landfill disposal represents a long-term threat to underground water supplies and must be considered much less acceptable than incineration. Landfill disposal is acceptable only when landfill sites are totally isolated from ground and surface water and thus meet the requirements of California Class 1 type landfills (62).

BENZOYL PEROXIDE

Chemical Degradation

Unusable benzoyl peroxide is best stored wet until disposal. The slurry of waste material can be destroyed safely and easily by adding it slowly with stirring to ten times its weight of 10% sodium hydroxide solution. The reaction is only mildly exothermic. The final solution containing sodium benzoate and benzoic acid, both biodegradable, may be flushed into the drain. Disposal of large quantities of solution may require pH adjustment before release into the sewer (64).

Incineration

An equal weight of noncombustible material such as vermiculite or perlite can be added to a slurry of benzoyl peroxide. The excess water is then filtered off and the wet paste is carefully burned in a slurry bed, auxiliary fired incinerator. This method is suitable only if quantities of material are small and the incinerator is equipped with particulate removal equipment to insure against noncombustibles being discharged to the atmosphere. All cartons and bags should be examined to ascertain that they contain no residual benzoyl peroxide. Residual material, if present, should be flushed out with water. The empty cartons should be crushed and placed in a special collection drum. Keeping the contents of the drum moist is a good safety precaution. Periodically, the contents of the drum should be carefully incinerated.

BENZYL CHLORIDE

Concentrated benzyl chloride should not be washed down the drain or burned in open pits. In the event it becomes necessary to dispose of a significant quantity of concentrated benzyl chloride, two disposal options are available (62).

Incineration

The first option is to incinerate the material. It is expected that a liquid combustor followed by secondary combustion and aqueous or caustic scrubbing would be an acceptable disposal method. Primary combustion should be carried out at a minimum of 1500°F for at least 0.5 second with secondary combustion at a minimum temperature of 2200°F for at least 1.0 second. The abatement problem may be simplified by insuring against elemental chlorine formation through injection of steam or methane into the combustion process. This is an adequate means of disposal.

Landfill

The second option is adding solid sodium carbonate or solid lime to the concentrated benzyl chloride and then burying the material in an approved California Class 1 type landfill.

BUTANETRIOL TRINITRATE (BTTN)

Open Burning

BTTN collected from spills and catch tanks, or considered unsuitable for use, should be disposed of by careful burning after absorption in sawdust, wood pulp or fuller's earth. If BTTN is spilled on the ground, the contaminated ground should be removed with low impact tools and burned. Ignition of BTTN is usually accomplished by igniting a black powder squib placed on the surface. All burning is performed in a remote area (64). Although the products of combustion contain considerable NO_x, processes to control pollution from the products of combustion of BTTN and similar compounds are not in wide use. Investigations are being conducted to develop better methods for the disposal of BTTN and similar compounds, including nitroglycerin, than the open-burning techniques currently used. These methods are based upon the use of scrubber-equipped incinerator systems for the controlled burning of explosive wastes, but this method is not available for wide use at this time. Additional research is required.

CARBON BLACK

Incineration

A process developed by M. Reichert et al (65) is a process for the oxidation of substances which are suspended or dissolved in a liquid resistant to oxidation. The problem of oxidizing waste substances presents itself, for example, in the removal of by-products in acetylene production. In the thermal cracking of hydrocarbons, for example in the production of acetylene by partial oxidation of hydrocarbons, by cracking hydrocarbons in an electric arc or with the aid of solid or gaseous heat carriers, carbon black is formed which becomes suspended in the quenching water injected into the hot gases, in wastewater from direct gas coolers and in the wash water from gas filters. The bulk of the carbon black is eliminated in conventional manner by allowing it to float to the surface and separating it in the form of a slurry which still contains a large percentage of water.

Since such carbon-black slurries cannot be discharged into drains or allowed to soak away into the ground, attempts have been made to remove the carbon black by means of a rotary filter and to burn it in the grate of a steam boiler. Difficulties arise especially in the oxidation of the separated moist substances, because the particulate solids are usually present as lumps with relatively large surface area and, during combustion of the same, uncontrollable high temperature peaks occur which strongly attack the combustion equipment, for example the grate of a boiler furnace. If the heat evolved is to be utilized, for example for steam generation, the uncontrollable heat of combustion which is occasionally evolved in the overheated regions makes regular operation impossible, especially when varying amounts of substance are to be oxidized.

In operation of the process it has been found that carbon black in aqueous suspension, which advantageously contains such an amount of water that the mixture is still pumpable, for example 80 to 99% by weight of water, can be burnt completely and without overheating. When suitable conveying and metering equipment, for example a screw conveyor, is employed, it is also possible to introduce a carbon-black slurry containing less than 80% by weight of water into the first zone of the combustion chamber and burn it without overheating. The apparatus used in this process is shown in Figure 103.

Into a vertical cylindrical chamber 1, which is subdivided into two zones 2 and 3 by an orifice plate 11 and provided with a superimposed smaller cylindrical antechamber 7, gaseous, liquid or solid fuels are supplied tangentially at 9 and gaseous oxidizing agents are introduced tangentially at 10 in such an amount that the fuel is burned completely and an excess of oxygen is present in the rotating hot combustion gas for oxidation of the substances to be oxidized. A suspension or solution of the substances to be oxidized is injected through an atomizer pipe 8 in the direction of the axis of the cylinder.

The atomized particles are entrained by the rotating hot combustion gas, and the liquid contained in the particles is substantially evaporated. Deposit formation on the wall of the chamber is thus prevented. Partial oxidation takes place already during evaporation of the liquid. In the first zone 2, the partly oxidized substance contained in the suspension is thrown into the vicinity of the wall by centrifugal force and entrained downwardly by the hot gas in a spiral path to the orifice plate 11.

FIGURE 103: INCINERATOR FOR CARBON SLURRIES IN WATER

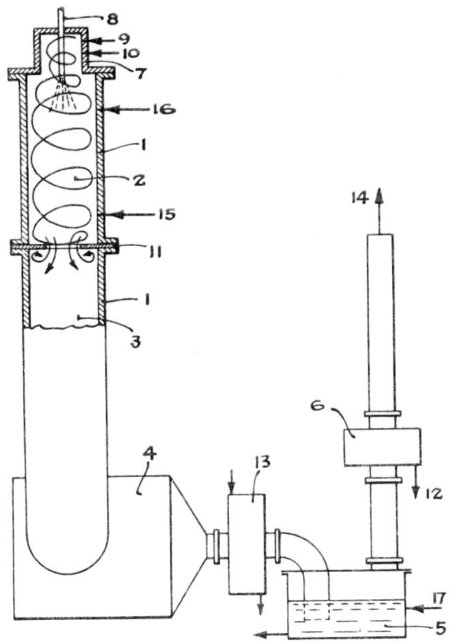

Source: U.S. Patent 3,145,076

When the tangential speed of the oxidizing agent is sufficiently high, combustion of the fuel takes place mainly in a ring of flame whose diameter is approximately the same as the internal diameter of the orifice plate 11. Excess oxidizing agent added in one or more separate planes, for example 15 or 16, rotates however mainly in the vicinity of the cylindrical wall and cools the same. At the inner edge of the orifice plate, the oxidizing agent streaming toward the center entrains the particulate solid into the center of the chamber and mixes with the same, mainly at the edge of the orifice plate. The particulate solid reacts with the oxidizing agent in the second zone 3, whose internal diameter is the same or, advantageously, smaller than that of the first zone 2.

In the second zone, a high temperature occurs, which is uniformly distributed over the cross section of the chamber. The chamber 1 discharges tangentially to a cylindrical afterburner 4 in which any incomplete oxidation is taken to completion. The suspension 17 is concentrated in an immersion vessel 5. The recovered liquid is withdrawn at 12 from a condenser 6. The condenser may be cooled with media which are used in the process. A heat exchanger 13 is provided for preheating the combustion of air and/or the fuel. The cold off-gases are withdrawn at 14. The chamber 1 may also be arranged horizontally or in any other position instead of vertically. The vertical position is, however, especially advantageous. Operation of the process will be further illustrated by the following example.

Example: Approximately 170 m³ (STP)/hr of coke-oven gas was introduced at 9 and approximately 1,300 m³ (STP)/hr of air at 10 into a cylindrical chamber 1 whose first zone 2 had a diameter of 680 mm and a length of 1,800 m. 372 kg/hr of wet carbon black with a water content of 84.7% was injected axially through an atomizer pipe 8 by

means of 150 m^3 (STP)/hr of air. Halfway up the combustion chamber, a temperature
of 630°C was measured in the vicinity of the wall. The carbon black rotating in the
vicinity of the wall was mixed with unconsumed air at the edge of an orifice plate **11**
of 350 mm inside diameter, and burnt completely in the second zone **3** of the chamber **1**, which had a length of 2,100 mm and a diameter of 550 mm. A temperature
of 960°C was measured at the lower end of this zone. The off-gas which had cooled
to 610°C and no longer contained any carbon black was passed to an immersion vessel
5, into which 945 kg of wet carbon black with a water content of 94% (=57 kg dry
carbon black + 888 kg water) was introduced per hour.

573 kg/hr of water was evaporated in the immersion vessel **5**. This is 64.5% of the water
contained in the wet carbon-black feed. The off-gas and the steam left the immersion vessel at **14** at a temperature of 82°C. 372 kg/hr of concentrated carbon-black suspension
with a water content of 84.7%, which also had a temperature of 82°C was withdrawn
from the immersion vessel and injected into the cylindrical chamber **1** as described above.
510 kg/hr of an aqueous slurry containing 14.1% of dry carbon black and 2.7% of polymer of higher acetylenes was burned in a similar manner.

Another process developed by M. Reichert et al (66) involves processing carbon black
(soot) suspensions obtained in the manufacture of petrochemical raw materials, such as
acetylene, ethylene and the like, from natural gas or petroleum or fractions of the same
by partial combustion of these starting materials followed by quenching of the reaction
gas formed with water. The suspensions of soot in water are creamed and separated from
the bulk of the water, the separated sludge is homogenized by stirring and again creamed,
and the concentrate thus obtained is withdrawn and filtered and then dried and ground
to fine-grained soot which is burnt after removal of the vapor. Figure 104 illustrates the
conduct of this process.

**FIGURE 104:   ALTERNATIVE INCINERATOR FOR DISPOSING OF CARBON (SOOT)
SLURRIES IN WATER**

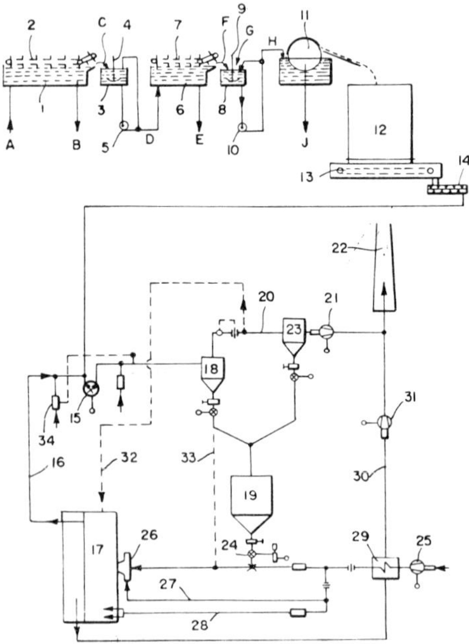

Source: U.S. Patent 3,401,654

The soot suspension A is first passed through an open channel 1 and the creamed soot concentrate, which has a foamy consistency, is removed with scrapers 2, the residence time of the suspension in the channel being dependent on the rate of supply and being such that adequate clarification of the water B is achieved, the water, after the concentrate (hereinafter called foam) has been separated, being used again for washing out soot from reaction gas. This water normally contains 20 to 40 grams of soot per cubic meter. The foam C is collected in a vessel 3. The soot content of this foam is up to 4 to 5% by weight with proper adjustment of the scrapers. It is not possible to achieve a higher concentration than this owing to the structure of the foam. Apparently the soot flakes form a framework which encloses 20 to 25 times their weight of water.

Surprisingly a further increase in the soot concentration may however be achieved by stirring the foam, so that it liquifies, and allowing it to cream again. Stirring of the foam may be effected by drawing it off into a stirred vessel designed according to the amount in question. The mean residence time of the foam in the vessel is from about 5 to 30 minutes. It is advantageous however to adopt the procedure illustrated in the drawing. The foam is first stirred for only a short time in the vessel 3. The foam during this short stirring treatment assumes a pumpable consistency and is supplied to a circulating pump 5 where it is stirred in consequence of the shear forces acting thereon. Some of the foam, for example at least 20 times the amount supplied to the stirred vessel 3, is pumped back into the stirred vessel after it has left the circulating pump and the remainder of the liquefied foam is supplied to another channel 6.

It has been found that liquefied foam treated in this way creams again, the creaming rate having been found to be between 0.2 and 1.0 mm per second. This rate is much below the creaming rate of the soot suspension A but this does not result in any difficulty because the amount of foam C to be processed is much less than the amount of soot suspension A. The channel 6 is entirely identical in construction to the channel 1 except that it may differ in size. This may be calculated from the amount of the supply C and the creaming rate. In general the second channel can be smaller than the first.

The clarified water E withdrawn from the channel 6 in general contains less than 40 grams of soot per cubic meter and may be returned to the process for the production of the petrochemical raw material. The enriched soot concentrate F is moved into container 8 by means of scrapers 7. Concentrations of 8 to 9% by weight are obtained in this way. The container 8 is provided with blade stirrer 9 and sludge pump 10, for example a two-rotor gear pump. It is also possible to meter additives G, particularly flocculants, into the container 8. The effect of the flocculant is described below. The concentrate H is moved direct into the trough of a rotary vacuum filter 11 by the sludge pump 10.

By means of the rotary vacuum filter 11 (which is equipped with a pressure band unit and scrapers for removing the filter cake), the concentrate is separated into a clear filtrate J having a maximum of 50 grams of soot per cubic meter of water and a filter cake K having 72 to 78% by weight of water. The capacity of the filter 11 is about twice as high as when using soot sludge D produced by a single creaming, when the above method is adopted.

Flocculating agents are generally polymers having a molecular weight of for example 50,000 to 80,000. Examples are starch products, such as the product known under the trade name Polyfloc, polyethylene oxides, such as the product known under the trade name Polyox, and proteins. Filtration may also be influenced by the temperature of the concentrate H supplied. At temperatures below 50°C, the filtering efficiency falls off sharply and the water content of the filter cake rises to more than 79% by weight so that further processing in a combined grinding and drying unit is rendered impossible because such a material is no longer compact and the conveyor and mill entrance become clogged. For practical reasons, the temperature of the concentrate supplied to the filter should not exceed 90°C.

The filter cake is supplied to a mill 15 through a bunker 12 by a feed scraper 13 and a double screw **14**. Drying of the cake and its comminution into fine-grained soot takes place in the mill. Hot gases, whose temperature is for example 750° to 850°C from the process itself may in general be used for drying; it has been found to be convenient however to use for the purpose hot gas from the boiler firing plant in which the soot is burnt; this may be sucked through the gas channel **16** from the steam generator **17** by the mill **15**. From the mill, the vapor and the comminuted soot pass first into a cyclone separator **18** from which the soot is drawn off into a bunker **19** while the vapor passes through vapor line **20** and is supplied by means of vapor blower **21** into the stack **22**. It is advantageous to provide a fabric filter **23** in the vapor line **20**, in addition to the cyclone separator **18**, for complete separation of the soot.

Soot deposited in the fabric filter **23** is combined with the soot from the cyclone separator **18** in the bunker **19** whence the appropriate amount of soot ready for burning is moved into the burner **26** of the steam generator **17** by means of a distributor **24** in correlation with the fresh air blower **25**. Pipes for secondary air **27** and tertiary air **28** are provided in the firing unit in the conventional manner. The fresh air is heated up in an air preheater **29** which is located in the gas line **30** in front of the suction line **31**. A line **32**, between the cyclone separator **18** and the fabric filter, passes from the vapor line **20** direct to the boiler firing plant into which vapor is supplied to the tertiary air jets. In this way an appropriate variant is provided for the fire of the steam generator **17**.

Another variant consists in providing a line **33** for direct injection of comminuted soot ready for burning while bypassing the bunker **19**. This provides for the case where the soot is to be supplied direct to the fire without intermediate bunkering. The temperature of the mill **15** is regulated by a control member **34**, cold air being introduced into the gas channel **16** behind the mill in dependence on the temperature. The process ensures a satisfactory destruction of soot occurring in various processes as a very dilute suspension in water, while utilizing its calorific value, the soot being first converted into compact material and dried and burnt as such.

CARBON DISULFIDE

Incineration

Unusable carbon disulfide solvents, as well as residues from manufacture, can be disposed of by controlled incineration. A sulfur dioxide scrubber is necessary when combusting significant quantities of carbon disulfide (62).

CARBOXYLIC ACIDS AND THEIR SALTS

Incineration

Bulk quantities of contaminated carboxylic acids that cannot be reprocessed or released by controlled dilution can best be disposed of by incineration. Since this will probably be an unusual occurrence, this is best done at industrial or municipal incineration sites (62).

A process developed by M. Akune et al (67) is one in which liquid waste containing a mixture of organic and inorganic substances or organic metallic compounds is, in turn, dehydrated, concentrated and combusted and the ash produced therefrom is recovered as an aqueous solution or a slurry. The process is shown schematically in Figure 105.

Referring now to the figure, a liquid waste storage tank **1** is in fluid communication via pump **2** and pipe a with a jacket **4** provided in the upper portion of a concentrator or evaporator **3**. Provided in the interior of the evaporator is a downcomer **5** to which the waste is introduced. The overflow of the concentrated solution from the evaporator **3** is delivered into a concentrated solution storage tank **6** by way of pipe b. A pump **7** conveys the concentrated solution from storage tank **6** via pipe c to a spray nozzle **9** provided on combustion furnace **8**.

FIGURE 105: INCINERATOR SYSTEM APPLICABLE TO SODIUM CARBOXYLATE DISPOSAL

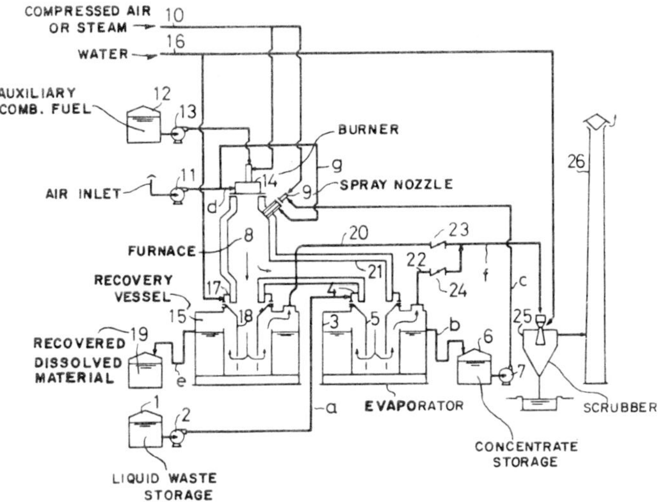

Source: U.S. Patent 3,912,577

Compressed air or steam under a medium pressure is introduced into the nozzle **9** via conduit **10** to accelerate the combustion of organic substances in the waste and the evaporation of the water contained therein. A blower **11** is provided so as to supply to combustion furnace **8** the air required for combustion of the waste concentrate, said blower being connected by way of a pipe **d** with a burner **14** and by way of a pipe **g** to the spray nozzle **9**. The auxiliary combustion fuel is contained in a storage tank **12**, and via pump **13** is delivered to burner **14** mounted on the furnace **8**. Beneath the furnace is a recovery vessel **15** to receive the combustion exhaust gases produced. A downcomer **18** is disposed inside the recovery vessel, which is provided also with a header **17** through which water introduced via pipe **16** is supplied to the recovery vessel **15**.

The recovery vessel **15** is connected to a recovered waste storage tank **19** through a pipe **e**, and the combustion gases are exhausted through a duct **20** located in the upper portion of the recovery vessel. Approximately one-half of the combustion gases discharged from the furnace **8** is delivered via duct **21** to downcomer **5** to heat the waste delivered by jacket **4**. Connected with the evaporator **3** is a duct **22** to discharge the combustion gases from the evaporator. Valves **23** and **24**, provided in ducts **20** and **22**, which ducts join in **f**, serve to adjust the volume of the combustion gases being delivered to the evaporator. Scrubber **25** is in communication with an exhaust tower **26** and with a water supply via pipe **16**.

In operation, the liquid waste contained in storage tank **1** is delivered by pump **2** to the downcomer **5** via jacket **4** provided on the upper portion of the evaporator **3** and then concentrated therein, and delivered to the waste concentrate storage tank **6**. The concentrate from the storage tank **6** is delivered by pump **7** to the spray nozzle **9** mounted on the furnace **8** and then spray-injected into the furnace for burning. It is preferable to use steam or compressed air as the atomizing source necessary for spraying. The compressed air is led through duct **10** to accelerate the combustion rate of the organic substances contained in the waste as well as the evaporation rate of the water in the concentrate. The air necessary to support the combustion of the organic substances is supplied by the blower **11** and the auxiliary combustion fuel from storage tank **12** is supplied by pump **13** to the burner **14** of the furnace.

In the event that the waste has self-sustaining combustibility, the auxiliary fuel is not required, but in general, aqueous solutions such as the waste of carboxylic acid solution do not provide self-sustaining combustibility, thus requiring the auxiliary fuel.

It is possible to make the furnace size compact and maintain a suitable flame by using a high-load short-flame burner. The temperature of the flame from the auxiliary flame burner is as high as 1600° to 1800°C, so that it may effect a rapid evaporation of the water as well as a complete combustion of the organic substances contained in the waste.

It is commonly accepted that the organic materials, in general, can be burned completely at a temperature of over 800°C. On combustion of the waste, the metallic compounds contained in the waste may be obtained as ash and if sodium carbonate is present, this will be released as a portion of the ash. The melting point of the sodium carbonate being about 850°C, the molten sodium carbonate tends to adhere to the inner surface of the furnace when combustion is carried out at a temperature below 850°C, thus causing clinkers.

In this case, therefore, the combustion temperature should be set at about 950°C. In case of inorganic materials such as sodium chloride, the melting point of the ash will be considerably lower, and thus no damage to the furnace will occur, when combusting at 850°C. One-half of the combustion gases thus produced is injected into the recovery vessel 15 and the remaining half is injected into the evaporator 3. Water is then supplied to the recovery vessel from the downcomer 18 via water pipe 16 and header 17.

Most of the sodium carbonate produced during combustion flows in the molten state along the inside of the furnace walls into the recovery vessel 15 and goes then into solution. Part of the sodium carbonate, in the form of particles, is entrained with the combustion gases into the recovery vessel directly. In this manner, 90% or more of the sodium carbonate may be absorbed into solution in the recovery vessel. The combustion gases are injected into the recovery vessel at about 950°C and then discharged therefrom through duct 20 at a temperature of 90° to 100°C.

As an illustrative example, a system was employed with 11,000 lb of waste/hr. The concentration of the solute was 20% and the major components to be treated were sodium carboxylate and inorganic materials, the lower calorific value based on anhydride being 5,035 Btu/lb. In this case, the rate of the steam produced from the recovery vessel was about 11,000 lb/hr. During combustion of the concentrated waste in the furnace, 2,640 lb/hr of water corresponding to the entire water contained in the concentrate was evaporated and the steam to atomize the waste amounted to 1,760 lb/hr so that the steam contained in the exhaust gases discharged from the recovery vessel was 15,400 lb/hr. The heat contained in this steam was utilized in evaporating the water contained in the waste, and 6,160 lb/hr of steam were yielded in the evaporator for use in concentrating.

CHLORINATED HYDROCARBONS

See Halogenated Hydrocarbons, page 348 and Polychlorobiphenyls, page 377.

CHLORINATION OFF-GASES

Incineration

A device developed by F.L. Gladney et al (68) is a burner unit which is particularly suitable for use with a furnace for burning carbonaceous waste gases. The air and fuel gas mixture for igniting the burner is mixed at the front end of the burner, to prevent flashback into the burner body. Mounted on the front end of the burner is a closure disc, which is connected into an air operator unit. The air operator unit moves the closure disc backward or forward in reponse to air pressure fluctuation in the mixing chamber. Moving the disc back and forth regulates the velocity of air which passes through an annular opening defined between the disc and the front end of the burner. Regulating the velocity of the air enables precise control of the firing rate of the burner. Various chemical products, such as chlorinated solvents, are obtained by the chlorination of methane or higher hydrocarbons. One of the waste products left over from the reaction process is a mixture of carbonaceous gases. Since it is undesirable to dishcarge the chlorinated gas waste into the atmosphere, it is usually disposed of by burning in a furnace with a sealed combustion chamber, preferably employing the present burner unit design.

CHLOROPICRIN

In the event it becomes necessary to dispose of a significant quantity of concentrated chloropicrin, three disposal options are available. The first option is to contact the manufacturer and determine if it is possible to return the materials. Niklor Chemical Company and Sobin Chemical Incorporated have indicated that recycling of chloropicrin is possible (64). The second option is incineration; the third is lagooning.

Incineration

Since chloropicrin is nonflammable, it must be added to fuel or sprayed into an incinerator at high temperatures. It is expected that a liquid combustor followed by secondary combustion and aqueous or caustic scrubbing would be an acceptable disposal method. Primary combustion should be carried out at a minimum of 1500°F for at least 0.5 second with secondary combustion at a minimum temperature of 2200°F for at least 1.0 second. The abatement problem may be simplified by insuring against elemental chlorine formation through injection of steam or the utilization of methane as a fuel in the combustion process. Incineration provides an adequate way to handle the concentrated waste but it is strongly recommended that recycling be used whenever possible.

Lagooning and Landfill Disposal

Dilute aqueous waste containing chloropicrin is lagooned and allowed to gas off or is incinerated with scrubbers to remove the hydrogen chloride from the effluent gas. It is recommended that the material to be lagooned be reacted with sodium sulfite or sodium bisulfite to suppress the noxious properties of the chloropicrin and any resulting sludge be incinerated or buried in an approved California Class I type landfill. Both lagooning and incineration are adequate means of disposing of chloropicrin provided the concentrations released to the environment do not exceed the provisional limits.

Dilute organic waste generated during the manufacture of chloropicrin are disposed of either by land burial or incineration. Incineration with proper treatment of the effluent gas to remove hydrogen chloride and nitrogen oxides is the primary option for disposing of dilute organic waste. An alternate method of disposal would be in an approved chemical California Class I type landfill (64).

CYCLOHEXANE OXIDATION WASTE LIQUORS

See Oxidation Process Wastes, page 371.

DIMETHYL SULFATE

Incineration

The Manufacturing Chemists Association (MCA) recommends that dimethyl sulfate be disposed of by incineration in properly designed and operated chemical waste incinerators. The MCA also recommends that the waste be diluted and neutralized before disposal, whenever conditions permit (58).

Incineration of dilute (1%), neutralized dimethyl sulfate waste is recommended as the best method for disposal of the material. The incineration must be performed by trained and experienced personnel using equipment properly designed to handle hazardous materials and equipped with efficient oxides-of-sulfur scrubbing devices. These incinerators should expose the waste material to a minimum temperature of 1800°F for at least 1.5 seconds. These conditions may be easily met through the use of afterburners. Disposal of concentrated dimethyl sulfate wastes by direct incineration is judged to be less acceptable for all but very small (laboratory) quantities of material. The danger of exposure to vaporized, but unburned dimethyl sulfate is significantly greater when the material is disposed of in concentrated form and, therefore, requires the employment of specially trained personnel operating properly designed and maintained equipment.

The use of biological wastewater treatments methods for the disposal of dimethyl sulfate is acceptable only for very dilute, neutralized waste streams. These procedures are less satisfactory than proper incineration because of the danger of exposure to incompletely hydrolyzed dimethyl sulfate.

DINITROBENZENE

Incineration

In the event it becomes necessary to dispose of a significant quantity of concentrated dinitrobenzene, incineration appears to be the only acceptable technique currently available. Qualified personnel familiar with handling toxic and explosive materials must be available to operate the facility. The dinitrobenzene must be diluted with other combustible materials which are not explosive and incinerated under controlled conditions where oxides of nitrogen are removed from the effluent gas by scrubbers, and/or catalytic or thermal devices (64). Combustion should be carried out at a minimum temperature of 1800°F over a minimum residence time of 2.0 seconds.

Biological Treatment

Dinitrobenzene appears as aqueous waste in its manufacturing process. Other sources of waste are from water used in the cleaning of equipment used in dinitrobenzene service. It is assumed that methods used to dispose of aqueous dinitrobenzene waste are similar to methods for dinitrophenol, i.e., secondary treatment utilizing activated sludges. The adequacy of this practice is in doubt due to the apparent difficulty of microorganisms to degrade aromatic nitro compounds. Until data are avilable to show that dinitrobenzenes can be degraded satisfactorily in secondary treatment facilities, it is recommended that aqueous waste streams be concentrated and treated as in the care of the concentrated dinitrobenzenes—by incineration.

DINITROPHENOL

Recycle

The first option is to contact the manufacturer and determine if it is possible to return the material. Southern Dyestuff Company has indicated a willingness to accept concentrated dinitrophenol for reprocessing (64). The option of recycling the material is adequate and should be used whenever possible.

Incineration

Incineration is the second choice option provided controlled combustion processes where the oxides of nitrogen are scrubbed from the effluent gas are used or where a thermal device is used to reduce the oxides of nitrogen to their elemental form. The material should be combusted at a minimum temperature of 1800°F for at least 2.0 seconds.

Landfill

Sanitary landfills are generally not recommended for dinitrophenol (64) because of the potential of ground and surface water pollution as well as possible occupational hazards resulting from on-site handling. However, California Class I type landfills are adequate when the material is contained and has low vapor pressure.

DIPHENYLAMINE

Incineration

In the event it becomes necessary to dispose of a significant quantity of concentrated diphenylamine wastes, and purification/recycling is impractical, then incineration of the diphenylamine is the recommended method of disposal. The material must be incinerated under controlled conditions where oxides of nitrogen are removed from the effluent gas by scrubbers and/or thermal or catalytic devices (64).

Landfill

Diphenylamine appears as dilute waste in the tars which appear as bottoms products in the manufacturing processes. Currently, these tars are disposed of in a plant landfill. Disposal of dilute waste in plant landfills is a satisfactory means of disposal provided the site is acceptable from a geologic and groundwater hydrology standpoint and meets California Class 1 landfill requirements. Land burial of concentrated diphenylamine wastes is a satisfactory means of disposal provided the site is acceptable from a geologic and groundwater hydrology standpoint and meets California Class 1 type landfill standards.

ESTERS

Incineration

Bulk quantities of contaminated esters that cannot be reprocessed or released by controlled dilution can best be disposed of by incineration. Since this will probably be an unusual occurrence, this is best done at municipal or industrial incineration sites (62).

ETHERS

Prime factors to consider in the disposal of ethers are: possible presence of peroxides in liquid or on container elements; flash points; vapor density with respect to air; liquid flammability; explosive limits with air; and propensity to accumulate static electrical charges that may ignite vapors. Waste materials containing lower molecular weight ethers should not be emptied into drains or sewers, as sewer explosions are likely to result. The vapors will travel long distances before dissipation, and a steam or process line may be hot enough to effect ignition (64).

Where there is certainty that peroxides are not present, disposal may be accomplished by discharging liquid (not vapor) at a controlled rate near a pilot flame. An inert gas line connected to the vapor phase and a valve controlled line from the bottom of the container will permit a gravity feed of the liquid to a pilot flame at a safe distance. The inert gas avoids contact of the liquid with air. The rate of burning can be valve controlled. Lines and containers should be grounded.

Where peroxides are known to exist, or it is suspected that they might exist, it is suggested that no attempt be made to open the containers. (Peroxides tend to form in the threads of bungs and caps, and are easily ignited by the frictional heat generated when these closures are turned.) Handling of the containers should be minimized.

The containers should be destroyed from an upwind position of 100 feet or more away, using rifle fire so aimed that most of the ether will be forced to spill on the ground and evaporate. This should be done at a remote location, away from habitation, where no harm to the public or surroundings can occur. It should be remembered that there is always the possibility that fire may ensue or the container might explode. Alternatively, a blasting cap or small explosive charge may be used to perforate the container from a safe distance. Dilute organic waste containing ethers can be safely incinerated provided the burning temperature is sufficiently high to reduce the waste to CO_2 and water.

Since ethers generally resist biological degradation, dilute aqueous waste containing ethers above preliminary limits for water and soil can be disposed of by incineration. It is recommended that dilute aqueous waste be thermally oxidized by spraying the aqueous solution into an incinerator which is at a temperature of at least 1500°F.

HALOGENATED HYDROCARBONS

Chemical Degradation

The use of chemical reagents to decompose concentrated ethylene bromide, methyl bromide or methyl chloride wastes to less toxic forms has not been specifically reported in the literature. Based on the known reactions of these compounds, however, it appears that: (1) metal alkoxides could be used to react with methyl bromide or methyl chloride leading to the formation of ethers and metallic bromides or chlorides according to the Williamson ether synthesis; (2) sodium in liquid ammonia could be used to react with methyl bromide or methyl chloride leading to the formation of methane, methylamine, and sodium bromide or chloride; and (3) zinc could be added to ethylene bromide with the formation of ethylene and zinc bromide.

Since these methods generally do not lead to the recovery of valuable halogens and little is known of the optimum conditions under which the chemical reagents could be applied, chemical degradation could not be recommended as a technique for the disposal of the halohydrocarbons (62).

Incineration

Badly contaminated solvents that cannot be reclaimed, as well as chlorinated and non-chlorinated residues and sludges, can be disposed of by incineration. Some of the materials, especially those with high chlorine content may not support combustion by themselves, but can be burned if an auxiliary fuel is used. Care must be exercised to assure complete combustion. An acid scrubber is necessary to remove the halo acids produced. In this manner, both long- and short-term environment effects are eliminated at the sources in one simple disposal step.

The complete and controlled high temperature oxidation of halohydrocarbons in air or oxygen with adequate scrubbing and ash disposal facilities offers the greatest immediate potential for the safe disposal of these compounds with the possibility of recovering the greater part of all of the halogens in some usable form. In particular Dow Chemical has designed (62) incinerators to dispose of the concentrated ethylene bromide and methyl bromide wastes at its Midland plant. Auxiliary fuel will be required in the combustion of ethylene bromide.

The off-gas from the incinerator will contain both hydrogen bromide and bromine, both of which are harmful pollutants if released to the atmosphere. The hydrogen bromide can be readily removed by scrubbing with caustic soda, but Dow claims to encounter some problems with bromine removal. However, it is not clear whether Dow has fully investigated the several methods available for the separation of bromine vapors from air: (1) contacting the air-bromine mixture with moist scrap iron leading to the formation of ferrous bromide; (2) absorption in ammonia solution with the formation of ammonium bromide; (3) absorption in ferrous bromide solution with the formation of ferric bromide; (4) alkaline absorption in which bromine reacts with sodium hydroxide or carbonate to form sodium bromate, sodium bromide, and either water or carbon dioxide; (5) lime absorption in which bromine reacts with calcium hydroxide to form calcium bromide and bromate; (6) reduction of bromine by means of sulfur dioxide giving rise to the formation of a spray of fine droplets of hydrobromic and sulfuric acids that could later be trapped by a solution of the same mixed acid circulating in an absorption tower; and (7) absorption with a concentrated solution of sodium bromide cooled below $0^{\circ}C$ in packed towers.

The absorbed bromine that is bound chemically in the form of bromide, bromate or hydrogen bromide can all be recovered by either acidification with sulfuric acid or reoxidizing with chlorine. For small quantities of methyl bromide and ethylene bromide, both compounds could be safely disposed of by dissolution in a flammable solvent and spraying the mixture into the fire box of any incinerator equipped with an afterburner and alkali scrubber. In the case of methyl chloride, combustion units designed for the disposal of chlorinated organic wastes and capable of recovering chlorine in the form of usable hydrogen chloride have been developed, and a 7,000 lb/hr plant has been built for E.I. DuPont de Nemours and Company in Victoria, Texas by Union Carbide Corporation. Properly designed and operated incineration is therefore considered as the best current and near future method for the disposal of concentrated ethylene bromide, methyl bromide, and methyl chloride wastes.

A device developed by J.A. Cull et al (69) is a special furnace capable of converting halogenated organic residue materials to recover hydrogen halide therefrom as a saleable product. Halogenated residues from chlorination and bromination processes, such as residues from the chlorination processes manufacturing hexachlorocyclopentadiene, benzoyl chloride, chlorendic acid, etc. have in the past presented a difficult disposal problem. They are toxic to plant and animal matter, so that they should not be sewered into rivers or lakes, nor should they be dumped on land where the drainage therefrom would reach waters being used for drinking purposes. Also, because they have a large amount of halogenated material in them, normal furnace disposal means have been unsuitable because the materials either would not burn easily or gave free halogen in the exit gases which corroded the equipment and contaminated the air. Further, the halogen content of the residues was lost.

Figure 106 shows a plan and elevation view of such a furnace. The furnace has an outer cylindrical firewall **1** vertically positioned on a foundation **2** at the bottom and covered at the top with refractory brick **3** and an insulated manhole cover **4** which also serves as an explosion relief means. The outer firewall is constructed on the inside of refractory material **5** which has a low heat conductivity and which is resistant to free halogen, oxygen and hydrogen halide, such as mullite, a composition of aluminum silicate. This is backed up by firebrick **6**. An air air gap **7** separates the refractory materials from a gastight steel shell **8**.

The furnace also has an inner cylindrical firewall **9** concentrically aligned with the outer firewall **1** so that an annular space **10** is formed between the inner and outer walls. The inner firewall is also positioned on the foundation **2** and is in communication at the bottom with an insulated outlet means **11**. The inner wall is in open communication with the top inner space **12** of the outer firewall **1**. The inner wall should be constructed of heat conductive refractory material such as silicon carbide brick, so that there can be an exchange of heat from the inside **13** of the inner wall cylinder through the refractory material to the annular space and thereby both preheat the incoming feed materials to their operation temperature and cool down the exiting reaction products. The height of the inner wall depends largely upon the amount of effective heat transfer across the inner wall.

The outer wall **1** has a residue feed inlet means **14**. This inlet is positioned in the outer wall preferably below the top of the inner wall **9** and as close to the bottom as possible without causing carbon build-up in the relatively cool bottom zone or a decrease in the effective mixing of the gases in space **10**.

FIGURE 106: HOOKER CHEMICAL CORP. INCINERATOR DESIGN FOR DISPOSAL
OF HALOCARBONS

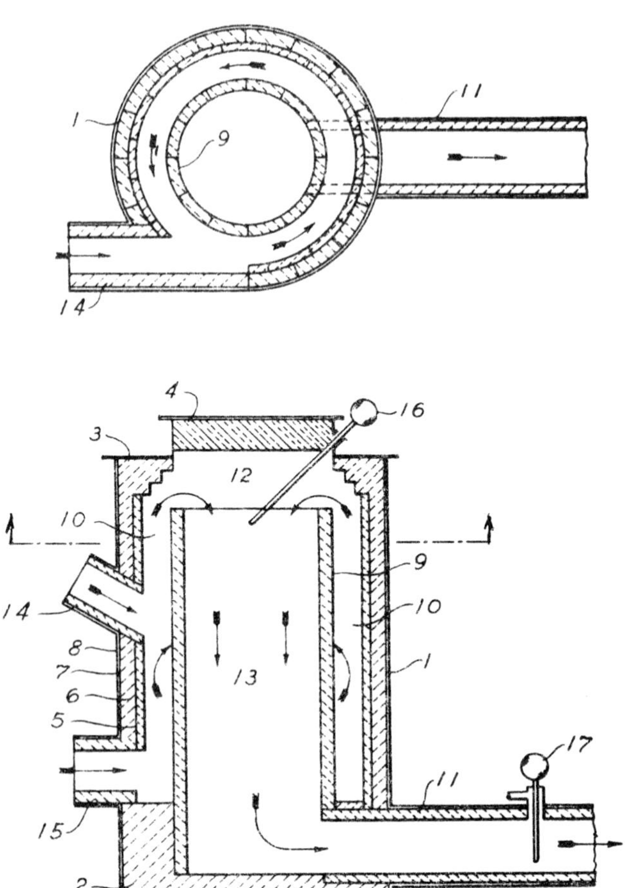

Source: U.S. Patent 3,140,155

The angle of the inlet **14** is preferably downward and tangential to the cylindrical outer
wall **1**. The incoming residue materials injected in this direction aid in creating a greater
amount of turbulence in the annular space **10**, which in turn increases the overall heat
transfer rate across the inner wall **9**. The residue inlet can also be used to introduce air
or other materials such as steam or hydrogen.

The outer wall **1** also has a second inlet **15** positioned to have materials enter near the
bottom of the annular space **10**. This second inlet means is preferably used to start up
the furnace by injecting air and hydrogen or fuel gas therein, and afterward to control or
regulate the composition of the final product by injecting steam or air or a hydrogen
source, etc., as required, as more fully described below. This inlet can also be used to bring
the furnace back up to the required operating temperature if it drops below that point.
Residue materials have also been fed into the furnace through this inlet and converted suc-

cessfully, but in general, the operation of the furnace has not been as smooth. This inlet **15** is preferably positioned in the outer wall **1** to be tangential to the annular space in the same direction as inlet **14** which aids in creating a greater amount of turbulence in the annular space and thereby increases the efficiency of preheating the reactant materials and overall heat transfer across the inner wall **9**. Although the inlet **15** is depicted as being directly below the first inlet means **14**, this need not be done. Successful conversions have been accomplished by feeding everything in inlet **14**, or everything in inlet **15**, or combinations of feeds in inlets **14** and **15**.

As shown the furnace preferably has a reaction temperature indicating means **16** positioned in the top inner space **12** of the outer wall **1**. The furnace preferably also has an exit temperature indicating means **17** in the outlet **11**. Suitable temperature indicating means are thermocouples or pyrometers.

In general, temperature variations within the furnace seemed to have little or no effect upon the overall reaction in the range of 900° to 1300°C. When burning perchlorinated materials, the furnace should be preheated to 900°C; however, lower operating temperatures have been found satisfactory when feeding more flammable materials. The upper temperature limit is dependent mainly on the materials of construction of the furnace and the temperature measuring devices, if employed.

Under normal conditions the temperature of the furnace at the top of the inner space **12** was about 300°C higher than the temperature of the exit gases. However, if feed rates are increased above the capacity of the furnace, more time will be required to bring the reactants up to the operating temperature. Therefore, the temperature of the inner space will decrease, and the exit temperature will increase, causing the furnace to be in a state of thermal unbalance.

Maximum furnace capacity of the furnace was found to be when the exit gas temperature **17** was equal to the furnace temperature at space **12**. However, this is also dependent on the size of the annular space **10**, the heat transfer efficiency across the inner wall **9**, the height of the inner wall and other factors. The following are specific examples of the operation of such a furnace.

Example 1: A furnace having an inside volume of approximately 60 cubic feet was preheated with a hydrogen-air flame at inlet **15** to a temperature of about 900°C. Residues from a process manufacturing hexachlorocyclopentadiene were then metered into the first inlet means **14** at a rate of about 425 pounds per hour.

This feed comprised substantial quantities of the following materials: CCl_4, C_2Cl_4, C_2Cl_6, C_4Cl_6, C_5Cl_6, C_5Cl_8 and C_6Cl_6. The overall average feed composition of the materials was 20% carbon and 80% chlorine. In addition to the residue fed, there was also fed 100 pounds per hour of steam through the second inlet means **15**.

Air, at the rate of 660 pounds per hour was also introduced, some being fed in with the residue at inlet **14** to keep the inlet feed tube cool, and the balance being fed in at the bottom inlet **15**. The furnace temperature at the top **12** was 930°C. The exit gas comprised 350 pounds per hour hydrogen chloride, 311 pounds per hour carbon dioxide, 507 pounds per hour nitrogen, and the balance was excess steam and oxygen plus traces of carbon monoxide. The gases contained no detectable free chlorine (Cl_2), carbon, or organics.

Example 2: In a furnace and manner after Example 1, hexachlorocyclopentadiene residues were fed in at a rate of 425 pounds per hour. The steam feed rate was varied from 40 to 82 pounds per hour with an air rate of 650 pounds per hour. The temperature at space **12** in the furnace varied from 955°C to greater than 1300°C.

There was no detectable carbon or organics in the exit gas. When the steam was fed at the lower rate of 40 pounds per hour, some free chlorine appeared in the exit gases; however, at the higher steam feed rates, no free chlorine was detectable.

Another device developed by J.A. Cull et al (70) is very similar to that described above (69). A device developed by R.G. Woodland et al (71) is similar to those described by J.A. Cull et al (69)(70) but incorporates a special mixer-nozzle apparatus in the inlet. The mixer-nozzle apparatus includes means for introducing halogenated organic material and gaseous medium into the mixer-nozzle, means for mixing the gaseous medium and the halogenated organic material, a constricted zone after the mixer and an expanded zone after the constricted zone, so that atomization of the mixture is caused in the expanded zone, from whence the atomized mixture passes into the combustion zone of the furnace.

A process developed by S. Ezaki (72) is claimed to be an economical process for recovering hydrogen chloride from a spent organochlorine compound comprising carrying out a combustion treatment of spent liquor or waste gas containing organochlorine compound, the discharge of which in the untreated form is undesirable from the viewpoint of minimizing public nuisance. The process involves converting almost all the chlorine contained in the organic spent liquor or waste gas to hydrogen chloride, recovering the thus converted hydrogen chloride as hydrogen chloride gas or hydrochloric acid having a concentration of higher than the azeotropic composition of HCl-H_2O system, i.e., approximately 20% by weight, and at the same time, discharging other components into the atmosphere in the form of completely harmless and odorless gases.

Figure 107 shows a suitable form of apparatus for the conduct of the process. The spent organic compound to be subjected to combustion treatment and the combustion air are both led to burner **3**, where the spent organic compound is burned, via lines **1** and **2** respectively. The hot combustion gas containing hydrogen chloride is injected into an extracting agent liquid bath (extracting agent solution concentrator) **4**, where the gas is cooled by the direct heat exchange between the gas and the liquid, and simultaneously, the water in the extracting agent liquid bath is caused to evaporate by the combustion gas and, thus the concentration of the extracting agent solution is carried out.

FIGURE 107: YAWATA PROCESS DESIGN FOR ORGANOCHLORINE WASTE INCINERATION AND HCl RECOVERY

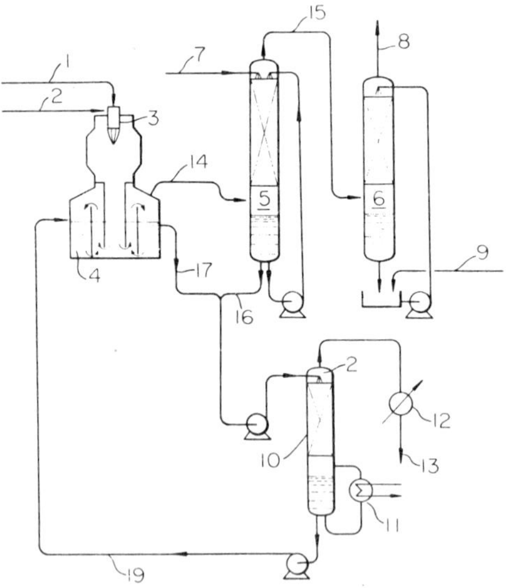

Source: U.S. Patent 3,589,864

The method by which a liquid is heated and caused to evaporate by contacting with a hot combustion gas injected directly into the liquid stored in a tank is generally called "submerged combustion method." The apparatus, wherein burner 3 is combined with extracting agent liquid bath 4, is a kind of "submerged combustion apparatus." The concentrated extracting agent solution is transported to the undermentioned extractive distillation step.

The gas cooled by heat exchange is introduced to the lower part of absorber 5 via line 14, and is contacted countercurrently with water provided via line 7 that joins the upper part of the column. The structure of the absorber does not play an essential role in the method and therefore, any suitable contacting apparatus can be used. Some of the column types usable therefor include plate tower such as bubble cap plate and perforated plate, and packed tower packed with Raschig ring, Lessing ring, cross-partition ring, Pall ring, Berl saddle, Intalox saddle, Tellerette, spiral ring, etc. Corrosive materials are involved in the operation, so a packed tower is preferable, since it is easier to maintain.

The hydrogen chloride vapor in the gas is absorbed by contacting, directly and countercurrently with a water spray, the water preferably being sprayed from the upper part of the column. As a result, a 16 to 18% hydrogen chloride solution, and in some cases, a solution approximating the azeotropic composition can be obtained from the bottom of the column. In case the concentration obtained is low, it is possible to recirculate a part of the hydrogen chloride solution obtained from the bottom of the absorber to the top. In case a packed tower is used as an absorber, the recirculation is required in some cases in order to secure the liquid load necessary for providing appropriate gas-liquid contact.

The gas released from the upper part of the absorber consisting mainly of nitrogen, carbon dioxide and steam can be transported to the lower part of scrubber 6 via line 15 in order to eliminate any remaining harmful materials. In the scrubber, the gas is countercurrently contacted with a dilute alkaline solution which is introduced via line 9, and circulated through the column. The dilute alkaline solution contains, for example, sodium carbonate, sodium bicarbonate, sodium hydroxide and calcium hydroxide or the like. The small amount of unrecovered hydrogen chloride can be removed in column 6 through neutralization. Thus, the gas is made completely harmless and odorless and is released into the atmosphere from the top via line 8.

Thereafter, in order to concentrate the dilute hydrogen chloride solution withdrawn from the bottom of the absorber via line 16 to a concentration of higher than the azeotropic composition, the dilute hydrogen chloride solution is subjected to a continuous mixing with the concentrated extracting agent solution passed from the liquid bath via line 17, and is provided to the top of the extractive distillation column 10. Reboiler 11 provided at the bottom of the extractive distillation column supplies the heat necessary for effecting the distillation, and thus the extractive distillation is carried out. The vapor withdrawn from the top of the distillation column is condensed as it is in condenser 12, and is recovered in the form of hydrogen chloride solution having a concentration higher than the azeotropic composition and high purity via line 13.

As a result of the extractive distillation, the extracting agent solution diluted with water is recirculated to the extracting agent liquid bath via line 19 from the lower part of the column, where the extracting agent solution is reused after effecting a direct heat exchange with the combustion gas and carrying out concentration as has been mentioned in the foregoing.

The following is one specific example of the conduct of the process. It involves the treatment of a spent liquor (calorific value of approximatley 7,500 kcal/kg) having as a main ingredient monochlorocyclohexane obtained from a hexane-recovering equipment in the photonitrosation method caprolactam production plant. Sulfuric acid was used in the treatment as an extracting agent.

(a) The spent liquid was burned by submerged combustion apparatus, and simultaneously an appropriate amount of water was added to the extracting agent liquid bath so as to avoid excessive concentration of the agent and to lower the gas temperature. Auxiliary fuel was required for effecting preheating of the furnace only at the time of "start-up" of the apparatus; the use of auxiliary fuel was not necessary after the combustion reached a normal running. At this point the temperature at the inside of the furnace was approximately 1600°C, while the concentration and temperature of the sulfuric acid solution discharged from the extracting agent liquid bath were approximately 71.5% and 160°C, respectively.

	Kg/hr
Spent liquor	500
Air for effecting combustion	4,960
Water added into the bath 4	3,350
Sulfuric acid solution discharged from the extracting agent liquid bath	4,180

(b) The packed section of the packed tower used as an absorber was divided into three separated parts; upper, middle and lower parts respectively, and a part of descending water led from each part thereof was recirculated to the top of each part in addition to the fresh absorbing water provided. The concentration and temperature of the hydrochloric acid solution discharged from the bottom were approximately 13% and 90°C respectively.

	Kg/hr
Absorbing water	1,720
Circulating water in each part	16,500
Hydrochloric acid solution discharged from the bottom	2,160

(c) In a packed tower used as a scrubber, slaked lime slurry having a concentration of approximately 5% was circulated, and as the basisity of the slurry weakened, the slaked lime was supplemented, or a part of the circulating liquid was discharged and was replaced by a fresh slurry. Hydrogen chloride in the spent gas was negligible. Chlorine in the amount of approximately 20 ppm was detected.

	Kg/hr
Circulating slaked lime slurry	18,000

(d) A packed tower was used as an extractive distillation tower, wherein the distillate was totally condensed. Hydrochloric acid having a concentration of approximately 35% was recovered from the condenser. Meanwhile, the concentration and temperature of the sulfuric acid solution discharged from the bottom were approximately 50.9% and 120°C respectively. In addition, hydrochloric acid in the amount of approximately 2.1% was contained in the solution.

	Kg/hr
Steam required for reboiler	450
Sulfuric acid solution discharged from the bottom	5,900
Recovered hydrochloric acid solution	420

A device developed by W. Sansom et al (73) is a chemical waste incinerator having a combustion chamber lined with a refractory material in communication with a spray quench chamber lined with a corrosion resistant material, such as carbon blocks, and an annular heat shield at the juncture of the combustion chamber and quench chamber, there is provided an improvement wherein the heat shield consists essentially of a corrosion resistant outer layer separated from a heat resistant inner layer by a resilient insulating barrier of high thermal efficiency.

Figure 108 shows such an apparatus and its operation will now be described with reference to that figure. Chlorinated hydrocarbon waste **1**, atomizing air **2**, and natural gas **3** are passed through atomizing nozzle **4** to form spray **5** which mixes with air **6**. The resulting mixture is ignited in burner **7** and passes into combustion chamber **8** where the mixture is burned.

**FIGURE 108: DU PONT INCINERATOR DESIGN FOR CHLORINATED CHEMICAL
WASTES**

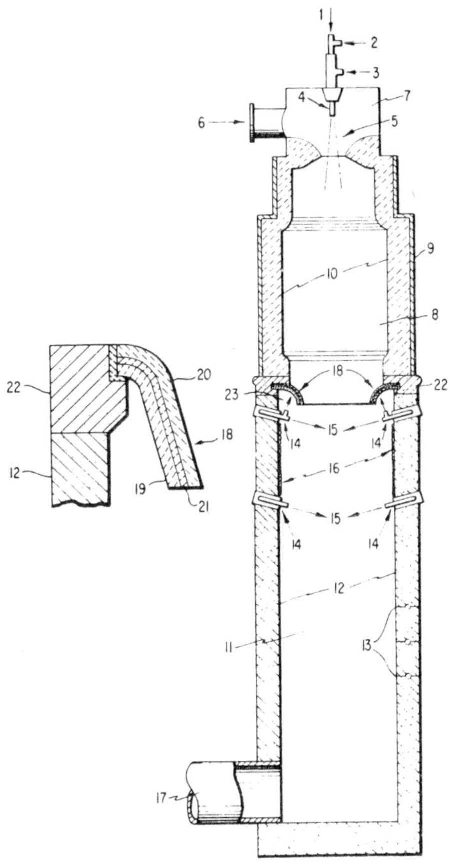

Source: U.S. Patent 3,712,796

The combustion chamber **8** is formed by a shell **9**, such as metal, lined with a refractory
material **10**, such as mullite or a high alumina refractory material. The actual refractory
materials used in the combustion chamber are not critical to the successful operation of
the heat shield, but rather are dependent upon the properties of the chemical waste and
temperature of operation. For example, when a chlorinated hydrocarbon waste product
is incinerated, the temperature in the combustion chamber is typically about 2000° to
3000°F. During combustion, hot HCl gas is formed. Thus, the refractory material in the
combustion chamber should be resistant to chemical attack by HCl gas up to at least about
3000°F.

By "resistant to chemical attack" it is meant that the structural material referred to should
be sufficiently resistant to chemical attack by the materials for which it was designed in
order to remain substantially unchanged over extended periods of operation of the apparatus. Products of combustion exit the combustion chamber and enter quench chamber **11**
where the products of combustion are cooled and HCl removed. The quench chamber is

lined with carbon blocks **12**, such as graphite or high density carbon blocks. The carbon block lining can be reinforced, such as by the block and rod joints shown at **13**. Other carbon block linings and methods of construction which can be used are well-known to those skilled in the art.

The quench chamber contains a number of spray nozzles **14** through which scrubbing liquid **15** is sprayed. The nozzles should be resistant to the chemicals which they contact. For example, graphite spray nozzles have been found to be particularly desirable with combustion products and scrubbing liquid containing HCl. Other materials of construction which can be used will be obvious to the art skilled.

The scrubbing liquid can be any liquid in which HCl or other corrosive gas is soluble under process conditions. Preferred liquids are water and dilute aqueous HCl. The dilute HCl solution is conveniently obtained by recycling scrubbing liquid removed from the quench chamber. The temperature of the scrubbing liquid is typically about 25° to 100°C in order to provide cooling for the hot products of combustion. The scrubbing liquid also forms a wet film **16** on the inside surface of the quench chamber and this cools carbon blocks **12** along the full height of the quench chamber.

The products of combustion stripped of corrosive gas (HCl) and scrubbing liquid are removed from quench chamber **11** at outlet **17**. The incinerator also contains a heat shield **18** at the juncture of the combustion chamber and the quench chamber. This heat shield of improved construction consists essentially of at least three layers, a corrosion resistant outer layer **19** separated from a heat resistant inner layer **20** by a resilient insulating barrier of high thermal efficiency **21**. The heat shield is shown mounted on a carbon block **22**, which in turn is mounted on the carbon block **12** of the quench chamber.

The corrosion resistant outer layer can be any material resistant to chemical attack by about 1 to 30% by weight aqueous hydrochloric acid solution. For example metallic or ceramic materials can be conveniently employed. Preferred metals are Hastelloy C and tantalum. Tantalum is conveniently used as a foil about 10 mils thick due to its high cost. Tantalum can be used to cover a structural element, such as steel, which provides support for the foil. Hastelloy C contains 2.5% maximum cobalt, 15.5% chromium, 16% molybdenum, 3.75% tungsten, 5.5% iron, 1% maximum silicon, less than 1% sulfur, vanadium and phosphorus, the balance being nickel. Preferred ceramic materials are fused silica, mullite or other aluminum silicate, alumina, silicon carbide, or other ceramic materials having the required corrosion resistance. The operating temperature of the surface of the corrosion resistant outer layer varies over a range typically about 70° to 150°C. The corrosion resistant outer layer should be self-supporting at the operating temperature.

As previously noted, it is necessary to keep the carbon blocks **12** in the upper portion of the quench chamber cool to prevent deterioration. This is accomplished by spraying a portion of the cool scrubbing liquid on the blocks and against the corrosion resistant outer layer as shown at **23**. Consequently the corrosion resistant outer layer must be resistant to chemical attack by HCl in the scrubbing liquid.

A process developed by J.M. Bruce, Jr. (74) is a process for the oxidation of halocarbons comprising chlorofluorocarbons which comprises contacting the halocarbon with oxygen and a member selected from the class consisting of oxides of calcium, aluminum, barium, magnesium, iron, nickel and mixtures thereof at from 750° to 1100°C.

Landfills

The total production of the major halocarbons exceeded 16 billion pounds in 1970. If one makes the conservative estimate that by-product production of chlorine-containing residues and sludges is equivalent to 1% of production, at least 160 million pounds (80,000 tons) of unusable material must be disposed of each year. Presently, this material is collected by various commercial scavengers and landfilled, often in remote places (62). The dry cleaning, degreasing and solvent recovery industries, as well as the industries that use halocarbons for raw materials, also generate sludge of waste solids but with very much lower chlorine content. These wastes, especially the smaller amounts, are often landfilled locally.

Deposit by landfill of halogen-containing waste, especially those which are generated in the production of chlorinated hydrocarbons, presents a serious long-term environmental hazard to man, animals and fish. These materials are toxic and to some extent water-soluble. Biodegradation will further increase their mobility. Thus, landfill disposal of these wastes represents an excessive, long-term threat to underground water supplies and must be considered much less acceptable than incineration. Landfill disposal is acceptable only when the landfill site is totally isolated from ground and surface water and meets California Class 1 standards.

Ocean Disposal

The chlorinated hydrocarbon waste category includes beta-chloropropylene, trichloropropane, isopropyl chloride, allyl chloride, miscellaneous chlorides, and other constituents discharged by chemical and petrochemical plants. The disposal operations discussed by Reed (75) are confined to the Gulf of Mexico. Although wastes containing various percentages of chlorinated hydrocarbons are discharged off the Atlantic and Pacific coasts, their environmental effects have not been investigated primarily because in those areas such studies are not a prerequisite for obtaining a disposal permit.

Four separate chlorinated hydrocarbon disposal operations in Gulf waters have been studied. The work included field and laboratory investigations of the effects of these wastes on various marine organisms, with emphasis on phytoplankton, zooplankton, and fish. Observed effects at the disposal site included sinking of the bulk of the waste, discoloration of the water, and the death of fish and plankton on direct contact with undiluted waste. After two to four hours, the surface waste field was found to be sufficiently diluted so that no effect on marine life was observed.

Because the inherent patchy distribution of the microorganisms at sea makes it difficult to obtain statistically significant results, the effects of various toxicity levels on phytoplankton and zooplankton were tested under controlled laboratory conditions. Inhibition to photosynthesis and respiration caused by the wastes was determined by using standard oceanographic procedures for measuring carbon-14 uptake and oxygen evolution. Comparison of the results with those obtained from uncontaminated control samples maintained under identical laboratory conditions provided an index of inhibition. Acute toxicity tests were also conducted on the plankton and various species of fish. Results of the laboratory toxicity studies showed that the inhibition of photosynthesis and respiration is a much more sensitive and meaningful measure of the toxic effects of organic contaminants in seawater than acute toxicity tests on organisms such as fish.

The effects of wastes on organisms endemic to the disposal sites was assessed in the field by examining the flora and fauna associated with the seaweed, Sargassum, present both within the outside of the disposal area. It was found that direct contact with the undiluted hydrocarbon wastes at the point of discharge was generally lethal to the small crabs, shrimp, fish, snails, etc., but in most cases samples taken in the disposal area showed conditions had returned to normal within three to eight hours.

With regard to diffusion and dispersion of the hydrocarbon wastes, it was estimated that about 7 mi^2 of surface area would be affected by a single disposal operation. The work further indicated that within 3 to 9 hours, the wastes in the surface layers were diluted to levels below the limit of detection by the analytical method used.

In contrast to this, slow dispersion of the wastes occurred at depths from 50' to 500'. For example, at one station the waste concentration at 500' was roughly equivalent to that observed at the surface in the actual disposal wake 42 hours earlier. Because of the possibility of these waste-laden waters working their way shoreward to bottom fishing areas or regions of upwelling, it was recommended that future disposal areas should be located seaward of the 400 fathom line.

It was concluded that chlorinated hydrocarbon wastes could be discharged at sea with only minor adverse effects on marine organisms within the dispersal area. It should be pointed out, however, that the work did not examine the possible effects of the wastes at depth or of bottom contamination on benthic organisms, even though samples of bottom mud within the disposal area contained from 10 to 100% of the original surface concentration of hydrocarbon waste.

Laboratory tests on the toxicity of two chlorinated hydrocarbon wastes produced by a chemical plant were carried out. This study was in preliminary support for an application for marine disposal off the Texas coast. The tests, which used top-water minnows, brine shrimp, and various phytoplankton, measured both the median toxicity level at 24 hours and the critical concentration range, defined as that dilution range between which there is no mortality on the one hand and no survival on the other. In the case of the test plants, 50% survived at concentrations of 0.05 to 0.5% by volume on 24 hour exposure to the hydrocarbon wastes, and the waste, which included dilute sulfuric acid, was somewhat more toxic, i.e., 50% survival at concentrations of 0.02 and 0.05% by volume.

This greater toxicity should be greatly reduced in a marine disposal operation because of the rapid neutralization of sulfuric acid which occurs in seawater. It was estimated that the waste would be diluted below the 0.006% concentration range within an hour by using a pumping rate between 8 and 15 tph while underway at 5 knots, and was concluded that no significant mortality would be expected from the disposal of either waste in the open ocean.

Recycling

The dry cleaning, refrigeration, degreasing and solvent extraction industries, as well as processors using halocarbons as raw materials, all recycle as much of their halocarbons as possible. For small users or companies with badly contaminated lots, commercial solvent reclaimers reprocess the materials for further use (62).

HYDROCARBONS

Many small and some larger companies contract with local scavengers who then dispose of the materials by methods such as landfill, desert burial, open incineration at remote sites, and ocean burial. These scavengers offer high potential for resources recovery since they could be capable of concentrating wastes at a few sites where they could be treated by sensible means such as recycling or incineration (62).

Incineration

Unreclaimable solvents, solvent mixtures or sludges containing hydrocarbons can be disposed of by incineration. Although not in wide use at present, this is an excellent method because hydrocarbons are easily combusted to carbon dioxide and water. Incineration, being easy and safe, can be performed on site, by scavengers, or at a municipal incineration site.

An apparatus developed by J.D. Hummell (76)(77) is a disposal apparatus for burning liquid fuel waste comprises a vaporizing surface disposed below a combustion zone and adapted to receive the liquid fuel. The vaporizing surface can be a movable channel extending outside of the apparatus. In operation, the fuel is introduced into the apparatus in liquid form and then vaporized and burned.

Figure 109 shows a plan and elevation view of such an apparatus. The apparatus includes a housing, indicated generally at **20**, which forms a chamber **22** having an upper combustion zone **25** generally defined by the upper portion of housing **20**.

FIGURE 109: PPG INCINERATOR DESIGN FOR LIQUID HYDROCARBON WASTE DISPOSAL

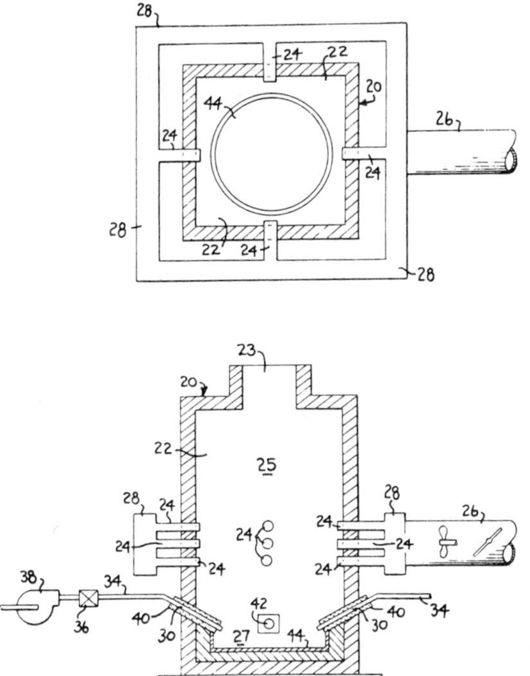

Source: U.S. Patent 3,748,081

A plurality of air inlet ports **24** are provided in the side walls of housing **20**, in the combustion zone, and are communicated via air duct **28** to a regulated source of combustion air, such as a conventional blower diagrammatically illustrated at **26**. A pair of fuel inlet passage means **30** are provided in the lower portion of housing **20** and are communicated via pipe lines **34** and valve means **36** to a conventional pump **38** which in turn is communicated with a storage tank or source of liquid wastes, not shown.

The pump and valve means mainly function to meter the flow of liquid wastes into the incinerator as the liquid wastes are not required to be preheated nor are high pumping pressures required. The dimensions of the inlet passages although not critical are relatively large as compared to restricted orifices or the like which function to divide the inlet fluid into a spray as employed in prior liquid waste disposal apparatus. Generally substantially all liquid wastes commonly incurred could function with merely a gravity feed and pump **38** is used primarily to control the flow rate of liquid entering the housing.

The inlet passages are preferably protected from the heat generated in the combustion chamber by a surrounding insulating material **40** to maintain the temperature of passages **30** low enough to prevent reaction of any of the materials or premature vaporization of the volatile constituents in the liquid waste being fed into the housing. Suitable insulating materials include water, air, ceramics, etc. Fuel ignition means, such as auxiliary burner **42**, is provided adjacent to the vaporizing surface **44**, which is in the bottom portion of chamber **22**. The vaporizing surface is usually the bottom of the housing along with the lowermost portion of the sidewalls. This area of the chamber 20, having the fuel inlets and the vaporizing surface, forms vaporizing zone 27.

Vaporizing surface 44 is preferably heated initially by burner 42 to cause an increased rate of vaporization of the volatile constituents present in the thin liquid layer of fuel on the vaporizing surface. The combustible vapors are then ignited by the burner as they become mixed with the combustion air entering inlet ports 24. This permits rapid and smoke-free start-up even with liquid waste fuels having relatively low volatility. After the start of the operation, the heat generated in the upper portion of housing 20 by the burning fuels and radiated down to the surface of the liquid waste and the vaporizing surface is generally sufficient to provide rapid enough vaporization of the volatile constituents to maintain smoke-free combustion. Then auxiliary burner 42 is no longer needed.

The above described arrangement provides ease of control because only a small volume of liquid fuel is present on the vaporizing surface at any time. This also permits the vaporizing surface to be heated by radiant energy, so as to provide a rapid rate of vaporization of the volatile constituents in the waste fuel, to obtain a maximum rate of complete combustion of the combustible constituents, but not so rapid as to cause any solids that might be suspended in the liquid waste to be carried out with the exhaust vapors.

The heat energy supplied to the vaporizing surface should not be sufficient to cause substantial film boiling of the liquid where a thin layer of vapor actually forms between the hot vaporizing surface and the liquid being introduced. This is particularly true when the waste fuel contains appreciable solids. However nucleate boiling of the liquid layer is desirable where the liquid layer actually lies on the vaporizing surface and the vapors bubble through the liquid to escape and any solids suspended in the liquid remain on the vaporizing surface.

The liquid waste is usually metered into the chamber through passages 30 at a rate consistent with the combustion air flow, which may be regulated with conventional controls, and the burning capacity of the structure. The volatile constituents vaporize in the vaporizing zone and burn in the combustion zone where by appropriate arrangement of the air pattern and control of the fuel vaporization, as described above, the air-vapor mixture is maintained in the proper proportions to provide continuous smoke-free combustion. The mixture within the chamber ranges from vapor-rich in the vaporizing zone, near the vaporizing surface, to ideal combustion proportions near the air ports, to air-rich near the exhaust outlet or flue 23.

In one example of operation, there is employed the apparatus illustrated with refractory-lined chamber 4½ ft^2 and 12 ft high. The fuel inlet passages are two one-half inch diameter pipes, through which liquid waste fuel is introduced. The exhaust gases are vented through a breeching and stack. The liquid fuel is organic solvent, mainly aliphatic and aromatic hydrocarbons, containing about 20 to 25% dissolved organic solids and 15 to 20% suspended organic solids. To initiate operation, an auxiliary burner is fired at about 3,000,000 Btu/hr for about 4 hours to preheat the refractories to a temperature of 1600° to 1800°F. The liquid fuel is then fed at a rate of up to 150 gal/hr, which combustion air is introduced from the blowers at 6,000 to 8,000 cfm. After the liquid fuel ignites the auxiliary burner is turned off.

Stable, self-sustaining and virtually complete combustion is achieved in this manner, with no flames visible in the stack and fully transparent exhaust gases. The rate of liquid fuel feed is varied (depending on heating value and viscosity) so as to maintain the temperature of the combustion gases entering the stack at the desired temperature, usually between 1700° and 2000°F, while maintaining the flow of combustion air constant.

Landfills

Presently hydrocarbon residues, sludges and in some areas hydrocarbon mixtures are landfilled, often at local municipal landfill sites. Since incineration is such an attractive alternative, and because landfill represents a long-term threat to underground water supplies, this option must be considered much less acceptable than incineration (62).

HYDROQUINONE

Incineration

Incineration of hydroquinone in a well designed and operated incinerator is an acceptable waste disposal method. Hydroquinone should be combusted at a minimum temperature of 1800°F for a minimum of 2.0 seconds. Care must be taken to avoid leakage of unburned hydroquinone vapors, and to remove harmful combustion products (e.g., SO_2, NO_x) produced by other components in the waste mixture. The incineration option is most applicable to concentrated hydroquinone wastes and in areas removed from ready access to appropriate municipal or private secondary treatment facilities (64).

Landfill

Landfill or deep well disposal of hydroquinone waste streams is not generally recommended because of the danger of release of this water-soluble, toxic substance to the environment. In addition to possible leaching of hydroquinone by rain or subterranean water, its possible oxidation to the more toxic, volatile, and less biodegradable quinone must be considered as contraindicative for burial disposal methods. However, a landfill meeting California Class 1 requirements is adequate (64).

Biological Treatment

The major portion of hydroquinone wastes are expected to be in the form of dilute aqueous solutions. Hydroquinone can be effectively removed from such waste streams by conventional secondary sewage treatment (biochemical oxidation) methods. Of those processes, activated sludge is the most efficient, with aerated lagoons or trickling filters equally effective if sufficient residence time for complete decompostion is available. Anaerobic processes (facultative ponds) are not recommended for hydroquinone waste streams which contain components incorporating sulfur atoms (e.g., photographic process effluents) because of the danger of H_2S production.

Concentrated hydroquinone wastes may also be treated by common municipal secondary wastewater treatment methods after adequate dilution. The required dilution will depend upon the capacity of the treatment plant in question; hydroquinone exerts a BOD of 1.12 lb/lb of waste. The direct introduction of untreated hydroquinone wastes into surface or underground waters is not recommended because of the known high toxicity of the material to fish and other fauna (64).

MALEIC ANHYDRIDE

Essentially, four options are available for the disposal of concentrated maleic anhydride wastes. These are recycle, incineration, landfill and deep sea burial (64).

Recycle

By far, the most appropriate method of disposal of concentrated maleic anhydride wastes is to recycle the material. This is accomplished either by locating a consumer willing to use low grade material or by reprocessing the material at the plant for captive use.

Incineration

If recycling is impossible, controlled incineration is the disposal method of choice. Incineration must be controlled to ensure that the waste maleic anhydride, as well as any other material in the waste stream, is completely oxidized to nontoxic combustion products. Incineration of the tar by-products containing maleic anhydride is an adequate means of disposing of this material particularly since the tars can be used as fuel in boilers. Care must be taken to ensure that complete oxidation of the maleic anhydride, as well as any other materials in the waste stream, is attained.

Landfill

The use of landfills for the disposal of concentrated maleic anhydride wastes should be considered only when the options of recycle and incineration are impractical and the landfill is of the California Class 1 type. Sanitary landfills are currently used to dispose of tars containing maleic anhydride and maleic acid (64). This method is recommended only when the landfill utilized meets California Class 1 landfill standards since this will minimize the possibility of water contamination.

Ocean Disposal

The statement has been published that the use of deep sea burial is not recommended since injury to aquatic life is possible and once dumped, control of the material is lost. However, it is also stated (64) that wastewater containing maleic acid from maleic anhydride equipment wash downs can be adequately handled in municipal sewers after neutralization of the dilute solution. Once neutralized by NaOH addition and in dilute aqueous solution there is no evidence of toxicity, taste or odor problems in the context of disposal to municipal or industrial treatment plants since the sodium maleate (in dilute form) is readily oxidized biologically.

MERCAPTANS

Incineration

It is recommended (62) in the handling and storage of mercaptans that incineration be used to burn vapors emitted during pressure reduction of storage tanks. This method is adequate provided some means are used to remove the sulfur components in the effluent gas. For low volume or intermittant discharge such as tank car venting, an activated-carbon scrubber may be used.

Recycling of the concentrated mercaptans is practical in some cases depending on the contaminating materials, but in most cases the waste is disposed of by incineration. In the current incineration disposal process, a Claus sulfur recovery unit is used to recycle the sulfur and reduce hydrogen sulfide and sulfur dioxide emissions to meet state and local regulations. Incineration followed by effective scrubbing of the effluent gas is an acceptable means of disposing of mercaptans.

Dilute mercaptan waste can appear as organic or aqueous waste. Dilute organic wastes are incinerated and the effluent gas scrubbed with a caustic solution. Dilute aqueous waste is thermally oxidized (incinerated) by spraying the aqueous solution into an incinerator at a temperature of 2000°F then scrubbing the effluent gas with caustic solution. These are adequate methods for the disposal of dilute mercaptan waste.

MISCELLANEOUS ORGANIC WASTES

Incineration

A process developed by G. Hegner (78) is a process for the purification of wastewater which mainly contains organic impurities in dissolved, suspended, or colloidal form. There are already many processes known which carry out wastewater purification in mechanical, chemical, or biological manners. In most cases flocculent chemicals or biological slurry are used. The slurry resulting there usually contains over 95% water and can in this condition, therefore be thrown on the dump. It is necessary to expend additional apparatus and work in order to obtain it in solid form or after further dehydration to be able to burn it. Since the effectiveness of the slurry formation and separation essentially depends on the duration of time, very large basins and ground surfaces are required. However, these large basins and ground surfaces are not always available and therefore make wastewater purification often very problematical.

In a known wet burning process, these difficulties are avoided to a certain extent. This process consists of the disintegration of the organic, dissolved material by oxidation with air at temperatures between 100°C and the critical temperature of the water, especially between 220° to 320°C, and at pressures between 50 and 150 atm excess pressure in an autoclave.

Since the process runs off in the liquid phase, it is necessary that one choose the operating pressure of the reactor at 20 to 50 atm excess pressure higher than the vapor pressure of the water at operating temperature, with reference to the air present and the other gases resulting from the oxidation. In this manner, the highest achievable temperature is limited by the vapor pressure, making impossible the oxidation of some compounds (e.g., methanol in the liquid phase). This, however, limits the utilization of this process for the purification of industrial wastewater.

According to the process, it is possible to purify wastewater having predominantly organic impurities of different compositions to the extent necessary in that the water is brought in its fluid phase to a reaction chamber by being overheated with hot, oxygen-containing gases or waste gases or immediately with burners working with excess air, until the desired oxidation temperature is reached. By the expression "fluid" is meant liquid and/or vaporous or as a water-vapor air mixture.

Since in this process the work range lies in the overheated vapor area, a very quick and complete disintegration of the organic impurities is achieved. As a result, no slurry is precipitated, while the ash residues can be removed in a simple manner. The process is carried out at normal pressure, so that the operating pressure of the equipment has no influence on the achievable temperature, and the latter can thus be adjusted to the requirements of the apparatus or the wastewater. In such manner, the introduction of corrosion and heat resistant materials, including ceramic linings, is also possible.

Since the oxidation processes are exothermic, per the degree of impurities in the wastewater, the corresponding amount of heat energy can be saved. Consequently, this process, contrary to biological equipment, operates most economically with wastewaters heavily contaminated organically. Its introduction is therefore advantageous where concentrated wastewater is precipitated immediately at the production operation without previous mixing with cooling water or rain water. With such introduction it is furthermore achieved that the amount of water to be treated is relatively small and therefore requires only small equipment. With reference to the balance of energy in the process, it is effective to use extensively, by skillful heat exchange, the heat brought in for preheating, evaporation and overheating of the untreated water.

Figure 110 is a flow diagram of the process. The water to be purified flows from the reservoir 1 to the preheated and vaporizer 2 where it is heated with the air for combustion 3 required for the burner 5. By the thus-achieved vapor pressure lowering of the water, it is possible to evaporate, corresponding to the preheating temperature chosen, a more or less large portion of the wastewater as well as the volatile constituents eventually contained there already at temperatures below 100°C. This vapor-air mixture then continues further in the overheater 4, where it is overheated and finally, mixed with fuel, it is conducted to a burner 5. The wastewater not vaporized in the preheater overflows to the reactor 6 and is there vaporized and overheated by the hot combustion gases. Since the burner operates with excess of air, the organic impurities of the wastewater are oxidized by the free air oxygen and disintegrated. The amount of heat set free can save on fuel energy.

The hot exhaust gases of the reactor are then conducted to the overheater where they give off a part of their heat content to the vapor air mixture to be heated, without cooling off so much that eventually vaporous, anorganic impurities could be precipitated or condensed from the water vapor. At 9 from the already purified wastewater or from other pure water available, such an amount of water is injected into the exhaust gas that its temperature is lowered to the thaw point temperature, and accordingly further eventual precipitation can be dissolved or washed away by the injection water so that the precipitation does not become lodged in the heat exchanger 2.

FIGURE 110: WASTEWATER OXIDATION PROCESS

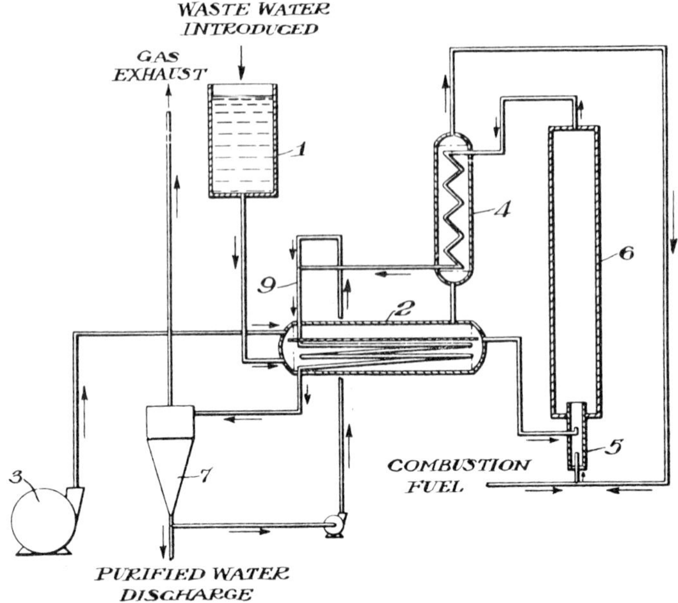

Source: U.S. Patent 3,296,125

Since the vapor content of the wastewater is always greater than the vapor air mixture, even without the additional water injection at **9** the thaw point temperature of the exhaust gas must be in any case above the boiling temperature of the wastewater in **2**. Thus, an effective heat exchange of condensing vapor with the boiling liquid at small exchange surfaces is possible as well as the recovery of a large portion of the heat brought in. The amount of heat available behind the exchanger **2** can also be used economically for other purposes or the mixture goes directly to a separator **7** where the condensate is separated from the exhaust gas.

An apparatus developed by H.C. Schutt (79) is a still and column apparatus for treating water contaminated with organic chemicals, e.g., hydrocarbons, alcohols, ethers, aldehydes and amino compounds. The apparatus comprises a column connected at the top with a reflux condenser and communicating at the bottom with a tank-still containing the contaminated water to be treated, the tank-still being equipped with submerged combustion burners arranged at opposite sides of the still and adapted to heat the water therein, the combustion products and volatilized materials being discharged upwardly of the column, the column comprising vertically stacked trays each formed by spaced apart parallel angle irons.

Figure 111 shows a suitable form of apparatus for the conduct of the process. There is shown the hollow, cylindrical exchange column **11** mounted above and communicating with the base tank **12**. Connected between a source of raw process water and the water inlet nozzle assembly **10** in the upper portion of the column is the feed line **13** which passes through the heat exchanger **14**. Positioned in the water removal line **15** is the pump **16** which withdraws purified water out of the base tank and through the heat exchanger. The fluid permeable decks **17** are stacked throughout the column in a spaced apart, parallel relationship.

FIGURE 111: SUBMERGED COMBUSTION BURNER FOR WASTEWATER TREATMENT

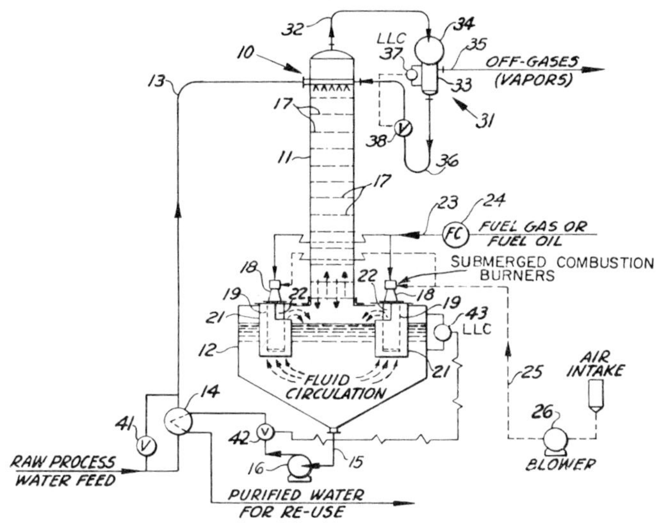

Source: U.S. Patent 3,432,399

The burners **18** are supported by and project into the base tank **12**. Extending downwardly from each burner is a hollow, open bottomed downcomer **19** surrounded by a concentric pipe **21** having in its upper surface a 180° weir opening **22** facing the column **11**. Operatively connected to the burners are both the fuel line **23** with the associated flow controller **24** and the air intake line **25** with the associated blower **26**. The burners are commonly known as submerged combustion burners and are of the same general type as shown in U.S. Patent 3,138,150.

The discharge apparatus **31** is connected to the top of the column by the discharge line **32**. Included in the discharge apparatus is the reservoir **33** mounted beneath the cooler-condenser **34**. Communicating with the reservoir are the offline **35** for removal of gases and vapors and the return line **36** connected to the nozzle assembly **10**. The controller **37** maintains a desired liquid level in the reservoir by controlling the valve **38** in the return line.

During operation, the waste process or raw water is preheated by the effluent or purified water in the heat exchanger **14**. Part of the raw water can by-pass the exchanger by passing through the by-pass valve **41** in order to control the temperature of the raw water entering the top of the column as required by the types and amounts of contaminants. The transfer to and retention of the solute, hydrocarbons or organics, in the vapor phase are effected by a carrier gas which is generated in the base tank on which the column is mounted. The carrier gas is derived from the combustion of fuel gas or oil which is burned with a minimum amount of excess air required for complete combustion in the burners.

The dry or inert combustion products should not contain more than 1.2% of oxygen. The amount of fuel burned is governed by the concentration and phase behavior pattern of the contaminant to be removed. The temperature in the base tank is controlled by adjusting the temperature of the raw process water entering the column, as described above. The proper temperature will vary from case to case, and only temperature ranges can be cited based on application of the process.

The vapors leaving the top of the column through the discharge line 32 are cooled and partially condensed in the cooler-condenser 34 and returned to the column 11 for combination with the raw water in its downward passage thereby retaining the maximum amount of water in the system. The condenser may be mounted above the column which permits gravity induced return of the condensate or may be located at any other appropriate elevation for separating the condensate which is then returned to the top of the column by means of a pump.

The uncondensed materials comprising the inert or carrier gas, the contaminants removed from the raw water, and some water vapors approximately equivalent to those produced in the combustion process are then discharged through the offline 35 into the plant flare system or the combustion chamber of a furnace or boiler where the contaminants are burned or completely oxidized and released with other inert materials into the atmosphere.

The liquid descending in the column which is equipped with trays or packing, is in intimate contact with the ascending gases and vapors to effect the required heat and mass transfer for driving the contaminants out of solution. It maintains a marked temperature differential between the top of the column and the base tank 12, in the range of 30° to 70°F, the exact temperature pattern being dependent on the composition and the pseudo-vapor pressure of the solute, to effect a high thermodynamic efficiency.

Transfer of hydrocarbons and other contaminants of the process water entering at the top of the column takes place at relatively high solute concentrations and low temperatures, in the order of 150° to 180°F, so that the amount of water vapor contained in the off-gases is kept at a minimum. The proper temperature at the top of the column is maintained by regulating the temperature of the raw water feed to the column in conjunction with the temperature maintained in the base tank.

The transfer of hydrocarbons in the lower part of the column is abetted by the large volume of steam and inert gases leaving the base tank. Most of the steam is recondensed in the column imparting its heat to the descending liquid, thus effecting a high thermal efficiency of the process system. The final disengagement or transfer of contaminants from the water takes place in the base tank where temperatures in the order of 185° to 225°F are maintained, depending again on the nature of the contaminants, by direct application of combustion heat. The amount of fuel burned and carrier gas generated is maintained at a constant, predetermined value by means of the flow controller 24 in the fuel gas supply line 23.

The burner assembly is of special design which induces a high rate of liquid circulation in contact with the hot combustion gases, resulting in a large interfacial area and transient high temperatures at the gas-liquid boundaries, thus intensifying the heat and mass transfer. The combustion gases, issuing from the burner muffle or downcomer 19 at relatively high velocity, pass upward through the annular space formed by a surrounding and concentric pipe 21 of approximately twice the diameter of the downcomer and having a 180° weir opening 22 at the top.

At the normal liquid level maintained in the tank, water enters at the bottom of the annular passage and is lifted over the weir opening which faces the liquid stream issuing from the column being in open communication with the base tank. The desired liquid level is maintained by emptying purified water from the quiescent zone at the bottom of the tank at a flow rate determined by the valve 42 controlled by the liquid level controller (LLC) 43. Two or more burner assemblies 18 may be applied depending on the amount of waste process water to be treated; they are then arranged concentrically to the column with overflow weir openings facing the downcoming liquid stream.

The pressure in the base tank should be as low as possible, and the means provided in the column for vapor-liquid contacting as well as the partial condenser 34 in the overhead vapor-stream are designed for relatively low pressure drop but are consistent with the required efficiency of the heat and mass transfer processes. For relatively small capacity units handling up to 150 gpm of raw water, a column equipped with packing, such as Stedman or Pall rings, may be provided. For larger units the column may be equipped with conventional fractionating trays but preferably with grid decks which are uniquely suited for this application because of their low liquid holdup and pressure drop. With these decks, the pressure in the base tank **12** would normally not exceed 20 psia for larger capacity units.

Hydrocarbons with atmospheric boiling points as high as 400°F which may be dissolved in the process water in measurable quantities can be removed at a temperature level of 180° to 210°F in the base tank and 150° to 160°F at the top of the column **11**, i.e., a temperature level throughout the system considerably below the boiling point of water. The residual hydrocarbons in the treated water amount to 5 ppm or less. Some organic chemicals, oxygenated hydrocarbons in particular, such as higher molecular weight alcohols, aldehydes, ethers, and ketones, which are partially soluble in water, also exhibit high activity coefficients in dilute solutions and therefore respond to this treating method.

A typical application of the process follows. The waste process water from a basic petrochemical plant involving pyrolysis of petroleum distillates, recovery of olefins and associated by-products usually contains 1,200 ppm of hydrocarbons in true solution after passing through an API separator. The concentration of various species of dissolved hydrocarbons is: benzene, 850 ppm; alkyl benzenes, 160 ppm; styrenes, 50 ppm; diolefins and olefins, 140 ppm. This raw water was processed at a rate of 500 gpm or 250,000 lb/hr, and its hydrocarbon content was reduced to less than 5 ppm under operating conditions shown below. If a lower temperature is maintained in the base tank and the heat released by the burner reduced, the residual hydrocarbons in the treated water increase to 8 to 9 ppm. By reducing the feed temperature, restoring the heat input of the burners and the amount of inert gases, this can be avoided.

	P.s.i.a.	° F.
Pressures and temperatures:		
Raw water feed (pH=6.2 to 6.4)		175
Base tank	19.8	205
Top of column	18.5	180
Effluent from overhead condenser	15.8	152
Fuel gas burned, s.c.f./hr	11,100	
Heat release (basis LHV), B.t.u./hr	10,200,000	
Efficiency in use of fuel gas, percent	97.1	
Treated water recovered, lb./hr	249,300	

NITRILES

Incineration

In the event it becomes necessary to dispose of a significant quantity of organic waste and purification/recycling is impractical, incineration of nitriles is the recommended method of disposal. The material must be incinerated under controlled conditions whereby oxides of nitrogen are removed from the effluent gas by scrubbers and/or thermal or catalytic devices (62).

A process developed by K. Kishigami et al (80) involves incinerating combustible industrial wastage containing organic nitrogen compounds of a peptide bond, cyan radical, etc. by using a so-called fluidized bed incineration furnace. This method uses steps of supplying, the aforesaid wastage together with combustible materials abundant in carbon content, such as pulverized coal, heavy oil, etc., in the fluidized bed to thereby maintain the interior of the fluidized bed in a reductive atmosphere, while reducing nitrogen oxide produced from the combustion of the wastage, whereby nitrogen oxide may be removed as innocuous nitrogen from waste-gas or stack gas.

Figure 112 shows a suitable form of apparatus for the conduct of the process. An incineration furnace **1** includes a perforated plate **2** and an air chamber **3** in the lower portion thereof, while dehydrated mud waste, which contains pulverized coal or fine granular coal, is supplied from the top portion of the incineration furnace therein from a mud supply device **4**. The air which is required for fluidizing and combustion is supplied through a blower **8**, damper **15** and duct **9** to the air chamber, thereby fluidizing by passing through the perforated plate 2 the fluidizing medium particles such as cement clinker and the like as well as the mud which contains pulverized coal and fine granular coal, the particles and mud being positioned on the perforated plate, and thus the fluidized bed is provided.

FIGURE 112: INCINERATOR FOR ORGANIC NITROGEN-CONTAINING WASTES
SUCH AS NITRILES

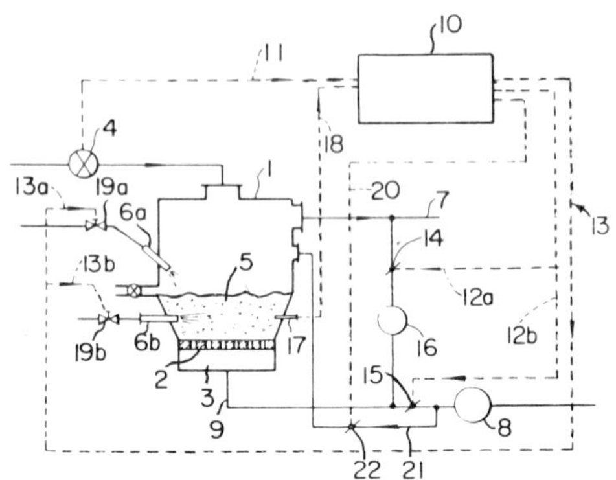

Source: U.S. Patent 3,888,194

The operation of the incineration furnace begins with heating the fluidized bed by using a burner **6a**, after which the incineration of the wastage proceeds due to the combustion of the pulverized coal (or fine granular coal) mixed with the muddy wastage. If the heat accruing from the combustion of the pulverized coal is insufficient, the burner is used as an auxiliary means. The waste gas is discharged through a duct **7** out of the furnace. When the increase in the quantity of fluidizing gas is required for forming a fluidized bed **5**, the waste gas is introduced from the duct to thereby supply the waste gas through a damper **14** and a blower **16** into the duct **9**.

For operation of the incineration furnace **1**, it is required to control the quantity of air and waste gas to be mixed therewith. To this end, a signal **11** which represents the supplied amount of muddy wastage and is obtained in connection to rpm of a rotary feeder of the muddy wastage supply device **4**, and a temperature signal **18** from a thermometer **1** adapted to measure temperatures at the fluidized bed, are fed into a control box **10**, whereby inlet dampers **15** and **14** of the blowers **8** and **16** are each controlled by means of signals **12a** and **12b**, while the control valves **19a** and **19b** for controlling the amount of oil for burners **6a** and **6b** are controlled by means of signals **13a** and **13b**.

The burner **6a** is primarily used for starting a furnace or for an auxiliary purpose, while burner **6b** is used for supplying combustible materials abundant in carbon content, such as heavy oil, to the fluidized bed and is normally used in case the muddy wastage to be charged in the furnace contains no pulverized coal or the like. On the other hand, a damper **22** provided in the secondary air supply duct **21** is controlled by means of a signal **20**.

For incineration of combustible industrial wastage containing organic nitrogen, particularly muddy industrial wastage of a high water content a fluidizing bed which permits the incineration at an extremely low excessive air ratio (1.1 to 1.3) is used. Carbon components of no less than 20% by weight, based on the weight of the wastage, are mixed with the solid matter such as muddy wastage thereby to form a reductive atmosphere within the fluidized bed while the temperature at the fluidized bed is maintained to be not less than 500°C to incinerate the wastage. Nitrogen oxides produced are reduced by CO, H_2, CH_4

and the like which are thus produced where nitrogen oxides may be removed from the waste combustion gas from the wastage in the form of nitrogen gas. According to another aspect of the process, there are used, as fluidizing-medium, cement clinker particles containing a great amount of Fe_2O_3, Cr_2O_3, MnO, Al_2O_3, chrominum ore particles and manganese ore particles for the purpose of accelerating the aforesaid reductive reaction, thereby effectively reducing and removing nitrogen oxides.

NITROANILINE

Incineration

In the event it becomes necessary to dispose of a significant quantity of concentrated nitroaniline, and purification/recycling is impractical, then incineration of nitroaniline is the recommended method of disposal. Qualified personnel familiar with handling toxic materials must be available. The material must be incinerated under controlled conditions where oxides of nitrogen are removed from the effluent gas by scrubbers and/or thermal or catalytic devices (64). Combustion should be carried out at a minimum temperature of 1800°F for at least 2.0 seconds.

NITROBENZENE

In the event it becomes necessary to dispose of significant quantities (55 gal drum or 10,000 gal tank car) of concentrated nitrobenzene, two disposal options are available (64).

Recycle

The first option is to contact the manufacturer and determine if it is possible to return the material. E.I. du Pont de Nemours has indicated a willingness to accept concentrated nitrobenzene for reprocessing provided the contaminant or contaminants in the nitrobenzene are compatible with their reprocessing system and the economic aspects of the situation are favorable (64).

Incineration

The second option is incineration since nitrobenzene is flammable and amenable to the treatment. Combustion should be carried out at a minimum temperature of 1800°F for at least 2.0 seconds. The open burning of nitrobenzene is not an adequate means of disposing of this material since oxides of nitrogen as well as incomplete combustion products may be generated during its combustion. Controlled combustion processes where the oxides of nitrogen are scrubbed from the effluent gas or where a thermal or catalytic device is used to reduce the oxides of nitrogen to their elemental form is acceptable (64).

NITROCHLOROBENZENES

In the event it becomes necessary to dispose of a significant quantity of concentrated nitrochlorobenzene, two adequate disposal options are available.

Incineration

The first option is to incinerate the material (64). It is expected that either a rotary kiln or liquid combustor, depending upon the form of the waste, followed by secondary combustion and aqueous or caustic scrubbing would be an acceptable disposal method. Primary combustion should be carried out at a minimum of 1500°F for at least 0.5 second with secondary combustion at a minimum temperature of 2200°F for at least 1.0 second. The chloride abatement problem may be simplified by insuring against elemental chlorine formation through injection of steam or methane into the combustion process. The nitric oxides may be abated through the use of thermal or catalytic devices. Concentration followed by incineration is the recommended method of disposing of dilute aqueous waste.

Landfill

The second option is to bury the material in a California Class 1 type landfill (64).

NITROPARAFFINS

Recycle

Presently, manufacturers who use nitroparaffins as solvents or chemical intermediates recycle all possible material. This is the method of choice (64).

Incineration

Unusable nitroparaffins that are unfit for recycling can be incinerated. The incineration of large quantities of material may require NO_x removal by catalytic or scrubbing processes.

NITROPHENOLS AND NITROTOLUENES

Incineration

Contaminated material that cannot be reclaimed is best disposed of by controlled incineration (64). The toxic nature of these compounds requires extreme care to maintain complete combustion at all times. Incineration of large quantities may require the use of scrubbers and/or thermal or catalytic devices to control the level of effluent oxides of nitrogen.

Landfill

This method is suggested by the Manufacturing Chemists Association for nitrobenzene and should also be applicable to the nitrotoluenes (64). This method of disposal is not considered adequate unless the landfills are of the California Class 1 type since these nitro compounds are toxic, mobile and slightly volatile. In addition, their solubility of 100 to 150 parts per million in water presents a long-term potential hazard to underground water supplies.

OXALIC ACID

Incineration

One common method for the disposal of industrial oxalic acid waste is to neutralize it with limestone to form insoluble calcium oxalate which is then incinerated (64). This is probably the best method for disposal of oxalic acid since the products of combustion are harmless carbon dioxide and calcium oxide which can be recycled to neutralize more acid waste. The efficient collection of calcium oxide requires either electrostatic or wet collection equipment downstream of the incinerator. Combustion of oxalic acid without neutralization is not recommended since toxic carbon monoxide and formic acid are among the materials produced. In addition, there is the strong possibility of the release to the atmosphere of some of the volatile oxalic acid.

Landfill

Many users of oxalic acid neutralize their waste streams with limestone and bury the recovered calcium oxalate in Class 2 landfill or deep wells (64). These methods are not generally recommended because of the danger of release of oxalic acid by the leaching action of acidic wastes. Another potential danger lies in inadvertent contact with strong oxidizing agents in the burial area. Disposal in California Class 1 type landfills is adequate if strong oxidizing agents are not present.

OXIDATION PROCESS WASTES

Incineration

A process developed by W. Kube (81) is a process for burning organic substances contained in aqueous effluents in a combustion chamber with an offgas cooler arranged downstream thereof, the aqueous liquid being passed through the offgas cooler on the coolant side with or without the supply of additional heat, heated under superatmospheric pressure and then fed into the combustion chamber.

The combustion of organic substances contained in aqueous effluents in prior art combustion plant has always been carried out with the addition of an auxiliary fuel such as oil or gas. The auxiliary fuel serves the purpose of maintaining a supporting flame which vaporizes some of the water present in the aqueous effluent so that the effluent becomes more concentrated and burns spontaneously owing to the improvement in the calorific value and ignitability of the organic substances. A high consumption of auxiliary fuel is associated with the maintenance of the supporting flame bacause the amount of heat required to evaporate the water if fairly large. If it is desired to recover at least some of the additional heat introduced into the combustion plant with the auxiliary fuel, it is necessary to provide downstream of the combustion plant a waste heat boiler for the production of heating stream.

Preliminary evaporation of the aqueous effluent is not possible in most cases because volatile constituents of the effluent (for example organic substances which owing to their odor or their toxicity have to be burnt and thus destroyed in the combustion chamber) would pass into the vapor phase together with the water fraction. Economical recovery of heat introduced with the auxiliary fuel is also excluded when using submerged burners because the aqueous effluents are usually vaporized at atmospheric pressure and the vapor formed can therefore reach a maximum temperature of only 100°C. The vapors formed are moreover not brought to the temperature necessary for complete combustion of readily volatile substances.

Another known method concerns the wet-air oxidation of sludge. In this method the sludge concentrate is heated up with live steam, oxidized by passing air through it for several hours and then drained. The maximum temperatures reached in the oxidation substances which are volatile in steam are not oxidized by the wet-air oxidation. It is an object of the development to provide a process for burning all organic substances contained in aqueous effluents in which little or no auxiliary fuel is supplied depending on the calorific value of the aqueous effluent, and a large part of the heat contained in the offgas from the combustion plant is recovered in an economical manner.

The water content of such effluents is as a rule from 50 to 95% by weight, and other solvents, particularly organic solvents, may be present as components. A specific example of a liquid which is suitable for combustion in accordance with this process is the waste liquor from the oxidation of cyclohexane which contains 40% by weight of dry substance consisting of sodium hydroxide and sodium salts of low molecular weight monocarboxylic and dicarboxylic acids and having a calorific value of about 200 to 300 kcal/kg of dry substance.

As shown in Figure 113, the combustion plant used for carrying out the process consists essentially of a combustion chamber **4**, a burner nozzle **3** for supplying and burning the aqueous effluent, an offgas channel **16**, a cyclone-type dust separator **8**, an offgas cooler **2** downstream of the cyclone **8**, a fine filter **9** and an offgas flue **10**. When auxiliary fuel is to be added, an additional burner nozzle **6** is provided in the combustion chamber.

FIGURE 113: APPARATUS FOR BURNING AQUEOUS ORGANIC WASTES

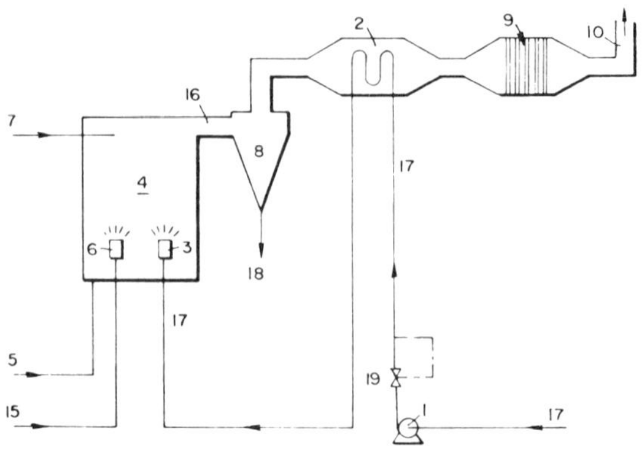

Source: U.S. Patent 3,583,339

In the simplest case the aqueous effluent **17** is brought to a pressure higher than the combustion chamber pressure by a pump **1**, heated by passage through an offgas cooler **2** generally to a temperature below the boiling temperature at the prevailing pressure and to at least the critical temperature to at least the critical pressure, and (after leaving the offgas cooler) is fed direct to the burner nozzle **3**. The aqueous effluent is flashed in this burner nozzle and some of the liquid is vaporized as a result of the pressure falling below the boiling pressure. The effluent/vapor mixture containing organic substances then passes in atomized condition into the combustion chamber **4** where the organic substances are burnt by means of the combustion air **5** also supplied to the combustion chamber.

The additional burner nozzle **6** provided in the combustion chamber serves for heating up the combustion chamber by means of an auxiliary fuel **15** before the commencement of the organic substances contained in the aqueous effluent and in the case of effluents having very high water contents can also be used to support the combustion process. The offgas formed in the combustion in the combustion chamber may be cooled by adding cooling air **7** prior to entry into the offgas channel **16** in order to cool below their melting point and to separate as dust any molten constituents present in the offgas which could cause soiling and corrosion of the heat-exchange surfaces of the offgas cooler. The offgas entering the offgas channel is given a coarse purification from solid constituents in the downstream cyclone **8** and supplied to the offgas cooler.

The solid constituents **18** are removed at the lower end of the cyclone. A large part of the heat contained in the offgas is given up in the offgas cooler to the aqueous effluent flowing on the coolant side through the offgas cooler and thus serves to heat up the aqueous effluent to a temperature near to its boiling temperature at the prevailing pressure and at least to the critical temperature. A fine filter **9** and a flue **10** for the offgas are provided on the offgas side of the offgas cooler, and the offgas leaves the combustion plant through these. If the heating up of the aqueous effluent in the offgas cooler is not sufficient to cool the offgas to flue temperature, it may be advantageous to interpose between the offgas cooler and the offgas flue a conventional waste heat boiler for the production of steam (not shown in the drawing).

The amount of aqueous effluent passed through may be regulated either by providing, downstream of the pump, a pressure control valve **19** which keeps the pressure of the aqueous effluent upstream of the burner nozzle constant, or by regulating the output of the pump or by keeping the pressure upstream of the burner nozzle constant by using an adjustable burner nozzle.

The plant described may be used to burn a waste liquor consisting of 30% by weight of organic sodium salts and 70% by weight of water. The waste liquor is brought by means of pump 1 to pressure of about 200 atm gauge and heated during passage through the off-gas cooler 2 from an inlet temperature of about 40°C to about 360°C. The waste liquid is flashed from 200 atm gauge to about 1 atm abs in the burner nozzle 3 so that about 62% of the water vaporizes and finely atomizes the residual liquid phase as it passes through the jet.

Oil is added as auxiliary fuel 15 through the additional burner nozzle 6 and the mixture is burnt by means of combustion air 5 in the combustion chamber 4 at about 1000°C. The offgas formed is mainly steam, soda and products formed in the combustion of the oil. Since soda does not solidify until the temperature has fallen to about 700°C, cooling air is blown into the offgas so that it is cooled to about 690°C. The offgas then passes through the cyclone 8 in which the soda is separated in powder form 18. The offgas is cooled in the subsequent offgas cooler to a flue temperature of about 200°C and is discharged through the offgas flue 10 after having passed through the fine filter 9.

A process developed by I.Y. Sigal et al (82) provides a method of thermal purification of waste gases from impurities, mainly organic ones, consisting in burning them directly in the furnaces of boiler units including those whose main function is generation of thermal and electrical energy. The waste gases to be purified are delivered under, or into the beginning of, the fuel burning zone instead of part or all of the air required for the combustion of the fuel.

PENTACHLOROPHENOL

The disposal of pentachlorophenol wastes as well as other chorophenoxy wastes in open pits, lagoons, unapproved landfill sites, by application to the soil surface, and by on-site burning or deep sea burial are not recommended practices because of the obvious potential contributions to air and water pollution (58).

Deep Well Disposal

Although pentachlorophenol is only sparingly soluble in water, its persistence and stability in water and the potential contamination of groundwater make deep well at best a questionable method for disposal. The method is not recommended (58) and should not be considered.

Incineration

The complete and controlled high temperature oxidation coupled with adequate scrubbing and ash disposal facilities offers the greatest immediate potential for the safe disposal of concentrated pentachlorophenol. The research on incineration of pesticides conducted by Kennedy et al at Mississippi State University (58) has led to the conclusion that chlorophenols approach complete oxidation when combusted at temperatures in the 600° to 900°C range with hydrogen chloride being the only pollutant liberated. If proper aqueous or caustic scrubbing systems are utilized, efficient abatement of the hydrogen chloride can be achieved. Therefore, properly designed and operated incineration is considered the best present and near future method for the disposal of concentrated pentachlorophenol.

PHENOLS

Biological Treatment

Phenolic compounds, with few exceptions, have been found to show very little resistance to acclimated microorganisms (62). The use of biological treatment processes, particularly activated sludge, is a recommended method for the disposal of aqueous wastes containing these materials and of concentrated phenolic wastes after they have been extensively diluted with water.

Incineration

Controlled incineration of phenolic wastes is currently used and is an acceptable method for the disposal of organic and aqueous waste containing these materials (62).

PHTHALIC ANHYDRIDE PLANT OFF-GASES

Incineration

A process developed by D.C. Ferrari et al (83) is one in which chemical effluent waste gases from chemical plants, particularly effluent waste gases from phthalic anhydride and maleic anhydride plants, are effectively water washed of residual organic matter (98 to 99% removal) in a wet scrubber using recycled water to concentrate the organic pollutants in the scrubber liquor. A concentrated liquid purge (blowdown) from the scrubber recycle circulating loop is directed to a thermal incinerator where the purge is vaporized and the organic pollutants are oxidized to nonpollutant products.

Figure 114 shows the essential elements of the process. With reference to the drawing, **10** represents a single shell two-stage, liquid-gas, slurry-handling scrubber and **11** represents a conventional aqueous incinerator for incinerating liquids, such as an incinerator known as Thermal Oxidizer. Switch condenser off-gas from the solid condensers of a phthalic anhydride plant containing organic pollutants, i.e., phthalic anhydride, maleic anhydride, benzoic acid and naphthaquinone, and at a temperature of 150° to 200°F enters the scrubber through line **30** and passes upwardly through the first scrubbing stage **12**, where it is contacted with a countercurrent, downwardly flowing scrubbing liquor, comprising an aqueous slurry of the organic pollutants removed from the off-gas during preceding cycles.

The aqueous scrubbing liquor is introduced into the top of the first scrubbing stage through line **13** and distribution manifold **13b**. Essentially all of the phthalic anhydride, benzoic acid and naphthaquinone, along with approximately 80% of the maleic anhydride, are removed from the off-gases by the scrubbing liquor in this way in the first scrubber stage. The organic anhydrides absorbed from the gas stream are hydrolized to the corresponding acids (one mol water per mol of organic anhydride), i.e., phthalic acid and maleic acid, in the scrubbing liquor by the water in the scrubbing liquor. It is also in the first scrubber stage where the temperature of the off-gas is reduced from 150° to 200°F to approximately 100°F as evaporation (of the water) cooling takes place and the gas stream becomes saturated with the evaporated water.

The organic pollutants are removed from the off-gas by absorption into the scrubbing liquor and also by precipitation thereof when the liquor becomes saturated therewith since they exist in a solid state at the scrubbing temperature. Accordingly, an aqueous slurry of solid organic pollutants in a saturated solution thereof in water is formed. The concentrated slurry at the bottom of the first stage is continuously recycled through line **13** to the upper portion of the first stage by pump **13a**. The scrubbed gas stream from the first stage then proceeds upwardly through the second scrubbing stage **15**, then through the conventional mist eliminator **16** and then out the stack **17** to the atmosphere at **35**.

In the second stage, the gas is scrubbed with a countercurrently and downwardly flowing scrubbing liquor, comprising a dilute aqueous solution of organic pollutants (chiefly maleic acid) which is introduced into the top part of the upper stage **15** from line **14** and spray nozzles **14b** and which is isolated from the first stage and the first stage scrubbing liquor by a conical shaped deflector plate **18** and collection tray **19** in the intermediate portion of the scrubber. The collected liquor on the tray supplies hold up through the line **14** for the second stage liquor recycle pump **14a** which pumps it to the upper end portion of the second stage through line **14**.

FIGURE 114: INCINERATOR FOR PHTHALIC ANHYDRIDE PLANT OFF-GASES

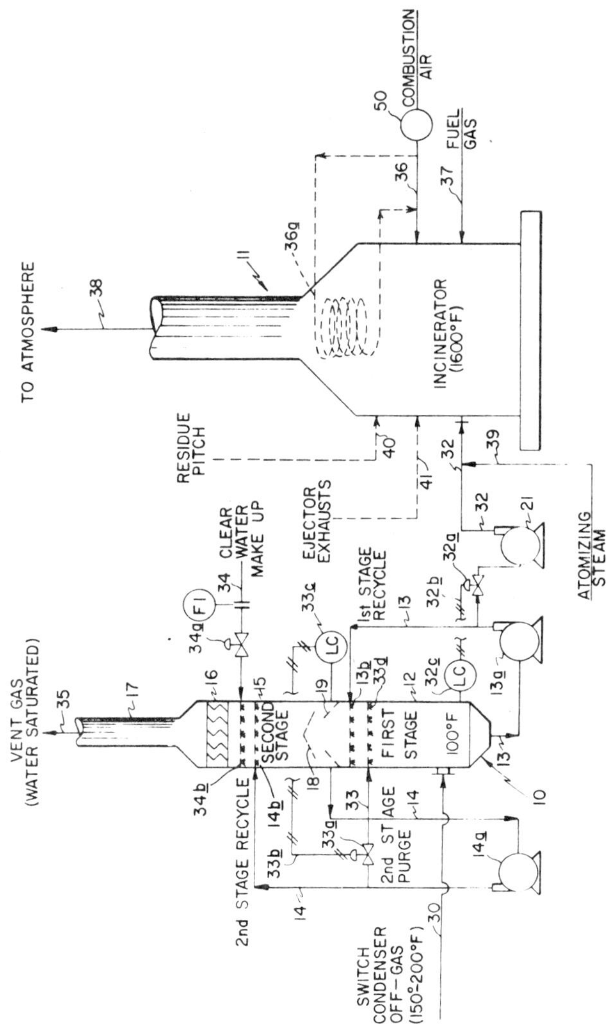

Source: U.S. Patent 3,624,984

The recycle liquor **14**, together with fresh makeup water **34**, which is introduced into the top of the second stage **15** at a point below the mist eliminator **16** and above the spray nozzles **14b** and through the line **34**, valve **34a** and spray nozzles **34b**, form the makeup scrubbing liquor for the second stage scrubber.

The scrubbed off-gas from first scrubber **12** flows upwardly past the deflector **18** and past the second scrubber stage hold-up liquor on tray **19** into the bottom of the second stage scrubber. A portion of the recycle liquor is continuously purged from the second stage recycle circuit through the second stage purge or blowdown line **33** to control the concentration of organic pollutants in the second stage scrubbing liquor, the purge liquor **33** being flowed to the upper end portion of the first scrubber stage through line **33** and spray nozzles **33d** to combine with the first stage recycle liquor **13** to make up the first stage scrubbing liquor. The rate of the second stage purge removed from the recycle line is controlled by a level controller device **33a**, **33b** and **33c**.

It is in the second stage where the remainder of the maleic acid is removed from the gas stream, thereby providing an exit gas stream **35** essentially free of organic pollutants. A controlled portion of the first stage concentrated recycle scrubbing liquor is continuously purged from the recycle circuit through first stage purge line **32** to control the concentration of organic pollutants in the scrubbing liquor of the first stage. The rate of makeup water introduced at **34b** is adjusted by valve **34a** to replace water removed from the scrubber system [1] in the saturated scrubbed gas **35** (this is the water which is evaporated in the scrubber and saturates the gas stream) by evaporation and [2] in the purge line **32**, these being the only exits from the system.

Each of the scrubber stages contains a fluid bed of packing on a supporting grid or grids to achieve intimate liquid-gas contact. The packing is made up of relatively large but light smooth-surfaced pieces such as spheres or rings or cylinders, e.g., Ping-Pong balls or Pall rings, which are large in bulk but very light in weight so that the bed can be easily fluidized but yet does not hold up the solid organic pollutant particles in the slurry. This type of packing and scrubber are conventional. One such scrubber construction using Ping-Pong balls is known as Turbulent Contact Absorber.

Purge stream **32** from the scrubber first stage is fed from the recycle line **13** through line **32** to the liquid incinerator **11** via pressure booster pump **21**. The rate of flow of purge **32**, which controls the concentration of organic pollutants in the first stage scrubbing liquor is controlled by a level controller device **32a**, **32b**, **32c**.

Atomizing steam **39** is introduced into the liquid purge stream to finely divide it immediately before it enters the combustion area of the incinerator. Preferably the atomizing guns discharge the atomized liquid directly into the combustion zone. It is preferred to atomize the purge immediately before incineration to eliminate the danger of converting maleic acid to the more insoluble fumaric acid before burning occurs. The purge can be atomized with pressurized air if desired.

Combustion air (through air pump **50**) and fuel gas, e.g., methane, are introduced into the incinerator at **36** and **37**, respectively, in conventional manner, to provide a temperature of 1400° to 1600°F. The rate of flow of combustion air is in excess of the stoichiometric amounts required to completely burn the fuel gas to CO_2 and H_2O and to completely oxidize the organic pollutants in the purge stream to CO_2 and water. A part of the air can also be introduced with the atomizing steam.

Additional organic pollutant-containing liquid waste streams from the PA plant, including the residue pitch **40** from the purification area or areas of the PA plant and PA distillation column ejector exhausts **41**, may also be fed to the incinerator, where, at 1400 to 1600°F, the organic pollutants there are 99.9% oxidized to carbon dioxide and water.

If desired, the combustion air **36** can be preheated by the hot combustion gases by leading it through heat exchange coil **36a** located at the top of the incinerator, as shown in broken lines in Figure 114, before passing it into the incinerator at **36**, to thereby provide recovery of heat. Also the coil **36a** can be used to generate steam to increase heat recovery. Preheating of the aqueous scrubber or blowdown stream **32** is inadvisable since the maleic acid may be converted to fumaric acid.

POLYCHLOROBIPHENYLS

Incineration

Contaminated polychlorinated biphenyls are reprocessed by Monsanto by passing the material through a series of clay filters (64). Concentrated materials that cannot be reprocessed in this manner are incinerated at 3000°F and the effluent gas scrubbed to remove any chlorine-containing products. These are adequate means of disposal/reuse.

Materials containing polychlorinated biphenyls such as plastics are placed in incinerators at relatively low temperatures for disposal. The polychlorinated biphenyls are not destroyed in this process and may be emitted to the atmosphere. High temperature (3000°F) incinerators with effluent gas scrubbers (see section on concentrated polychlorinated biphenyls) are also used. The high temperature incinerators are adequate for the disposal of these materials. The low temperature incinerators are not recommended because of incomplete destruction of the polychlorinated biphenyls.

An apparatus developed by T. Nakamura et al (84) is one for incinerating a mixture containing a random mixture of organic and inorganic materials. The apparatus includes an upright cylindrical furnace, a refuse chute having at the uppermost part of the furnace doors which are constructed so as to be always able to shut off the outside atmosphere, combustible gas outlet located in the middle of the furnace, a heat molten material bath located at the lowermost part of the furnace, a means for directly heating the heat molten material in the tank, exhaust openings formed respectively in the middle and lower parts of the tank.

Materials chiefly composed of metallic oxides are kept molten in the molten material bath shut off from the outside atmosphere, the combustible gas by-product of the pyrolyzed organic materials in the refuse dropped into the bath is taken outside the bath and the silicious by-product and metallic by-product of the pyrolyzed inorganic materials in the refuse are taken outside the bath respectively from above and below the molten material. Since the bath is operated at superhigh temperatures, there is no possibility of clinkers being produced, and deleterious material, such as polychlorinated biphenyl, which is considered difficult to decompose, is completely decomposed into a harmless material, and a nitrogen oxide attendant upon high-temperature incineration is not produced because of pyrolysis in a nonoxygen atomsphere.

QUINONE

Incineration

Incineration in a properly designed and operated incinerator is recommended as the most satisfactory method for the disposal of waste quinone (64). It is necessary that the incinerator be designed to burn the quinone completely (minimum of 1800°F for at least 2.0 seconds) and that leakage of unburned quinone vapors be eliminated. This option is applicable to both concentrated and dilute quinone wastes provided that an appropriate incineration method is used.

Biological Degradation

Biochemical wastewater treatment processes are less satisfactory than incineration for the disposal of quinone waste streams. Quinone is toxic to most microbiota, is relatively resistant to oxidation since it is itself an oxidizing agent, and is sufficiently volatile to be a potential odor and toxicity problem in open lagoons. Treatment by the activated sludge process using well acclimated seed is the most satisfactory biochemical treatment method for chemically unmodified quinone wastes. Quinone wastes can be rendered significantly more biodegradable by prior reduction to hydroquinone or a hydroquinone derivative. Sodium sulfite and sulfur dioxide are suitable reducing agents. Reduction to hydroquinone, in addition to providing a more easily destroyed compound, also eliminates the problems posed by the volatility of quinone.

Landfill

Land burial disposal processes are generally less satisfactory than incineration or biological degradation for disposal of quinone wastes. The high sublimation pressure of this material poses the threat of release of toxic quinone fumes to the atmosphere. Another possible route to environmental contamination is the leaching of this water-soluble material by rain or liquid wastes. A properly located and operated burial facility meeting California Class 1 requirements may limit the impact of leaching but is unlikely to completely eliminate the danger of sublimation (64).

TETRAALKYLLEAD COMPOUNDS

Dilute, aqueous solutions of organic lead compounds generated in the manufacturing process are the major type of TEL and TML wastes. The contaminated water is produced during steam distillation, washing, purification and maintenance operations. The aqueous wastes are collected in settling pits to recover lead and insoluble lead salts. The water from the settling pits is then treated to adjust its pH to 8 to 9.5 in the presence of precipitating agents such as ferrous sulfate or sodium carbonate. Precipitated lead salts are recovered for recycling to lead metal, and the water is allowed to return to the environment.

Despite the ubiquity of the disposal method described above, it is now generally agreed that it does not reduce the lead concentration in effluent streams to acceptable levels. In view of the inadequacy of the current disposal process, the DuPont plant in Antioch, California has been forced to store its lead-containing wastes pending the completion of an incineration facility to convert them to a form more amenable to recycle by lead smelters. The problem of lead recycling is especially complicated at the Antioch plant, which uses a continuous rather than batch TEL synthesis (64).

The Baton Rouge plant of the Ethyl Corporation has also experienced difficulty in maintaining acceptably low lead effluent concentrations. They are pursuing conventional (unspecified) methods of reducing those concentrations.

The scale and sludge which form in TEL and TML storage, mixing and transport vessels are another source of tetraalkyl lead waste. The former practice of dumping the old tanks containing lead residues at sea is no longer used. The lead sludges are buried or sold to secondary smelters for recycle. Petroleum refineries follow the practice of exposing the sludges to the air (weathering) until they are "inactive," after which they are landfilled. Presumably, the weathering process results in the slow oxidation of the lead compounds to insoluble PbO. However, it seems likely that the evaporation of toxic organic lead compounds cannot be avoided in this essentially uncontrolled exposure and that this method may result in local atmospheric organic lead pollution. This method of disposal is not adequate (64).

The disposal processes outlined in the following subsections for tetraalkyl lead compounds do not appear to be in actual use at this time. They are currently under development or have been reported in the literature (64).

Incineration

An incineration process to convert organic and inorganic lead wastes to PbO for recycle to lead metal is now under development (64). The use of this or a similar process for treatment of TEL and TML waste streams is recommended as the most desirable method of reducing their deleterious effects on the environment. An acceptable lead waste burning installation must be fitted with efficient scrubbing devices to prevent the contamination of stack effluents with toxic lead or lead-containing particulates and/or vapors.

Chemical Degradation

A number of methods designed to improve the removal of lead-containing compounds from TEL and TML aqueous wastes are now under study. One such method involves the treatment of aqueous effluents from tetraalkyl lead manufacture by: [1] adjusting the pH of the effluent to between 8.0 to 9.5, [2] intimately contacting the aqueous effluent with an ozone-containing gas in the presence of a soluble carbonate, [3] precipitating the converted inorganic lead compounds, and [4] separating the precipitated compounds. Reduction of the dissolved organic lead content to below 5 ppm is claimed. Any improved precipitation should be coupled to a process for conversion of the precipitated lead compounds to metallic lead for reuse.

Strongly acidic cation exchange resins have been shown to effect almost complete removal of organic lead compounds from typical TEL aqueous waste streams. The inorganic lead compounds were removed by conventional means before the stream was introduced to the ion exchange column and the final concentration of lead in the effluent water was less than 1 ppm. After elution from the column with caustic soda the eluate was subjected to oxidative chlorination, affording almost complete conversion to recoverable inorganic lead compounds. Although a judgment with respect to economic practicality of this process is not possible with the data in hand, the removal of organic lead compounds from TEL wastes with ion exchange resins appears to be technically attractive. A complete disposal system incorporating this process must also provide for rendering the inorganic lead obtained suitable for conversion to metallic lead.

TETRANITROMETHANE

The manufacturer, Hummel Chemical Company, will buy back for recycle any excess tetranitromethane in storage (64). If it becomes necessary to dispose of excess or contaminated tetranitromethane, the procedure recommended by JANAF Hazards Working Group for liquid nitro compounds is presumably followed. The procedure calls for open burning, and may be summarized as follows: Tetranitromethane, collected or stored in drums, cans or carboys, is destroyed by open burning at remote burning sites. Container tops or drum bungs are removed, combustible material carefully placed around the containers (avoiding any spillage) and the tetranitromethane is ignited with a black powder squib.

This procedure is not entirely satisfactory, since it makes no provision for the control of the toxic effluents, NO_x and HCN. Suggested procedures are to employ modified enclosed pit burning, using blowers for air supply, and passing the effluent combustion gases through loosely packed earth (as an adsorbent), or through wet scrubbers. No engineering data on tetranitromethane combustion product characteristics is available for use in design of modified enclosed pit burning systems.

TRICRESYL PHOSPHATE

Landfill

TCP is used as a lubricant and hydraulic fluid additive as well as being a major constituent in many synthetic lubricants and special purpose liquids. In these applications, small amounts are used at many scattered sites. These fluids are periodically changed due to breakdown, contamination, equipment failure or routinely in critical applications. Their exact fate is not known but it is generally thought that most material is landfilled at local municipal sites. The small amounts landfilled from this source along with the low solubility of TCP (20 ppm) probably presents no excessive long-term hazard to underground water supplies. However, TCP bulk soil concentration should be maintained below 0.005 parts per million (64). Refined TCP is produced by distilling technical material under vacuum. The chemical composition of the distillation bottoms is similar to TCP but of higher molecular weight. Landfill of concentrated TCP wastes is recommended only in sites meeting California Class 1 standards.

Incineration

TCP can be destroyed with proper incineration. This is adequate as long as phosphorus oxide emissions are not significant.

XYLOLS

Incineration

A process developed by C. Belliot et al (85) for the oxidative destruction of gaseous organic compounds consists essentially of passing the organic compounds and an excess of air at a temperature from about 180° to about 800°C over a catalyst. The catalyst consists essentially of from about 50 to about 85% by weight of manganese oxides, primarily in the form of rhombohedral manganese dioxide; up to about 30% by weight of manganese carbonate; from about 3 to about 10% by weight of cupric oxide; and up to about 15% by weight of nickel oxides in a refractory excipient. Organic compounds destroyed include xylenols, pyridine, methylpyrrolidone, thiophene, and the like.

REFERENCES

(1)   API Petrochemical Waste Questionnaire (1958).

(2)   N.V. Chalov and L.P. Volskaya, "Decontamination of Waste Liquors Containing Phenols, Aldehydes and Methanol," *Zh. Prikl.*, 25, 1082-8 (1952); *Chem. Abstr.*, 47, 3497b (1953).

(3)   E.R. Strong and R. Hatfield, "Treatment of Petrochemical Wastes by Superactivated Sludge Process," *Ind. Eng. Chem.*, 46 [2] 308-16 (1954).

(4)   H. Heukelekian, et al., "1957 Literature Review," *Sewage Ind. Wastes,* 30 [6] 717-873 (1958).

(5)   J.R. Cushman and J.R. Hayes, "Pilot Plant Studies of Pharmaceutical Wastes at the Upjohn Company," *Purdue Univ. Eng. Bull. Ext. Ser.,* 91, 62-72, Lafayette, Ind. (1956).

(6)   R.E. McKinney and J.S. Jeris, "Metabolism of Low-Molecular-Weight Alcohols by Activated Sludge," *Sewage Ind. Wastes,* 27 [6] 728-35 (1955).

(7)   J.L. Reagan, "Pilot Study of Synthetic Organic Waste Disposal on Trickling Filters," paper presented at Fourth Ann. Regional Conf. on Indus. Health, 103, Houston (1951).

(8)   H.F. Elkin, "Condensates, Quenches and Wash Waters as Petrochemical Waste Sources," *Sewage Ind. Wastes,* 31 [7] 836-40 (1959).

(9)   K.L. Schulze and B.N. Raju, "Studies of Sludge Digestion and Methane Fermentation: II. Methane Fermentation of Organic Acids," *Sewage Ind. Wastes,* 30 [2] 164-83 (1958).

(10)   D. Tarvin and A.M. Buswell, "The Methane Fermentation of Organic Acids and Carbohydrates," *J. Am. Chem. Soc.,* 56, 1751-5 (1934).

(11)   R.V. Green and D.V. Moses, "Destructive Catalytic Oxidation of Aqueous Waste Materials," *Sewage Ind. Wastes,* 24 [3] 215-21 (1952).

(12)   W.W. Eckenfelder, R. Klefferman and J. Walker, "Some Theoretical Aspects of Solvent Stripping and Aeration of Industrial Wastes," *Purdue Univ. Eng. Bull. Ext. Ser.,* 91, 14-25, Lafayette, Ind. (1956).

(13)   R. Hatfield, "Biological Oxidation of Some Organic Compounds," *Ind. Eng. Chem.,* 49 [2] 192-6 (1957).

(14)   S.J. Paradiso, "Disposal of Fine Chemical Wastes at the Upjohn Company," *Purdue Univ. Eng. Bull. Ext. Ser.,* 89, 49-60, Lafayette, Ind. (1955).

(15)   C.B. Lamb and G.F. Jenkins, "B.O.D. of Synthetic Organic Chemicals," *Purdue Univ. Eng. Bull. Ext. Ser.,* 79, 326-39, Lafayette, Ind. (1952).

(16)   E.J. Mills, Jr., and V.T. Stack, Jr., "Acclimation of Microorganisms for the Oxidation of Pure Organic Chemicals," *Purdue Univ. Eng. Bull. Ext. Ser.,* 87, 449-64, Lafayette, Ind. (1954).

(17)   E.J. Mills, Jr., and V.T. Stack, Jr., "Biological Oxidation of Synthetic Organic Chemicals," *Purdue*

Univ. Eng. Bull. Ext. Ser., 83, 492-517, Lafayette, Ind. (1953).

(18) L.D. Dougan and J.C. Bell, "Waste Disposal at a Synthetic Rubber Plant," *Sewage Ind. Wastes,* 23 [2] 181-7 (1951).

(19) R.F. Rocheleau, "Incineration of Organic Wastes," *Proc. Air Water Pollution Abatement Conf.,* 89-98, Mfg. Chemists Assoc., Washington, D.C. (1957).

(20) D.W. Hood, B. Stevenson, and L.M. Jeffrey, "Deep Sea Disposal of Industrial Wastes," *Ind. Eng. Chem.,* 50 [6] 885-8 (1958).

(21) I. Gellman and H. Heukelekian, "Biological Oxidation of Formaldehyde," *Sewage Ind. Wastes,* 22 [10] 1321-5 (1950).

(22) H. Heukelekian and M.E. Rand, "Biochemical Oxygen Demand of Pure Organic Compounds," *Sewage Ind. Wastes,* 27 [9] 1041-53 (1955).

(23) G. Taylor, "Chemical Oxidation: Some Laboratory Experiments in the Treatment of Chemical Trade Waste," *Water San. Engr.,* 3 [1] 31-3 (1952).

(24) M.M. Wells, "The Reactions and Resistance of Fishes in Their Natural Environment to Salts," *J. Exptl. Zool.,* 19 [3] 243-83 (1915).

(25) N.H. Kirchgessner, "Treatment of Phenolic Plastics Manufacturing Wastes," *Sewage Ind. Wastes,* 22 [10] 1314-20 (1950).

(26) F.J. Ludzack, R.B. Schaffer, R.N. Bloomhuff and M.B. Ettinger, "Biochemical Oxidation of Some Commercially Important Organic Cyanides," *Sewage Ind. Wastes,* 31 [1] 33-44 (1959).

(27) A.B. Cherry, A.J. Gabaccia, and H.W. Senn, "The Assimilation Behavior of Certain Toxic Organic Compounds in Natural Water," *Sewage Ind. Wastes,* 28 [9] 1137-46 (1956).

(28) C.E. Renn, "Biological Properties and Behaviors of Cyanogenic Wastes," *Sewage Ind. Wastes,* 27 [3] 297-310 (1955).

(29) J.T. Garrett, "Multipurpose Incineration," *Ind. Wastes,* 2 [5] 111-3 (1957).

(30) B.W. Dickerson, "High-Rate Trickling Filter Operation on Formaldehyde Wastes," *Sewage Ind. Wastes,* 22 [4] 536-45 (1950).

(31) N.H. Kirchgessner, "Phenol Recovery by Use of Isopropyl Ether," *Sewage Ind. Wastes,* 30 [2] 191-8 (1958).

(32) T. Waldmeyer, "Treatment of Formaldehyde Wastes by Activated Sludge Methods," *Surveyor Municipal County Engr.,* III, 445-7 (July 12, 1952).

(33) B.W. Dickerson, "A High-Rate Trickling Filter Pilot Plant for Certain Chemical Wastes," *Sewage Works J.,* 21 [4] 685-93 (1949).

(34) "Anaerobic Fermentations," *Illinois State Water Surv. Bull.,* 32, 51-3, Urbana, Ill. (1939).

(35) G.E. Eden, A.M. Freke and K.V. Melbourne, "Treatment of Waste Waters Containing Hydrogen Peroxide, Hydrazine, and Methanol," *Chem. Ind. (London),* 1104-6 (Dec. 15, 1951).

(36) I.E. Wallen, W.C. Greer and R. Lasater, "Toxicity to Gambusia Affinis of Certain Pure Chemicals in Turbid Waters," *Sewage Ind. Wastes,* 29 [6] 695-711 (1957).

(37) G. Gutzeit, "Treatment of Complex Chemical Wastes Resulting from Tank Car Cleaning Operations," *Sewage Works J.,* 21 [1] 91-9 (1949).

(38) H.O. Henkel, "Surface and Underground Disposal of Chemical Wastes at Victoria, Texas," *Sewage Ind. Wastes,* 25 [9] 1044-9 (1953).

(39) A.N. Heller, E.W. Clark and W.M. Reiter, "Some Factors in the Selection of a Phenol Recovery Process," *Purdue Univ. Eng. Bull. Ext. Ser.,* 94, 103-22, Lafayette, Ind. (1957).

(40) D.I. Macht and H.P. Leach, "Pharmacological Studies of Twenty-three Isomeric Octyl Alcohols," *J. Pharmacol. Exp. Therap.,* 39, 71 (1930).

(41) R.G. Edmonds and G.F. Jenkins, "Recovery of Phenolics from Waste Effluents," *Chem. Eng. Progr.,* 50 [3] 111-5 (1954).

(42) B.A. Adams, "The Lethal Effect of Various Chemicals on Cyclops and Daphnia," *Water and Water Eng. (London),* 29, 361-4 (Sept. 20, 1927).

(43) I.S. Wilson, "Treatment of Chemical Wastes," *Surveyor Municipal County Engr.,* 113, 315-8 (Apr. 17, 1954).

(44) W.W. Mathews, "Treatment of Ammonia Still Wastes by the Activated Sludge Process," *Sewage Ind. Wastes,* 24 [2] 164-80 (1952).

(45) H.W. De Ropp, "Chemical Waste Disposal at the Victoria, Texas Plant of the DuPont Company," *Sewage Ind. Wastes,* 23 [2] 194-7 (1951).

(46) H.H. Black and V.A. Minch, "Industrial Waste Guide, Wood Naval Stores," *Sewage Ind. Wastes,* 25 [4] 462-74 (1953).

(47) N. Federgreen and A.J. Weinberger, "Methyl Styrene—A Case Study in Spent Acid Catalyst Treatment," *Ind. Eng. Chem.,* 49 [1] 46-8 (1957).

(48) F. Majewski, "The Treatment of Wastes at the Rohm and Haas Company," *Purdue Univ. Eng. Bull. Ext. Ser.,* 83, 328-45, Lafayette, Ind. (1953).

(49) B.W. Dickerson, "Treatment of Powder Plant Wastes," *Purdue Univ. Eng. Bull. Ext. Ser.,* 76, 30-42, Lafayette, Ind. (1951).

(50) D.T. Laurie, "Combustion and Bio-Oxidation Combined in Waste Plant at Chemstrand Corp." *Power Eng.,* 63 [4] 80-2 (1959).

(51) S.D. Faust and H.E. Orford, "Reducing Sludge Volume with Crystal Seeding in Disposal of Sulfuric Acid Wastes," *Ind. Eng. Chem.*, 50 [10] 1537-8 (1958).

(52) A.D. McRae, "Disposal of Alkaline Wastes in the Petrochemical Industry," *Sewage Ind. Wastes,* 31 [6] 712-8 (1959).

(53) H.D. Lyon, "Disposal of Synthetic Organic Wastes," *Chem. Eng. Progr.,* 46 [8] 388-94 (1950).

(54) "Biological Waste Treatment," *Petrochem. Ind.,* 1 [4] 16-9 (Oct. 1958).

(55) H.W. Eustace and R.J. McVeigh, "Design of a Sedimentation-Type Waste Treatment Plant," *Proc. Air Water Pollution Abatement Conf.,* 99-112, Mfg. Chemists Assoc., Washington, D.C. (1957).

(56) Booz-Allen Applied Research, Inc., *A Study of Hazardous Waste Materials, Hazardous Effects and Disposal Methods,* Report PB 221 467, Springfield, Va., Nat. Tech. Information Service (July 1973).

(57) P.S. Sharpe, U.S. Patent 3,892,190; July 1, 1975; assigned to Brule' C.E. and E., Inc.

(58) R.S. Ottinger, J.L. Blumenthal, D.F. Dal Porto, G.I. Gruber, M.J. Santy and C.C. Shih, *Recommended Methods of Reduction, Neutralization, Recovery or Disposal of Hazardous Waste,* Vol VIII, *Misc. Inorganic and Organic Compounds,* Report PB 224 587, Springfield, Va., Nat. Tech. Information Service (August 1973).

(59) W.O. Fitzgibbons, E.M. Schwerko and A.H. Brainard; U.S. Patent 3,734,943; May 22, 1973; assigned to The Standard Oil Co. (Ohio).

(60) H.R. Sheely; U.S. Patent 3,895,050; July 15, 1975; assigned to The Badger Co., Inc.

(61) J.L. Callahan, H.F. Hardman and R.K. Grasselli; U.S. Patent 3,804,756; April 16, 1974; assigned to The Standard Oil Co. (Ohio).

(62) R.S. Ottinger, J.L. Blumenthal, D.F. Dal Porto, G.I. Gruber, M.J. Santy and C.C. Shih, *Recommended Methods of Reduction, Neutralization, Recovery or Disposal of Hazardous Waste,* Vol X, *Organic Compounds,* Report PB 224 589, Springfield, Va., Nat. Tech. Information Service (August 1973).

(63) D.L. Miller; U.S. Patent 2,707,171; April 26, 1955; assigned to Sharples Chemical, Inc.

(64) R.S. Ottinger, J.L. Blumenthal, D.F. Dal Porto, G.I. Gruber, M.J. Santy and C.C. Shih, *Recommended Methods of Reduction, Neutralization, Recovery or Disposal of Hazardous Waste,* Vol XI, *Organic Compounds (Continued),* Report PB 224 590, Springfield, Va., Nat. Tech. Information Service (August 1973).

(65) M. Reichert and L. Unterstenhoefer; U.S. Patent 3,145,076; August 18, 1964; assigned to Badische Anilin- & Soda-Fabrik AG.

(66) M. Reichert, U. Wagner, W. Heitmuller and G. Huebner; U.S. Patent 3,401,654; September 17, 1968; assigned to Badische Anilin- & Soda-Fabrik AG.

(67) M. Akune and K. Yoshii; U.S. Patent 3,912,577; October 14, 1975; assigned to Nittetsu Chemical Engineering Co.

(68) F.L. Gladney and M.L. Turner; U.S. Patent 3,764,254; October 9, 1973; assigned to The Dow Chemical Co.

(69) J.A. Cull and T. Hooker; U.S. Patent 3,140,155; July 7, 1964; assigned to Hooker Chemical Corp.

(70) J.A. Cull and T. Hooker; U.S. Patent 3,220,798; November 30, 1965; assigned to Hooker Chemical Corp.

(71) R.G. Woodland and M.C. Hall; U.S. Patent 3,445,192; May 20, 1969; assigned to Hooker Chemical Corp.

(72) S. Ezaki; U.S. Patent 3,589,864; June 29, 1971; assigned to Yawata Chemical Engineering Co. Ltd.

(73) W. Sansom and F.W. Thompson; U.S. Patent 3,712,796; January 23, 1973; assigned to E.I. du Pont de Nemours & Co.

(74) J.M. Bruce, Jr.; U.S. Patent 3,845,191; October 29, 1974; assigned to E.I. du Pont de Nemours & Co.

(75) A.W. Reed, *Ocean Waste Disposal Practices,* Park Ridge, N.J., Noyes Data Corp. (1975).

(76) J.D. Hummel; U.S. Patent 3,748,081; July 24, 1973; assigned to PPG Industries, Inc.

(77) J.D. Hummel; U.S. Patent 3,834,855; September 10, 1974; assigned to PPG Industries, Inc.

(78) G. Hegner; U.S. Patent 3,296,125; January 3, 1967; assigned to Friedrich Uhde GmbH.

(79) H.C. Schutt; U.S. Patent 3,432,399; March 11, 1969; assigned to The Fluor Corp., Ltd.

(80) K. Kishigami, H. Kobayashi, T. Sente and K. Sugiyama; U.S. Patent 3,888,194; June 10, 1975; assigned to Babcock-Hitachi KK.

(81) W. Kube; U.S. Patent 3,583,339; June 8, 1971; assigned to Badische Anilin- & Soda-Fabrik AG.

(82) I.Y. Sigal, G.F. Naidenov, N.A. Gurevich, J.I. Danilevich, A.M. Shiman, V.F. Shishkin and N.I. Goncharenko; U.S. Patent 3,782,302; January 1, 1974.

(83) D.C. Ferrari and C.G. Bertram; U.S. Patent 3,624,984; December 7, 1971; assigned to The Badger Co., Inc.

(84) T. Nakamura and Y. Iwasaki; U.S. Patent 3,841,239; October 15, 1974; assigned to Shinmeiwa Kogyo KK.

(85) C. Belliot, E. Cheylan, S. Madelaine and J. Ebbing; U.S. Patent 3,925,535; December 9, 1975; assigned to Compagnie Francaise Thomson Houston-Hotchkiss Brandt.

PAINT INDUSTRY WASTES

The U.S. paint industry is a source of waste sludge that contains a variety of heavy metal compounds including three based on lead, cadmium and chromium. The manufacture of paint contributes a relatively modest amount of sludge for disposal. In total, this probably approximates six million gallons per year. On the basis of 1970 pigment consumption by the coating industry, this sludge contains over 600,000 pounds of lead pigments, 150,000 pounds of chromium pigments, and about 5,000 pounds of cadmium pigments. About 1,700 establishments, most relatively small, make paint.

The amount of waste generated by the production of paint is modest compared to the paint sludge that originates with spray paint operations which utilize production finishes. It is estimated that close to 100 million gallons per year of paint sludge are generated through spray painting. The pigment content per gallon is lower than that of the sludge from paint manufacture both by virtue of the method of generating the sludge and the types of paints used for industrial product finishes. Nevertheless, the total poundage of heavy metal pigments produced as wastes from spray painting operations is quite probably an order of magnitude larger than the pigment contained in wastes from paint manufacture, as described by Arthur D. Little, Inc. (1).

The inorganic pigments—which include compounds of lead, chromium and cadmium—are principally responsible for the toxicity of paints and paint sludges. Paint waste also contributes modest amounts of other toxic materials added to improve stability or performance. These include phenyl mercury compounds, used as a mildewcide. These compounds are used virtually only in water base paints in very low concentration. During 1971, about 600,000 pounds of mercury were consumed by the paint industry. In addition, modest amounts of other toxic materials, such as lead-based dryer additives, are used in the manufacture of solvent-based paints. The contribution of heavy metal compounds to the paint by these additives is modest, however, compared to the amount of heavy metal compounds present in the paint as pigments.

As a consequence of paint manufacture, toxic wastes are produced in the form of [a] finished paint which is insoluble for whatever reason, and [b] waste from washing and cleaning operations. Waste contained in wash solvents is eventually concentrated either by settling or, in the case of solvent paint, by settling and distillation.

Industry sources indicated a very wide range of waste in actual practice—from one gallon of sludge per 60 gallons of finished paint for a large plant with varied output to as little as one gallon of sludge per 500 gallons of paint for small plants manufacturing a high proportion of the same product. There is a strong inclination, particularly among the smaller producers, to utilize every bit of raw material possible and, with continuous batch production of the same product, washings normally go into subsequent batches.

Generally speaking, there is more waste in the production of latex paint; it has been estimated that one gallon of the sludge is obtained from the decantation and solvent recovery operations for every 120 gallons of solvent based paint produced.

INCINERATION

The ultimate disposal of hydrocarbons and odor-producing substances from the paint and varnish industry has been described by E.J. Dowd (2). The only control technique currently being used that has proven effective for all cases is combustion. Three general methods are employed to combust waste gases, as shown below.

> [1]   Flame combustion
> [2]   Thermal combustion
> [3]   Catalytic combustion

All of the above methods are oxidation processes. Ordinarily, each requires that the gaseous effluents be heated to the point where oxidation of the combustible will take place. The three methods differ basically in the temperature to which the gas stream must be heated.

Flame combustion is the easiest of the three to understand, as it comes the closest to everyday experience. When a gas stream is contaminated with combustibles at a concentration approaching the lower flammable limit, it is frequently practical to add a small amount of natural gas as an auxiliary fuel and sufficient air for combustion when necessary, and then pass the resulting mixture through a burner. The contaminants in the mixture serve as a part of the fuel. Flame incinerators of this type are most often used for closed chemical reactors. They are not used on resin reactors at present. They may be an ideal solution some day, however, when methods of operating a closed, pressurized resin reactor are developed.

It is far more likely that the concentration of combustible contaminants in an air stream will be well below the lower limit of flammability. When this is the case, direct thermal combustion is considerably more economical than flame combustion. Direct thermal combustion is carried out by equipment such as that illustrated in Figure 115. In this equipment, a gas burner is used to raise the temperature of the flowing stream sufficiently to cause a slow thermal reaction to occur in a residence chamber.

Whereas flame temperatures bring about oxidation by free radical mechanisms at temperatures of 2500°F and higher, thermal combustion of ordinary hydrocarbon compounds begins to take place at temperatures as low as 900° to 1000°F. Good conversion efficiencies are produced at temperatures in the order of 1400°F with a residence time of 0.3 to 0.6 seconds.

Catalytic combustion is carried out by bringing the gas stream into intimate contact with a bed of catalyst. In this system, the reaction takes place directly upon the surface of the catalyst, which is usually composed of precious metals, such as platinum and palladium. While thermal combustion equipment brings about oxidation at concentrations below the limits of flame combustion, catalytic combustion operates below the limits of flammability and below the normal oxidation temperatures of the contaminants. The reaction is instantaneous by comparison to thermal combustion and no residence chamber is required. Catalytic combustion is carried out by equipment such as that illustrated in Figure 116.

In general, catalytic afterburners are less expensive to operate; however, they depend directly on the performance of the catalyst for their effectiveness. It will not function properly if the catalyst becomes deactivated. Because of this, catalytic units are not inherently functional when operated at design conditions. In many areas, means for ensuring adequate performance of the catalyst on a long term basis will be required by environmental control offices.

FIGURE 115: THERMAL AFTERBURNER SYSTEM FOR RESIN REACTOR OR CLOSED KETTLE

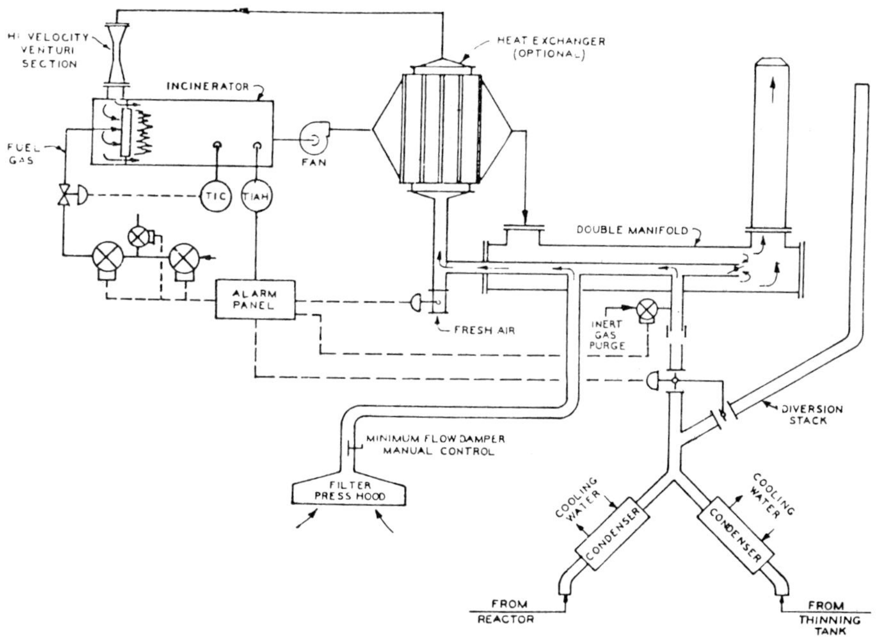

FIGURE 116: SCHEMATIC DIAGRAM OF A CATALYTIC INCINERATION SYSTEM FOR VARNISH KETTLES

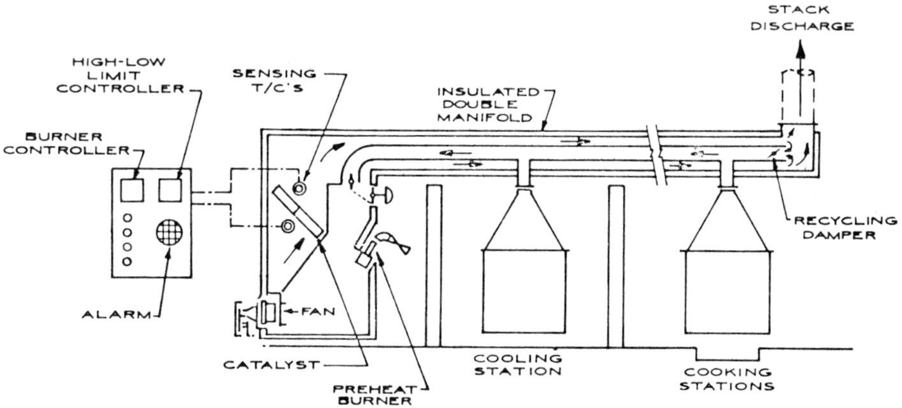

Source: PB 238 058

The basis for design of either catalytic or thermal combustion is the hydrocarbon concentration of the exhaust gases handled by the incinerator. The maximum hydrocarbon level is set by most insurance companies at one-quarter of the lower explosive limit (LEL) which is equivalent to 13 Btu/scf of exhaust gas. As outlined earlier, the quantity of emission may vary significantly with cooking time and the type of cook. There is also likely to be a large variety of different hydrocarbons emitted. For this reason, theoretical calculations of emissions for design purposes are not satisfactory. On site emission measurements are required.

Once the rate of emission is determined, it is then necessary to calculate the dilution air required to meet ¼ LEL and set up the duct work system to provide for this dilution. When possible, dilution air should be utilized to help capture as many fugitive fume emissions as possible. For example, this can be accomplished by taking the dilution air from a hood positioned over the resin filter press and venting the thinning tanks and product rundown tanks into the same system.

A concentration of ¼ LEL or 13 Btu/scf will give a temperature rise of about 600°F in the afterburner. This is too high if a heat exchanger is to be used, and in these cases, it would be necessary to dilute to a maximum concentration of 12 Btu/scf. In all cases, the heat exchangers will be the parallel flow type having a thermal efficiency of 42%. This is required due to the high emission concentration to assure temperature balance and control.

The major problem with catalytic or thermal afterburners as applied to open or closed resin and varnish kettles is the danger of fires and/or explosions. This has happened in numerous occasions in the past due primarily to excessive hydrocarbon emission from kettles. These problems have been all but eliminated on newer units by assuring that the design was based on actual emission measurements of the highest emitting cook and the addition of some of the following system safety features:

[1] High limit temperature alarm to shut off burner and activate a diversion system.

[2] High velocity duct section to assure gas flow to afterburner substantially exceeds flame propagation velocity of hydrocarbons being burned.

[3] Double manifolding or hot gas recycle to prevent condensation of heavy hydrocarbons or phthalic anhydride.

[4] Diversion system to block off hydrocarbon emissions to unit, by-passing them directly out of separate exhaust and introducing fresh air to purge the unit.

[5] Pneumatic operation of the diversion system to assure fast positive action and provide a fail-safe system in the event of either air or electrical failure.

[6] Purging with inert gas in the event of power failure.

The above general requirements are applicable to all types of afterburner control.

The envisaged process for the incineration of solvent recovery sludges, reaction waters and latex wastes is illustrated in Figure 117. Capacity was based on the production of 125,000 gallons annually of solvent recovery sludge which had an average heating value of 120,000 Btu per gallon. Operating conditions were selected on the basis of experimental runs made during the study, (i.e., bed temperature 1750°F, superficial velocity 2.0 fps).

Calculations made to determine material and energy balances indicate that an incinerator of about 6 feet in diameter would be required for this sytem. The results also indicate that approximately 54 gallons per hour of latex wastes and/or reaction water could be disposed of by incineration concurrently with the waste sludge.

FIGURE 117: SIMPLIFIED FLOWSHEET OF INCINERATOR FOR PAINT WASTES

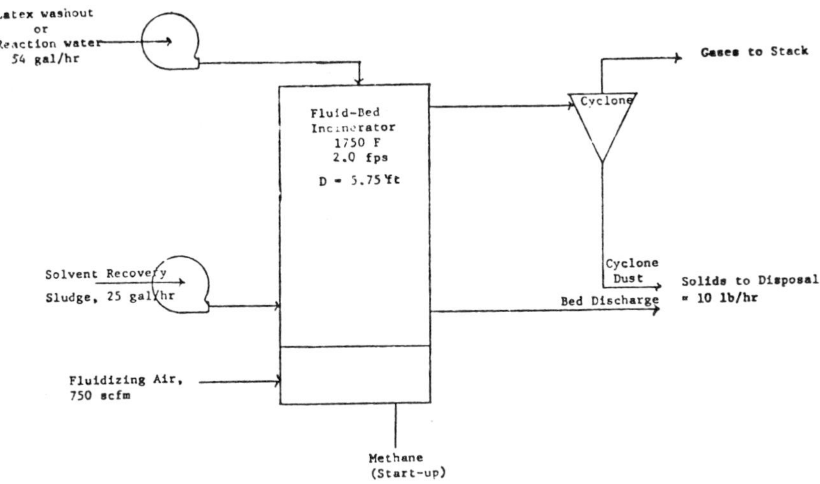

Source: Report 12020 FYF 03/72

LANDFILL DISPOSAL

Wastes from paint manufacture, as sludge and finished paint, have been drummed and disposed of in landfill operations, but local regulations are making this practice difficult. There is increasing dependence on disposal contractors paid to haul wastes away and assume responsibility for disposal. Small paint plants often store wastes for future use (1). This practice has probably grown because of the problems presented by disposal. Storage is used primarily with solvent paint systems since latexes do not mix and store well.

The process of G. Slikkers, Jr. (4) involves recovering landfill suitable solids from sludges composed of industrial waste solids and organic solvents. The process comprises mixing an alkali or alkaline earth metal salt into a sludge comprised of industrial waste solids selected from epoxy resins, polyester resins, phenolic resins, cellulosic fibers, polyolefins, vinyls, acrylics and organic and inorganic ink and paint pigments, and being dissolved or suspended in industrial organic solvents selected from aliphatic or aromatic esters, ketones, aldehydes, alcohols, glycols, ethers and halogenated hydrocarbons. Sufficient alkali or alkaline earth metal salt is added in stoichiometric excess based on the industrial waste solid constituents of the sludge to create a mixture having a pH above 7. The mixture is gradually added to boiling water in a still and steam distilled. The industrial organic solvent is recovered through the steam distillation step; the still-bottom residue, comprising the organic-solvent-free industrial waste solids is removed from the still and disposed of in a landfill.

REFERENCES

(1) Arthur D. Little, Inc., *Alternatives to the Management of Hazardous Wastes at National Disposal Sites,* Vol. II, Report PB 237 264, Springfield, Va., Nat Tech Information Service (1973).
(2) E.J. Dowd, *Air Pollution Control Engineering and Cost Study of the Paint and Varnish Industry,* Report PB 238 058, Springfield, Va., Nat Tech Information Service (June 1974).
(3) T.L. Tewksbury, A.K. Reed, H. Nack and G.R. Smithson, Jr., *Fluidized Bed Incineration of Selected Carbonaceous Industrial Wastes,* Report 12020 FYF 03/72, Wash., D.C., U.S. EPA (March 1972).
(4) G. Slikkers, Jr.; U.S. Patent 3,929,586; December 30, 1975; assigned to Organic Chemicals Co., Inc.

PESTICIDE INDUSTRY WASTES

Control of wastes in the manufacture of pesticides involves strict process controls, to reduce to a minimum the wastes which must be treated, and advanced waste treatment measure, to successfully treat toxic wastes prior to their release to the air or water (1).

Many methods are used to recover plant wastes. Curbs and collecting sumps are placed around pumping areas. Tanks are used to collect pump drippings and accidental losses. Such lost material may be returned to the process. Drain tiles, connected to a collecting sump, catch contaminants which may have seeped into the ground. Filters and scrubbers recover particulates. Industrial vacuum cleaners are used to immediately clean up spills of dry materials.

Frequently, a buffer unit is placed between the processing plant and the final waste treatment unit. It is a method of control, equalization, and stabilization. If the maximum capacity of the intermediate buffer unit is reached, the waste generating plant is shut down. The buffer unit serves also as a trap so that high process losses can be given additional or different treatment. Buffer units used may be tanks, receiving ponds, or sumps.

Empty containers are never abandoned or allowed to accumulate in an area accessible to humans or animals. They are either burned, if combustible, or decontaminated, if noncombustible. If not contaminated, rinse solutions are burned in an isolated area away from water supplies (1).

TYPES OF PESTICIDE WASTES

Virtually every pesticide production process produces aqueous or gaseous streams and frequently solid wastes which contain unreacted ingredients, unrecovered products and solvents, and unavoidable or undesired by-products. Extensive efforts are usually made to minimize by-products and to recover, recycle, or otherwise prevent these process losses from occurring. For each process, however, a balance point is eventually reached between the expense of recovery and the value of the recovered product.

In the past, the economic considerations were frequently dominant and process losses were included as unavoidable costs. Under the recent emphasis on environmental contamination further efforts have been made to recover many previously lost materials, even when economics indicated that it was more expensive to do so, and most pesticide manufacturers have invested in or are in the process of building extensive waste treatment facilities wherein those wastes which cannot be recovered are degraded to acceptable levels or disposed by state approved methods. A summary of the principal waste generated and the disposal method employed by the producers of the key pesticides is shown in Table 33.

TABLE 33: SUMMARY OF MANUFACTURING WASTES AND DISPOSAL

Pesticide	Liquid Wastes	
	Source	Disposal
DDT	Processing solutions	Evaporative basin
Aldrin	Floor washings, etc.	Evaporation basin
Dieldrin	Process solutions	Evaporation basin
Chlordane	Process solutions	Deep well
Toxaphene	Pinene-Camphene plant	Bio-treatment plant
	Process solutions	Neutralize, hold, discharge
Disulfonton	Process solutions	Secondary treatment plant
Malathion	Process solutions	Barge to deep sea
Phorate	Process solutions	Barge to deep sea
Parathions	Process solutions	Waste treatment plant
Carbaryl	Process solutions	Secondary waste treatment
Aldicarb	Process solutions	Neutralize, secondary waste treatment
2,4-D (Dow)	Process solutions	Trickling filter; biological waste treatment plant
2,4-D (Chipman)	Process solutions	Charcoal absorption/ filtration treatment
2,4,5-T (Dow)	As per 2,4-D	
2,4,5-T (Thompson-Hayward)	Process solutions	Oxidation pond, discharge
Atrazine	Process solutions	Most to river; some to deep well
Trifluralin	Process solutions	Biological waste treatment
Alachlor	Process solutions	Discharge
Captan	Process solutions	Hold, discharge
Methyl Bromide		
Pyrethrin	Aqueous still bottoms	Sewer
Bacillus _t._ (Abbott)	Process solutions	Sterilized; biological waste treatment
Bacillus _t._ (Nutrilite)	Process solutions	Evaporation pond
$HgCl_2$-Hg_2Cl_2	Process solution	Hg-recovery; discharge to sewer

Pesticide	Solid or Other Wastes	
	Source	Disposal
DDT	Reactor solutions	County Dump
Aldrin	Lime slurry	Lime pit
Dieldrin	Filter solids	Incinerate
Chlordane	Filter solids	Clay pit
Toxaphene		
	Filter solids	Solid waste
Disulfonton	Filter solids, etc.	Commercial landfill
	H_2S	Flare
Malathion	Filter solids	Landfill (with lye)
Phorate	Filter solids	Landfill
	Mercaptan losses	Flare
Parathions	H_2S; S	Flare; incinerate
Carbaryl	H_2, $COCl_2$, amine	Flare
	Heavy residues	Incinerate
Aldicarb	Process vents	Flare

(continued)

TABLE 33: (continued)

Pesticide	Solid or Other Wastes	
	Source	Disposal
2,4-D (Dow)	Filter solids and still bottoms	Incinerate, Scrub
2,4-D (Chipman)	-	-
2,4,5-T (Dow)		
2,4,5-T (Thompson-Hayward)	Solids	Landfill
Atrazine		
Trifluralin	NO_x	Scrubber
Alachlor	Solvent	Fuel
Captan	Gas streams	Scrub, vent
Methyl Bromide	Gaseous wastes	Scrub, waste treatment plant
Pyrethrin	Process solids	Storage
Bacillus t. (Abbott)	Filter solids	Landfill
Bacillus t. (Nutrilite)	Process air	Incinerate or filter
$HgCl_2$-Hg_2Cl_2	Filter solids NO_x, H_2S	Hg recovery Recovery

Source: PB 213 782

Ground disposal and incineration seem to predominate in Table 33 but ocean dumping, deep well disposal and other methods are used as will be discussed later in this chapter. Generally speaking, there are five methods of ground disposal used.

Deep Well: dug far from fresh water sources. Noxious fluid wastes are disposed of.

Sanitary Landfill: refuse is reduced to smallest possible volume, then covered with earth after each day's operation. There exists here the possibility of ground and surface water pollution.

Disposal Pits: excavations or dumps for the disposal of waste materials. They are left open to the air for extended periods of time. In addition, disposal of waste, both solid and liquid, is not controlled. Waste is dumped haphazardly, is not compacted and covered, and is left exposed to the elements.

Lagoons: shallow excavations or natural topographic depressions used as retention basins or ponds. The waste is oxidized or degraded biologically, suspended solids settling to the bottom, and evaporation then reduces the amount of effluent to be disposed of. However, there exists the great potential of ground and surface water pollution because the wastes are already in a fluid state.

Surface: liquid and solid wastes are evenly distributed on selected soil surfaces for pesticide degradation by oxidation, microbial metabolism, or photochemical transformation. If the pesticides are not rapidly metabolized, the possibility for reappearance of the unaltered pesticide in the environment is high.

Although pesticides are designed for widespread use on land, careful control of plant effluents is essential to the health of nearby streams. Unless care is taken to decompose these compounds by chemical, biological or thermal means, the persistency of some pesticides may permit them to leach into the underground water system.

Pesticides which enter streams may also enter the ecological chain by being taken up by plankton and subsequently eaten by higher levels of aquatic life until concentrated into edible fish species. Regulatory agencies and public interest will serve to advance the management and handling systems for these compounds. The disposal of small quantities of unused pesticides by various techniques has been reviewed by E.W. Lawless et al (2).

Organophosphorus Wastes: The treatment of organic phosphorous wastes can vary from lagooning to incineration. The system used by a manufacturer of several varieties of organic phosphorous pesticides was developed to treat a waste stream containing unreacted raw materials, partially reacted materials, cleaning products, solvents and other plant wastes. The steps involved include:

> pH adjustment with lime to form calcium phosphate
> Primary settlings
> Activated sludge processing
> Final settlings
> Sludge thickening
> Sludge dewatering

The deceleration times may extend to seven to ten days to ensure the destruction of the toxic compounds by the aerobic digestion system. The off-gases from the production of sigma phosphorus compounds (which include hydrogen sulfide and mercaptans) are incinerated. Residual sludges may be landfilled, sea dumped, or incinerated. In some cases, they are diluted and control fed to streams.

Organochlorine Wastes: Chlorinated organic compounds produce pollutant solids (dust concentrates and powders), liquids (waste solutions), or gases (vapors and mists). Open burning is not used since the hydrogen chloride (HCl) gas released would cause atmospheric pollution. The HCl volatilized inorganics or other acid gases formed are removed by scrubbing towers and/or activated carbon towers. Solid waste materials are either buried in an area designated for disposal of toxic materials, placed in a permanent stockpile, or sent to a settling pond. Liquid disposal may involve the concentration and incineration of combined wastes. Other pesticides and their disposal methods include:

Carbonates: These are somewhat easier to decompose. They are lower in toxicity, decompose quickly in soil, are insoluble in water, breakdown rapidly in alkali, and burn readily. Mixing with alkali decomposes carbonates. Reaction products may be sent to a sewage treatment facility. Caustic treatment in a settling tank is sufficient for water-soluble wastes. Solid wastes, which cannot be easily treated, are buried using landfill techniques. Empty containers are burned in unpopulated areas or else are buried.

Phenoxy Acids, Salts, and Esters: Disposal methods include incineration, chemical treatment (chlorination or precipitation), or biological treatment (trickling filters, activated sludge, or sewage lagoons). Deep well disposal is also used.

Inorganics: Three disposal methods are used: [a] burial, [b] incineration, or [c] municipal sewage system.

ORGANIC PESTICIDE WASTE DISPOSAL

The disposal of pesticide wastes in open pits, lagoons, unapproved landfill sites, and by on-site burning or deep sea burial are not recommended practices because of the obvious contributions to air and water pollution.

The only process recommended for treating concentrated chlorinated or DNOC pesticide wastes is incineration (3). The only adequate methods for the disposal of concentrated 2,4-D wastes are incineration, and soil surface application. The adequate methods for the disposal of concentrated organophosphorus pesticide wastes are: incineration, chemical degradation with 2 N to 8 N sodium hydroxide solution, and approved sanitary landfill.

Chemical Degradation

The use of chemical reagents to decompose concentrated chlorinated pesticide wastes to less toxic forms has been investigated. Mississippi State work (3) showed that liquid ammonia and ammonia and metallic sodium or lithium would completely decompose DDT, dieldrin, or 2,4-D but the reagents are dangerous to use and the toxicity of the degradation products is not known. The investigators demonstrated that the degradation of DDT can be effected by a mildly acidic solution of zinc powder or zinc-copper couple. The principal degradation product of DDT is bis(p-chlorophenyl) ethane, or DDT with all three aliphatic chlorines removed; however, although this material has been stated to be void of the neurotoxic effects of DDT, it nevertheless possesses insecticidal properties and thus detoxification of DDT is not complete.

In the same studies, it was also found that lindane could be completely degraded by the mildly acidic reduction of zinc powder, although the degradation products have not been identified. It has also been shown (3) that polychlorocyclodiene insecticides could be substantially degraded by the mildly acidic reduction of zinc powder or zinc-copper couple; however, the degradation products have not yet been identified and the soluble zinc ions formed are toxic and pose another environmental problem.

Caustic alkalies have been used to decontaminate chlorinated pesticide containers, but their actions on the concentrated pesticide wastes at ambient temperatures are relatively slow and involve the evolution of noxious hydrogen chloride. Based on the results to date and the effectiveness of incineration by comparison, chemical degradation could not be recommended as a method for the disposal of concentrated chlorinated pesticide wastes, especially when large quantities of these pesticide wastes are involved.

The use of chemical reagents to decompose concentrated organophosphorus pesticide wastes to less toxic forms has also been investigated (3) and shown that [1] sulfuric and nitric acids are not effective in destroying the organophosphorus pesticide malathion; [2] sodium hydroxide from 2 N to 8 N concentration will break down malathion sufficiently to yield inorganic phosphates; and [3] liquid ammonia and metallic sodium or lithium will completely decompose malathion, but the reagents are dangerous to use and the toxicity of the degradation products is not known. Based on the results obtained, treatment with sodium hydroxide is the only recommended chemical method for the disposal of concentrated organophosphorus pesticide wastes.

The use of chemical reagents to decompose concentrated DNBP pesticide wastes to less toxic forms has also been investigated (3) and showed that [1] DNBP was altered structurally by sodium hydroxide but the spectrum produced was not resolved; [2] 80% of the DNBP were decomposed when treated with the sodium biphenyl reagent prepared by heating a mixture of metallic sodium, anhydrous toluene, and the dimethyl ether of ethylene glycol; [3] 93.8% DNBP degradation were obtained when the pesticide was treated with liquid ammonia and metallic sodium. Although liquid ammonia and metallic sodium would probably also completely decompose DNOC, the reagent is dangerous to use and the toxicity of the degradation products is not known. Based on the results to date, chemical degradation could not be recommended as a method for the disposal of concentrated DNOC wastes.

A composition developed by B.C. Wolverton (4) is a solution for decontaminating and converting selected organophosphorus insecticides into products which can be water washed and discharged into streams without the usual toxic effects to fish normally exhibited by unaltered insecticides.

More specifically, the process involves the steps of [1] exposing the insecticide in a volume-to-volume ratio of ten parts of a decontaminant to one part of the insecticide, the decontaminant consisting essentially of 12.5 to 25% by volume of monoethanolamine in dipropylene glycol monomethyl ether as a solvent and [2] continuing the exposure for at least 30 minutes.

A process developed by K.H. Sweeny et al (5) is one in which halogenated organic compounds, especially chlorinated organic pesticides are decomposed by reaction with metallic couples in a mildly acidic environment.

Previous work with the zinc degradation of DDT and similar pesticides, disclosed in U.S. Patent 3,640,821, tended to indicate that reduction of DDT may proceed by way of a complex, two-path reaction. A portion of the p,p'-DDT appears to be rapidly converted to the ethane derivative, 1,1-bis(p-chlorophenyl) ethane referred to as DDEt, losing all three aliphatic chlorine atoms. Another portion appears to be reduced in a stepwise and much slower fashion. This reaction appears to progress from p,p'-DDT to p,p'-DDD [2,2-bis(p-chlorophenyl)-1,1-dichloroethane] to DDMS [2,2-bis(p-chlorophenyl)-1-chloroethane] and finally to DDEt. The latter product, DDEt, is a preferred reaction product since it appears to have far less physiological activity than DDT or the other intermediate products.

It has been found that use of particular metallic couples as the reducing agent for halogenated organic compounds, particularly such pesticides as DDT, results in substantial advantages compared with use of zinc alone. A marked increase in reaction rate is observed but, more importantly, the proportion of relatively innocuous decomposition products increases.

The preferred couples are zinc-copper, iron-copper and aluminum-copper, but other couples showing substantial reductive activity include zinc-silver, magnesium-copper, and cadmium-copper. Mildly acidic reaction conditions are necessary with the preferred reaction pH ranging from 1.5 to 4. Reaction temperature may range from ambient to moderately elevated or within the general range of $0°$ to $100°C$. Even higher temperatures may be employed but would require use of pressurized reaction equipment. When reacting low concentrations of pesticide in wastewaters, it is preferred to operate at the temperature of the waste stream which will usually range from $20°$ to $80°C$. The reaction medium may comprise water, organic solvents or mixtures of the two.

The metallic couples which have been found useful may be described as catalyzed metal reductants. They may be prepared by depositing, preferably by chemical means, a thin film or layer of a catalytic metal such as copper or silver upon a metal reductant such as zinc or they may alternatively be prepared in the form of an alloy. The effect appears to be truly catalytic since the copper or silver may be recovered in metallic form at the conclusion of the reaction.

It is expected that this process will prove most useful in the decomposition of halogenated organic residues to produce less highly halogenated products. Specific uses contemplated include the decomposition of chlorinated organic pesticides such as DDT found in process wastewaters and even in soils.

Deep Well Disposal

The question of deep well disposal of pesticide manufacturing wastes is touched upon by Atkins (9), who points out that since deep well disposal places the waste in a remote location where its ultimate fate cannot be adequately determined, the practice of subsurface injection is receiving critical review in many agencies. The uncertainties associated with this type of disposal can make it questionable, especially for long-lived pesticides and pesticidal wastes.

Although all the chlorinated insecticides are practically insoluble in water, their persistence and stability in water and the potential contamination of groundwater thus make deep well at best a questionable method for the disposal of these pesticides. Deep well disposal of chlordane manufacturing wastes has, however, been practiced at Velsicol's Marshall, Illinois plant (3). The method is not recommended and should be considered only under special situations where hazards would be nonexistent.

Although 2,4-D is only sparingly soluble in water, its persistence and stability in water and the potential contamination of groundwater make deep well at best a questionable method for the disposal of 2,4-D. Incidents of groundwater contamination that persisted for over 3 years leading to the damage of lawns, shrubs and crops, as a result of the penetration of the 2,4-D wastewater through permeable sediments after being discharged to rivers or lagoons have been reported. Deep well disposal of 2,4-D wastes has also been practiced at Dow Chemical Company, Midland, Michigan (3).

Although a properly planned, properly designed, properly constructed, and properly operated deep well disposal installation is an expensive investment, it may still be the only economic alternative for pesticide manufacturers and large formulators to dispose of large volumes of organophosphorus pesticide wastes from productive operations. However, the method is uneconomical for the occasional disposal of small volume pesticide wastes. The solubility in water of four organophosphorus insecticides is: parathion, 24 ppm; methyl parathion, 50 ppm; demeton, 60 ppm; Guthion, 30 ppm. Because of the potential contamination of groundwater, deep well disposal of organophosphorus pesticide wastes is not recommended (3) and the method should be considered only under very special situations where hazards would be nonexistent.

Although DNOC itself is only sparingly soluble in water, the potential contamination of groundwater by the water-soluble ammonium, potassium, sodium, and calcium salts of DNOC makes deep well, at best, a questionable method for the disposal of DNOC. Deep well disposal is not recommended (3) and should be considered only under special situations where hazards would be nonexistent.

Incineration

The disposal of pesticide wastes by incineration has been described by R.S. Ottinger et al (3). The incineration of pesticide wastes has also been discussed by T.L. Ferguson (6). He cites a waste tar burner developed by Bigelow-Liptak jointly with Dow Chemical Co. and operated by Dow Chemical Co. at their Midland Michigan plant which is used for incinerating pesticide wastes. It is shown in Figure 118. For additional discussion of this unit, see the earlier chapter of this volume on Incineration and more specifically the section on Liquid Waste Combustors, page no.

Incineration is a popular disposal method for concentrated pesticide waste streams. Most organic compounds (liquid and gaseous) can be effectively destroyed by the method, but other components of the waste streams often cause significant problems. Nitrogen compounds present in many pesticide manufacturing wastes invariably produce oxides of nitrogen (NO, NO_2, etc.) when burned. Since these gases are of importance in the atmospheric reactions which produce photochemical smog, regulations against their production are often severe and will no doubt become more restrictive in the future. Sulfur-bearing compounds may produce significant quantities of sulfur dioxide and sulfuric acid. Regulations restricting SO_2 emissions are already in effect in almost all states.

These problems make it difficult but not impossible to burn many pesticide wastes. Temperature control, excess air control, fuel and waste feeding techniques, and proper incinerator design can all reduce the magnitude of the emission problems. A well-designed and operated air cleaning device (which will normally constitute a major capital investment) can remove many of the gaseous pollutants and eliminate virtually all of the particulates. These devices should be included in the original design of the incineration equipment and not merely added at a later date when air pollution problems become obvious.

Table 34 summarizes results of a combustion study made by Kennedy, et al (7) on nineteen pesticide formulations at five different temperatures. Most pesticides showed a high percent loss on combustion. However, the data shown indicate that care must be taken to provide adequate combustion temperatures and contact times for complete disposal. The tests were made in controlled temperature chambers where adequate oxygen was available. Commercial formulations of the pesticides were burned rather than waste streams. The data indicate that most pesticide compounds are destroyed effectively by burning at 800° to 1000°C. Exceptions are atrazine, zineb, bromocil, dalapon, DSMA, and Sevin.

FIGURE 118: PESTICIDE WASTE INCINERATOR DESIGN

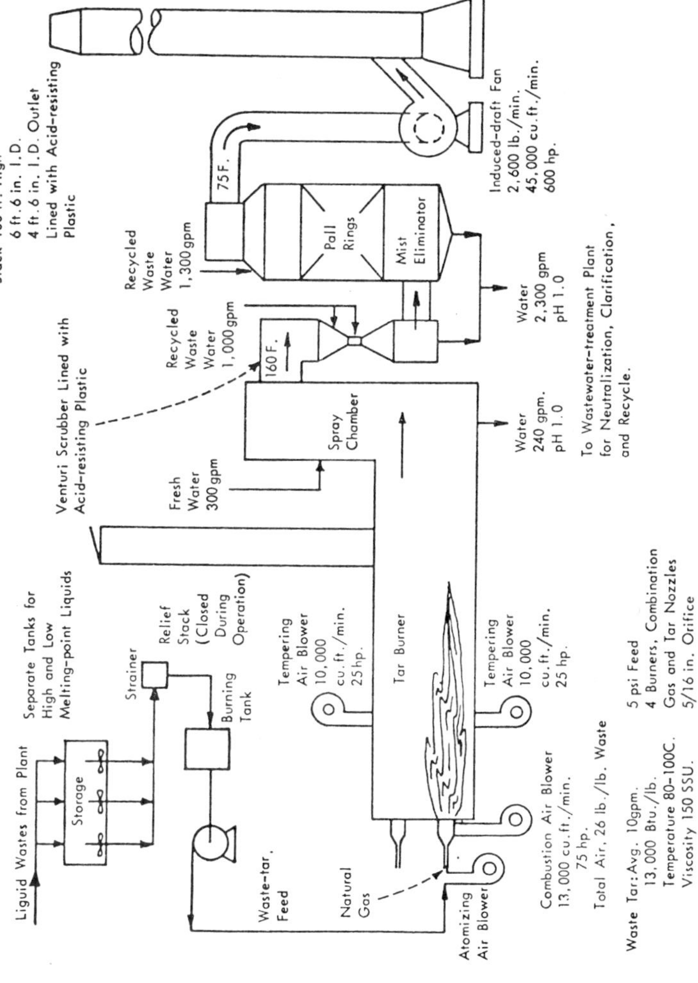

Source: PB 241 001

TABLE 34: PERCENT LOSS ON COMBUSTION OF COMMERCIAL FORMULATIONS
OF PESTICIDES AT FIVE TEMPERATURES

Commercial Formulation	Loss (%) at				
	$600^\circ C$	$700^\circ C$	$800^\circ C$	$900^\circ C$	$1000^\circ C$
Picloram	90.8	91.8	95.6	98.7	99.2
Atrazine	87.8	88.1	88.8	88.9	89.0
Nemagon	99.6	99.6	99.6	99.6	99.6
Trifluralin	99.7	99.8	99.8	99.8	99.8
Malathion	95.3	96.0	96.3	96.4	96.7
2,4,5-T	99.9	99.9	99.9	99.9	99.9
Zineb	70.1	71.3	71.5	72.7	72.8
Vernam	99.6	99.6	99.6	99.6	99.6
Paraquat	98.3	98.6	99.0	100.0	100.0
Dicamba	98.6	98.7	98.9	99.0	99.4
Bromacil	88.8	89.1	89.4	90.5	91.3
Dieldrin	99.1	99.4	99.5	99.5	99.5
DDT	99.2	99.3	99.7	99.9	100.0
Dalapon	64.3	64.3	67.8	73.8	91.0
2,4-D	99.8	99.9	99.9	99.9	99.9
Diuron	94.6	95.0	95.4	95.5	95.7
DNBP	99.8	99.8	99.8	99.8	99.8
DSMA	80.6	80.7	80.7	81.2	81.2
Sevin	88.7	88.8	88.8	89.1	89.5

Source: *Residue Reviews* 29, 89-104 (1969)

In a combustion chamber, the toxic compounds may undergo decomposition by oxidation, pyrolysis, isomerization or polymerization. If the combustion temperature is too low, the residence time in the chamber is too short, adequate air is not present, or the waste is not well dispersed to expose a large amount of surface area, then complete destruction of the toxic compounds may not occur. In the case of organophosphate insecticides, toxic isomers of the original compound may be produced (9).

Many of the major problems associated with incineration of pesticide wastes in the past resulted from failure to consider the constituents of the waste other than toxic materials. Pilot studies should always be conducted to provide information on effluent gas and particulates from the incinerator exhaust. In a well-designed, well-operated incinerator, phosphorus-containing compounds produce P_2O_5 and phosphoric acid, which can result in a heavily persistent white plume visible for long periods of time. High potassium or sodium levels in a waste can produce hygroscopic salt nuclei that cause visibility problems in humid atmospheres. In addition, the salt particles may be air pollutants of consequence, especially in agricultural areas.

The complete and controlled high temperature oxidation of chlorinated insecticides in air or oxygen with adequate scrubbing and ash disposal facilities offers the greatest immediate potential for the safe disposal of pesticides, however the research on incineration of pesticides conducted at Mississippi State University (3) has led to the conclusion that temperatures at or near 1800°F will be sufficient to degrade 99% or more of most reagent grade pesticides and commercial pesticide formulations. It is expected that either a rotary kiln or liquid combustor, depending upon the waste form, followed by secondary combustion and scrubbing would be an acceptable disposal method.

Primary combustion should be carried out at a minimun of 1500°F for at least 0.5 second with secondary combustion at a minimum of 2200°F for at least 1.0 second. The volatile products identified from burning of the formulations at 1650°F include carbon monoxide, carbon dioxide, chlorine and hydrogen chloride. As the same combustion products will be obtained in the incineration of other polychlorocyclodiene insecticides, an adequate gas clean up system to remove chlorine and hydrogen chloride must be installed to alleviate the air pollution problem. The abatement problem may be simplified by ensuring against elemental chlorine formation through injection of steam or methane into the combustion process.

Incineration is also one of the principal means of disposing of heptachlor and endrin manufacturing wastes at Velsicol's Memphis, Tennessee plant (3). In addition, combustion units designed for the disposal of chlorinated organic wastes and capable of recovering chlorine in the form of usable hydrogen chloride have been developed, and a 7,000 lb/hr plant has been built for E.I. du Pont de Nemours & Company in Victoria, Texas by Union Carbide Corporation (3). Properly designed and operated incineration is therefore considered as the best present and near future method for the disposal of concentrated chlorinated pesticide wastes.

The complete and controlled high temperature oxidation of organophosphorus insecticides in air or oxygen with adequate scrubbing and ash disposal facilities offers the greatest immediate potential for the safe disposal of these pesticides. The research on incineration of pesticides conducted at Mississippi State University (3) has led to the conclusion that temperatures at or near 1800°F will be sufficient to degrade 99% or more of most reagent-grade pesticides and commercial pesticidal formulations. It is expected that either a rotary kiln or liquid combustor, depending upon the form of the waste, followed by secondary combustion and scrubbing would be an acceptable disposal method.

Primary combustion should be carried out at a minimum of 1500°F for at least 0.5 second with secondary combustion at a minimum temperature of 2200°F for at least 1.0 second. The results indicate the possible formation of objectionable combustion products such as hydrogen sulfide and phosphorus oxides. A higher air/fuel ratio will lead to the formation of sulfur dioxide instead of hydrogen sulfide.

As the same combustion products will be obtained in the incineration of other organophosphorus insecticides, an adequate gas clean up system must be installed to alleviate the air pollution problem. Monsanto's Anniston, Alabama parathion and methyl parathion manufacturing facility uses incineration to dispose of its semisolid residue wastes, and has proven that an aqueous scrubbing system followed by a mist eliminator is effective in recovering 99.9% of the phosphorus pentoxide (3).

The other major air pollutant of concern, sulfur dioxide, is undoubtedly removed by the same system. Properly designed and operated incineration is therefore considered as the best present and near future method for the disposal of concentrated organophosphorus pesticide wastes.

The complete and controlled high temperature oxidation of dinitro-o-cresol (DNOC) in air or oxygen with adequate scrubbing and ash disposal facilities offers the greatest immediate potential for the safe disposal of the pesticide. The research on incineration of pesticides conducted at Mississippi State University (3) has led to the conclusion that DNBP approached complete combustion at temperatures as low as 600°C and identified carbon monoxide, carbon dioxide, and ammonia as the volatile products from burning of a dinitro-sec-butyl phenol (DNBP) formulation at 900°C. Because of the similarity in chemical structure of DNBP and DNOC, temperatures not far above 600° C should also be sufficient to degrade DNOC and the same combustion products should be obtained. It is expected that either a rotary kiln or liquid combustor, depending on the waste form, followed by secondary combustion and scrubbing would be the best current and near future method for the disposal of concentrated DNOC wastes. Again, primary combustion should be carried out at a minimum of 1500°F for at least 0.5 second with secondary combustion at a minimum temperature of 2200°F for at least 1.0 second.

As described by Ferguson (6), wastes should be collected in suitable containers, properly labeled for transfer to incineration, unless of course, the incinerator is an integral part of a manufacturing or formulating operation. Depending upon quantity, containers should preferably be metal pails, drums or tanks, or in some instances, fiber packs if the wastes are dry solids. Identification should include trade name, type of material, active concentration, weight. If the wastes are solids, the unit quantity will need to be compatible with the incineration facility. If the wastes are liquid and represent a continuing and sizable consideration, receiving tanks at the incineration site will be necessary for proper segregation and blending.

Further, as noted by Ferguson (6), with few exceptions, facilities required for the proper incineration of pesticides cannot be justified for small-scale operations. The applicator, distributor, and many formulators will find it expedient to contract for the services of a municipal or commercial incinerator equipped to meet all governmental regulations.

For the incineration of flammable liquid wastes in those cases where a gas scrubber is not required it may be quite satisfactory to mount an oil burner and refractory burner block (such as supplied by National Aeroil Burner Company) on a pedestal in the open with or without a pile of rubble as a target. Conventional vertical refractory lined incinerator units capable of handling liquids having a high water content and operating at rates up to several thousands of pounds per hour are supplied among others by Thermal Research and Engineering Corporation, John Zink Company, Prenco Division, Pickards Mather and Company, and Surface Combustion Division, Midland-Ross Corporation.

Present practice in the combustion of solid wastes employs multiple-chamber incinerators where the combustion proceeds in two stages. The primary or solid phase combustion occurs in an ignition chamber which is followed by a secondary or gaseous phase occurring in a secondary combustion zone. Heat release is ordinarily in the range of 25,000 to 50,000 Btu per hour per cubic foot of combustion volume. The Bigelow/Phillips Incinerator represents a large-scale application of this configuration. The Cleanaire Radicator furnished as a package unit by the Midland-Ross Corporation has a capacity of several hundred pounds per hour and with adaption for vapor scrubbing might be suitable for pesticide formulation or small-scale manufacture.

For large-scale operation a rotary kiln incinerator has certain advantages, among which is the ability to handle a large variety of feed including packs and drums of solid waste chemicals. The Dow Chemical Company operates an incinerator of this type and a Dow subsidiary has designed a similar installation for another company.

A further type of incinerator coming into prominence is the fluid bed unit which has special application when an inorganic residue such as sodium sulfate or sodium chloride will be produced. Dorr-Oliver and Copeland Systems, Battelle Memorial Institute, the Dow Chemical Company, and others, have been active in this field.

Fumes and air-dispersed mists can be incinerated in equipment similar to that used for liquids. The concentration of combustible contaminants will usually be well below the lower limit of flammability, but it may be possible to pass the entire stream through a burner after adding a small amount of natural gas as auxiliary fuel and adjusting the air to effect complete combustion. Fuel oil can, of course, be accommodated as supplementary fuel in fume incinerators as well as in any other type.

The two most critical factors in pesticide waste incineration are temperature control and retention time. The incinerator must be capable of maintaining a set temperature with minimal fluctuation. Retention time is fixed by operating parameters such as amount of excess air and by the volume of the combustion chambers which should be large enough to provide several seconds residence for the gaseous products of combustion. The temperatures required for the complete destruction of most pesticides will be about 1800°F. An extensive study in this regard has been reported by M.V. Kennedy, et al (7) as a contribution from the Mississippi Agricultural Experimental Station under a grant from the U.S. Department of Agriculture.

Conventional fuel oil burners are used for burning waste liquids. All the air necessary for combustion can be supplied as primary air, or primary air may be used to control flame character while secondary air is introduced through openings in the burner block to provide additional requirements for complete combustion. To achieve proper burning, the waste liquid, or supplemental fuel when required, must be finely atomized. High-pressure air atomization, low-pressure air atomization, steam atomization and mechanical atomization have all been used satisfactorily.

Temperatures in the separate combustion zones are controlled by strategically located thermocouples which cause fuel (or waste) and air flows to respond to the temperature settings. Draft gauges, flowmeters, and stack gas analyzers are most useful in properly controlling incinerator operation. Ignition and flame detection systems operating in conjunction with purge arrangements and other safety features and interlocks are necessary accessories which should be provided in a manner to meet Factory Insurance Association requirements.

Liquid wastes are strained, segregated and blended as necessary and are preheated if required for proper atomization in an oil type burner. The feed pump should be a positive displacement type such as a vane, gear or screw pump and it should be equipped with a variable speed drive actuated by furnace temperature control. When handling solids, it is likely that provisions would be made for directly injecting pails and drum quantities of solid wastes into the incinerator. Typically, refuse other than full drum or pails would be dumped into a refuse pit from which an overhead hoist would transfer the material to a charging hopper.

The disposal of halogen-bearing pesticides and solvents presents complications of no small matter. These materials release the corresponding acid vapors when burned, thus precluding the use of simple pits or furnaces. The incinerator stack gas must be scrubbed with water or an alkaline medium, and the scrubber effluent should be sent to a wastewater treatment plant employing secondary treatment. If the wastes contain sodium salts, a simple packed tower will probably be inadequate and a high-energy venturi-type scrubber may be required. Under any circumstances, appropriate governmental approval must be obtained for release of gases to the atmosphere or discharge of liquids to municipal sewers or to other waterways.

A process developed by S.J. Yosim et al (11) provides for the ultimate disposal of organic pesticides with negligible environmental pollution by feeding the pesticide and a source of oxygen into a molten salt containing an alkali metal carbonate and preferably also an alkali metal sulfate to pyrolytically decompose and at least partially oxidize the pesticide.

Some of the resulting decomposition products react with and are retained in the melt; remaining gaseous products pass through the melt to the atmosphere or are conducted to a second reaction zone where oxidation of any combustible matter present is completed. Certain organic pesticides may be completely combusted in the first reaction zone by using an excess of oxygen or air in this zone during the decomposition reaction. Thereby the final gases vented to the atmosphere as a result of oxidative treatment in only the first zone or in both zones include only such gases as carbon dioxide, water vapor, oxygen and nitrogen.

Details of the construction of this apparatus have been presented earlier in this volume in the chapter on Incineration and the section on Molten Salt Incinerators.

Where containers of plastic, paper, glass or metal having adherent pesticide residues are to be disposed of, it is preferred to use a molten salt mixture consisting essentially of the Na_2CO_3-K_2CO_3 eutectic (MP 710°C) and containing from about 1 to 25 weight percent sodium sulfate. The process may conveniently be practiced for the disposal of such pesticide-containing containers by combusting the containers in such a melt at a temperature between 700° and 1000°C in a portable disposal unit mounted on a truck bed.

Where relatively large quantities of pesticides are to be disposed of, or higher temperatures to insure total disposal are required, it is preferred to maintain the molten salt mixture at a temperature between 850° and 1000°C, particularly between 900° and 950°C. Such a molten salt mixture preferably consists essentially of sodium carbonate containing from about 1 to 25 weight percent sodium sulfate. An amount between 5 and 15 weight percent sodium is particularly preferred. Where only partial combustion of organic matter takes place in the first reaction zone, then in order to complete oxidation of this combustible matter, which may include carbonaceous particulate matter as well as gases that are further oxidizable, it is fed to the second reaction zone.

A second source of oxygen, preferably air, is also fed into the second reaction zone. For certain applications a metal mesh, e.g., stainless steel, or a ceramic-coated metal mesh, e.g., aluminized or alumina-coated stainless steel, is added to this second zone to serve as a source of ignition and also to demist the gaseous effluent of any particles of the molten salt mixture present in it prior to venting this effluent to the atmosphere.

Depending upon the decomposition products of the pesticide that are retained in the melt, it is preferable and frequently required that the melt be further treated so as to be nontoxic and noncontaminating when ultimately disposed of in a body of water or in the soil. The melt containing the retained decomposition products is treated with air to oxidize any residual sulfide to sulfate. It is then cooled for disposal in a dry lake bed or treated with water and lime for disposal in an approved dump site. Where the melt essentially contains only chlorides, such as sodium chloride resulting from the decomposition of chlorinated hydrocarbon pesticides, the entire residue may be directly disposed of into a large body of water such as the ocean.

A wide variety of organic pesticides may be rapidly and conveniently treated by the process with relatively minor modification in treatment techniques because the process basically involves destruction of the organic compound whether the initial step is partial or complete combustion. The process is particularly suitable for disposal of those organic insecticides and herbicides of large-scale manufacture and which require frequent and convenient disposal.

Thus particularly suitable for treatment by this process are the chlorinated hydrocarbon insecticides such as chlordane, DDT, dieldrin, heptachlor, and aldrin; the phenoxy acetic acid, toluidine, and nitrile herbicides, illustratively, trifluralin, 2,4-D, 2,4,5-T, dichlobenil, and MCPA; and the phosphorus-containing insecticides such as diazinon, disulfoton, phorate, malathion, and parathion. These compounds are rapidly destroyed at the decomposition temperatures used, and the resulting acid products and gases are absorbed and neutralized by the molten alkali metal carbonate. Any toxic gases and carbonaceous particulate matter evolved are passed through the melt and completely combusted therein or are further oxidized in the second reaction zone.

Where chlorinated hydrocarbons constitute the pesticides being treated, sodium chloride is formed in the melt. In the case of organic phosphate compounds, sodium phosphates are formed. Where sulfur is present in the material, sodium sulfate will be produced. All of these formed inorganic compounds are retained in the melt. When the capability limit of the salt to react with pesticides has been reached, the salt is removed and fresh make-up salt is added. The spent salt is recovered or otherwise disposed of. The following are some specific examples of the operation of the process.

Example 1: A representative pesticide used for evaluation consisted of 50% chlordane $C_{10}H_6Cl_8$, equivalent to 30% octachloro-4,7-methanotetrahydroindane and 20% related compounds. The tests were carried out in such a manner that most of the pesticide was reacted while it was in submerged contact in the melt. An upright mullite tube which was about one-half to two-thirds full of a molten mixture of sodium carbonate (about 85%) and sodium sulfate (about 15%) at a temperature of about 980°C was contained in an electric furnace.

Downstream from the mullite reactor tube were located a glass wool trap for particulate carbonaceous material, a port for taking gas chromatograph samples, an analyzer for carbon monoxide and hydrocarbons, a water scrubber and a flow meter. Small quantities consisting of a few tenths of a gram of 50% chlordane were intermittently added as a powder directly into a stainless steel air inlet tube of the mullite reactor. An air flow through the air inlet tube forced the pesticide and its pyrolysis products through about one foot of molten salt in order to permit better contact between the melt and the pesticide.

Monitoring of the gas evolution, followed by analysis of the water scrubber and of the small amount of soot evolved for chlorides and of the gas phase for hydrocarbons showed that virtually all of the solid pesticide, in excess of 99.9%, was rapidly decomposed. The exit gases contained products of the reaction between the carbonaceous material and sulfate, i.e., carbon dioxide and carbon monoxide. By increasing the ratio of air to pesticide, the presence of carbon monoxide in the evolved gases was markedly decreased. The hydrocarbon content of the gas phase was below 10 ppm. Even when the air to pesticide ratio was relatively low permitting a CO concentration in the gas phase of about 1% by volume, the hydrocarbon content in the gas phase was only 100 to 150 ppm. These low hydrocarbon levels indicate that the maximum amounts of chlorinated hydrocarbons in the gas are quite small. No chlorides were detected in the water scrubber when silver nitrate was added.

The melt was also analyzed for total chlorine. The results showed that amounts substantially in excess of 80 weight percent were retained by the melt.

Example 2: In a second type of apparatus designed for continuous disposal of larger quantities of pesticide, the molten salt consisted of 10 pounds of 80 weight percent sodium carbonate and 20 weight percent sodium sulfate contained in a 6-inch i.d. by a 30-inch long alumina tube closed at one end. The alumina tube was placed in a vertical position in a 9-inch inner diameter by 30-inch long tube furnace into which was inserted a closed-end thick-walled 7-inch i.d. by 30-inch long stainless steel vessel flanged at the top. Polyethylene packets containing 5- and 10-gram amounts of the pesticide chlordane were repeatedly added to an airstream which was bubbled through the melt through a concentric 1½-inch i.d. by a 48-inch long alumina tube, open at both ends, which extended 6 inches below the melt level.

All gases evolved due to pyrolysis of the pesticide in the addition tube were bubbled through about 5½ inches of molten salt where the noxious gases were chemically absorbed. Air used to oxidize the sulfide to sulfate was added through an auxiliary tube. Mixing of the smaller amount of molten salt within the addition tube and the main body of salt in the wider alumina tube was obtained by agitation of gases escaping from the melt.

It was found that substantially complete reaction of 10-gram quantities of chlordane took less than two minutes, indicating that reaction of one pound per hour is readily attainable in the described apparatus. A secondary combustion chamber was used to insure complete combustion of evolved gases. It was found that most of the particulates were removed by the melt. None were visually observed in the exit gas downstream of the secondary combustion chamber.

Example 3: Using the same apparatus as employed in Example 2, a melt consisting of 90 weight percent sodium carbonate and 10 weight percent sodium sulfate was maintained in the molten state at an initial temperature between 950° and 1000°C. No attempt was made to maintain the temperature of the melt constant during the actual course of the reaction.

Malathion, a representative organophosphorus-type pesticide which also contains organic sulfur, was placed in polyethylene bags and periodically dropped into the melt by way of an additional port. The port was closed so that all emission passed through the salt, through a secondary combuster where heated air was drawn into the system, through a glass-wool particulate trap, through the water scrubbers, and out of the system. Benzene extracts of the particulates in the glass wool and water scrubbers were evaporated to dryness and the residue analyzed. During the course of the tests, carbon monoxide and hydrocarbon emissions were determined using infrared detectors.

Ten 5-g packets of malathion were added at three-minute intervals. By monitoring the carbon monoxide and hydrocarbon emission, it was determined that 5 g of the sample was destroyed within about 30 seconds. The secondary combustion unit consisted of a heated air feed and a hot mesh over which the gaseous emission was contacted. No carbon monoxide or hydrocarbons were detected downstream of the secondary combustion. The malathion samples contained 2.42 g sulfur and 1.21 g phosphorus. Analysis of the benzene extracts of the resulting emissions showed only 1.2 mg S and 1.8 mg P, indicating that at least 99.9% of the pesticide was destroyed. Analysis of the melt after the test showed a phosphorus content of $80 \pm 15\%$.

Example 4: Using the procedure and apparatus of Example 3, the representative herbicide Weed B Gon was used. This consists of 17.8 weight percent of the isooctyl ester of 2,4-D and 8.4 weight percent of the isooctyl ester of silver in a solution of kerosene. Six 5-g polyethylene-bagged packets of the herbicide were added to the system, the emitted particulates were extracted with benzene, and the benzene residue was analyzed for chlorine. Of the 1.48 g chlorine added, only 0.77 mg was found in the total emissions. This indicated a pesticide destruction of at least 99.96%. Analysis of the melt before and after reaction indicated that the chlorine content of the melt (as chloride) had increased sufficiently to account for virtually all the chlorine in the pesticide added.

Example 5: Using the procedure and apparatus of Example 3, Sevin a typical carbamate pesticide identified as carbaryl (1-naphthyl N-methyl carbamate), which contains 7 weight percent nitrogen, was evaluated. Five 5-g polyethylene-bagged packets of the pesticide were added to the molten salt. Of the 0.875 g nitrogen added, less than 0.075 mg was found in the benzene extract. This indicated that more than 99.99% of the pesticide had been destroyed. As expected, no nitrogen was found in the melt because of the very rapid carbon-nitrate-nitrite reduction reaction occurring in these melts at these temperatures. Excessive nitric oxide emissions were not observed in the tests, peak emissions showing only about 20 ppm.

Example 6: Liquid chlordane pesticide (72% emulsifiable concentrate) was completely combusted in a continuous feed system using excess air. The molten salt consisted of a mixture of 90 weight percent sodium carbonate and 10 weight percent sodium sulfate. This molten salt mixture was contained in an alumina tube placed in a stainless steel retainer vessel. The liquid chlordane pesticide was pumped into the system at a feed rate of about 1.6 lb/hr. The air and pesticide were injected into the molten salt reactor by way of an alumina feed tube. The air feed rate was about 4 scfm. The run was carried out for a total time of 75 min, the feed rate being adjusted during the first 20 min. A steady-state condition was achieved during the last hour of the run with about 75% excess air.

The exit gas had a superficial linear velocity of about 1.6 ft/sec. The temperature of the melt was maintained at about 1000°C. Besides providing the excess oxygen for the combustion, the injected air provided cooling of the melt so as to maintain its temperature below 1050°C. The total pesticide consumption was about 2 lb.

A particular filter and a water scrubber were used to trap any evolved organic chlorides. No organic chlorides were detected in either fraction. The detection limits indicate that less than 0.04% of the original pesticide escaped as organic chlorides from the system. Thus greater than 99.96% of the pesticide was destroyed by molten salt combustion.

The evolved exhaust gases were analyzed for the presence of possible gaseous pollutants. Without using a secondary combuster, the NO_x content of the gaseous effluent was found to be less than 70 ppm. The CO and unburned hydrocarbon emissions were also low (less than 0.1% and less than 25 ppm, respectively). Gas chromatographic analysis showed about 12% CO_2, 8% O_2, and 80% N_2 in the exhaust gas.

The process is considered to make feasible the disposal of pesticides in a continuous manner at a rate up to 1,000 lb/hr. It will of course be realized that many variations in reaction conditions may be used in practice, within the limits of the parameters set forth, depending upon the organic pesticide used, whether in liquid, solid, or solution form, whether associated with packaging material, the temperature required for complete pyrolysis, and the desired rate of feed of pesticide and of oxygen to the system. Where the pesticide is in the form of a liquid or a solution, a screw conveyor will generally be replaced by a continuous duty pump.

Land Surface Application

Research on the reduction of 2,4-D waste process liquors into biologically inactive compounds by means of application to and degradation in the soil surface has been conducted at Alkali Lake in eastern Oregon under the direction of R.L. Goulding of Oregon State University (3). The results of the study here suggested subsurface injection of 2,4-D as a useful economic approach to control airborne losses of the waste component, and indicated clearly the apparent degradation of 2,4-D by soil microorganisms when applied at the rate of 250 pounds per acre equivalent of 2,4-D.

Data derived from samples taken from the actual test plots showed that the 2,4-D concentration in the soil layer subjected to subsurface injection has declined from an initial 135 ppm to 30 ppm after a 480-day period. The Oregon State work thus provides strong support to the adequacy of the soil surface application as a disposal method for concentrated 2,4-D wastes.

Ocean Disposal

Pesticide (insecticide, herbicide, and fungicide) manufacturing operations produced about 7% (290,000 tons) of the total industrial wastes disposed of at sea in 1968. Of this amount, about 170,000 tons were barged down the Mississippi from the Memphis upriver area and dumped 100 miles offshore from Beaumont, Texas into 1,200 feet of water.

The probable environmental effects of the proposed marine disposal of wastes from a Gulf Coast plant that manufactures herbicides and fungicides have been reported (8). The wastes included chemicals of the following types: anilines (primarily chloroaniline and aniline with small amounts of monochlorobenzene); liquid organic solvents (methanol, p-xylene and chlorobenzene); and dry chemicals including Thiram, Thiram-E, Thionex, Zineb, Ferbam, Monuron, and carbon disulfide. Because of their toxic nature, these wastes are disposed of in weighted steel drums. To facilitate discharge and dispersal along the bottom, a small air space is left in each drum to ensure deformation and rupture of the drum on the ocean floor.

Toxicity data on these waste materials on the basis of the available literature were evaluated, and it was concluded that: The herbicides and fungicides are generally more toxic than the liquid organics. These dry chemicals, however, have a very low solubility so that once in solution a relatively low dilution ratio (on the order of 100:1) will reduce the concentration below the median tolerance limits (TL_M). In addition, the susceptibility to biodegradation and chemical instability in an aqueous environment would further reduce localized toxic conditions.

Sanitary Landfill

Burial of pesticide wastes is a common practice (9). Solid, semisolid and liquid wastes are often buried in landfill or placed in open pits. The liquid and semisolid wastes are usually sealed in containers for transport to landfill sites or placed in the open pit areas in bulk. Therefore, the two methods differ considerably and will be discussed separately.

The landfill type disposal method normally involves opening a trench, often fifteen feet or more in width and up to twelve feet deep, depositing the waste material in the trench, compacting the material (thus rupturing any containers present) and covering the waste daily with a minimum of six inches of compacted earth. The final fill cover should be at least 18 inches deep in order to isolate the waste material. Where topography permits, the waste material can be used to fill low areas without opening a trench. Borrow pits are necessary to provide the necessary cover material. However, if this procedure is used, special measures must be taken to provide for surface drainage around the fill site.

Subsurface pesticide movement is possible and should be considered especially in highly permeable soils. In tight, clay material and high organic content soils, most pesticides will move slowly as rainwater and landfill leachate pass through the soil. Studies indicate that chlorinated hydrocarbons including dieldrin and aldrin migrate only a few feet in natural soils over periods of years.

The USGS reports little horizontal movement of dieldrin, endrin, heptachlor and heptachlor epoxide from a landfill located at a site underlain by a nearly horizontal strata composed of deposits of sand, silt and clay. However, it was reported that the subsurface disposal had produced a zone of contamination directly beneath and peipheral to the landfill. Laterally the limits of the zone of contamination appeared to be confined to 25 feet or less from the trench margins. The depth of the zone, however, extended nearly to the water table, 90 feet beneath the land surface. In addition, surface waters in a nearby stream contained traces of the four pesticides probably from surface runoff.

Organophosphate pesticides also move slowly in soils. In a gravelly sand soil (5% clay) six to eight feet of eluted water was required to move 25% of the applied thiamate through one and one-half inches of soil. In a heavy clay soil (37% clay), after 28 feet of elute water, only 5% of the same pesticide had moved through the soil.

These studies indicate that, if proper care is taken, landfill area can be a suitable disposal method. The soil type should be such that waste migration is minimized (i.e., high clay or high organic content). The water table should not be near the disposal site (a minimum of 10 feet between trench bottoms and the water table should be maintained). Surface flow should be controlled to prevent erosion and surface water contamination. Site management should be strict enough to prevent spills, leaks and improper dumping or covering. A monitoring program should be instituted which would include surface water sampling, groundwater sampling below the fill, and water sampling at any nearby drinking water sources. Table 35 shows in generalized fashion the relative mobility of various pesticides in typical soil.

It should be realized that a natural soil system is never a simple homogenous medium that can be modeled with assurance. Undetected layering, faults and fractures, and drainage channels to groundwater can cause unforeseen problems. Since the rate of travel of chemicals in a soil-water system is dependent on the solubility and sorption characteristics of the pollutant and the physical characteristics of the aquifer, detailed study and sampling should be conducted before a fill site is chosen.

Pesticides buried in a landfill may be degraded biologically or chemically. Unsaturated, branched chain compounds of high molecular weight are less susceptible to degradation than the saturated short chain compounds. Aromatics are also resistant to degradation in soils. The principal factors that govern the stability of pesticides and related halogenated and phosphorylated compounds include soil texture, humus content, temperature, moisture and pH. The half-life of these materials may vary from a few weeks to several years. A listing of a number of representative pesticides and their persistence in typical soil is given by Atkins (9).

The open dump, or open pit disposal method, is also used by many pesticide manufacturers. The pit disposal method consists of placing wastes in an open pit or trench without compaction or cover. The wastes are exposed to the elements until the pit is filled. When full,

TABLE 35: RELATIVE MOBILITY OF PESTICIDES IN SOILS*

Immobile	Slightly Mobile	Mobile
Aldrin	Atrazine	2,4-D
Chlordane	Simazine	2,4,5-T
DDT	Prometryne	MCPA
Dieldrin	Azinophosmethyl	Picloram
Endrin	Carbophenthion	Fenac
Heptachlor	Diazinon	
Toxaphene	Ethion	
TDE	Methyl Parathion	
Lindane	Lindane	
Heptachlor Epoxide	Heptachlor Epoxide	
Trifluralin	Parathion	
	Phorate	
	Diuron	
	Monuron	
	Linuron	
	CIPC	
	IPC	
	ERTC	
	Pebulate	

*Mobilities are based on soil thin-layer chromatography – mobile compounds move between R_f 1.0-.65, slightly mobile .64-.10, and immobile .09-.00. (R_f = "relative to fructose").

Source: EPA Report 12020 FYE 01/72

the pit is covered with earth and abandoned. These methods are used because of their relative cost. Landfill operation requires labor and equipment costs of at least $1.00 per ton of waste plus the cost of haul. Pit disposal can be accomplished for almost no cost if the pit is located on or near the plant site.

Pit disposal can produce the same types of contamination as a landfill. Since abandoned sand and gravel pits, rock quarries and other naturally occurring topographic features are often utilized as disposal pits for reasons of availability and economy, the probability of large scale subsurface contamination is increased. In general, the problems, both real and aesthetic, associated with pit disposal make it undesirable. Odors, volatile toxic gas releases, and increased likelihood of water contamination are potential problems.

In addition, the probability of accidents near the pit is high, since haphazard dumping is usually employed. Even after final cover, the disposal site is subject to significant settlement, cracking, and perhaps surcharge during heavy rains. For these reasons, it is recommended that this method of pesticide waste disposal should be discouraged except under extremely favorable geologic conditions and where the site is very remote from producing wells.

Burial practices are receiving close scrutiny by various regulatory agencies. The degree of control and severity of the regulations will continue to increase and care must be taken when choosing the method as a long term disposal practice. A regulation to control waste burial proposed by the state of California is typical of regulations which may be expected in the future.

Soil burial of chlorinated pesticide wastes, because of their extended persistence of up to six years, is not a satisfactory means of disposal (3). In the case of aldrin and heptachlor, their activity is extended through the formation of their respective metabolites, dieldrin and heptachlor epoxide. Both dieldrin and heptachlor epoxide are more chemically inert and persistent than the pesticides they derive from and heptachlor epoxide is also more toxic than heptachlor to all organisms. Sanitary landfill should therefore be only considered for the disposal of small quantities of pesticide wastes, and only at approved sites that are acceptable from a geologic and groundwater hydrology standpoint.

Soil burial or organophosphorus pesticide wastes, because of their relatively short persistence time in soil of one to three months, is a satisfactory means of disposal provided the site is acceptable from a geologic and groundwater hydrology standpoint and has been approved as a sanitary landfill by appropriate authorities. The practice of disposing large quantities of concentrated pesticides at any one sanitary landfill site, however, is not recommended (3).

Although data are not available on the persistence of DNOC in soil, the pesticide readily forms water-soluble ammonium, sodium, potassium, and calcium salts and poses the problem of potential ground and surface water pollution. Sanitary landfill should therefore be considered only for the disposal of small quantities of DNOC wastes, and only at approved sites that are acceptable from a geologic and groundwater hydrology standpoint.

ARSENICAL PESTICIDE WASTES

As described by R.S. Ottinger et al (3), various options exist for ultimate disposal of arsenical pesticide wastes.

Incineration

Compounds containing heavy metals including lead, mercury, and arsenic should be burned only if proper precautions are taken to control atmospheric emissions. Lead and mercury salts as well as arsenic can escape the combustion chamber and travel great distances, contaminating large areas. Low levels of these compounds may produce recognizable effects in plants or animals as well as people. Compounds containing chlorine produce hydrochloric acid, which can cause serious corrosion problems in the incinerator and in some cases significant downwind pollution problems.

Land Application

Some users, when faced with unwanted stocks of some of these pesticides, use land spreading as a means of disposal. This approach involves spreading the waste materials over large amounts of land in very light applications so as not to significantly affect any of the crops or plant life growing in the vicinity. The technique might be used for rare or infrequent disposal of small amounts of the material but in light of the fact that these compounds will build up in the soil by repeated application, this method cannot be considered as completely adequate and other methods should be considered first.

Landfills

The disposal of cacodylate wastes in sanitary landfills is generally not acceptable because of the potential danger of ground and surface water pollution, as well as possible occupational hazards resulting from on-site handling. There are, however, certain approved sites located over nonwater-bearing sediments or with only unusable groundwater underlying them and are completely protected from flooding and surface runoff and drainage such as those designated as Class 1 sites in California. The disposal of small stocks of cacodylate wastes or empty containers contaminated with cacodylates in Class 1 sites is considered as adequate, provided special handling techniques are employed to protect site personnel.

The solid waste salt by-product from cacodylate manufacture contains sodium chloride, sodium sulfate, and 1 to 1.5% cacodylate contaminants. The volume of this waste material currently in storage amounts to 60,000,000 lb. There is a definite need for the development of an economical process capable of extracting the cacodylate constituents from the salt waste so that the latter can be readily and safely disposed of in municipal landfills (3).

The disposal of arsenate or arsenite compounds in sanitary landfills is generally not acceptable because of the potential danger of ground and surface water pollution as well as possible occupational hazards resulting from on-site handling. There are, however, certain approved sites located over nonwater-bearing sediments or with only unusable groundwater underlying them and are completely protected from flooding and surface run-off or drainage such as those designated as Class 1 sites in California. The disposal of small stocks of arsenate or arsenite pesticides or empty containers used to ship arsenates or arsenites in Class 1 sites is considered as adequate, provided special handling techniques are employed to protect site personnel.

Long-Term Storage

Storage of cacodylates and cacodylate wastes, until they can be used or reprocessed, is a satisfactory waste management option. These materials are stable compounds and require the minimal storage precautions needed for any toxic material. Storage in the original containers is recommended but they should be periodically checked for corrosion or breakage of the containers. Bulk quantities of waste, such as the salt by-products from cacodylate manufacture, can be stored in concrete vaults or weatherproof bins.

The use of large, weatherproof and siftproof storage bins or silos is currently being used for the storage of arsenic compounds, especially arsenic trioxide (3). Waste arsenate or arsenite pesticides already packed in multiwall paper bags could also be safely stored in these large weatherproof bins. This approach is relatively expensive. Land and storage equipment are required for long, unknown periods of time before these materials can be moved by sale or disposal to other outlets. Considering the fact that these arsenic compounds are poisonous in all forms, long-term storage has to be considered as adequate and currently practical.

REFERENCES

(1)　Booz-Allen Applied Research, Inc., *A Study of Hazardous Waste Materials, Hazardous Effects and Disposal Methods,* Vol. II, Report PB 221 466, Springfield, Va., Nat Tech Inf Serv (July 1973).

(2)　E.W. Lawless, T.L. Ferguson and A.F. Meiners, *Guidelines for the Disposal of Small Quantities of Unused Pesticides,* Report PB 244 557, Springfield, Va., Nat Tech Inf Serv (June 1975)

(3)　R.S. Ottinger, J.L. Blumenthal, D.F. Dal Porto, G.I. Gruber, M.J. Santy and C.C. Shih, *Recommended Methods of Reduction, Neutralization, Recovery or Disposal of Hazardous Waste,* Vol. V, *Pesticides & Cyanides,* Report PB 224 584, Springfield, Va., Nat Tech Inf Serv (August 1973)

(4)　B.C. Wolverton; U.S. Patent 3,725,269; April 3, 1973; assigned to the U.S. Secretary of the Air Force

(5)　K.H. Sweeny and J.R. Fischer; U.S. Patent 3,737,384; June 5, 1973; assigned to U.S. Secretary of the Interior

(6)　T.L. Ferguson, *Pollution Control Technology for Pesticide Formulations and Packages,* Report PB 241 001, Springfield, Va., Nat Tech Inf Serv (January 1975)

(7)　M.V. Kennedy, B.J. Stovanovic and F.L. Shuman, Jr., "Chemical and Thermal Methods for Disposal of Pesticides," *Residue Reviews* 29, 89-104 (1969)

(8)　A.W. Reed, *Ocean Waste Disposal Practices,* Park Ridge, N.J., Noyes Data Corp. (1975)

(9)　P.R. Atkins, *The Pesticide Manufacturing Industry—Current Waste Treatment and Disposal Practices,* EPA Report 12020 FYE 01/72, Washington, D.C., U.S. EPA (January 1972)

(10)　Midwest Research Institute, *The Pollution Potential in Pesticide Manufacturing,* Report PB 213 782, Springfield, Va., Nat Tech Inf Serv (June 1972)

(11)　S.T. Yosim, D.E. McKenzie, L.E. Grantham and J.R. Birk; U.S. Patent 3,845,190; October 29, 1974; assigned to Rockwell International Corporation

PETROLEUM INDUSTRY WASTES

Some of the general disposal problems concerning petroleum industry pollutants have been reviewed by H.R. Jones (1). While that book did discuss some ultimate disposal techniques such as flare disposal, most of its emphasis, like that of many other books on pollution control, was on pollutant reduction rather than pollutant elimination.

CATALYTIC CRACKING OFF-GASES

Incineration

A device developed by J.S. Zink, et al (2) is an assembly for carrying out the combustion of waste gases, such as those released during catalytic refining of petroleum so that the heat developed during such combustion may be employed for useful purposes and at the same time complete and continuous combustion may be attained and maintained with minimum quantities of auxiliary heat.

Large volumes of gases are released during the refining of petroleum by the catalytic process. These products are often discharged to atmosphere as waste gases. Such gases contain carbon monoxide, methane, traces of oil vapors, nitrogen and water. While the calorific value of the combustible materials of such waste gases is relatively low and in the order of fifteen to thirty British thermal units per cubic foot the temperature of such waste products as they are released is in the range of approximately 1000°F.

Apparatus is available for partial recovery of the sensible heat but the combustible materials of the waste products are not readily convertible into useful heat because the temperature of the waste gases must be elevated in the presence of oxygen to a level that will enable the combustible materials to burn to completion. The temperature for carrying out such combustion is in the range of about 1500° to 1800°F. Such temperatures are not produced by the combustion of the waste gaseous products so as to maintain continued burning of the combustible materials. It is an object of the apparatus to carry out complete burning of the combustible materials contained in waste gaseous products and to provide for delivering the heat produced by both combustion processes into a space to be heated.

COKE FROM REFINERY PROCESSES

In a variety of refinery operations, circulating contact materials are coated with tars or cokes. Failure to remove such contaminants and permit reuse of such contact materials would be uneconomic in the first place and would produce a vast volume of pollutants in the second place. Hence various regeneration incineration processes have been developed for the recovery of the active contact materials and for the ultimate disposal of the contaminant deposits.

Incineration

A process developed by E.V. Bergstrom (3) relates to the conversion of fluid reactants in the presence of a subdivided solid contact material on which deleterious combustible deposits are formed. The solid particle-form material may be catalytic or noncatalytic. It may partake of the nature of Fuller's earth, natural or treated clays, or various synthetic associations, such as, for example, silica or silica with additions of alumina or zirconia or chromia. Some processes utilize granular particles of inert refractory materials, such as corhart, mullite or even iron balls. Sufficient heat is stored in these materials so that when the materials are contacted with hydrocarbons, suitably prepared for conversion, the heat is released to effectively convert the hydrocarbons. The device described by Bergstrom is a simplified kiln for regeneration of fouled granular contact material in which the contact material will not be heat damaged and afterburning will be prevented.

LNG VENT GASES

In recent years the use of liquefied natural gas as a source of fuel in areas where natural gas is unavailable has increased. Commonly, the natural gas is liquefied at the area where it is produced and then shipped over land or by sea to the area of use by means of special refrigerated vessels. A problem encountered in such shipment of liquefied natural gas is the disposal of vent gas continuously formed due to the vaporization of small portions of the transported liquefied natural gas. In order to eliminate a fire hazard and to prevent pollution of the atmosphere, such vent gas must be incinerated, and particularly with respect to the transport of liquefied natural gas by ship, the incinerator apparatus must be inexpensive, easily and quickly operated and cannot include a high stack.

Incineration

A system developed by H.O. Ebeling (4) is one in which the vent gas is intimately mixed with a stream of combustion air and the resulting air-gas mixture is combusted so that the stream of vent gas is converted to inert products of combustion. The products of combustion are quenched and released to the atmosphere.

OIL SPILLS ON BEACHES

The problem of restoration of oil-contaminated beaches by incineration has been reviewed at some length by F.N. Rubel (5). This topic has also been covered by Sittig (6).

Incineration

The details of incineration design and construction for decontaminating beach sand and ultimately disposing of oil residues on the sands have been covered by R.M. Roberts et al (7).

OIL SPILLS ON WATER

Open Burning

One way of disposing of oil on water is to burn it. The success of this method depends on supplying the blaze with sufficient oxygen and keeping it hot enough. One problem is that the thin layer of oil is cooled by the water, making it nearly impossible to ignite. A wicking agent, such as polyurethane foam, helps insulate the oil from the cool water below and keep the oil ignited.

There are disadvantages to burning. It results in air pollution, and it may create a fire hazard. For these reasons, burning is not usually practical in sheltered waters, although it may be desirable to burn oil on the open sea.

Burning of oil on water or land by special methods and materials seems to offer an attractive and perhaps inexpensive means of eliminating large amounts, providing of course, the many significant hazards are also recognized. Freshly spilled oils and crudes containing volatile components are relatively ignitible. If a thick layer of oil on water is present, the oil will sustain burning until the volatiles and a portion of the heavier fractions are combusted. Conversely, with fresh oil spills within a harbor or confined area, a significant fire danger exists when the level of hydrocarbon vapors is within the range of flammability, (e.g., gasoline, aviation fuels, or light crude oils).

Wood, debris or other material caught within an oil slick can serve as a wick to start or sustain an oil fire. The cooling of a layer of oil by the water body beneath will greatly deter burning. However, the wick will withdraw the oil and insulate the burning oils from the cooling action of the water, and at the same time provide a renewal and vaporization surface for combustion.

Experience by certain investigators has indicated that floating oils on the sea with thicknesses less than 3 mm (0.12 inch) will not burn. It is also reported that layers of kerosene, gas oil, lubricating oil and fuel oil on water will not burn at all without a wick. In one instance, attempts were made to ignite fresh Iranian crude oil five minutes after a spill without success. Once an oil spillage has spread, the material quickly loses its volatile components and ignition is extremely difficult. Weathered oils are consequently reported to present almost no fire hazard.

In experiments carried out at the Edison, N.J. laboratory of EPA, a heavy fuel oil and a lightweight crude oil were placed atop a layer of water contained in metal tanks with 24 ft^2 of exposed surface area. Attempts were made to combust the oils using different burning agents. It was concluded from these experiments that the light crude, freshly applied in a floating thickness of 2.5 mm (0.1 inch) required external support with burning agents and an ignition source to burn near completion.

When the No. 6 fuel was used for testing, one of the agents used would not sustain burning at a thickness of ½ to ⅔ inch. Another agent generously applied over the surface of the No. 6 fuel caused sporadic burning and the minimum required oil thickness to sustain burning was ⅓ to ½ inch. It is important to note that a thickness of ⅓ to ½ inch is equivalent to an oil slick of 7 million gal/sq mi. It was evident after this burning that appreciable oil was still remaining.

A third agent used performed well with the No. 6 fuel at a thickness of $\frac{1}{10}$ to ¼ inch. The manufacturer producing this material, however, suggests that in order to have complete burning of all the oil, the oil layer must be completely covered with the material with no broken patches. Required amounts of this material (very low bulk density) may be as high as 1 lb for each 12 to 15 ft^2 of oil slick.

Field scale oil slick burning experiments, with and without special agents were undertaken in 1970 by both the U.S. Navy and the Edison Water Quality Laboratory. Preliminary data from these experiments indicate the following:

[1] Burning of free-floating or uncontained oil slicks is extremely difficult unless the thickness of oil is 2 mm or greater.

[2] Adequate automated seeding methods for both the powder and nodule type burning agents are lacking. Spreading of the burning agent on the oil slick had to be accomplished by hand. This conclusion was also reached by the Navy, which conducted burning experiments in May 1970.

[3] Contained South Louisiana crude oil was successfully burned (80 to 90%

reduction) without the application of burning agents and/or priming fuels.
Bunker C could not be ignited under the same conditions.

[4]   Bunker C was successfully burned (80 to 90% reduction) when the slick
was seeded with burning agents and an appropriate priming fuel. It was
discovered that South Louisiana crude oil performed better as a priming
agent than did gasoline or lighter fluid.

[5]   Use of magnesium type flares and gasoline torches to ignite the burning-
agent-treated slick proved unsuccessful. Success was achieved, however,
using a blow torch once it was learned how to manipulate the torch in
such a manner that the torch gas pressure did not push aside the oil and
seed material so as to expose the water surface.

Burning agents were considered when the Torrey Canyon was in the final stages of destruction off the British coast in March of 1967. As a last resort, the British Government attempted simultaneous and complete burning of the 15,000 to 20,000 tons of oil remaining
in the badly broken tanker by aerial bombing, incendiaries, and catalyst-oxidizing devices.
The major objective was to penetrate and lay open the decks by explosive surgery, exposing the oil in the storage compartments to large amounts of oxygen required for burning.
High-explosive, 1,000 lb bombs filled with aluminum particles, thousands of gallons of aviation fuel, napalm bombs, rockets, and sodium chlorate devices served to produce a massive
fire, if not a sustained fire. The British concluded that no appreciable amounts of oil escaped burning, but if any did, it was lost to the open sea.

Controlling the burning oil mass and ensuing air pollution problems would appear to preclude intentional burning except where the oil mass is distant from the coastline, offshore
facilities, vessels, etc. The safety and welfare of all parties, however, remote from the burning site are of utmost importance. The possible loss of additional oil, and the loss of a
drilling platform or vessel at the source of the spill must be recognized. Because of potential merits in controlled burning, further research is desired on new methods, techniques
and procedures both in the laboratory and in the field, together with additional guidelines
for burning.

Commercial burning agents are available for promoting combustion of an oil slick. However,
in most cases it is apparent these agents are designed for a relatively thick oil layer which
is sufficiently contained. These agents are intended to serve one or more functions such
as providing increased surface area exposed to burning; addition of catalysts, oxidizers and
low-boiling volatile components; absorbency and entrapment of the oil; or creating a wicking mechanism via surface diffusion and capillary action by the material added. According
to EPA, burning agents are available from the following sources:

[1]   Eduard Michels GmbH, Essen, Germany: Kontax, the commercial name
for this agent, ignites spontaneously when it comes in contact with water.
In 1969, the Dutch conducted field experiments which included burning
of oil on beaches, and in open waters. Results of this investigation indicated that the quantity of agent required was dependent upon wind, condition of sea and continuity and thickness of oil.

Dutch engineers also reported that a method should be developed to jetti-
son, hurl or catapult the agent from a vessel in such a way that there is not
the slightest risk of having the agent come in contact with rainwater or
spray. Dropping from an airplane is worth considering provided special
packaging requirements are met.

[2]   Pittsburgh Corning, Pittsburgh, Pa.: Known as Seabeads, these cellulated
glass nodules are available in sizes from ⅛ to ¼ inch in diameter. By
capillary action, the nodules become coated with oil. Depending upon the
type and age of the spilled oil, combustion is accomplished by using an incendiary device alone, or in combination with a primer fluid such as gasoline. After burning, Seabeads still remain, therefore, they must be collected
or left to break up by abrasion. During the Arrow incident in 1970, this

burning agent was used, with varying degrees of success, on small patches of spilled oil.

[3] Guardian Chemical Corporation, Long Island City, N.Y.: Pyraxon, a powder material, is a catalyst containing small amounts of oxidizing materials. Pyraxon liquid is used as the priming agent or starter fluid, with the powder being sprayed, blown or dropped onto the oil, in and around the flame.

[4] Cabot Corporation, Boston, Mass.: Cab-O-Sil ST-2-0 promotes combustion by acting as a wicking agent. Produced from fumed silica, this powdery-like material, is surface treated to render it hydrophobic. It may be applied directly to the slick or by spraying in a stream of water. Combustion is best accomplished by using an incendiary device in conjunction with a primer fluid. Reportedly, this product was used successfully for handling a 2,000 gal spill at Heard Pond, Wayland, Mass.

A process developed by A. Molin and O. Carlsson (8) involves destroying drifting oil layers on the surface of water basins by sustained combustion in a zone contiguous to the oil layer and in relative motion with respect thereto. In the method for thus combating drifting oil, a plurality of jets of combustion sustaining gas, in particular compressed air, are blown against the oil layer in the zone for sustaining combustion therein.

In the combating means a hollow element is connected to a source of pressure gas, in particular compressed air, and kept afloat at the surface of the water with longitudinally spaced discharge openings on the element blowing the pressure gas against the oil layer in the zone for sustaining combustion therein. Figure 119 shows one version of the application of the process, using shore-based equipment for the combustion of a spill adjacent to the shore. As shown there, lines of interconnected oil booms **17** are moored by means of weights **40** to the bottom of the sea near the shore **43**.

FIGURE 119: ATLAS COPCO PROCESS FOR BURNING OFF WATER SURFACES

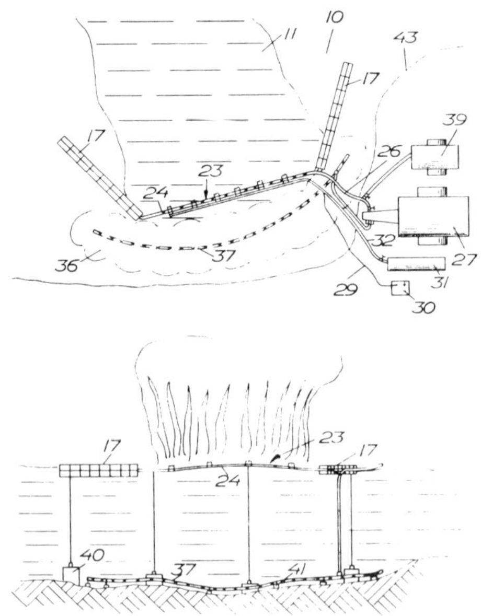

Source: U.S. Patent 3,586,469

An oil patch drifting with wind or currents **10** is caught by the oil booms and guided towards an element **23** laid out between the lines of the booms and provided with discharge openings spaced along the length of the element. More particularly, the element consists of a single floating tube **24** of plastic material provided with an asbestos covering.

By means of weights **41** there is furthermore immersed and moored at the bottom of the sea a perforated tube **37** for forming, when supplied with compressed air, an air bubble curtain and thereby an above-surface barrier **36** above the tube which barrier guides the oil layer **11** towards the combustion zone at the element. For purposes of increasing the combustion intensity gaseous oxygen additive may be led over the compressed air hose **26** from an oxygen tank **39** for liquid oxygen placed ashore, whereby the oxygen content in the compressed air may be increased.

In case of need fuel agents such as gasoline, spirit, coal gas and the like may be supplied to the combustion zone via special supply conduits, not shown, or as a dosage to the compressed air in the hose in case the combustion tends to become sluggish when encountering large accumulations of heavy oil in the combustion zone.

From the compressors **27** air under pressure is supplied to the element for sustaining by the oxygen contained therein, the combustion in the zone above the tubes when the oil layer reaches the latter and is washed thereover by drifting with wind and water currents towards the element. In the zone above the tubes the oil layer is subjected to a plurality of fine gas jets from the discharge openings hitting the layer transversely from below and vehemently mxing the oil with gas and conditioning the zone for continuous sustained combustion.

The oil zone above the tubes, thus effectively blown through and oxygenized, is ignited for example by means of torches or matches. Preferably there are arranged ignition coils on the tubes which coils are fed by an electric heating current via a cable **29** from a current source **30**.

If an increase of the intensity of combustion is desired, a fuel gas container **31** may be provided. The container supplies fuel gas such as coal gas via a conduit **32** to the fuel nozzles arranged at the tubes adjacent to the ignition coils. The element is displaced relative to the layer preferably at such rate that the layer flowing on will have time to burn away continuously and substantially completely in the zone above the tubes.

A process developed by P.R. Tully, W.J. Fletcher and H. Cochrane (9) is one in which particulate solids are applied to the spill and the resulting system is thereafter fired. Such treated spills are more easily ignited and the combustion thereof is more complete than experienced with untreated spills. When certain conditions pertaining to the type and amount of treating agent applied to the spill are met even further benefits accrue to the process. The benefits reside in improved physical character of the burned residue which is more amenable to physical removal thereof from the water or land mass than the burned residuum of untreated spills.

The particulate solid materials useful as the treating agents are generally any substantially hydrocarbon and water-insoluble particulate solid having an average ultimate particle diameter of less than about 250 mμ (preferably less than 100 mμ), an apparent density of less than about 50 lb/ft^3 (preferably less than about 15 lb/ft^3) and a specific surface area (as determined by the BET-N$_2$ method) of greater than 10 m^2/g (preferably greater than 50 m^2/g).

Specific examples of such materials are: carbons such as carbon black, activated carbons, chars and the like; metal and metalloid oxides produced by way of various precipitation, arc, plasma or pyrogenic processes such as silica, titania, alumina, silica-alumina, silica gel, alumina gel, iron oxide, zirconia, vanadia, chromia, magnesia, zinc oxide, copper oxides, and the like.

In addition, there exist various naturally occurring siliceous clays and minerals such as chrysotile asbestos which can be specifically treated so as to fall within the ambit of the above-recited solubility, particle diameter, density and surface area criteria. Certain of such naturally occurring materials, in particular, asbestos, are usually acicular, lamellar or fibrous in form rather than roughly spherical. However, such acicular or fibrous materials are to be considered as meeting the particle diameter criteria set forth hereinabove when the average cross-sectional dimension of the ultimate particle thereof is less than 250 mμ.

Additionally, particularly when it is desired that a substantial amount of physical integrity be imparted to the combusted residuum resulting from the firing of the treated oil spills, it is preferred that the particulate solid utilized be possessed of a substantial degree of structure. Structure is that property of a particulate solid which signifies the extent to which primary particles thereof tend to form into a chainlike network.

Accordingly, the higher the structure of a particulate solid, all other factors being equal, the greater the reinforcement capabilities thereof will normally be when dispersed into a suitable matrix. Generally, the structure of a particulate solid is roughly proportional to the ability of the solid to absorb oils, such as linseed or mineral oil or dibutyl phthalate.

The term oil absorption factor is to be considered as the minimum amount of dibutyl phthalate (in cc) required to cause the coalescence of 100 grams of a given particulate solid into single spherical structure by working incremental amounts of the oil into the solid by means of hand stirring with a spatula. Thus, those particulate solids which, in addition to meeting the limitations previously set forth, also bear oil absorption factors of greater than 100 cc/100 g of solid are normally greatly to be preferred.

Further, when such preferred solids are utilized, it is advantageous in terms of eliciting the maximum reinforcing function in the burned oil slick residuum, to treat the oil slick with at least about 2% by weight thereof of the highly structured particulate solid and even more preferably with between 4 and 10% by weight thereof.

In practice the particulate solids utilized are also preferably rendered hydrophobic prior to application thereof to the oil spill. When such particulate solids are utilized, potential loss of the solid material into the water phase is normally vastly lessened. In fact, water may be effectively utilized as a carrier liquid for the conveyance of such hydrophobic particulate solids to the oil slick. In particular, metal and metalloid oxides, such as silica or titania, which have been rendered hydrophobic by treatment thereof with various organo-silicon compounds have been found to be especially preferred treating agents.

The application of the treating agents to the oil spill may be undertaken in any suitable manner. For instance, the particulate solids may be applied by hand, by air drops, or other procedures well known to the particulate material application arts. When, however, the much preferred hydrophobic particulate solids are utilized the application to water-borne spills can be made advantageously either from beneath or above the surface of the spill.

A water stream can be utilized (in accordance with the Bernoulli pump principles) to aspirate the hydrophobic solid from its storage area and convey same to the spill. When this stream is played upon the upper surface of the slick the water carrier sinks therethrough leaving, however, the preponderance of the hydrophobic particulate solid in or on the slick.

Further, again referring to water-borne spills, the hydrophobic solids can be conveyed into the water phase from beneath the oil slick and thereafter released. Due to the hydrophobic/oleophilic nature of the particulate solid the solid rises to the oil/water interface and is entrained substantially entirely into the slick proper. Where air or other gas is also entrained in this subsurface application method, the gas tends to rise through the slick, thus providing an additional mild and often desirable dispersing function.

Additional benefits to be derived from the use of hydrophobic particulate solids and application thereof to the oil slick from beneath the water surface also reside in the substantial obviation of potential losses of the materials due to convection, breezes, wind gusts, etc. which are often encountered in the environment above the surface of the slick as well as providing improved capability of precision and uniformity of application.

Subsequent to the application of the treating agents to the oil spill, the resulting oil/solid system is ignited in any suitable manner. Such ignition may be expeditiously achieved by applying to a localized area of the treated slick a small amount of a highly flammable liquid such as lighter fluid, kerosene, or the like and igniting this so-called initiator. The resulting flame thereafter progresses into the treated slick proper and propagates across the surface of the oil/solid system.

A process developed by N.T. Castellucci and N.C. Krouskop (10) involves spreading a layer of substantially spherical ceramic nodules on the upper free surface of the layer of combustible liquid. The nodules are wetted with the combustible liquid and the combustible liquid is ignited on the upper surface of the nodules until combustion is self-sustaining. The combustible liquid on the upper surface of the nodules consumed by combustion is continually replaced with combustible liquid from the layer until substantially all of the combustible liquid in the layer is consumed. The cellular ceramic nodules have a multiplicity of separate closed cells and the outer surface of the nodules has a plurality of cup shaped recess portions.

The cellular ceramic nodules suitable for use in this process may be prepared in accordance with the process described in U.S. Patent 3,354,024 from a pulverulent glassy material and a cellulating agent or from other pulverulent materials and a cellulating agent in accordance with the process described in U.S. Patent 3,441,396. A description of the process for providing a textured surface on the nodule may be found in U.S. Patent 3,493,218, and entitled "Tower Packing Element". The cellular ceramic nodules enable and enhance combustion of the combustible liquid to be removed from the body of water through interaction of the physical characteristics of the nodules, such as the surface morphology, the density, the impermeability, the chemical composition, the thermal characteristics, and the like.

The nodules may have an apparent density of between about 6 and 30 lb/ft^3 and a thermal conductivity of between about 0.40 to 0.50 Btu/hr/ft^2/°F/in at 75°F. The nodules can be made in many different sizes. Nodules of a size between one-eight and one-half inch with an apparent density of between 10 and 20 lb/ft^3 were found suitable.

In U.S. Patent 3,354,024 the nodules are made by admixing relatively fine pulverulent glass with a cellulating agent such as carbon black or the like. A binder is then added to the mixture which is then pelletized and subsequently coated with a parting agent that serves to maintain the pellets discrete during the cellulation process. The coated pellets are heated in a rotary funace or kiln to a cellulating temperature and the pellets cellulate to form substantially spherical cellular ceramic nodules with a continuous outer skin. Although pulverulent glass is a preferred constituent of the cellular ceramic nodules, other glassy materials as described in U.S. Patent 3,441,396 may be used. The term ceramic is intended to encompass both pulverulent formulated glass and other suitable pulverulent glassy materials.

The cellular ceramic nodules thus produced have a core of individual completely closed cells of ceramic material and a continuous outer skin of ceramic material. For use with the herein described process, it is preferred that the cellular ceramic nodules produced as described above be abraded or otherwise treated to remove the relatively thin continuous outer skin and a portion of the layer of underlying closed cells to expose, over the entire surface of the nodule a portion of the layer of cells therebeneath. The cells on the abraded surface are opened to form a surface having a plurality of contiguous individual cup-like recessed portions or cell fragments. With their low thermal conductivity, the nodules function as thermal insulators during combustion thereby preventing loss of heat to the underlying water and confining and concentrating the available heat to the region of combustion in the thin film of liquid on the surface of the nodules.

A process developed by D.D. Sparlin (11) is one in which particles of foamed water-soluble and dispersible alkali metal silicates are distributed over oil slicks to absorb the oil. The oil is then burned after which the water-soluble and dispersible particulate foamed alkali metal silicate particles solubilize and disperse.

A process developed by R.J. McGuire et al (12) is one whereby oils floating on the surface of open bodies of water can be removed by burning them in situ in the presence of an oleophilic particulate material such as vermiculite which has been treated with a metallo cyclopentadienyl compound such as dicyclopentadienyliron.

A process developed by R.F. Rensvold (13) is one whereby an oil slick is destroyed by applying thereto finely divided particles of a compound capable of generating a combustible gas, upon contact with water, allowing the particles to contact the underlying body of water so that bubbles of combustible gas rise through the oil film and admix therewith, so as to enhance the combustibility of the oil, and then igniting the oil-gas mixture to burn and destroy the film, e.g., calcium carbide to form acetylene gas.

A process developed by F.J. Shell (14) involves removing hydrocarbons from the surface of a body of water by placing a polypropylene fabric sheet over and in contact with the hydrocarbons and combusting those hydrocarbons passing onto the upper surface of the sheet.

It has been found that initial combustion of hydrocarbons on the upper surface of the polypropylene sheet does not destroy the sheet and during the combustion, hydrocarbons located beneath and in contact with the sheet pass upwardly through the sheet and are combusted on the upper surface. The polypropylene sheet thereby functions as a wick and causes the hydrocarbons to be removed and separated from the surface of the water and supports the hydrocarbons in a spaced relationship from the water surface during combustion. Ignition of the hydrocarbons can be by several methods whereby a flame is brought into contact with the hydrocarbons.

A process developed by J.W. Marx (15) is one in which petroleum oil floating on the surface of water is removed therefrom by adsorbing the oil on a treated cellulose sponge and then burning the adsorbed oil from the sponge while it remains in contact with the water. During the combustion, the treated cellulose sponge continues to adsorb oil and deliver it to the combustion zone.

A process developed by W.D. Johnston (16) is a process for the substantially complete combustion of a combustible liquid including the combustion of a layer of the combustible liquid floating on a body of water. Cellular ceramic nodules having a core of separate closed cells and substantially continuous outer surfaces or skins with a coating of particulate alumina parting agent adhering thereto are treated to remove a substantial portion of the particulate alumina parting agent. The nodules are preferably washed in a water or dilute acid bath with mild agitation, to loosen and remove the particles of alumina parting agent agent adhering thereto without substantial abrasion or fracturing of the continuous outer skin.

The treated nodules are relatively dust free and have a smoother outer surface. A layer of the treated cellular ceramic nodules is formed on the upper surface of the combustible liquid with a substantial number of the nodules in contiguous relation with adjacent nodules in the layer. The upper exposed surfaces of the nodules are wetted with the combustible liquid to form a film or layer thereon and the wetted films on the exposed surfaces of the nodules are ignited until combustion is self-sustaining. The combustible liquid films on the exposed upper surfaces of the nodules consumed by combustion, are continually replaced with combustible liquid from the bulk of the liquid until substantially all the combustible liquid is consumed.

Another process developed by W.D. Johnston (17) is one in which cellular ceramic nodules are prepared by coating uncellulated pellets with a particulate carbonaceous parting agent and cellulating the coated pellets in a rotary furnace or kiln. The cellular ceramic nodules obtained by the above process have a relatively thin coating of the carbonaceous parting agent thereon and a relatively smooth continuous outer skin. These particles are used in the same manner as those described just above which were prepared using an alumina parting agent.

A process developed by A.J. Shaler et al (18) is one in which an elongated porous carbon member is impregnated with a combustible fluid and then floated in a generally upright position in a layer of combustible fluid on a body of water, with the lower portion of the porous member extending down in the water and with its upper portion projecting above the fluid layer. The fluid carried by the upper end of the porous member is ignited to produce a flame that is thereafter fed by combustible fluid moving up through that member by capillary action from the fluid layer, whereby to remove the fluid from the water and burn it.

Since oils wet carbon better than water does, carbon being hydrophobic and oleophilic, the rod acts as a selective wick for the oil. The rod extends above the layer of oil and is tapered upwardly to more or less of a point. Oil will move upwardly through this tapered portion by capillary action, whereas water will not. Before the device is put in use, it is impregnated with a combustible fluid. Then, after being dropped into the water where it floats with its tapered upper portion above the oil layer on the water, the tip of the device is ignited. This produces a flame that is confined to the tip and that will be fed by oil moving up through the porous carbon as long as a layer of the oil remains on the water in contact with the device.

The outer surface of the rod, where it is in contact with the oil, should have a sufficiently large perimeter so that the rate of entry of the oil into the pores in the carbon is not so rapid as to entrain foreign particles and other debris which might otherwise clog the pores. Also, the device should be as stable as possible in agitated waters and the upper portion should project above the oil far enough to prevent the wave motion from sinking its tip from time to time below the oil layer.

By tapering the upper portion of the rod, the volume of carbon at the apex is so small that it can easily be brought to the ignition temperature of the oil. Since the thermal conductivity of the carbon-oil composite is low to begin with and the tip is distant from the cool oil layer on the water, this ignition temperature is maintained long enough, even if the device is splashed or lightly sprayed with oil or water or even rained upon, to reignite the oil flowing upwardly through the carbon in case the flame is temporarily extinguished. Thus, the tip of the device should be at a substantial height above the level of the oil layer so that it is not cooled appreciably by conduction and also so that in agitated waters, spraying, splashing and the up and down motion of the device caused by wave motion will not extinguish the flame more than momentarily.

Although porous carbon is the preferred material, other light porous materials such as foamed plastics, pumice and other ceramics, textiles and metals may be utilized. If necessary, the walls of their pores may be coated, as with silanes or other hydrophobic but not oil-repelling substances. However, none of these porous materials are as satisfactory, all things considered, as porous carbon.

One form of such a device is shown in Figure 120. The device illustrated is formed from two intersecting vertical slabs **12** and **13** of oil impregnated porous carbon that will float. The two slabs are assembled by providing one of them with a central slot extending downwardly from its upper end and providing the other slab with a central slot **14** extending upwardly from its lower end. This second slab **13** is inserted in the slot in the other one and moved downwardly to straddle the first slab. The result is a device in the form of a cross in horizontal section, as shown. To keep the device floating upright, the lower end of slab **12** carries a small weight **15**. The portion of the device extending above the oil is tapered upwardly substantially to a point.

FIGURE 120: POROUS WICK TO ASSIST IN BURNING OIL FROM WATER

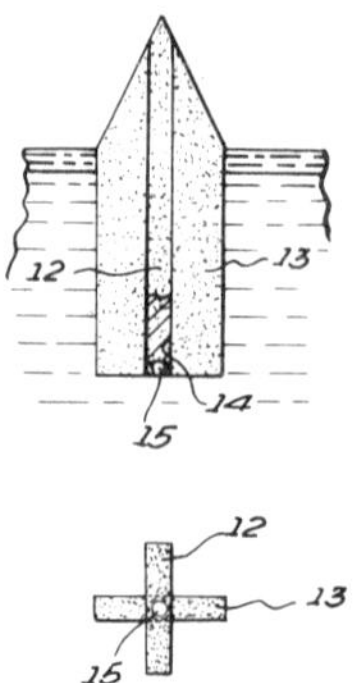

Source: U.S. Patent 3,659,715

This device has a greater surface area for entry of oil per unit of length than other designs, and generally is less costly to manufacture. It also has greater stability in high winds because its cone angle is less and the ratio of its greatest width to its length is greater. Large numbers of the components of these devices can be oil-impregnated at the factory and sealed into an efficient packing for storage on the deck of a ship. They can be readily assembled when needed by inexperienced crewman. A single such device is capable of burning off more than 50 lb of crude oil, lubricating oil, or even of a heavy fluid fuel oil per day from the surface of an ocean, lake, or pond in heavy seas and weather, and it will continue to burn for days until the oil slick has become a film only some thousandths of an inch thick.

It can easily be lighted and relighted if it is accidentally extinguished. There is no danger of the fire spreading away from the upright slab of the device, since the flame is confined to the upper portion of the latter by the cooling action of the water on its lower portions, as well as by the action of the wind. If a number of the devices are scattered over a spreading oil slick, they will tend to gather from wind drift along the downwind boundary of the slick where it is most desirable to burn off the oil and where any wind-caused water-spray is least, since upwind, the device faces a large expanse of oil-covered water. The wind does not tend to cause the fleet of devices to drift past this boundary into water that is not covered by oil, because the slick is also being wind-drifted in the same direction and also because capillary attraction holds the devices to the slick boundary.

A process developed by S. Kraemer et al (19) is a process for absorbing and burning away oil or other combustible liquids on water or other noncombustible liquids wherein absorbent and/or surface active noncombustible inorganic foamed particles are spread out over the combustible liquid, the combustible liquids are absorbed by the particles and the liquid absorbed by the particles is ignited.

Suitable substances of which the foamed glass particles are produced are siliceous substances such as sodium, calcium or aluminum silicate, alone or in mixtures, expanded mica or similar expanded natural products, but also oxidic materials such as Al_2O_3 and clay, and even foamed metals, especially aluminum and its alloys when they function as absorbers and remain floating on the surface of the liquid, which means that in addition to the open pores which give the material wick-like properties, it also has enough closed air cells to keep it floating.

The foamed materials which are especially suitable for absorbing and burning oil that is floating on a water surface are disclosed in U.S. Patents 3,184,371 and 3,261,894. During the production of these materials the ratio of the open pores to the closed inner cells is controlled within wide limits by the use of varying amounts of foaming agent.

These foaming agents include calcium carbonate, aluminum sulfate, zinc sulfate, carbon compounds such as glycerol, sugar and others in the presence of sulfates, and also compounds containing chemically bound water. The melting point of the products thus produced are varied within wide limits, at least between 400° to 1000°C.

The melting point of the foamed glass particles is brought up to at least 400°C by addition of lead, boron etc. compounds to the aqueous silicate solution. Lead compounds suitable for this purpose are preferably lead oxides PbO, PbO_2 and Pb_3O_4, while suitable boron compounds are, for example borax and other borates. The melting points of the foamed glass are brought to a maximum of 1000° to 1100°C by the addition of alkaline earths and earth metals to the foamable silicate solution. Reference is here made to compounds of the alkaline earths and of the elements of main Group III of the Periodic Table, such as aluminum, etc., and especially in the oxide form.

By suitable adjustment of the porosities and melting points it is possible to produce products that are suitable for absorbing and burning away of oils, etc., and which after completion of the burning away process become sintered at their surfaces and in this manner acquire sufficient density to sink in the water. In a similar manner other products are obtained which will retain their closed air cells in sufficient abundance to keep them floating on the surface after the burning has ended. The products of combustion are preferably allowed to escape into the air while some of the ash remains in the foam. After the burning, the foam remains floating on the surface if it retains sufficient porosity to keep the densities of the floating particles less than the densities of the liquids.

This is accomplished by using glasses, metal oxides or expansible clay of high melting point for the production of foamed particles. The foamed material will sink if there is sufficient pore loss to increase the density above that of the noncombustible liquid. This result is produced by the use of low melting glasses as starting materials. It is preferred that the foamed particles remain floating after the combustion because they can then be brought on land and collected without soiling the coast or acting unfavorably upon the marine fauna. If, however, economic considerations are controlling, then those foamed particles should be preferred which will sink after the burning, because the original raw materials are very cheap.

The treatment of the foamed particles to render them hydrophobic is accomplished in many ways with industrial silicone oils (produced from organosiloxanes derived from alkylhalogen silanes) which are dissolved in organic solvents, the foamed particles being sprayed with the silicone solution or immersed therein. The silicone oils are also applied as aqueous emulsions, or in the form of their precursors (e.g., mono-, di- and trichlorosilanes) which by hydrolysis with steam are coated upon the foamed particles. The water-repelling agents are also vaporized upon the foamed particles under vacuum.

From this it is seen that silicone oils or their precursors are applied in many ways. The foamed particles may be also rendered water-repellent by other known methods, e.g., by coating them with oils such as heating oil, asphalt, mineral fats, waxes such as montan wax or ozocerite, but also with salts of fatty acids, e.g., calcium stearate in solution, suspension, emulsion or dispersion. The substances used in the process have the following characteristics, they are inorganic silicious, oxidic or metallic foam. They are incombustible, buoyant, insoluble or only slightly soluble in water, insoluble in oil, absorbent for liquids that do not dissolve in water, stable in the presence of water and only slightly swellable or not swellable at all. Their affinity for combustible liquids is greater than their affinity for water.

Incineration

A process developed by R.B. Heagler (20) is one in which a generally U-shaped, buoyant, self-propelled vessel floats partially submerged in a body of water and has a longitudinal channel portion with a front opening. The vessel has an open bottom portion beneath the longitudinal channel portion. As the vessel advances into a body of water, a band of water with the layer of combustible liquid floating thereon, enters the channel portion of the vessel. The rate at which the combustible liquid, as a layer, enters the channel portion of the vessel is dependent on the forward speed of the vessel and the speed is controlled so that substantially all of the layer of combustible liquid is removed by burning before the band of water passes under the rear portion of the vessel.

As the vessel advances, the band of water with the layer of combustible liquid moves through a mixing chamber within the channel portion where a monolayer of cellular ceramic nodules is positioned on the top surface of the layer of combustible liquid. The layer of combustible liquid with the nodules floating thereon, moves rearwardly with the forward advance of the vessel into a combustion chamber where the layer of combustible liquid is ignited and burned. The modules within the combustion chamber are recycled to the mixing chamber where they are again positioned as a monolayer on the upper surface of the layer of combustible liquid.

Combustion air is provided for the combustion chamber and the combustion gases may be subjected to a secondary burning in the stack to remove the combustible materials in the combustion gases and provide a substantially smoke-free waste gas. Apparatus is provided to seal the combustion chamber and mixing chamber if the burning of the combustible liquid tends to spread beyond the receiver. The reader is referred to Figure 121 for plan and elevation drawings which give more details of the apparatus and process.

FIGURE 121: FURNACE SHIP FOR BURNING OIL SLICKS FROM WATER

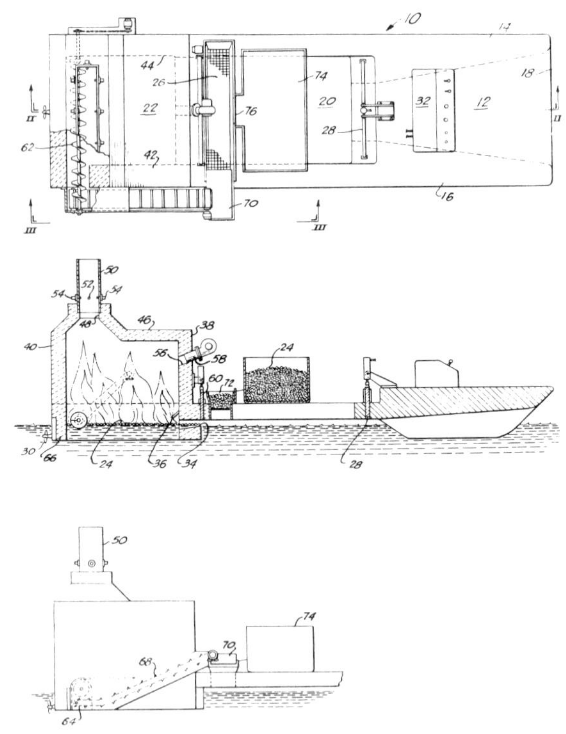

Source: U.S. Patent 3,663,149

There is illustrated a vessel **10** having a generally U-shaped configuration with a central longitudinal channel portion **12**. The vessel preferably has a pair of buoyant side portions **14** and **16** with an open bottom portion below the channel portion to provide an area within the vessel that confines a layer of combustible liquid floating on the surface of the body of water. The channel portion has a front opening **18** which permits the entry of the layer of combustible liquid floating on the surface of the water into the channel portion.

The channel portion may be described as having a mixing chamber **20** and a rear combustion chamber **22**. The mixing chamber is located in front of the combustion chamber so that the layer of combustible material may be prepared for burning before it enters the combustion chamber.

The vessel, particularly the side portions and the enclosure for the combustion chamber is preferably fabricated from a light material to permit the vessel to float on the body of water without displacing the latter and layer of combustible liquid within the central channel portion. The vessel may be fabricated from an admixture of cellular ceramic nodules and Portland cement in a ratio of 5 parts by weight of cellular ceramic nodules to 1 part by weight of Portland cement. Other materials having sufficient buoyancy may also be employed.

When floating in a body of water, the vessel is partially submerged as illustrated so that a band of the layer of combustible liquid floating on the top surface of the body of water will enter the front opening and move rearwardly through the channel portion as the vessel advances in the body of water. The layer of combustible liquid moves rearwardly into the mixing chamber within the confines of the vessel and within the mixing chamber. A layer of cellular ceramic nodules **24** is positioned on the top surface of the layer. The cellular ceramic nodules are distributed onto the upper surface of the combustible liquid within the mixing chamber by means of a distributor device **26**, such as, for example, a vibrating screen.

A gate member **28** extends along the front portion of the mixing chamber and is arranged to close the channel portion and isolate the mixing chamber from the forward portion of the channel portion. The gate is arranged to prevent the combustion of the layer of combustible liquid from spreading into the forward portion of the channel and possibly to the layer of combustible liquid beyond the confines of the vessel. Propelling means **30** are provided for the vessel and are preferably arranged adjacent the rear portion of the vessel to maintain the layer of combustible liquid within the channel portion in a quiescent state as the layer moves rearwardly within the channel portion. Suitable controls **32** are provided for the propelling means and the other apparatus for recycling and distributing the cellular ceramic nodules.

The layer of combustible liquid within the mixing chamber having the layer of cellular ceramic nodules positioned thereon, moves rearwardly in the mixing chamber and through an opening **34** into the rear combustion chamber. Suitable igniting means **36** are provided within the rear combustion chamber to ignite the layer of combustible liquid with the cellular ceramic nodules positioned thereon.

The rear combustion chamber is preferably enclosed with a front vertical wall **38**, a rear vertical wall **40**, a pair of spaced vertical side walls **42** and **44**, and a top wall or roof **46**. With this arrangement, an enclosure is provided for the rear combustion chamber that has a bottom opening which permits the layer of combustible liquid floating on the surface of the body of water to be moved into the enclosure as a layer floating on the body of water and burned while remaining as a layer on the surface of the body of water.

The roof has a stack opening **48** in which a preferably metallic stack **50** is positioned. The metallic stack has a plurality of radial openings **52** arranged to receive burner outlets **54**. Although not illustrated, the burner outlets are attached to a suitable source of combustion gas and may be utilized to ignite and burn the combustible material in the combustion gas and provide a relatively smoke-free gaseous product of combustion.

The combustion chamber front wall has an opening **56** in which a blower **58** is positioned to supply combustion air to the rear combustion chamber. It should be understood that other suitable means may be provided to supply combustion air to the combustion liquid within the combustion chamber. A gate **60** is provided to close the opening into the rear combustion chamber in the front wall to again prevent the spread of the combustion of the layer of combustible liquid beyond the confines of the vessel. It should be understood that the gates are safety devices and remain in a normally open position since the layer of combustible liquid without the layer of cellular ceramic nodules thereon, is difficult to ignite and it is seldom that the layer of combustible liquid will sustain ignition.

The layer of cellular ceramic nodules on the surface of the body of water, because of the forward motion of the vessel, tend to collect against the base of the combustion chamber rear wall. A suitable conveying means such as the screw conveyor **62** moves the layer of cellular ceramic nodules adjacent to the rear wall transversely into an enclosed receiver **64** that is positioned outside of the rear combustion chamber. It should be noted that the combustion chamber rear wall has a bottom end **66** beneath the top surface of the water so that the nodules which float on the surface of the water remain within the rear combustion chamber and only the water previously present within the confines of the rear combustion chamber is displaced beneath the combustion chamber rear wall.

The nodules are conveyed transversely by the screw conveyor and disposed in the receiver in which there is positioned a rib-type conveyor **68**. The conveyor conveys the nodules upwardly onto a vibrating feeder **70** that conveys the nodules transversely into a hopper **72** positioned above the mixing chamber. The hopper has a vibrating screen base portion **26** that feeds the nodules onto the surface of the combustible liquid within the mixing chamber. A suitable storage bin **74** has an outlet **76** that supplies makeup nodules to the hopper.

It should be understood that the size of the vessel may be varied to permit the complete combustion of the layer of combustible liquid material within the channel while the vessel moves through the body of water at a relatively fast speed. The relative position of the distributor device within the channel portion may also be advanced forwardly within the channel portion to permit the layer of cellular ceramic nodules to be positioned on the layer of combustible liquid within the channel and provide sufficient residence time of the nodules on the surface of the combustible liquid for the combustible liquid to move upwardly on the nodules.

Also, to ensure complete combustion of the layer of combustible liquid within the combustion chamber, the combustion chamber may also be elongated to permit a longer residence time within the chamber for burning before the nodules are removed therefrom by the screw conveyor.

In operation, the vessel floats on the body of water and is propelled into the portion of the body of water that has the layer of combustible liquid floating thereon. A band of the layer of combustible liquid enters through the front opening into the vessel channel portion and moves rearwardly therein as the vessel continues to move forwardly in the body of water.

In the mixing chamber a layer of cellular ceramic nodules is positioned on the upper surface of the layer of combustible liquid by means of the vibrating screen type distributor. The layer of combustible liquid floating on the body of water with the layer of cellular ceramic nodules floating thereon, thereafter moves into the combustion chamber through the opening in wall **38**. Within the combustion chamber, the layer of combustible liquid with the nodules thereon is ignited by means of the igniter.

It should be understood, however, that it may be necessary to only ignite the first portion of the layer of combustible liquid with the nodules thereon upon entering the combustion chamber. Combustion continues to spread to other portions of the layer of combustible liquid with the cellular ceramic nodules thereon as the combustible liquid enters the combustion chamber.

As the layer of combustible liquid with the cellular ceramic nodules thereon moves rearwardly in the combustion chamber, because of the forward advance of the vessel, the layer of combustible liquid is removed from the body of water by burning. The nodules adjacent the rear wall of the combustion chamber are removed from the surface of the body of water by means of the screw conveyor and are deposited in the receiver. A conveyor then conveys the nodules to a transverse vibrating conveyor of chute **70** where the nodules are again deposited in the hopper for subsequent distribution on the layer of combustible liquid therebeneath.

Combustion air is supplied to the combustion chamber by means of the blower and the secondary burners are positioned in the stack to burn any of the combustible material remaining in the combustion gases. With this arrangement, relatively smoke-free gaseous products of combustion are emitted to the atmosphere.

With the above described method and apparatus, it is possible to quickly, efficiently and substantially completely remove a layer of combustible liquid from a body of water. With the configuration illustrated, it may be necessary to make several parallel passes through the layer of combustible liquid floating on the body of water to burn substantially all of the combustible liquid in the oil spill.

Another process developed by R.B. Heagler (21) is one in which oil residues and emulsions floating on a body of water are burned by confining the layer of residue within a furnace chamber. The furnace has combustion air inlet means adjacent the upper surface of the residue and a stack with inlets for a combustible gas. The combustible gas burns the combustible material in the gases evolved from the combustion of the liquid residue to provide a relatively smokeless combustion process. The furnace is fabricated from a refractory material having insulating properties so that a substantial portion of the heat given off by the combustion of the residue is retained within the furnace to propagate further combustion of the residue and aid in the complete combustion of the difficult-to-burn portions of the residue.

The furnace is preferably fabricated from a material that permits the furnace to float partially submerged in the body of water and may be easily transported from one location on the body of water to another location thereon. The furnace may be supported from suitable pilings and the residue conveyed directly into the furnace chamber. For certain types of difficult-to-burn residues, a layer of cellular glass nodules with a textured outer surface is positioned to float on the upper surface of the residue within the furnace chamber. Figure 122 is a diagram showing such a furnace.

The furnace **10** is preferably fabricated from a mixture of concrete and cellular glass nodules so that the furnace will float partially submerged in a body of water. The furnace has a plurality of concentric rings **12, 14, 16, 18, 19** and **22** positioned in overlying relation with each other. The rings are preferably fabricated from an admixture of nodules and Portland cement in a weight ratio of 5:1. A typical furnace having a total height of 25", a bottom opening of 26" and a top opening of 10", weighs approximately 130 lb with an average wall thickness of 4". Where desired, suitable reinforcing wires may be used on the lowermost rings. Also where desired, the top rings may be fabricated from an admixture of cellular glass nodules and asphalt as a binder.

When positioned on a body of water, the lower ring **12** and a portion of the adjacent ring **14** are submerged below the water level. For example, a furnace weighing approximately 130 lb having the above discussed dimensions will have the lower rings submerged and the remainder of the furnace will be above the water level.

The intermediate ring **16** has a plurality of air passages **26** and an opening **28** is provided for the outlet of blower **30**. Although the air passages and the opening are illustrated as extending radially through the ring, it is preferred that the air passages and the opening for the blower extend angularly to the radius to provide a circular motion for the air entering the chamber **24**.

FIGURE 122: FLOATING FURNACE FOR BURNING OIL SLICKS FROM WATER

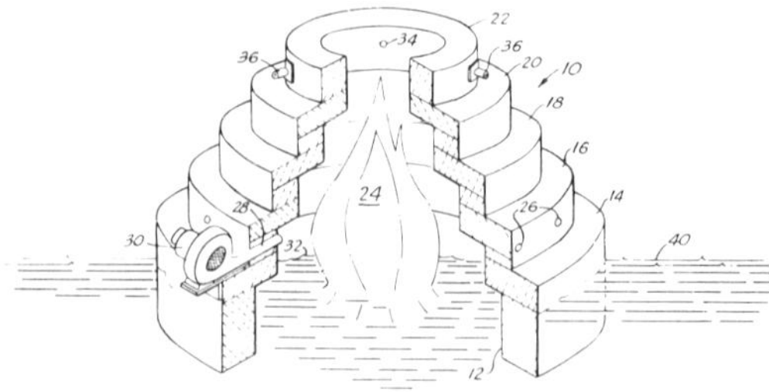

Source: U.S. Patent 3,695,810

With this arrangement, during combustion within the chamber, combustion air is drawn into the chamber through the air passages and, where it is desired, the blower provides air under pressure through the opening to the furnace chamber adjacent the top surface **32** of the liquid within the chamber. Suitable inlet openings **34** may also be provided in the top ring **22** for conduits **36** that are connected to a source of combustible gas such as methane, or the like. The openings in the top ring are arranged to provide for combustion of the combustible materials in the gases evolved during the combustion of the liquid within the chamber to provide relatively smoke-free combustion gases.

As indicated above, where the layer of combustible liquid is difficult to burn and where complete combustion of the combustible liquid is desired, a layer of cellular ceramic nodules may be positioned in the chamber to float on the upper surface of the combustible liquid and enhance the combustion of the combustible liquid.

The furnace is positioned on a body of water **40** to float thereon with a portion of the furnace submerged below the upper surface. The difficult-to burn combustible liquid is lighter than water and floats on the upper surface. To ignite the combustible liquid, a small amount of a primer such as Varsol or other suitable low flash point igniting agent, is poured onto the surface and is ignited by any suitable means. It should be understood that the furnace could include an automatic or remotely controlled resistance type igniter or the like to ignite the liquid on the surface.

After the combustible liquid is ignited combustion continues until substantially all of the combustible liquid within the confines of the chamber floating on the surface of the water is burned. Little, if any, residue remains. Since the furnace is floating on the body of water, it may be easily moved or transported thereon by any suitable means to be positioned with a new inventory of the combustible material within the chamber.

The furnace being fabricated from Portland cement and cellular glass nodules has refractory properties in that a minimum of spalling occurs within the furnace chamber during the combution of the combustible liquid. Also, the nodule-concrete mixture provides insulating properties for the furnace so that the heat generated during the combustion of the combustible liquid is retained within the chamber and it is believed, contributes to the substantially complete combustion of the combustible liquid so that little, if any, residue remains on the surface of the body of water within the chamber.

With the furnace, it is now possible to confine the combustion to the circumferential area of the chamber and to substantially and completely burn the combustible liquid floating on the surface of the body of water with a generally smokeless flame. Although the configuration of the floating furnace is generally described as circular, formed from a plurality of rings, it should be understood that the furnace may have other configurations, as, for example, rectangular or a polysided pyramidal shape.

A process developed by B.P. Martinez et al (22) involves filtering oil from a liquid containing oil by passing the liquid through a filter containing melt-spun thermoplastic synthetic fibers for absorbing the oil. When the fibers become saturated with oil, they are heated at an elevated temperature until they become liquid. The resultant liquid is then drawn from the filter for use as fuel. Figure 123 is a schematic showing the operation of this process.

FIGURE 123: APPARATUS FOR ABSORPTION OF OIL FROM WATER FOLLOWED BY COMBUSTION OF MOLTEN POLYMER-OIL MIXTURE

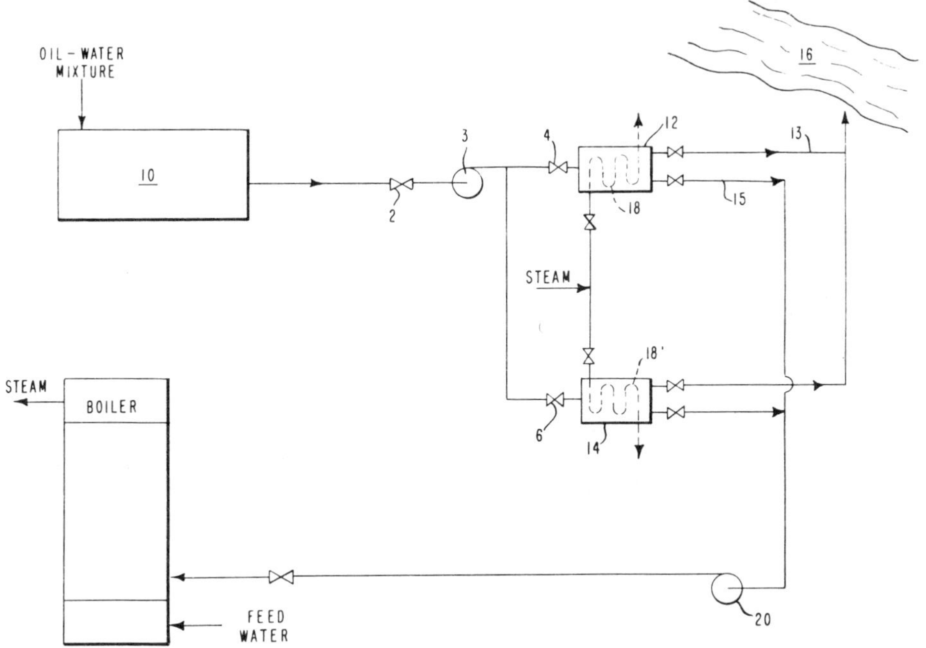

Source: U.S. Patent 3,923,472

A mixture of oil and water from a settling basin **10** is directed through valves **2, 4** and **6** by pump **3** into two filter boxes **12, 14**. The filter boxes are arranged so that one is on line while the other is on a standby basis. For example, the oil-water mixture passes through the filter box **12** and the oil is absorbed by synthetic fibers (less than 100 denier per filament) in the filter box and retained, while essentially oil-free water is returned to the stream **16** through pipe **13** attached to one outlet of filter **12**. When the first filter **12** becomes saturated with oil, the oil-water stream is directed into the second filter box **14** by means of the valves **4, 6**, i.e., by closing valve **4** and opening valve **6**.

After removal of the oil-saturated filter **12** from service, the entire filter is heated with steam by means of steam coil **18** in filter box **12** to convert the fiber-oil mixture into a liquid. Advantageously, the temperature may be maintained on the liquid to insure that it will remain pumpable. This liquid is then pumped by pump **20** through pipe **15** from the other outlet of filter **12** to a boiler where it is used to generate steam. The procedure is repeated using steam coil **18** to heat filter **14** when it becomes saturated with oil.

Synthetic fibers will absorb many times their own weight of oil, depending on type and composition. Typical numbers are 30 to 35 for polyester and 10 to 170 for polyolefins. The oil polymeric material mixture may be converted to a liquid by heating at atmospheric pressure or above, using temperatures in the range from 110° to 300°C. The process may be used with a wide variety of oils, including crude oil, fuel oil, used motor oil and light oils such as textile finishes. The process also may be useful in the disposal of waste synthetic fibers.

A similar process can be envisioned for collection of oil from oil spills at sea. In this case, the fibers, after absorption of oil, would be transferred to a holding tank on board ship. Heating the oil-fiber mixture would convert the mass to a liquid where it could either be used as fuel on board ship or transported back to shore and pumped into a tank for subsequent use.

PRODUCTION WASTES

Incineration

A system developed by P. Griffin III et al (23) is one in which an incinerator for removing hydrocarbon residues and other organic and inorganic components from oil and gas well drill cuttings is submerged in water to make the assembly explosive-proof with respect to the drilling rig. One embodiment uses a basket with a removable bottom in the combustion chamber to retain the cuttings during the burning. An alternate embodiment uses a series of metal plates for such retention.

After burning, the pollution-free cuttings are discharged from the combustion chamber beneath the surface of the water to eliminate sparks which might otherwise also cause an explosion. A belt-driven scoop located beneath the combustion chamber provides a means of testing the burned cuttings for unburned pollutants.

A device developed by W.J. Davies et al (24) is one for disposing of unwanted fluids having combustible constituents, such as petroleum wastes commonly collected in pits in oil-field operations. A screen or sheet of high velocity air is directed across an open-topped chamber of an incinerator, the air forming the sheet being directed downwardly at an angle relative to the horizontal so that it enters a combustion zone beneath the sheet. A mixture of air and waste is sprayed into the combustion zone from the same side as air which is supplied to form the sheet and in a direction substantially parallel to the direction of the travel of air in the sheet.

A process developed by R.K. Lewis et al (25) is a continuous combustion process for the substantially complete removal of the hydrocarbon content of cuttings from subterranean wells by combustion means which do not emit air-fouling gases. A feature of this process is the provision of means for retarding the movement of cuttings through an elongated combustion apparatus to subject them to conditions of high temperature and an excess of oxygen to obtain complete combustion of the hydrocarbon content thereof.

Solids, such as sand, are often entrained in the oil flow from an oil well. In such a case, it is conventional to pass the oil flow to a desander which consists of a cone wherein the oil flow is conducted through the cone in a whirling motion allowing the clean oil to exit through the top of the cone and the sand to fall by weight to the bottom of the cone.

The sand is held in a receiver at the bottom of the cone and is dumped under pressure as the sand accumulates. However, this sand still has a coating of oil or hydrocarbon on the surface and, as such, consititutes a pollutant. There is no acceptable method of treating the sand to remove the hydrocarbon coating in a safe and economical manner.

The device developed by G.O. Culpepper, Jr. (26) provides improved means for burning a hydrocarbon coating, such as an oil coating from the surface of particulate solid material such as sand in a pressurized firing zone or firing chamber where the solids are introduced under pressure into the firing zone. An air-fuel mixture is introduced into the firing zone and burned therein. Immediately downstream of the firing chamber is a doughnut restriction which confines the combustion to the firing chamber.

The ratio of air to fuel in the air-fuel mixture is always such that the fuel will be completely burned. The system will be provided with controls to shut the unit down when the temperature goes above or below a predetermined temperature and also to shut the system down when a lack of complete combustion of the fuel is indicated.

In the oil industry, crude oil pumped from oil wells is normally placed in storage tank means prior to distillation for producing a number of refined oil products, such as kerosene, gasoline, etc. While the crude oil is being held in the storage tanks, a certain amount of material will settle to the bottom of the tank and is not removed therefrom in the refining process. The oil sediment in a storage tank is referred to as tank bottom.

After a period of time, the sediment or tank bottom accumulated in the storage tanks must be removed. In the past, the oil sediment or tank bottom was removed from the storage tank and transferred to remote sediment ponds. After the sediment or tank bottom has been transferred to a sediment pond, the material is ignited and burned. However, due to increasing amounts of combustion, the atmosphere is being heavily polluted, causing many areas to adopt strict burning codes to prevent the burning of waste products in sediment ponds.

A number of attempts have been made to provide incinerator means which would burn liquid waste materials such as sediment from crude oil. These prior art incinerator means produced an excessive amount of combustion by-products which would not satisfy strictly enforced burning codes.

A device developed by J.M. Lang (27) is claimed to be capable of effecting maximum combustion of such flammable liquid waste material introduced therein with a minimum amount of exhaust. The incinerator requires a pressurized flame means to be directed at a detailed location across a flue entrance opening thereby producing a restrictive air block of the flue opening to increase the pressure within the combustion chamber. The pressurized flame means includes a mixing valve means for receiving and atomizing a supply of liquid waste material to be burned and includes means for directing the atomized liquid waste material past flame ignition means for igniting and burning the atomized liquid waste material.

The incinerator includes means operable for preheating the liquid waste material prior to introducing into the mixing valve means. Steam generating means is operatively associated with the combustion chamber for generating a predetermined amount of steam in response to burning the liquid waste material. The steam generating means includes nozzle dispersing means operable for directing an amount of steam into flue means associated with the combustion chamber for filtering the fly ash from the exhaust passing through the flue means.

Natural gas obtained at sea and on land during testing, and later during production, and which is separated from the oil at the well-head and in storage centers, is usually burned by means of a flare, where no market outlet exists.

The volumes of gas involved can amount to several hundred thousand cubic meters daily. Volumes of several tens of thousands of cubic meters can be burned by means of flares on the production platforms. For larger amounts, the heat discharged becomes intense, and variations in wind direction can endanger production installations and workers. The flare then has to be installed some distance from the production platform, which means that a support structure has to be built, usually another platform, the cost of which increases very quickly where the water is deep.

When the gas cannot be reused on the field, or collected and dispatched, it is generally injected into the sea, in the hope that it will be dispersed more effectively than by being simply discharged into the atmosphere, and also that it will be dissolved to some extent by the seawater. In fact, the process is a dangerous one, leading in calm weather to the possible formation of blankets of an explosive gaseous mixture; it also pollutes the sea. Another method consists of diluting the gas in air, using a mixing appliance, so that the mixture finally discharged into the atmosphere is noncombustible, remaining below the level of flammability.

This method removes the risk of explosion, but it involves cumbersome equipment with large installed capacity, and the need to extend and strengthen production platforms, where a second platform is not built. Moreover, the risk of atmospheric pollution remains, particularly when the gas contains a significant proportion of heavy hydrocarbons. When the proportion of hydrocarbon gas in the air is less than 3%, the risk of explosion is practically nil, but the toxicity threshold has to be taken into account: 0.2% for propane and 0.05% for pentane, hexane and heptane.

A system developed by F. Lazarre et al (28) overcomes these difficulties, allowing large amounts of gas to be eliminated, regardless of the position of the production platform, without endangering plant or workers.

This installation, used particularly to eliminate gas discharges from hydrocarbon production wells, is characterized by the fact that the submerged, watertight part of the installation comprises at least one burner consisting of at least one combustion chamber and at least one combustible mixture feed chamber, communicating with each other by means of apertures or slits, and at least one burner feed pipe, leading to the open air and comprising a mixing zone, consisting of a divergent passage possibly preceded by a convergent passage, in the inlet of which is placed one discharge passage leading to the open air, and means of adjusting the flow of combustible gas and air. Such a system is shown in Figure 124, which shows a hydrocarbon production platform **1** installed at sea or in a lake or pond.

FIGURE 124: SUBMERGED BURNER DESIGN FOR COMBUSTION OF WASTE GASES AT SUBSEA WELL HEAD

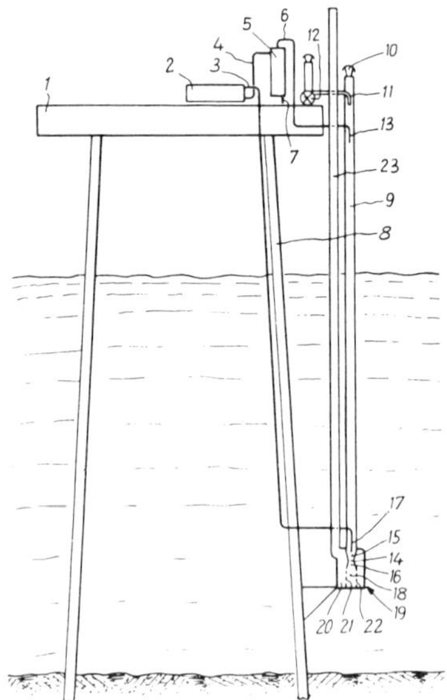

Source: U.S. Patent 3,913,560

Beside the wellheads and collection installation (not shown here) is the processing equipment, comprising a high-pressure separator **2**, with high-pressure gas outlet **3** and an outlet for incompletely degassed oil **4**, and a low-pressure separator **5**, with a low-pressure gas outlet **6**, and a storage oil outlet **7**. The submerged combustion installation is fixed to one supporting leg **8** of the platform. There is an approximately vertical supply pipe **9**, leading to the open air through an aperture **10**.

A duct **11** feeding in air supplied by an auxiliary starting blower **12** opens into this pipe, a short distance below the aperture **10**; further down is a passage **13** conveying gas from the low-pressure separator. At the lower end of this pipe is a mixing zone **14**, comprising two passages, a converging one **15**, followed by a diverging one **16**. At the inlet, and aligned with the convergent passage, is a nozzle **17** to which the gas from the high-pressure separator is injected; the position of this injector can be adjusted sideways, depending on the axis of the pipe.

This feed pipe **9** opens at the bottom into a feed chamber **18** for the burner **19**. This burner also comprises a combustion chamber **20**, alongside the feed chamber, from which it is separated by a partition **21** containing a number of apertures **22**. The combustion chamber **20** is connected by a set of collection passages (not shown here) to a pipe **23** discharging burnt gas into the atmosphere.

Ocean Disposal

The rock chips produced by the bit in drilling an oil well are termed drill cuttings. On offshore drilling platforms along the southern California coast, these cuttings are generally washed and discharged below the rig. The resulting solid waste accumulation is volumetrically of little significance. The cuttings were found to have no effect, either adverse or beneficial, on the environment. If the cuttings were to be disposed of several hundred feet away from the drilling platform and capped with stones or other rubble, they might provide an artificial habitat for sport fishing.

The disposal of drilling muds and cuttings in deep water from a barge has also been investigated. Observations on the disposal operation consisted of aerial photographs and visual inspection at the sea surface to assess the extent of the surface contamination. On the basis of the brief study involved, it was concluded that the disposal operation posed no threat to the environment, according to Reed (29).

REFINERY SLUDGES

Incineration

A fluid bed incinerator for refinery wastes which is the largest in the petroleum industry (although larger units for fluidized sludge incineration exist in the paper industry) has been described in the magazine, *Hydrocarbon Processing* (30), and is shown in Figure 125.

It burns liquid wastes generated in refinery operations which were formerly placed in sanitary landfills or processed in the refinery water treating system. The unit has a design capacity per day of 316 tons of sludges and 56 tons of spent caustics, making possible the disposal of these wastes without causing environmental pollution. Heaviest usage so far has been 212 tons of liquid sludges in one day and 19 tons of spent caustics. Air—forced through a grid in the bottom of the unit—fluidizes an 8-foot-deep bed of sand, which has been heated to 1240°F. Waste sludges and spent caustics are added to the bed and burned, being converted to carbon dioxide, water and inorganic ash.

An article by K.P. Becker and C.J. Wall (31) discusses in some detail the design and operation of fluid bed incinerators for refinery sludges.

FIGURE 125: LARGE FLUID BED INCINERATOR FOR REFINERY WASTES

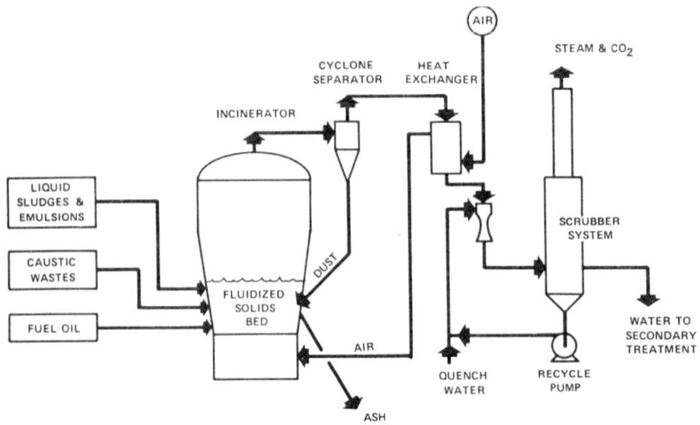

Source: *Hydrocarbon Processing*

A device developed by F. Vier et al (32) is one for burning combustible waste product sludges containing solid particulate contaminants, including a nozzle chamber adapted to mix the waste sludge with pressurized air. The static air pressure in the nozzle chamber is maintained at a predetermined pressure head in order to regulate the inlet of sludge to the chamber, so as to thereby effect control over the ratio of sludge to air in the mixture formed for burning.

A device developed by G.A. Vorms et al (33) is an apparatus for atomization and combustion of fluid industrial wastes having low and variable calorific value, such as petroleum sludge. It is a long-felt need for oil-refining and petrochemical plants to solve the problem of the disposal of petroleum sludge formed during the purification processing in the equipment such as in tanks, slurry ponds and oil traps, sewage settling basins and the like.

Such wastes comprise a nonstratifying mixture of dirty water and resinified oil products which form an emulsion with water and envelop solid impurities such as clay particles, scale, sand, oil coke grains and the like. The presence of even low-active detergents, such as waste alkali in this mixture additionally stabilizes it. The more pronounced are the colloid gel properties of such a mixture, the greater and more abnormal is the viscosity of the mixture, such mixture being difficult in batching and control. Even where constant volumetric or weight batching of such mixture of variable composition is available, it is difficult to ensure stable burning conditions, since it is very important for the combustion process to know the amount of the fuel component introduced into the combustion chamber per unit of time, the amount of air to be supplied, degree of combustion and other parameters being controlled on the basis of this amount.

In order to ensure reliable ignition of such wastes, still another factor is important, namely the content of water in the mixture to be burned since water is evaporated at the very beginning of the combustion process so that a large amount of heat is consumed for that purpose just at the time, when the amount of heat released is still very small. As a result, such wastes have a trend to spontaneous extinction and to absolutely unsatisfactory degree of combustion despite the fact that a total calorific value of wastes may be considerably greater than the lowest limit corresponding to self-sustained combustion.

The device comprises a housing accommodating a hollow shaft journalled in bearings and supporting a rotary atomizer at the end thereof facing the combustion chamber, and an annular air duct for supplying air for combustion, which is arranged around the rotary atomizer.

The hollow shaft accommodates a pipe for feeding the industrial wastes being burned to the inner surface of the rotary atomizer, as well as a pilot burner which is adapted to form an independently controlled flame along the axis of the spatter cone of the wastes being burned.

Such additional and controlled flame inside the very beginning of the spatter cone of the waste being burned formed by means of the pilot burner arranged coaxially with the apparatus, imparts the following new properties to the rotary burner as a whole: a thermal impact is obtained at the very beginning of the spatter cone, and considerably greater fraction of drops of the waste being atomized, and which is important, of the finest drops which move closer to the spatter cone axis will get into the pilot flame.

These fine drops are rapidly heated, evaporate their water, and after they have been ignited and changed their path under the action of almost radially flowing gases from the pilot burner they turn away from the axis and get mixed with the remaining mass of drops of the spatter cone. As a result, the main mass of the spatter cone is ignited not only by using valuable fuel burned in the pilot burner, but also by means of the ignited fine drops of the waste itself. All these phenomena take place at the very beginning of and inside the spatter cone, whereby the effect of the burning of small amounts of valuable fuel is multiplied.

A system developed by E.J. Roberts et al (34) is one in which oil refinery wastes containing alkali metal chlorides and sulfur and/or sulfates are incinerated in a fluid bed reactor by a process in which formation of low melting eutectics is avoided. The temperature of the fluid bed is maintained substantially above the vapor saturation temperatures of the alkali metal chlorides so that they are to a great extent carried out with the combustion gases.

By insuring the presence of sufficient reactive silica and a sufficiently high temperature, alkali metal sulfates can be prevented from accumulating. Any sticky alkali metal silicate glasses which tend to form are prevented from causing defluidization by insuring the presence of sufficient CaO, MgO, Fe_2O_3 and/or Al_2O_3 so that the glass is converted into high melting silicate compounds.

In operating certain fluid bed installations, it has been found that the organic waste streams are contaminated with chlorides. This is particularly prevalent in installations associated with refineries at coastal locations. The ground water near the seacoast is often briny and inevitably intrudes into the waste lines. The salts in seawater comprise from about 75 to 85% NaCl, small amounts of $CaSO_4$, $MgCl_2$ and $MgSO_4$ and lesser amounts of other salts.

Certain industrial wastes contain chlorides which are present as the natural consequence of reactions used to produce commercial products. Tanker ballast water is another waste material containing both organics and chlorides. Such chloride-contaminated waste streams or sludges are difficult to incinerate because chlorides in general, e.g., the alkali metal chlorides, have low melting points and give rise to formation of low-melting eutectics at the required combustion temperatures, which causes the ash to become sticky and, when this occurs in fluid bed units, the bed defluidizes.

Figure 126 shows the apparatus for the conduct of the process. The reactor **20** comprises an outer shell **10** which is capped by a cover **12**. Within the shell **10** the reaction chamber **14** is separated from the windbox **16** by the constriction plate **18**. A pump **21** supplies fluidizing gas to the windbox **16** through conduit **23**. A short down-coming exhaust conduit **25** conducts the exhaust gases to the venturi scrubber **65**.

In the reactor **20**, a bed **31** of finely divided solids is supported on the substantially horizontal constriction plate **18**. The reactor illustrated is arranged for the introduction of a pumpable feed such as partially dewatered refinery wastes. The pump **33** forces the waste feed from conduit **35** into conduit **37** which is connected to the annular distribution pipe **39** which surrounds the reactor shell **10**. A plurality of feed guns **41** are positioned about the periphery of the reactor **20** and penetrate shell **10** for direct injection of the waste feed into the fluidized bed **31**.

FIGURE 126: APPARATUS FOR FLUID BED INCINERATION OF WASTES CONTAINING
ALKALI METAL CHLORIDES

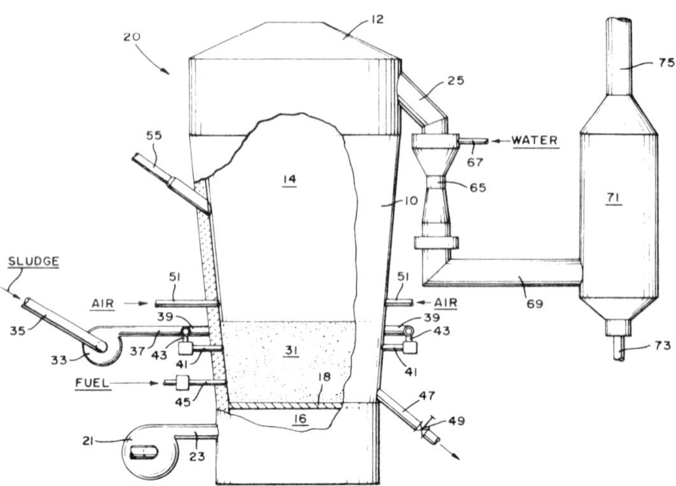

Source: U.S. Patent 3,907,674

Alternatively, feed guns **41** can be located just above the upper surface of the expanded fluidized bed so that the waste feed drops onto the top surface of the fluidized bed. Each feed
gun **41** is connected to the annular distribution pipe **39** by a feed tube **43** through which
it is supplied with waste feed. Also provided are a number of fuel guns **45** which are positioned about the periphery of the reactor shell **10** for direct injection of auxiliary fuel into
the fluidized bed. A discharge pipe **47** is provided to permit withdrawal of excess bed material.

The constriction plate **18** comprises a flat steel plate containing numerous orifices, perfora-
tions or apertures (not shown) to allow the fluidizing gas access to the reaction chamber **14**.
At a level just above the top of the expanded fluidized bed **31**, a plurality of air injection
tubes **51** penetrate the reactor shell **10** for the purpose of injecting secondary air into a
zone close above the bed. These air injection tubes **51** are supplied with pressurized air
from a bustle pipe (not shown) surrounding the reactor shell **10**. An inlet **55** is provided
for introducing additional sand into the bed as required.

The exhaust gases, as mentioned above, leave the reactor chamber **14** through the exhaust
conduit **25** which may be provided with a clean-out port (not shown). The exhaust conduit is as short as possible because the gas is cooling rapidly and condensation of at least
some vaporized constituents can be anticipated. The longer the exhaust conduit, the heavier
the deposit is likely to be and accumulations of such deposits must, sooner or later, be removed, with consequent downtime. The gases flow into the venturi scrubber **65** and there
contact scrubber water provided through conduit **67**.

Dust particles in the gases are picked up by the water, while the soluble chloride compounds
become dissolved in the water. The gases and liquid from the venturi scrubber **65** next flow
through conduit **69** to the liquid-gas separator **71** in which the liquid flows to the bottom
of separator **71**, exiting through the conduit **73** for disposal and the gas leaves through the
stack **75**.

The waste feeds on which the process operates may contain up to 19,000 ppm of chloride (full strength seawater) or more although this may require temperatures above

1000°C. It should be understood that the Fe_2O_3 addition should not be used for feeds containing much in excess of 5,000 ppm of chloride.

This limit comes about through the observation that at 10,000 ppm of chloride, a temperature of about 950°C is required for volatilization of the chloride, while a eutectic between SiO_2, Fe_2O_3 and acmite ($Na_2O \cdot Fe_2O_3 \cdot 4SiO_2$) forms at 955°C. Prudence dictates that a wide berth be given to this temperature in actual operation and this is best accomplished by not attempting to push the Fe_2O_3 addition into the treatment of feeds containing chlorides at the high end of the scale.

While certain of the metal oxides may be employed together, it is rather pointless to make additions of Al_2O_3 and CaO together because they tend to react with each other forming $CaO \cdot Al_2O_3 \cdot 2SiO_2$ in preference to the desired reaction, and losses of the additives are sustained in this fashion.

An apparatus for incinerating salt-containing liquid sludges such as refinery sludges has been described by R. Menigat et al (35). A liquid sludge containing salts such as NaCl is injected into such a fluidized-bed incinerator containing a bed of corundum or iron oxide which is fluidized by an oxygen-containing gas at a superficial velocity of at least 3 m/sec. The bed is therefore kept in constant rapid agitation heated to a temperature sufficient to vaporize the salts. The thus vaporized salts are thereafter scrubbed out of the exhaust gases of the incinerator with a water spray.

Landfill Disposal

A process developed by W.K. Lorenz et al (36) comprises dewatering and deoiling refinery sludges by filtering through a filter press at a temperature of from about 100° to 200°F, washing of the filter cake with hot water, and subjecting the solid filter cake thus obtained to soil biodegradation.

Ocean Disposal

Petroleum processing operations produced about 12% (about 560,000 tons) of the total industrial wastes dumped in the ocean in 1968. Refinery wastes include spent caustic soda solutions, sulfuric acid sludges, dilute process water solutions, spent catalysts, waste petrochemicals, and chemical cleaning wastes. The solid wastes (spent catalysts and sludges) are frequently slurried with the liquid wastes prior to shipment to sea for disposal. Varying amounts of sulfides, phenolates, naphthenates, cyanides, heavy metals, mercaptides and chlorinated or brominated hydrocarbons are among the hazardous minor contaminants of the refinery wastes.

SHALE PROCESSING WASTE

Some years ago in the early days of shale processing research, one somewhat cynical researcher at a northern New Jersey laboratory commented that making oil from shale was no great process problem but the real problem was plant location. Such volumes of waste would be produced he said, that the plant would have to be built atop the Rocky Mountains with a conveyor bringing up shale to be processed and a big shute down which the waste could slide into the Pacific Ocean! While this might have been something of a grandiose overstatement; the problems of ultimate disposal of such wastes are formidable.

Mine Fill Disposal

A process developed by B.D. O'Neal (37) is one in which waste spent shale resulting from gas combustion retorting operations is formed into an aqueous slurry and pumped into a mined-out area to deposit the waste spent slurry therein. A portion of the waste spent slurry is converted to cement which is pumped into the deposited waste spent shale to fill voids in and compact the deposited waste spent shale.

The scheme of operations for such a process is shown in Figure 127. Referring now to the drawing numeral **11** designates a gas retort such as one of the Bureau of Mines type or one having serially interconnected preheat, retorting, combustion, and cooling zones into which is fed by line **12** oil shale containing kerogen. This oil shale may be obtained from any of the western states of the United States such as Colorado or from any source. In gas combustion retort **11**, the kerogen is converted into shale oil and is removed by means not shown.

The spent oil shale which may be defined as waste spent shale is removed from retort **11** by line **13** and introduced thereby into a mechanical grader or particle separator **14** wherein the small particles having particle diameters within the range of from 5 microns to 0.5" are removed from large particles of waste spent shale.

The small particles of waste spent shale are withdrawn from particle separator **14** by line **15** and introduced thereby into a waste spent shale storage hopper **16**. The small particles of waste spent shale may suitably comprise from 75 to 90 weight percent of spent shale withdrawn from retort **11** by line **13**.

FIGURE 127: SYSTEM FOR DISPOSAL OF SPENT SHALE

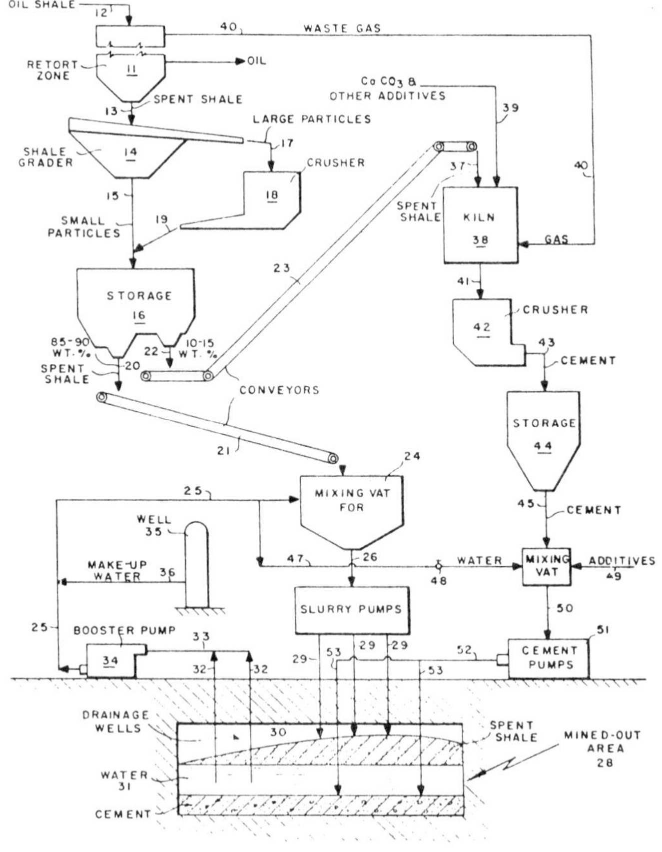

Source: U.S. Patent 3,459,003

The large particles of waste spent shale are removed from particle separator **14** by line **17** and introduced thereby into a crusher **18** wherein the large particles of waste spent shale are reduced in size to at least the size of the small particles of waste spent shale. The small particles from crusher **18** are withdrawn by line **19** and conducted into line **15** for introduction of small particles of waste spent shale into spent shale storage hopper **16**.

A portion of the waste spent shale in the storage hopper **16** is discharged by line **20** onto a conveyor or other mechanical transporting means **21**. A second portion of the waste spent shale is discharged from the storage hopper **16** by line **22** onto a conveyor or other mechanical transporting means **23**.

About 85 to 90% by weight of spent shale is discharged onto conveyor means **21** while about 10 to 15% by weight is discharged onto conveyor means **22**. The waste spent shale discharged onto conveyor means **21** is transported thereby to a mixing vat **24** into which water from a source which will be described further hereinafter is introduced by line **25**. Mixing vat **24** may be provided with mixing means such as a jet nozzle for introduction of water under pressure or with paddle or propeller type agitator means suitably powered by power means.

In any event, a slurry of spent shale is formed in mixing vat **24** comprised of about equal parts by weight of water and spent shale; however, the respective amounts of water and spent shale may vary depending only on the amount of water needed to provide a pumpable slurry. Thus, it is contemplated that greater or lesser amounts of water may be used.

The slurry in mixing vat **24** is withdrawn by line **26** by spent shale slurry pumps, generally designated by numeral **27**, for pumping the slurry of spent shale in water into a mined-out area **28** through lines and/or wells **29** depending on whether the mined-out area **28** is a confined area such as an underground (a room and pillar) mine or a cut-and-fill mine. In either event, the spent shale is deposited in the mined-out area **28** as a slurry, the spent shale being deposited to form body **30** of spent shale having substantial porosity or substantial percent of voids.

The water in the slurry separates to form a body **31** of water. This water is continuously withdrawn from body **31** by means of drainage wells or lines **32** connecting to line **33** containing a booster pump **34** to which is connected line **25**. Make-up water may be introduced into line **25** from a water well **35** or other source which connects to line **25** by line **36**.

The 10 to 15% by weight of the shale withdrawn from shale hopper **16** by line **22** and deposited on conveyor means **23** is transported thereby and deposited by line **37** into a vertical or rotary kiln **38** into which is fed by line **39** calcium carbonate in the form of crushed limestone and other additives such as calcium oxide, clay, or gypsum. For formation of Portland type cement, vertical or rotary kiln **38** may be heated by hot waste gas withdrawn by line **40** from gas combustion retort **11** and introduced in contact with kiln **38** to heat same and cause calcination of the spent shale, the limestone and other additives to form cement clinker which is withdrawn from kiln **38** by line **41** and deposited into a crusher means **42** where the clinker is crushed or finally ground to form cement which is withdrawn by line **43** to cement storage **44**.

The cement in storage **44** may be withdrawn by line **45** into a mixing vat **46** where it is mixed with water introduced thereto by line **47** which connects into line **25**, line **47** being controlled by valve **48**. Suitable additives such as calcium chloride, potassium hydroxide and sodium silicates may be used to accelerate the setting and hardening of the cement.

Additives such as gypsum and sodium tannate may be used to retard the setting and hardening of the cement. These additives, as needed, are introduced into mixing vat **46** by line **49**. Suitable, although not shown, the cement withdrawn by line **45** may be used for structural and other construction purposes as may be desired. Cement slurry is withdrawn from mixing vat **46** by line **50** and is pumped by cement pumps **51** by way of line **52** which connects to lines or wells **53** for introduction of the cement slurry into the spent shale body **30** to fill at least a portion of the voids and to cause compaction and cementation to the body **30** of spent shale.

By virtue of the operation as has been described hereinbefore, the spent shale is introduced into a mined-out area by pumping thereby disposing of it readily and by virtue of adding the cement material, it is caused to be compacted to give it substantially the strength it had prior to mining and making it competent to support considerable weight. The spent shale introduced as a slurry settles in the presence of water and under fluid conditions to enhance the compaction and cementing. Actually, the waste spent shale discharged from the retort sets up to form a low grade cement solid when exposed to water.

Thus, transporting it to the mined-out area as a slurry also provides a means for compacting. However, heating with a kiln further provides decomposition and a better quality of low grade cement. The additional limestone or calcium oxide may be used to upgrade cement if necessary; however, in special instances, additional additives may not be required.

Although the drawing illustrates the mined-out area incompletely filled, it is possible to fill completely an area such as an underground mine and cement by pressure pumping which in the case of underground mines prevents further subsidence in the mined-out area. As an example, disposing of the waste spent shale from above ground by gas combustion retort, in a 100,000 bpd operation requires mining about 180,000 tons of oil shale per day which represents a bulk volume of 50 acre feet per day.

After being crushed, processed and stacked, the solid waste or spent shale waste would weigh in the vicinity of 160,000 tons, but because of the porosity of the stacked material it would represent a bulk volume of about 60 acre feet. Stating this another way, the waste spent shale from a 100,000 bpd plant covers about 22,000 acre feet for each year of operation. This large volume of material would represent a blight on the countryside, and it is hence necessary to dispose of it underground. The process allows this to be done.

If the waste spent shale were returned to the mined-out areas in accordance with prior art techniques, the waste to be disposed of from a 100,000 bpd plant would be about 10 acre feet per day, more than can be disposed of in a mine by placing it dry. The waste spent shale alone amounts to 3,650 acre feet of surface waste per year for a 100,000 bpd plant.

In accordance with the process, all of the waste spent shale may be disposed of by taking advantage of the porosity of the waste spent shale which may be in the order of 20 to 30%, probably of 25%, which can be reduced to 10 to 12% by utilization of the cement manufactured from the waste spent shale allowing deposition of all of the waste spent shale in the mined-out area.

In accordance with the process water may be obtainable from surface sources or from subsurface sources through wells at shallow depths. Hence in oil shale mining operations using this technique, the deepest shale would be mined first to allow disposal of the water encountered at shallow depths. In a cut-and-fill mining operation, the process provides means for rebuilding a competent floor from which to conduct future operations.

SPENT SULFURIC ACID SLUDGES

Spent sulfuric acid sludges from such refinery processes as alkylation or such petrochemical process as olefin hydration are simultaneously concentrated in H_2SO_4 content and rid of oily impurities by direct-fired concentrators such as the Chemico concentrator.

TANKER DISCHARGES

Ocean Disposal

Some very significant work has been carried out on the wastes from tank cleaning operations conducted by oil tankers at sea (29). Based on earlier work regarding the oxidation of hydrocarbons by marine bacteria, the importance of discharging these wastes in a finely divided form in order to optimize bacterial degradation was shown. Calculations showed that the full quantity of waste oil (at an average concentration of 0.13 mg/l) resulting from cleaning the tanks of a 45,000 dwt tanker underway at 16 knots would be oxidized by bacteria in 2 to 4.5 days, depending on whether the wastes were discharged continuously over a 24-hour period; or discharged as fast as the tanks were cleaned.

VAPORS FROM PRODUCT TRANSFER OPERATIONS

Incineration

A process developed by H.W. Husa et al (38) involves collecting hydrocarbon vapors before they are dispersed in the atmosphere surrounding a site where loading or unloading of flammable fuels occur. The collected vapors are monitored to determine when the air-hydrocarbon vapor mixture is near or in the combustible range. When it is determined that the mixture is near or in the combustible range, a hydrocarbon gas such as propane is injected into the mixture to remove the danger of explosion or fire. The collected vapors are held in a holding tank and, preferably, are periodically burned.

A system developed by L.T. Cavallero et al (39) is a pollution-free system for disposing of gasoline vapors at a tank truck loading station, in which a booster pump energized in response to pressure within a vapor saver tank which receives the trunk tank vapors feeds the vapors from the vapor saver tank through a control valve to the supply line of the hydrocarbon oxidizer and in which means responsive to the temperature in the oxidizer operates the control valve to regulate the vapor flow to the oxidizer to maintain the temperature therein below a predetermined temperature and in which various safety devices are incorporated to ensure the safety of both the installation itself and of personnel working at the installation.

As shown in Figure 128, this system, indicated generally by the reference character **10**, is adapted to be installed at a station at which gasoline tank trucks such, for example, as a truck having a tank **12**, are to be refilled with gasoline for distribution. While there is illustrated only one tank truck in the drawings, it will readily be appreciated that there are facilities for handling a multiplicity of such trucks at the normal refilling station.

FIGURE 128: APPARATUS FOR INCINERATION OF WASTE VAPORS AT GASOLINE TRUCK LOADING STATION

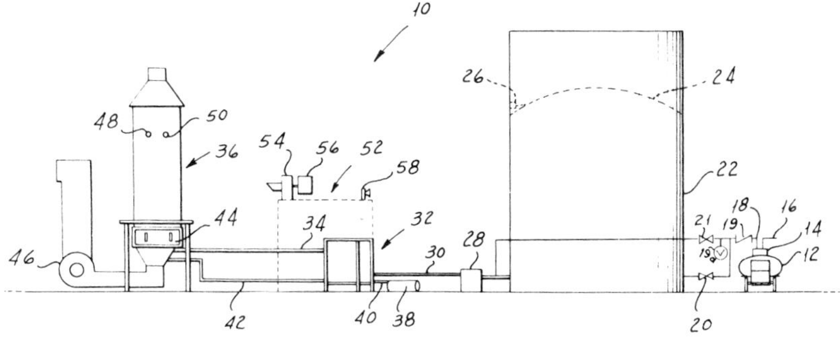

Source: U.S. Patent 3,817,687

At such a station, gasoline is fed into the tank **12** through a line **16** leading into the refill fixture **14** on the tank. At the same time, vapor from within the tank **12** is forced outwardly through a line **18** and past a check valve **19**. A gate valve **20** is adapted to be opened to pass the vapor to a vapor saver tank **22** of a type known in the art. Further as is know in the art, the tank **22** includes a bladder **24** which expands much in the manner of a balloon as the vapor pressure builds up within the tank **22**.

The tank **22** is provided with a switch **26** adapted to be operated by the bladder **24** when the pressure within the tank reaches a predetermined value to activate a booster pump **28** to feed vapor from the tank **22** through a line **30** leading to a piping rack, indicated generally by the reference character **32**.

In some instances it may be desirable to eliminate the tank **22** and to feed the vapor from the tank **12** directly to the booster pump **28**. In such a case the vapor comes from the truck through line **18** and past a check valve **19** to a specially designed valve **21**. The valve **21** is activated by a specially designed pressure switch **19A**. As the valve **21** is activated the booster pump **28** is also activated to feed vapor through a line **30** leading to a piping rack, indicated generally by the reference character **32**.

The arrangement is such that it continuously detects the difference between an actual measurement and its desired value and converts this difference to an air signal between 3 pounds and 15 pounds per square inch. It is this air signal which operates the valve. The actual measurement, the desired measurement, and the output signal are indicated on a controller.

The arrangement is such that an air supply passes through a regulator, then through a pneumatic controller, and then to the control valve. As the valve rotates to open position, it actuates a rotary switch (not shown) to make electrical contact to activate the booster **28**. Vapor from the rack **32** is carried by a pipe **34** to the supply line of a hydrocarbon oxidizer indicated generally by the reference character **36**, whereat the vapor is ignited and burns.

Any suitable type of pilot fuel may be used such, for example, as butane contained in a tank **38**. A line **40** conducts pilot fuel from tank **38** to the rack **32**. Another pipe **42** leads from the rack **32** to the pilot line of the oxidizer **36**.

Oxidizer **36** includes a door **44** which affords access to the interior of the oxidizer. A combustion air fan **46** is adapted to be activated to supply combustion air to the oxidizer **36**. A pair of thermocouples **48** and **50** sense the temperature within the stack of the oxidizer. The system includes a control house indicated generally by the reference character **52**, which houses the control panel (not shown) and associated equipment for the system.

A purging fan **54** driven by a motor **56** is mounted on housing **52** to ensure that the interior of the housing is at all times free of combustible vapors. A horn **58** on top of the housing **52** is energized each time the system starts up to warn personnel in the area. More specifically, the booster **28** may be, for example, a three-quarter horse-power turbine rated at 100 cubic feet per minute at eight ounces driven by a three-quarter horse-power explosionproof motor. The pressurizing or purge fan **54** may be any suitable type of blower driven by a one-third horse power explosionproof motor **56**. The supply fan may, for example, have a capacity of 6,500 cubic feet per minute driven at 1,368 rpm by a five horse-power motor which also should be explosionproof.

REFERENCES

(1) H.R. Jones, *Pollution Control in the Petroleum Industry,* Park Ridge, N.J., Noyes Data Corp. (1973).
(2) J.S. Zink and R.D. Reed; U.S. Patent 3,207,201; September 21, 1965; assigned to John Zink Co.
(3) E.V. Bergstrom; U.S. Patent 2,936,221; May 10, 1960; assigned to Socony Mobil Oil Co.

(4) H.O. Ebeling; U.S. Patent 3,917,796; November 4, 1975; assigned to Black, Sivalls & Bryson, Inc.

(5) F.N. Rubel, *Incineration of Solid Wastes,* Park Ridge, N.J., Noyes Data Corp. (1974).

(6) M. Sittig, *Oil Spill Prevention & Removal Handbook,* Park Ridge, N.J., Noyes Data Corp. (1974).

(7) R.M. Roberts and T.S. Hoyt, Report PB 198 227; Springfield, Va., Nat. Tech. Information Service (Nov. 1970).

(8) A. Molin and O. Carlsson; U.S. Patent 3,586,469; June 22, 1971; assigned to Atlas Copco AB.

(9) P.R. Tully, W.J. Fletcher and H. Cochrane; U.S. Patent 3,556,698; January 19, 1971; assigned to Cabot Corp.

(10) N.T. Castelluci and N.C. Krouskop; U.S. Patent 3,661,497; May 9, 1972.

(11) D.D. Sparlin; U.S. Patent 3,698,850; October 17, 1972; assigned to Continental Oil Co.

(12) R.J. McGuire, E. Mitchell and J.P. Pellegrini; U.S. Patent 3,696,051; October 3, 1972; assigned to Gulf Research & Development Co.

(13) R.F. Rensvold; U.S. Patent 3,705,782; December 12, 1972; assigned to Halliburton Co.

(14) F.J. Shell; U.S. Patent 3,607,791; September 21, 1971; assigned to Phillips Petroleum Co.

(15) J.W. Marx; U.S. Patent 3,677,982; July 18, 1972; assigned to Phillips Petroleum Co.

(16) W.D. Johnston; U.S. Patent 3,661,495; May 9, 1972; assigned to Pittsburgh Corning Corp.

(17) W.D. Johnston; U.S. Patent 3,661,496; May 9, 1972; assigned to Pittsburgh Corning Corp.

(18) A.J. Shaler and W.E. Clancy; U.S. Patent 3,659,715; May 2, 1972; assigned to Stackpole Carbon Co.

(19) S. Kraemer, A. Seidl and M. Seger; U.S. Patent 3,589,844; June 29, 1971; assigned to WASAG Chemie AG.

(20) R.B. Heagler; U.S. Patent 3,663,149; May 16, 1972; assigned to Pittsburgh Corning Corp.

(21) R.B. Heagler; U.S. Patent 3,695,810; October 3, 1970; assigned to Pittsburgh Corning Corp.

(22) B.P. Martinez and M.D. Zeisberg; U.S. Patent 3,923,472; assigned to E.I. Du Pont de Nemours and Co.

(23) P. Griffin III and W.C. Phillips; U.S. Patent 3,658,015; April 25, 1972; assigned to Dresser Industries, Inc.

(24) W.J. Davies, B. Hatton, G.C. Boyd and H.E.W. Hanlan; U.S. Patent 3,704,676; December 5, 1972; assigned to Kanting Oilfield Services, Ltd.

(25) R.K. Lewis and M.G. Bingham; U.S. Patent 3,570,420; March 16, 1971; assigned to Milchem Corp.

(26) G.O. Culpepper, Jr.; U.S. Patent 3,734,774; May 22, 1973; assigned to J.W. Kendall and J.W. Pitts.

(27) J.M. Lang; U.S. Patent 3,756,170; September 4, 1973; assigned to Care, Inc.

(28) F. Lazarre, G. Blu and J. Rozand; U.S. Patent 3,913,560; October 21, 1975; assigned to Societe Nationale des Petroles D'Aquitaine.

(29) A.W. Reed, *Ocean Waste Disposal Practices,* Park Ridge, N.J., Noyes Data Corp. (1975).

(30) *Hydrocarbon Processing* 51, No. 12, 17 (December 1972).

(31) K.P. Becker and C.J. Wall, *Hydrocarbon Processing* 54, No. 10, 88-93 (Oct. 1975).

(32) F. Vier and B. Winkeler; U.S. Patent 3,659,786; May 2, 1972; assigned to Wintershall AG.

(33) G.A. Vorms, P.I. Kuznetson and V.B. Volkov; U.S. Patent 3,862,821; Jan. 28, 1975.

(34) E.J. Roberts and P.A. Angevine; U.S. Patent 3,907,674; Sept. 23, 1975; assigned to Dorr-Oliver Inc.

(35) R. Menigat and W. Fennemann; U.S. Patent 3,921,543; Nov. 25, 1975; assigned to Metallgesellschaft, AG.

(36) W.K. Lorenz, E.C. Sebesta and C.L. McClellan; U.S. Patent 3,835,021; September 10, 1974; assigned to Sun Oil Co. of Pennsylvania

(37) B.D. O'Neal; U.S. Patent 3,459,003; Aug. 5, 1969; assigned to Esso Research & Engineering Co.

(38) H.W. Husa and I. Ginsburgh; U.S. Patent 3,783,911; Jan. 8, 1974; assigned to Standard Oil Co.

(39) L.T. Cavellero and J.J. Elnicki; U.S. Patent 3,817,687; June 18, 1974; assigned to Aer Corp.

PHARMACEUTICAL INDUSTRY WASTES

The following tabulation developed by Booz-Allen Research Inc. (1) shows the waste output from one pharmaceutical manufacturer which may be considered fairly representative of industry practice.

Solid Wastes: Solid wastes, exclusive of solid product material but including cafeteria garbage, construction rubbish, filters and filter aids are drummed and taken to the municipal dump. No estimate of quantity was available.

Production Wastes: Production wastes including dusts, off-quality products, packaging materials, returned goods are handled in one of two ways. A SOMAT process of wet pulping is used for all materials except plastic which is demolished by dry-grinding. Approximately 3,000 cubic yards are compacted monthly and hauled to a publicly owned sanitary landfill by company-owned trucks.

Process Water Wastes: Process water wastes are emptied into the sewage system, approximately 1.6 million gallons per day.

Radioactive Wastes: Radioactive wastes (process washing) are pumped directly to 10,000 gallon storage tanks for decay. Manufacturing residues and product rejects are given to an AEC disposal service. From 6 to 12 drums (30 to 55 gallon capacity) are disposed of each month.

Syringes and Needles: Syringes and needles are destroyed and taken to a landfill; 20 to 50 pounds per month appear as waste.

Hazardous Solvents: Hazardous solvents are reclaimed. Approximately 5,000 gallons per month may be given to a contract service for disposal. Additional quantities of solvents (research) are taken by a service company for reclamation. The quantity involved is less than 2,000 gallons per month.

Mycelia and Fermentation Residue: Mycelia and fermentation residue is emptied into the municipal sewer system. However, the company pays for special treatment required to reduce the BOD and COD load at the treatment site.

Figure 129 is a flow diagram of solid wastes in the pharmaceutical industry as taken from a report by D.M. Shilesky et al (2).

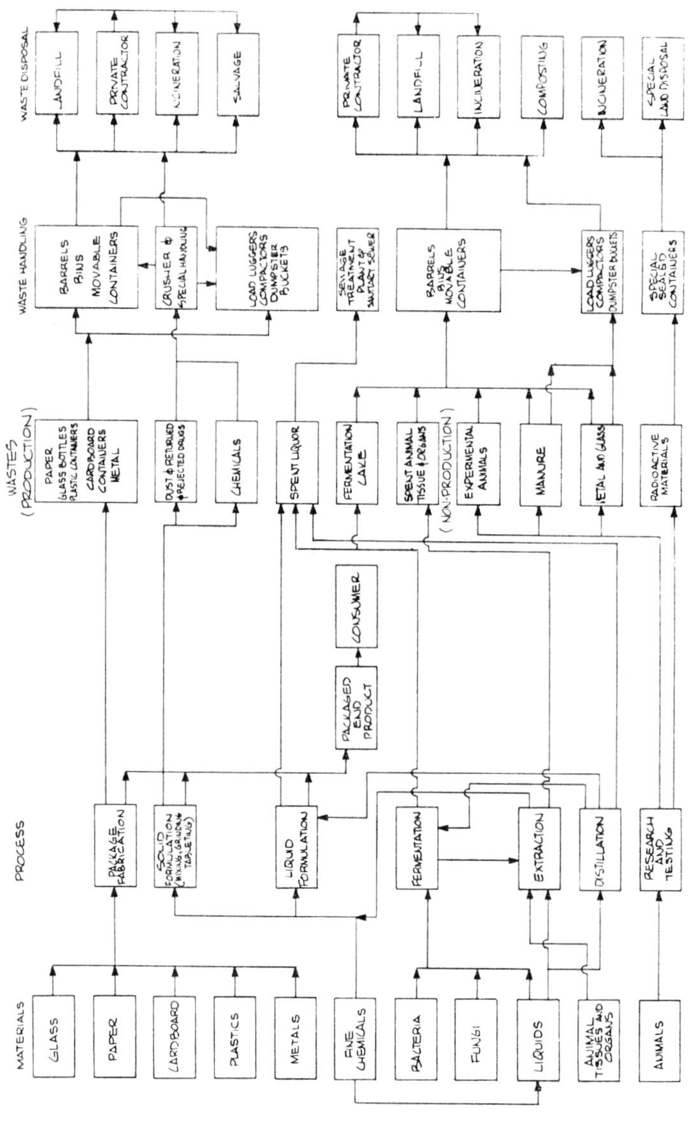

FIGURE 129: FLOW DIAGRAM OF SOLID WASTES IN THE PHARMACEUTICAL INDUSTRY

Source: PB 225 333

INCINERATION

As described in *Nickel Topics* (3), an incinerator-scrubber unit (Figure 130) may be used to avoid environmental pollution in pharmaceutical manufacturing.

FIGURE 130: DIAGRAM OF INCINERATOR-SCRUBBER UNIT

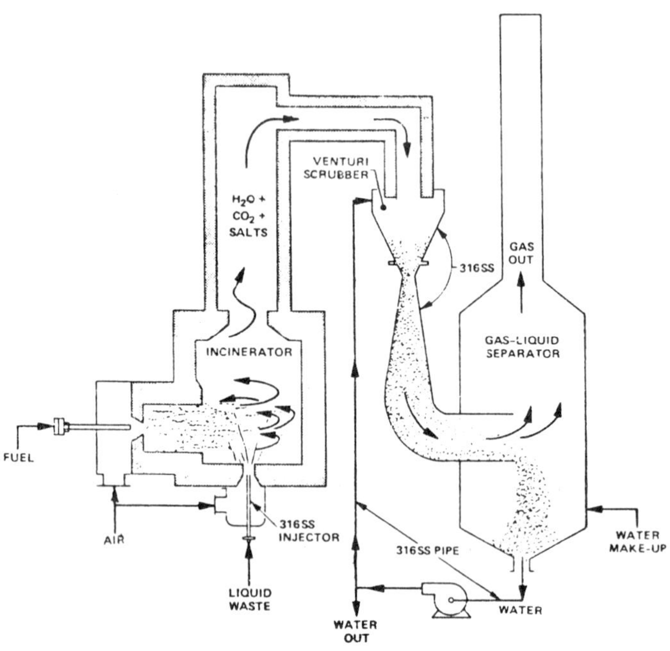

Source: *Nickel Topics*

A unit built by Pfizer, Inc. at their plant at Barcelonetta, Puerto Rico handles approximately 7,600 liters (2,000 gallons) per day of the end products of pharmaceutical production. The stream contains aqueous wastes, inorganic salts, and organic materials, including sulfates, phosphates, chlorides, phenols, and other corrosive chemicals. To provide the resistance required, all wetted areas in contact with these chemicals are fabricated in Type 316 stainless steel.

Incineration alone is not enough because the resultant gaseous effluent would still contribute to pollution in the form of fly ash fallout. To complete the pollution cleanup a venturi-type wet scrubber was included. This takes the gas from the incinerator and literally washes visible constituents from it. In operation, the aqueous wastes and chemicals are introduced into the incinerator through spray nozzles of Type 316 stainless steel. These nozzles atomize the liquid and direct it at an oil-fired flame front which generates temperatures in the incinerator of between 870° and 980°C (1600° and 1800°F). The organic wastes are turned to CO_2 and H_2O almost instantly.

A refractory-lined thimble feeds the 870° to 980°C (1600° to 1800°F) gaseous stream to the scrubber unit, consisting of a 3 meter (10 ft) long venturi with adjustable gate valve and a cyclonic separator. The W-A venturi and interconnecting piping are fabricated in

Type 316 stainless steel. The scrubbing liquor is introduced at the top section of the venturi by Type 316 stainless steel tangential piping inlets which cause the liquor to spin on an open shelf and cascade down the converging venturi walls toward the neck. At this point, the liquor is deflected across the high velocity gaseous flow. The liquor is atomized and solid particulates colliding with the droplets are effectively scrubbed from the gaseous stream, ultimately being carried to a slurry recirculation tank.

The gaseous stream, now considerably reduced in temperature but heavily laden with vapor and fine particulate-carrying mist, is accelerated by internal deflection vanes and caused to spin in a cyclonic separator. Centrifugal force combined with gravity separates the entrained liquids from the gas stream. Clean gases meeting EPA requirements exit from the stack at near ambient temperature and humidity. The liquids that were spun out of the steam drain into the reservoir or settling tank where they can be recirculated or occasionally bled off.

Since the introduction of the slurry at the top of the W-A venturi requires neither nozzles nor injection devices, which can plug up, reintroduction of the slurry is done easily, without loss of efficiency. However, since the recycled slurry liquor containing increasing concentration of salts is highly abrasive and corrosive, a suitably corrosion resistant material is required. Type 316 stainless steel is specified for the interconnecting piping.

The incinerator-scrubber, designed to meet OSHA standards, has been in use since mid-1973, with little downtime for modification. Combined with other programs, it has played a major role in helping Pfizer Inc. gain community acceptance and remain a "good neighbor" in the community of Barcelonetta.

REFERENCES

(1)   Booz-Allen Research, Inc., *A Study of Hazardous Waste Materials, Hazardous Effects and Disposal Methods,* Vol II, Report PB 221 466, Springfield, Va., Nat. Tech. Information Service (July 1973).
(2)   D.M. Shilesky, K.W. Krause and R.J. Sullivan, *Solid Waste Management in the Drug Industry,* Report PB 225 333, Springfield, Va., Nat. Tech. Information Service (1973).
(3)   *Nickel Topics,* 28, No. 2, 3-4 (1975).

PLASTICS INDUSTRY WASTES

GENERAL INDUSTRY WASTES

Incineration

Ultimate disposal of plastic wastes is often accomplished by incineration as described by M. Sittig (1). Incineration is basically a high temperature oxidation process. Thus, plastics, which contain primarily carbon and hydrogen, are oxidized to carbon dioxide and water vapor provided the conditions of time, temperature, and turbulence are right. When thermoplastics are heated, they become progressively softer and gradually melt; when the temperature is increased further, the viscosity of the melt decreases until it becomes a fluid. This generally occurs between 150° and 250°C.

In some cases, as with polystyrene, the polymer depolymerizes and often degrades or decomposes at or near its melting point. On the other hand, when thermosetting resins are heated, they generally do not deform but simply degrade. Because thermal decomposition starts from the outside, all materials with high ratios of surface to volume melt and decompose fairly rapidly. Conversely, the larger the article the more slowly it burns in the incineration process. Most plastics have a high calorific value, often equal to that of high grade fuel oil. For example, polyethylene has a fuel value of 18,000 Btu/lb; in contrast, PVC, which has chlorine in addition to carbon and hydrogen, has a value of only 9,500 Btu/lb.

The polyolefins burn relatively easily and give off no noxious gases; however, they tend to drip, and the drops often continue to burn. When burned in the open, polystyrene, which has a self-ignition temperature of 900°F, burns with a yellow flame throughout and develops dense smoke. PVC, on the other hand, burns with difficulty; burning has to be sustained by an outside heat source. PVC softens and chars, and if no plasticizer is present, it does not drip. On burning, PVC generates 58 lb of hydrogen chloride per hundred pounds of polymer.

Results of tests in which the normal refuse was supplemented with 3.25% PVC have been reported. It was noted in this case that only a small portion of the theoretical amount of HCl expected could be detected in the gases. It was also observed that if the incinerator operated at sufficiently high temperatures (above dew-point conditions), PVC presented virtually no risk of corrosion. On the other hand, the results of another study indicated that when industrial PVC waste was added to the urban refuse in an incinerator, inordinately high amounts of HCl were present in the flue gases (2,000 ppm). Also, severe corrosion occurred in the high temperature region of the incinerator. A clear statement concerning the corrosive effect of HCl from PVC is difficult to make, because corrosion can be caused by poor operating conditions, uneven addition of plastics to the burning refuse, other acid producing wastes, and other artifacts. Professor Elmer R. Kaiser, in a study sponsored by the Society of the Plastics Industry, added double and triple amounts of plastics to the normal solid waste loads and burned them in a relatively recently constructed municipal incinerator. He added PVC, polystyrene, polyethylene and polyurethane, and found that large amounts of plastics (up to 6%) in the solid waste mix did not have a corroding effect on the incinerator during the operating period. The short operating period of these experiments makes these findings questionable.

The incinerator was one that employed a conventional continuous feed system with oscillating grates and processed an average of 200 tons of mixed refuse a day. Prof. Kaiser also reported that the incineration of plastics in these large amounts caused no significant increase in smoke levels and did not increase pollution in the atmosphere. During the tests, the plastics did not drip through the incinerator grates. In fact, the higher burning temperatures of the plastics improved the incinerator performance, particularly when the normal refuse was wet.

As the discussion above indicates, PVC is the most troublesome plastic in an incinerator. Polystyrene also can be difficult, although it does not produce corrosive or noxious gases. Polystyrene burns easily in an incinerator if the rate of heat release, which is approximately 18,000 Btu/lb as opposed to 4,000 Btu/lb for normal refuse, is controlled. The very high rate of heat release must be retarded to insure complete combustion. Common methods to control this heat release are to cut the air supply and/or maintain a small ratio of polystyrene to normal refuse. The thick black smoke which can be generated in burning polystyrene is eliminated by using an efficient after burner.

As a treatment process, incineration is relatively expensive; incineration costs range from $2 to $8 per ton. The average is about $5.30 per ton. Incinerators require a large capital investment; construction costs vary from $1,500 to more than $6,000 a ton at a rated 24-hr burning capacity. Upkeep costs are also relatively high, and more skilled employees are required. Labor costs are a signficant part of the operating costs. Plastic wastes can affect these economics by requring more resistant materials of construction or heightened maintenance due due to the more rapid corrosion of refractory and metal parts.

The envisaged process for the incineration of primary treatment sludge from plastic manufacture is illustrated in Figure 130. Capacity was based on the production of 30,000,000 pounds annually of sludge containing an average of 10% solids. The heating value of these solids was estimated at 15,000 Btu/lb. Operating conditions were selected on the basis of experimental runs which indicated a bed temperature of 1325°F and superficial velocity of 1.8 fps were required. The design results indicate (2) that an incinerator of about 12 feet in diameter would be required for incineration of this waste sludge. Added fuel must be supplied (i.e., 156 scfm of methane or 625 lb/hr of waste polymers).

A device developed by B. Faurholdt (3) is an incinerator for the combustion of waste products and particularly plastic materials comprising two combustion chambers, the primary combustion chamber comprising refractory ceramic elements located adjacent to one another so as to form inclined continuous guide surfaces and means for supplying primary combustion air to the primary combustion chamber opening into the chamber at different levels above the bottom thereof. In the prior art incinerators the destruction of plastic materials presents serious problems because these plastic materials burn in a manner which is very different from that of household refuse.

The problems which are encountered during the incineration of plastic materials in prior art incinerators comprising a grate supporting the plastic materials are due to the fact that the grate quickly stops functioning in the normal manner because the holes in the grate through which the air is supplied become clogged. This clogging has the effect of initiating a destruction of the grate if it is made from commonly used materials, such as cast steel alloys. Other problems are due to the high soot content in the flue gases leaving incinerators in which plastic materials are incinerated. This soot content which is due to an incomplete combustion strongly contaminates the atmosphere and is a nuisance in the

area surrounding such incinerators. The reason for these difficulties is that plastic materials possess the property of becoming soft and melting before they are ignited and burn.
If refuse containing plastic materials is introduced into and ignited in an incinerator in which the grate consists of perforated grate elements of cast iron or grate lamellae separated by passages for supplying air, the openings or passages in the grate become clogged whether it is constructed as a step grate, a plane grate, or as an inclined grate because of the melting of the plastic materials.

FIGURE 130: SIMPLIFIED FLOWSHEET OF INCINERATOR FOR PLASTIC WASTES

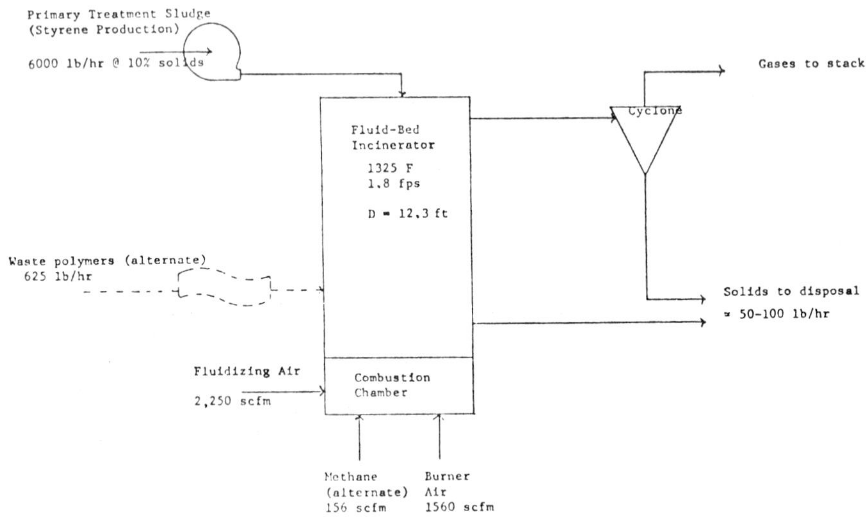

Source: EPA Report 12020

During the combustion which takes place on the basis of the combustion air which is supplied to the refuse without passing through the grate, very high temperatures are developed on the upper surface of the grate. Because of the clogging of the passages in the grate the latter is not subjected to the cooling which normally takes place during the passage of the cool combustion air through the grate passages. Thus, the grate, e.g., made from cast iron, is heated to very high temperatures and the heating causes an irreversible increase of volume.

Furthermore, the plastic materials possess the property that even when heated to a moderate temperature large amounts of gases are developed and the amount of the gases increases exponentially with increasing heating temperatures. Thus, if refuse containing plastic materials is introduced into and ignited in the above mentioned incinerators, large amounts of gases containing carbon particles are quickly formed in the zone above the refuse layer and the carbon particles escape from the combustion chamber together with the flue gases. If it is attempted to avoid the formation of these carbon particles by introducing more combustion air in the combustion chamber, the increased amount of heat from secondary combustion causes still more gases and sooty carbon particles. Thus, the problems are progressively increased until the gases and carbon particles have been completely burned away. Then the supply of a large amount of secondary air effects a strong cooling of the combustion chamber, thus decreasing the combustion temperature of the material which has not been completely burned out. This temperature decrease also leads to soot which leaves the incinerator together with the flue gases. As stated the rate of generation of volatile gases and soot particles within the primary combustion chamber depends on the

generation of heat and consequently the temperature within that chamber. However, by
having the combustion in two thermally separated chambers, the effects of the secondary
combustion (heat generation and temperature increase) do not influence the combustion
process in the primary chamber. In this development there is an incinerator with a primary
combustion chamber which comprises inclined continuous guide surfaces consisting of guide
elements made from a refractory ceramic material and in which the means for supplying primary combustion air to the primary chamber open into it at different levels above the bottom.

When incinerating plastic materials in the incinerator according to the process, the plastic
material while melting flows down along the inclined guide surfaces and during the movement the temperature is increased to a value such that a gasification takes place. Due to
the fact that the primary combustion air is supplied at different levels above the bottom
of the chamber, there is effected an intimate mixing of the primary combustion air and
the compounds formed during the gasification which compounds subsequently burn. Figure 131 shows the essentials of the incineration design.

FIGURE 131: TWO-CHAMBER INCINERATOR WITH INCLINED REFRACTORY GRATES

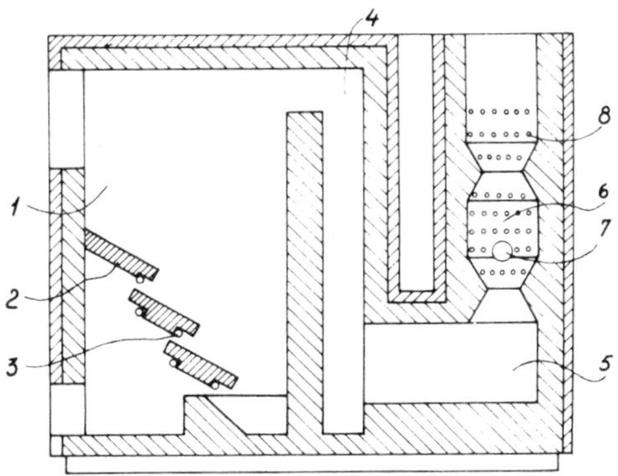

Source: U.S. Patent 3,670,667

In this figure **1** is a primary combustion chamber containing inclined guide elements **2**
of a refractory ceramic material. These guide elements **2** are supported by hollow metal
pipes **3** provided with holes and connected to means for supplying primary combustion
air to the combustion chamber. The primary combustion chamber is connected to a secondary combustion chamber **6** through a duct **4** and a flow-turning chamber **5**. In an
opening in the wall of the secondary combustion chamber **6** there is provided an oil burner
7 and a large number of air supply nozzles **8**.

A process developed by F. Ensslin et al (4) is a means for eliminating plastic waste and
recovering the metals and metal compounds contained in the waste. The plastic waste,
whether in a solid, pasty or liquid form, is converted into a combustible fluid-like form
and is directed through a burner into a combustion chamber. The combustion of the
waste results in combustion gas and fly ash which is cooled and directed through a separator to obtain the metals and metal compounds from the fly ash in the form of oxides.

Figure 132 shows the overall installation as well as a detail of the conveyor. As shown
there, the overall installation consists of a melt container **1**, a combustion system consisting of a burner **2** and a combustion chamber **3**, a heat exchanger **4**, a cooler **5**, a separator **6** for separating the fly ash and a conveyor **7**. The detail shows the conveyor **7**
embedded in a sand jacket. The conveyor **7**, for example a pump, and a pipe **8** for the
plastics waste are both embedded in the sand jacket **11**, which is itself enclosed in insulating

material **9** supported by a sheet metal jacket **10**. The sand is heated by a heater **12**. There is an outlet pipe **13** for removing the sand.

FIGURE 132: PLASTICS INCINERATOR USING HOT SAND-JACKETED CONVEYOR
TO FEED MOLTEN PLASTIC TO BURNER

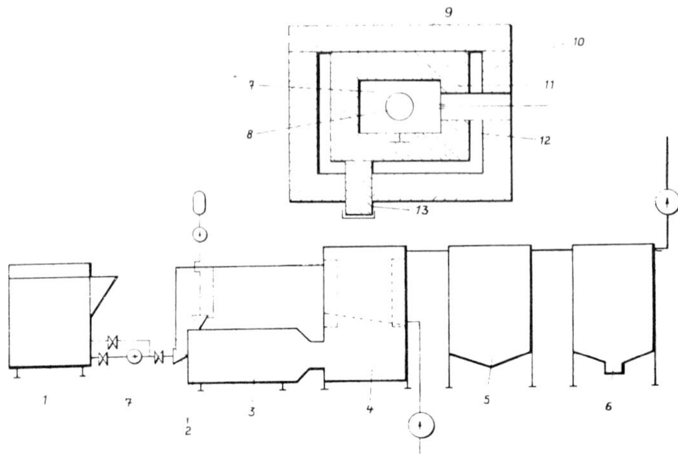

Source: U.S. Patent 3,776,147

The method of operating the apparatus will now be described on the basis of an example. In this example solid plastics waste is fed, either directly or with the help of a feed apparatus, to the melt container **1**. The melt container is constructed to allow liquid or pasty plastic waste to be introduced, either by gravity flow or by means of a feeding device. The container **1**, containing the plastic waste, is heated in the known way, for example electrically, to melt the waste and/or to bring it to a desired temperature and viscosity. If the charge is a solid waste the melting can be accelerated, and a constant temperature can be ensured, by agitating the melt and/or by using circulating pumps. During the agitating combustion promoters can be added. Circulating pumps may be used, arranged so that they project high speed jets of molten material against solid waste situated in front of a grating, melting and/or size reducing the solid waste so that it can flow.

As soon as the waste has acquired the desired temperature, which depends on its melting point and on the viscosity of the melt, the liquid or molten waste is conveyed through the feed pipe **8** to the combustion system consisting of the burner **2** and the combustion chamber **3**. To prevent the molten mass from freezing in the pipe **8**, this pipe together with all the necessary conveying, control and screening devices is embedded in a sand jacket **11** which is heated in a controlled manner. This arrangement ensures that the plastics waste cannot coke up due to local overheating. A further advantage of the sand jacket is that if a leak occurs in the conveying system the fire hazard is minimized.

In a particular advantageous modification the liquid waste is sprayed under a specified pressure by the burner **2** and is burnt in a current of preheated primary air. Simultaneously secondary air can be blown in centric-symmetrically in the direction of the flame, to provide the excess air necessary for ensuring complete combustion. Assuming that the liquid waste is fed at a constant rate, the flame temperature can be adjusted by precise adjustment of the primary and secondary air. If desired the combustion system can itself be heated, or preheated, by means of an auxiliary burner, for example an oil or a gas burner. This provides an igniting flame for initiating combustion and for sustaining combustion. The combustion chamber **3** consists of a steel jacket suitably lined, the lining being of a kind which does not react with the fly ash under the prevailing operating conditions. Suitable materials for the combustion chamber lining are for example refractory bricks made of chromium-magnesite, fireclay, Dynas, silica or the like. Downstream of the combustion chamber there is a heat exchanger **4**, connected to the combustion chamber either directly or through a connecting channel. The mixture of combustion gases and fly ash is cooled in the heat exchanger **4**, which utilizes the heat to warm the primary and secondary air. The heat exchanger can if desired also contain a further tube bundle through which a heat transfer medium flows, so that the combustion heat taken from the mixture of combustion gases and fly ash can be utilized elsewhere for heating purposes.

In the heat exchanger **4** the mixture of combustion gases and fly ash is cooled down far enough so that the downstream cooler **5** can be constructed of ordinary steel. The cooler **5** delivers the mixture at a controlled temperature to the separator **6**, for example a cloth-tube filter. The temperature should be such that the cloth-tube filter cannot itself cool down to the prevailing dewpoint. On the other hand of course the temperature must not be high enough to damage the filter. The fly ash separated by the filter contains the metals present in the initial plastics waste, the metals appearing in the fly ash in the form of oxides, mixed oxides or other inorganic compounds, whose nature depends on the nature of the waste material. The functioning of all the parts of the apparatus, including the filter, can be controlled by conventional devices.

A process dveloped by E.H. Swartz (5) is one in which plastic-coated scrap metal is heated within a vented combustion chamber to cause decomposition of the plastic without adversely affecting the metal core with respect to its subsequent separation from the decomposed plastic and recovery as a substantially pure metal product. A fuel gas mixture is fed at a controlled rate to the combustion chamber to initially elevate the temperature of the chamber to an optimum value that is then maintained constant. The products of combustion form a nonoxidizing atmosphere in the chamber within which the desired decomposition of the plastic occurs.

A process developed by R.F. Atkin (6) utilizes an incinerator which is capable of disposing of combustible waste containing a substantial percentage of plastic materials without creating attendant atmospheric pollutants. This is accomplished by provision of a controlled, stabilized process of combustion within the incinerator, wherein the solid wastes are volatilized at a controlled rate in the presence of a flame, the resultant combustion gases are combined with excess oxygen and are burned in an oxidizing flame; the remaining combustion products are then thoroughly mixed with the remaining excess oxygen at an elevated temperature and then accelerated into another oxidizing flame where the residue of the plastic materials is completely consumed.

A device developed by J.F. Straitz (7) is a smokeless portable incinerator for solid waste materials such as plastic, foam and straw, liquid waste materials, and other combustible wastes. It includes a cylindrical shell open at the top, with tangentially oriented pilot gas inlets in the side wall of the shell and tangentially mounted forced air inlets of the positive or air inspirating pressurized type.

Inlet pipes are provided tangentially oriented to the side wall of the shell for delivery of liquid combustible wastes for combustion. The shell is mounted to an insulated base plate which rests on the floor or other location where wastes are to be burned. The incinerator may be provided with a bottom drain from removal of noncombustible liquids. A water ring can also be provided for delivery of low combustible content liquid wastes, for highly water saturated wastes, and for water delivery to enhance combustion and for smoke elimination.

In one approach to the recovery of oil spilt on a body of water, pieces of sponge-like styrofoam plastic are scattered on the water in the vicinity of the spill where they absorb the spilt oil. The foam is collected and passed through rolls where the oil is squeezed out and recovered. The plastic sponges are reused until they are worn out and are ready for disposal. The disposal of the waste plastic, which is preferably accomplished on board the ship collecting the spilt oil, must be accomplished without causing atmospheric or other pollution with the preferred method of disposal being by burning. The device is claimed to be particularly applicable to the disposal of such plastic wastes.

An apparatus developed by T. Ishikawa et al (8) is a solid plastic waste incinerator incorporating two stages of jet scrubbing for exhaust gas treatment. The first stage is for HCl removal and the second stage for cleaning up the effluent after first stage treatment.

Landfill Disposal

Plastics make a suitable component of sanitary landfill (1). Since they do not decompose readily, they release no odors, gases or liquids that pollute surrounding land, air or water. The Los Angeles County Sanitation District Office has given the following statement in a letter to the Society of the Plastics Industry dated December 7, 1970.

> The District staff does not subscribe to the opinion or suggestion that sanitary land-fill is not a proper disposal method for plastic material because they are nonbiodegradeable. If we consider for discussion purposes a sanitary landfill in which only inert materials have been placed, the settlement anticipated would occur only through mechanical means. Nonbiodegradeable plastic wastes disposed of in such a landfill would seem to be as suitable as dirt, broken concrete, bricks and such like materials which also do not undergo decomposition. Since plastic materials do not degrade, they can be considered the equivalent of any other inert material in this respect. The economics of plastics products versus decomposable materials in a sanitary land-fill are insignificant.

Compacted plastic wastes or residues from incineration are normally delivered to a landfill or, in a few instances, transferred to barges for subsequent ocean dumping. More than 90% of the household, commercial, and industrial solid wastes are handled by some type of landfill. Costs include capital investment of about $2,000 per ton and one to two dollars per ton as operating costs.

The practice of landfill can range from an open burning dump to a truly sanitary landfill operation. This latter operation is basically one of spreading refuse on the ground; in some instances, the refuse is densified by burning or by mechanical means using a compaction vehicle, and afterward covering the refuse with a layer of clean earth. In the case of the open burning dump, many plastics are undoubtedly subjected to some measure of volume and mass degradation with an accompanying emission of air pollutants. Where compaction vehicles are used to densify the refuse, many plastic items are ruptured and reduced in volume.

Plastic materials generally do not decay or rot even after prolonged soil burial and therefore, represent some of the most persistent components in the fill. For example, plastic bottles are not easily squashed, and only those in the lower strata of the refuse become distorted, flattened, or compressed by the overlying weight. Those with less weight or overburden are likely to remain intact and can, therefore, contribute to the springiness of the top layer. Accordingly, whatever process is used to compress refuse on-site or off-site, the results will be a reduction, if not a complete elimination, of the deleterious effects that plastics may have in landfill. If plastic bottles are properly squashed, they lose much of their springiness, although some shapes tend to unsquash more than others.

The public has often attacked plastics because they are not biodegradable. However, as recently pointed out by some officials of the county sanitation district of Los Angeles, there are no meaningful differences between plastics and decomposible materials when used in a sanitary landfill. In fact, they state that because plastics can be considered the equivalent of an inert material such as broken concrete, plastics are better suited than decomposible materials for landfill disposal, because they provide immediate stability. Here again, plastics are most suitable if they are in a pulverized form. Biodegradation is dependent on time and the geometry of the article in the disposal area. In a study of plastics that were buried in a dump in Sweden for five years, it was found that while heavy plastic parts were found, thin plastic films had already disappeared. It was also reported that after five years the molecular weight of the polyethylene in refuse was ¼ of the original, whereas the molecular weight of PVC in refuse was ⅔ of the original (1).

POLYOLEFIN WASTES

Incineration

A device developed by R.C. Pryor (9) is one in which effective incineration of waste products including plastic products of various types is obtained by means of a two-stage grateless incinerator including a first stage vaporization zone with burner and forced air supply, and a second stage combustion zone with burner and forced air supply. The two-stage vaporization-combustion incinerator provides effective oxidative incineration, with gaseous products being smokeless, and any nonvaporized noncombusted solid matter is essentially retained in the first stage vaporization chamber.

POLYVINYL CHLORIDE WASTES

Incineration

A device developed by W.R. Fuller et al (10) permits reducing the amount of halogen emitted during the incineration of halogen-containing plastics. The method involves applying an alkali to the plastic before it is burned. The method can reduce the emission of halogen by greater than 75% when properly employed.

A process developed by M. Shigaki, S. Kido and Y. Chiba (11) is a device for the disposal of scrap vinyl polymers containing chlorine and hydrogen. The apparatus includes a rotary kiln that is adapted to receive the scrap plastic and heat the same to a temperature at which the plastic partially decomposes, and the major portion of the chlorine escaping therefrom as hydrogen chloride gas, which gas together with any chlorine which results from the decomposition is recovered in a confined space for subsequent usage. The partially decomposed scrap plastic is sequentially removed from the rotary kiln in a dried state and crushed. The crushed plastic is discharged by a blower onto a grate in a furnace where the scrap plastic is burned.

The hot gases of combustion from the burning together with fines and dust arising from the burning operation are conducted through a passage to heat air that is subsequently used in the heating of the rotary kiln. The hot gases of combustion, together with entrained solid materials, are discharged into a dust collector to remove the fines and dust therefrom. The gases after being freed of solid materials are subsequently discharged through a chimney to the ambient atmosphere. The gases discharged to the ambient atmosphere are substantially free of chlorine and hydrogen chloride gases.

The apparatus, as shown in Figure 133 includes a vertically extending storage pit **1** into which scrap plastic **P** may be dumped by gravity through an opening **1a**. A wall structure **1b** extends upwardly from the pit **1** and defines a confined space **1c**. The wall structure **1b** supports a traveling crane **2** that is movably supported on a horizontal rail **2a** that has the ends thereof secured to the wall structure **1b**. The traveling crane **2** includes vertically movable buckets or claws **2b** that may be raised and lowered by power means (not shown) to sequentially grasp portions of the scrap **P**, and raise the same to a position where the scrap **P** can be dropped onto a horizontal power driven inclined vibrating feeder **3** that is of conventional design. A power driven crusher assembly **4** is shown in the drawing that is of conventional design, and has an opening **4a** in the upper portion thereof that sequentially has scrap plastic **P** discharged therethrough from the vibrating feeder **3**. The crusher **4** has a discharge opening **4b** in the lower portion thereof through which crushed scrap plastic **P** is discharged into an enclosed bunker **5**.

FIGURE 133: APPARATUS FOR INCINERATION OF PVC

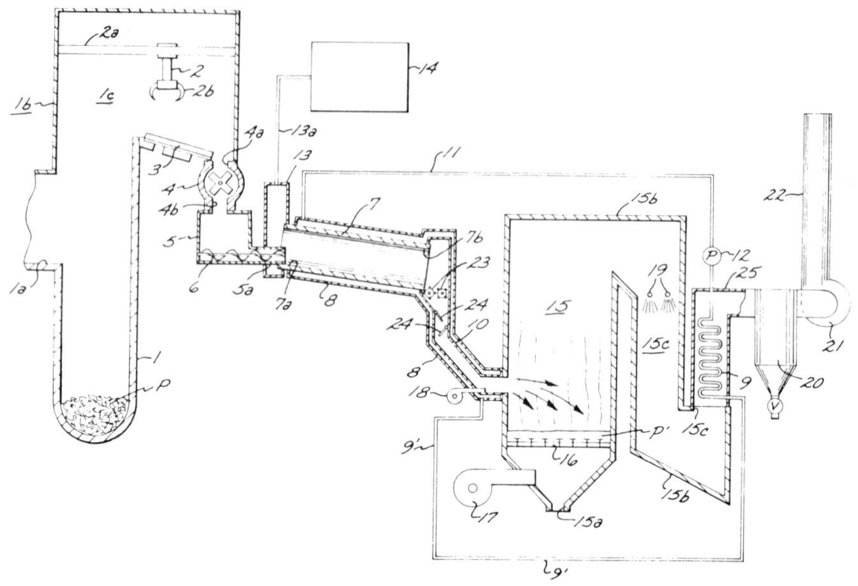

Source: U.S. Patent 3,716,339

The bunker **5** has a power driven screw conveyor **6** situated in the lower portion thereof, which screw conveyor when operating discharges plastic **P** from the bunker **5** through a conduit **5a** into a first end **7a** of a horizontal inclined power driven rotary kiln **7**. The kiln **7** as may be seen in the drawing has a second end **7b** that is of lower elevation than the first end **7a**. The rotary kiln **7** has a jacket **8** extending therearound and outwardly spaced therefrom. The jacket **8** extends not only around the kiln **7**, but also around a downwardly extending chute **10** that is in communication with the second end **7b** of the kiln **7**. An air heater **9** is provided as may be seen in the drawing which is in the form of a heat exchanger, and has a conduit **9'** extending from a first end thereof of the lower portion of the jacket **8** as shown in the drawing.

A second conduit **11** extends from the upper portion of the jacket **8** as may be seen in the drawing to the suction side of a power driven blower or fan **12**, which blower is connected to the air heater **9**. When the blower **12** is actuated, air is continuously circulated through the conduit **9'**, jacket **8** and conduit **11** to transfer heat from the air heater **9** to heat the chute **10** and kiln **7**. The plastic material **P** discharged into the kiln **7** is heated to not over 280°C whereby the scrap plastic **P** in the kiln partially decomposes, with the chlorine escaping therefrom in the form of hydrogen chloride gas. The gas escaping from the plastic material **P** in the kiln **7** flows into a collecting chamber **13** that has a conduit **13a** extending therefrom to an enclosed structure **14**. The hydrogen chloride gas discharging into structure **14** is easily recovered due to its ready solubility in water and may subsequently be used for industrial purposes, such as in the manufacture of chemicals, fertilizers and the like.

The plastic material **P** after being partially decomposed in the kiln **7** solidified, and is reduced in size prior to passage into the chute **10** by power driven crushers **23** as may be seen in the drawing. Flow of the crushed plastic through the chute **10** is by gravity, the flow of the crushed plastic being controlled by pivotally movable baffles or other flow controlling means 24 as shown in the drawing. A furnace 15 is provided as may be seen in the drawing. A furnace 15 is provided as may be seen in the drawing, which is vertically disposed, and has an apertured grate 16 located within the lower confines thereof. A power driven blower 17 discharges a current of air into the confines of the furnace 15, with the current of air flowing upwardly through the grate 16 and the partially decomposed scrap plastic P' that rests on the grate. The plastic P' is discharged onto the grate by use of a second power driven blower 18 shown in the drawing that discharges a current of air into the lower portion of the chute 10 and as a result directs the crushed scrap plastic P' onto the grate to cover the latter. The furnace 15 has an opening 15a in the lower portion thereof through which particles of solid material that do not burn on the grate 16 may drop downwardly therethrough by the action of gravity.

The furnace 15 is defined by a wall structure 15b that provides a downwardly extending passage 15c that has at least one water spray unit 19 mounted therein. The water spray units 19 are used to control the temperature of the gases of combustion from the furnace 15, prior to these gases contacting the air heater 9. The walls 15b of the furnace 15 have an opening 15c therein that is in communication with an upwardly extending inverted L-shaped conduit 25 that has the heater 9 situated within the confines thereof.

The hot gases of combustion from the furnace 15 as they pass through the conduit 25 contact the heat exchanger on air heater 9 and heat the air that is circulated therethrough by action of the blower 12. The hot gases of combustion and solid materials entrained therewith such as fly ash and the like discharge therefrom into a dust collector 20 of conventional design as shown in the drawing, which dust collector is connected to a power driven ventilator unit 21. The ventilator unit 21 draws the gases of combustion through the conduit 25 and dust collector 20 and discharges these gases free of fly ash and particles of solid material up a chimney 22 for discharge to the ambient atmosphere. The gases of combustion so discharged to the ambient atmosphere are substantially free of chlorine and hydrogen chloride gases and hence are not obnoxious or detrimental to the area surrounding the apparatus.

THERMOPLASTIC WASTES

Incineration

A device developed by R.F. Stockman (12) provides for incinerating a meltable plastic material by means of first reducing the plastic to a liquid, heating the liquid to vaporization, and then burning the fumes that are given off from the surface of the liquid before they may be exhausted to the surrounding atmosphere. Figure 134 shows details of the construction and operation of such a device. 10 refers to a housing or a container for the incineration of plastic material or the like. The housing has substantially vertical side walls 12 and an apertured top 14 with an inlet duct 16 which extends therethrough and terminates above the floor of housing 10 for adding solid plastic material into the container while an outlet port 17 at the top of the housing exhausts the fumes given off therefrom.

The port 17 is provided with an outlet duct 18 of cylindrical design which terminates a short distance above the housing 14 and is centrally positioned within the housing in a manner that permits a reasonable freedom from the congestion of fumes given off from the molten plastic. The stack 22 exhausting from the container 10 may be of any approved construction, however one of modified venturi configuration performs best in mixing the fumes with air for combustion and then subjecting the mixture to a secondary process of combustion.

At the lower end of the stack air supply ports 25 are provided whereby the upward flow of fumes from the container 10 will induce the flow of air therethrough. As the air mixes with the fumes it is compressed slightly by passing through the converging section of the venturi-shaped stack. Adjacent its point of greatest restriction of a burner 42 including an independent source of gas and air is injected into the stack and ignited whereby the fumes from the plastic together with the air supplied through ports **25** are then subjected to combustion. As the gases ignite in the stack they expand and draw in still more air through the ports **44** whereby combustion is nearly complete in the remaining section of the stack before the spent gases are exhausted to the atmosphere.

FIGURE 134: PLASTIC BURNER

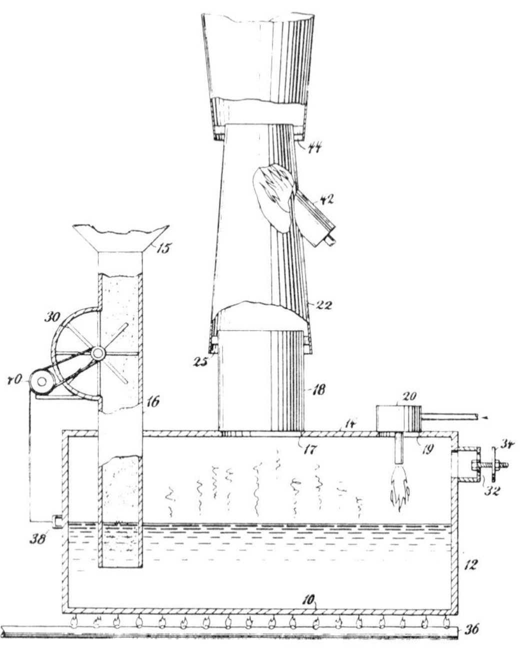

Source: U.S. Patent 3,572,265

The top **14** of the container includes a third opening **19** for a start up burner **20** which is inserted therethrough to initiate combustion of the plastic fumes. An air supply port **32** in the upper portion of the housing **12** includes an adjustable valve **34** therefor that may be opened a predetermined degree to partially burn the fumes being given off from the surface of the liquified plastic and produce sufficient heat in the combustion process to insure that the mass of plastic material in the housing is maintained in a liquid state.

In operation a mass of plastic material is loaded into the hopper **15** leading to container **10** through the inlet duct **16**. The size of the individual pieces of plastic will determine the size of duct **16**, thus for a mass of particle sized plastic an inlet duct **16** of smaller diameter is adequate while for larger chunk sized pieces of solid plastic an inlet of larger diameter is required. A feeding device such as a star wheel **30** may be required to feed the plastic evenly and to preclude the back flow of excess quantities of vapor or fumes given off from the molten plastic which lies within the inlet tube **16**. A temporary source of heat such as a burner **36** is applied to the container **10** adequate to heat the contents thereof and thus reduce the solid plastic to a liquid state. As the plastic is heated it liquifies and its rate of vaporization increases so that it increasingly gives off fumes and vapors which are in turn ignited by the burner or other ignition means in the opening **20**. Fumes given off from the incomplete combustion of fumes in the housing **10** are exhausted through the outlet **18** to the stack **22**. The fumes from the incompletely burned plastic are then mixed with air entering through ports **25**, mixed in the throat of the venturi and subjected to the incinerating action of the secondary burner **42**. Such incineration causes the inspiration of additional air through ports **44** so that combustion is complete before the gases are exhausted to the atmosphere. The step of injecting air and combustible fuel can be located at as many points along the stack as may be required to provide for the complete elimination of combustible products in the exhaust gas.

Heat from the plastic burning in the container **10** is sufficient to melt additional plastic material being added thereto so that an auxiliary or temporary source of heat such as the burner **36** is not required for continuous operation. A liquid level sensor **38** on the container **10** may be made to control operation of the motor **40** and star wheel feeder **30** so that the level of plastic supplied through tube **16** to container **10** is maintained above the lower end opening to the inlet tube **16** so that little vapor from the molten plastic is allowed to escape up the tube **16** to the atmosphere. The star wheel **30** provides an additional barrier in tube **16** preventing still further the escape of vapor from the container.

The amount of combustion air permitted to enter the container **10** through supply valve **34** is controlled carefully to provide heat for the combustion of plastic within the container **10** adequate only for the continuous reduction to a molten state of the solid plastic being added through inlet **16**, it being understood that level of the plastic within the container is continuously controlled by the liquid level controller **20** so that it remains substantially constant at a point lying between the open bottom end of tube **16** and the top of the container.

Therefore the surface area of the body of molten plastic remains substantially constant, its temperature too is almost constant, and the air for its combustion is allowed to vary little so that an almost constant rate of combustion is permitted. Inasmuch as wide fluctuations in the amount of fuel available for burning are substantially eliminated, combustion is nearly complete and little smoke or obnoxious exhaust gas escapes to the atmosphere.

A device developed by H. Liu (13) is a waste incinerator for burning a variety of sludges, liquids and solid materials that can be reduced to a combustible liquid by exposure to heat. Such materials are fed into an incinerator and fall onto the uppermost of several superposed perforate beds. As the waste material is exposed to heat it becomes less viscous and slowly drips through the perforation of each bed to a subjacent bed where it is exposed to additional heat and ultimately gasified and burned.

A process developed by E.O. Ohsol et al (14) provides for the disposal of thermoplastic materials utilizing a fuel oil as a solvent for the thermoplastic, by dissolving from about 1 to 20%, by weight, of a thermoplastic material or a mixture of the materials in from 99 to 80%, by weight, of a fuel oil, in the presence or absence of a cracking catalyst, usually from 1 to 10%, by weight, at elevated temperatures under atmospheric or superatmospheric pressure and, thereafter, burning the fuel oil containing thermoplastic material with resultant minimal residues and other solid, liquid, or gaseous pollutants.

A device developed by J.G. Roy (15) provides a means for the self-supportive incineration of thermoplastic materials wherein the thermoplastic material serves as the fuel and wherein the heat from the incineration is utilized to melt the thermoplastic material to permit rapid gravity flow onto an essentially vertical heat conductive surface maintained at the vaporization temperature of the thermoplastic material. This method and apparatus are claimed to provide a solution to the ecological problem of bulk waste disposal of thermoplastic materials. Figure 135 shows an overall section of such a device as well as a detail at the lower part of the figure showing the feed-melt-combustion transition area.

First, in the operating sequence for the apparatus, there is the feed zone wherein solid thermoplastic **M-1** material is transferred to the feed hopper **8**. The solid thermoplastic material resides on the top surface of the melting plate **4** and (in operation) on molten thermoplastic material M-2. Secondly, there is the melt zone wherein solid thermoplastic material M-1 receives heat from orificed melting plate **4** and is thereby melted to form M-2 and whereafter the molten thermoplastic material M-2 flows by gravity onto the ignition plate **3**. Thirdly, there is the combustion zone wherein the flowing molten thermoplastic material rapidly spreads on ignition plate **3** in the presence of oxygen supplied by supply means **10** and ignition means (not shown) so as to ignite and incinerate, whereinafter the incineration hot gases flow upward to the bottom surface of melting plate **4** and parallel to the ignition plate **3** before exiting the combustion chamber **1** through flue means **11**.

FIGURE 135: INCINERATOR FOR THERMOPLASTIC WASTE MATERIALS

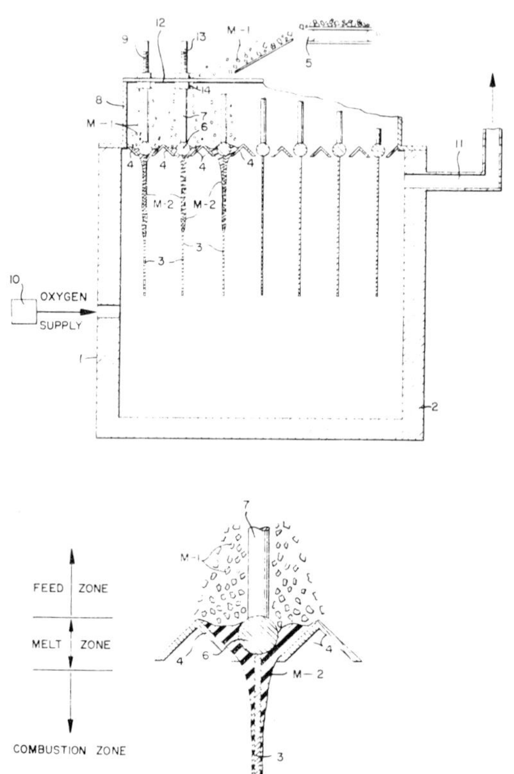

Source: U.S. Patent 3,710,739

The apparatus includes material feed means **5** depicted as a conveyor. Any suitable solid handling system is operable including pneumatic, mechanical as well as manual systems. The apparatus also includes a feed hopper **8** into which the thermoplastic material M-1 is fed. The feed hopper **8** is mounted above the combustion chamber **1**. There is shown an adjustment support member **12** mounted to the feed hopper **8**. The adjustment support member **12** as the nomenclature suggests supports a flow adjustment member **7** having flow regulating adjustment means **9** attached at the top thereof and a flow directing member **6** attached at the bottom thereof. It is to be noted that the flow regulating adjustment means **9** is depicted as a screw **13** and lock nut **14** assembly, however any controlling means for moving the adjustment member **7** in a vertical mode is suitable. The vertical motion of the adjustment member **7** results in a variation in the orifice represented as the space between the flow directing member **6** and the orificed melting plate **4**. This variable orifice permits control of the flow rate of the molten thermoplastic material **M-2** from the feed zone through the melt zone onto the combustion zone as depicted in Figure 135. It is most preferred that the apparatus have a variable orifice for controlled flow of molten thermoplastic material **M-2**. Apparatus of fixed orifice are suitable but as stated are not preferred.

The apparatus further comprises an orificed melting plate **4** or a plurality of the melting plates wherein the melting plate is mounted in concert with the feed hopper **8** and the combustion chamber **1**. The orificed melting plate **4** is of any heat conductive material preferably a metal and serves to transfer the heat involved in the combustion chamber **1** to the solid thermoplastic material **M-1** to melt same. The apparatus further comprises an ignition plate **3** having an essentially vertical heat conductive surface or a plurality of same. As shown in the lower detail view, the ignition plate **3** is mounted to the flow directing member **6**. This mounting arrangement is advantageous in that heat from the ignition plate **3** is conductive to the flow directing member **6** to ensure molten material at the surface of the flow directing member **6** thereby providing a free flowing system more responsive to control.

The ignition plate **3** is housed in combustion chamber **1** wherein the chamber has suitable insulation **2** to minimize the heat losses. The ignition plate **3** may be of any suitable material that is heat conductive, preferably a metal. Mounted in concert with the combustion chamber is oxygen supply means **10** of either a compressed cylinder supply, an induction fan or any suitable gaseous supply source. Regulatory valve and safety valve means may of course be added to the oxygen supply means **10** as desired. The oxygen may be supplied neat or in mixture with an inert gas such as nitrogen, argon and the like or in ambient air. The supply gas may of course be preheated.

Also, mounted in concert with the combustion chamber is flue means **11** whereby the waste gas from the combustion of the molten thermoplastic material **M-2** passes across the ignition plate **3** and the bottom surface of the orificed melting plate **4** transferring heat thereto and thereafter exiting the combustion chamber **1** by flue means **11**. The ignition plate **3** (specifically the major surfaces thereof) are shown at a right angle to the direction of the oxygen supply **10** flow direction. This need not be so and the major surfaces of the ignition plate **3** may be and are preferably parallel to the oxygen supply **10** flow direction.

A parallel arrangement of the oxygen supply **10** flow direction, the ignition plates **3** and the flue means **11** permits the oxygen to fill a maximum volume of the combustion chamber and is a preferred arrangement. In the event the waste gas contains undesired gaseous products, scrubbing or collecting means may be provided in connection with the flue means **11**. In the event the waste gas contains incomplete combustion products, an afterburner and gas recycle system may be further provided in connection with the flue means **11**.

A device developed by J.A. Boyd and D.E. Boyd (16) provides an improved incineration system for thermoplastic materials wherein a combustion chamber is provided with an inlet or charging port. A moving stream of air is generated outside the incineration combustion chamber and is directed toward and through the charging port into the interior of the combustion chamber. The thermoplastic materials to be incinerated are introduced into the air stream at a location remote from the charging port and are conveyed along and supported by the air stream and are thus introduced through the charging port into the combustion chamber where they are burned. By using the air stream for supporting and conveying the thermoplastic materials, such materials are maintained out of contact with the elevated temperature portions of the incinerator itself and hence will not melt and stick to surfaces of the incinerator.

Instead, the thermoplastic materials will be carried into the central portion of the combustion chamber where they are to be burned, and the air stream which carries the thermoplastic material additionally creates turbulent flow conditions within the combustion chamber to promote and facilitate burning of the materials.

Another device developed by J.A. Boyd and D.E. Boyd (17) is an incinerator specially adapted for burning plastic materials in such a manner that the materials are substantially completely burned with a minimum of smoke and odors. Such plastic materials are introduced into a primary combustion chamber where they are subjected to initial burning, and thereafter, the products of combustion from the primary combustion chamber are directed to at least one secondary combustion chamber which includes a diffusion chamber means therein having at least one refractory wall with spaced openings therein. A burner in the secondary combustion chamber means heats the refractory wall to an incandescent condition and preheated air from the diffusion chamber means is mixed with the products of combustion to promote additional burning thereof. Turbulent heated air is introduced into the primary combustion chamber to promote combustion therein.

REFERENCES

(1) M. Sittig, *Pollution Control in the Plastics and Rubber Industry,* Park Ridge, N.J., Noyes Data Corp. (1975).

(2) T.L. Tewksbury, A.K. Reed, H. Nack and G.R. Smithson, Jr., *Fluidized Bed Incineration of Selected Carbonaceous Industrial Wastes,* Report 12020 FYE 03/72, Washington, D.C., U.S. Environmental Protection Agency (March 1972).

(3) B. Faurholdt; U.S. Patent 3,670,667; June 20, 1972.

(4) F. Ensslin, G. Frenzel and E. Löbermann; U.S. Patent 3,776,147; December 4, 1973; assigned to Preussag AG-Metall.

(5) E.H. Swartz; U.S. Patent 3,821,026; June 28, 1974.

(6) R.F. Atkin; U.S. Patent 3,828,701; August 13, 1974; assigned to Pyrocom, Inc.

(7) J.F. Straitz III, U.S. Patent 3,869,995; March 11, 1975; assigned to Combustion Unlimited, Inc.

(8) T. Ishikawa, T. Endo, T. Masuyama, F. Uruma, M. Morishita and K. Takeshi; U.S. Patent 3,927,986; December 23, 1975; assigned to Nippon Carbon Co. Ltd.

(9) R.C. Pryor; U.S. Patent 3,680,500; August 1, 1972; assigned to Phillips Petroleum Co.

(10) W.R. Fuller, B.H. Bieler and D.C. Morgan; U.S. Patent 3,556,024; January 19, 1971; assigned to The Dow Chemical Co.

(11) M. Shigaki, S. Kido and Y. Chiba; U.S. Patent 3,716,339; February 13, 1973; assigned to Takuma Kikan Mfg. Co.

(12) R.L. Stockman; U.S. Patent 3,572,265; March 23, 1971; assigned to The Air Preheater Co., Inc.

(13) H. Liu; U.S. Patent 3,780,674; December 25, 1973; assigned to The Air Preheater Co., Inc.

(14) E.O. Ohsol and A. Perlmutter; U.S. Patent 3,750,600; August 7, 1973; assigned to American Cyanamid Co.

(15) J.G. Roy; U.S. Patent 3,710,739; January 16, 1973; assigned to Union Carbide Corp.

(16) J.A. Boyd and D.E. Boyd; U.S. Patent 3,490,395; January 20, 1970; assigned to Washington Incinerator Sale and Service, Inc.

(17) J.A. Boyd and D.E. Boyd; U.S. Patent 3,495,555; February 17, 1970; assigned to Washington Incinerator Sales and Service, Inc.

PULP AND PAPER INDUSTRY WASTES

Among the disposal operations conducted by the pulp and paper industry (1) are sludge disposal, strong waste disposal, and by-product recovery.

Sludge Disposal: The handling, dewatering, and disposal of primary and secondary sludges, either together or separately. Several dewatering methods are used: vacuum filtration using conditioning chemicals, centrifugation, sludge presses, drying beds, and sludge lagoons. The selection of method depends on sludge characteristics, land availability, ultimate disposal considerations, and proportion and secondary sludges. Landfilling or incineration of primary and secondary sludges is common.

Strong Wastes Disposal: Deep well disposal of liquor and pulp washer water. The effectiveness is dependent on geological formations at the mill location and the amount of waste discharged. Another method is dilution of spent liquors for land application, having similar factors as in irrigation disposal to define its effectiveness, and their spray application to land. However, runoffs cause significant damage to receiving stream conditions because of high BOD and solid concentrations.

Waste Treatment By-Products: Several organic by-products can be recovered from the cooking liquor resulting from various types of pulping operations. They include turpentine, oil, yeast, alcohols and dimethyl sulfoxide (DMSO). Bark by-products include roofing felts, thermal insulation materials, and wrapping paper. The sulfite spent liquor can be used to produce insecticides, tanning agents, reinforcing agents, reinforcing agents in rubber, cement dispersing agents, etc.

In addition, there is cooking liquor disposal (or recovery, sometimes involving techniques which dispose of contaminants while recovering valuable chemicals) and also the ultimate disposal of odorous gases from pulp cooking operations.

BLACK LIQUOR

Incineration

A process developed by G. Steinert (2) involves drying and burning so-called black liquor, a concentrated pulp digester residual liquor derived from various pulping processes such as the sulfate, soda or sulfite process practiced in the paper making industry. The process more specifically relates to a method of drying and burning black liquor where the concentrated liquor is sprayed into a hot gas stream, dried and then introduced with combustion air into the furnace of a steam generator for burning. As hereinbefore practiced in the industry, the concentrated liquor is sprayed into a duct carrying hot gases for drying and for delivery of the dried particles into a furnace chamber while they are suspended in the drying gas. This method, however, suffers from the disadvantage that a large amount of liquor thus introduced adheres to the duct walls, where drying proceeds but slowly and unevenly.

Since the amount of dried liquor particles dropping off the duct walls continually fluctuates, the quantity of dried particles transported by the gas stream being discharged into the furnace also fluctuates. This is highly undesirable, especially when the dried material is directly introduced into the furnace in a continuous stream, since under these circumstances it is practically impossible to maintain the constant fuel and air ratio which is so much desired for economical operations.

In accordance with the process these disadvantages are avoided by introducing the concentrated liquor into the hot gas duct in the vicinity and in front of the fan, and into the fan inlet opening preferably in a direction which is generally coaxial with the axis of rotation of the fan rotor. During operation a cushion of compressed gas is produced surrounding the rotor blades of the fan.

This cushion then moves at high velocity either in radial or axial direction depending on the type of fan. Accordingly, the finely divided liquor is successfully prevented from adhering to the fan blades and passes through the fan with the gas mass as an integral part thereof. Drying of the small black liquor droplets is therefore achieved in a uniform and speedy manner.

This technique finds an important application in installations in which the black liquor is burned together with pulverized coal. The method is specifically useful in a system where a mill is employed performing the function both of a coal pulverizer and of a blower. In such a system hot furnace gases and crushed coal are fed to the suction inlet of the mill.

Figure 136 shows the system which is capable of burning black liquor either as a single fuel or in combination with powdered coal. The mill **10** comprises a beater wheel **12** having radially extending beater blades **14** supported between an annular disc **16** and a wheel disc **18**. The beater wheel **12** is carried at the end of a shaft **20** journaled at **22** and driven by means not shown, to cause rotation of wheel **12** at high velocity within a mill housing **24**.

The housing is of a well-known shape which spirally expands in direction of mill rotation. Precrushed coal is fed from coal bin **26** by way of duct **27** and coal gate **28** into a gas duct **30** through which flow hot furnace gases from combustion chamber **32**. They intermix in a vertical shaft **34** and are introduced into the axial mill inlet **36** by way of mill door **38**. The nozzle **40** projects through the front wall of mill door **38** in the vicinity and directly in front of the mill inlet **36**.

This nozzle is in adjustable manner fastened to the wall, such as by way of flange **41** and is connected to a source of black liquor not shown for discharging liquor of suitable concentration into the descending coal and gas stream in a direction which is preferably coaxial with the axis of rotation of the mill wheel **12**. A shield **43** is provided above the black liquor nozzle **40** to protect it against abrasion by the descending coal.

The preferred shape of the liquor spray is that of a cone having a configuration which protrudes into and substantially fills the circular axial inlet opening **36** of the beater mill housing **24**. This shape and manner of spraying result in an even distribution of the liquor droplets throughout the hot gas and coal stream at the point of entry into the mill. The compressed gas cushion formed on the surface of the beater wheel parts during operation of the mill successfully prevents adherence of the liquor particle on the wheel. The liquor droplets instead are moved about in the hot gas stream and are speedily dried while being carried through the mill by the rotating gases.

FIGURE 136: APPARATUS FOR DRYING AND BURNING BLACK LIQUOR

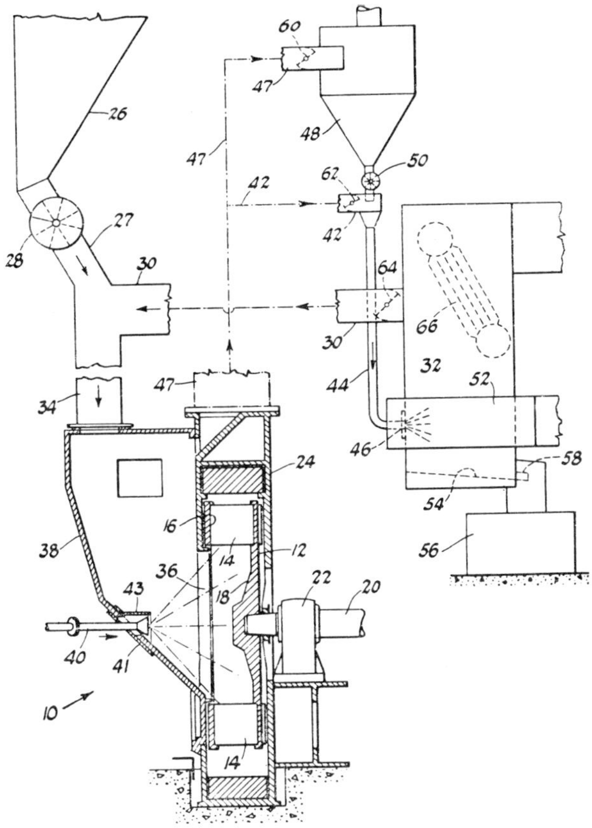

Source: U.S. Patent 3,053,615

This method of drying and burning black liquor is equally applicable to a system where only black liquor is dried and burned, or to a system where a mixture of pulverized coal and black liquor is dried and burned. The same apparatus could be used equally well to dry black liquor and deliver it to a furnace chamber for burning.

For this purpose beater mill **10** is only required to perform the function of a mixer and transport fan by drawing hot gases out of the furnace chamber **32** via conduits **30, 34** and **38**. The black liquor is sprayed into the thus created stream of hot gases by way of nozzle **40** in the manner earlier described, the liquor droplets being uniformly and speedily dried as they are carried at high velocity through the fan by the rotating hot gases.

The dried liquor particles, if black liquor alone is burned, or the liquor particles and coal dust mixture, can be carried directly in the gas stream to the furnace chamber **32** by the way of duct **42**, pipe **44** and burner **46**. Or the gas and fuel mixture can be discharged into the cyclone separator **48** via duct **47** for separation of the fuel particles from the moisture-laden gases.

The fuel collected in the cyclone **48** is then fed into gas duct **42** by way of valve **50** for delivery to burner **46**. Combustion air in a conventional manner is furnished from a source not shown to the burners via duct **42**. While the combustible material in the black liquor is burned off, the residual chemicals accumulate on the hearth **54** of the furnace chamber and form a smelt which periodically is discharged into a smelt tank **56** by way of spout **58**.

Dampers **60, 62** and **64** are provided in ducts **47, 42** and **30** respectively, for regulating the flow of the combustion gases. Heating surfaces **66** for generating or superheating steam are provided in the path of the combustion gases, with the gases being released into the atmosphere after having given up a major portion of the heat contained therein.

A process developed by J.E. Greenawalt (3) involves the recovery of chemicals from waste black liquor, a by-product obtained in the production of pulp from wood; and more particularly a method for not only recovering the chemicals by a burning or combustion operation but also for utilizing the calorific value of the black liquor for producing steam, all in a manner which eliminates objectionable odors in the residual gasesous products which, after passing through the apparatus, are discharged into the atmosphere and dissipated.

Figure 137 illustrates the application of this process. The apparatus, in general, comprises a reduction furnace **A** in which the concentrated black liquor is initially burned in a reducing atmosphere, an oxidizing furnace **B** connected with furnace **A** in which the combustible particles entrained in the combustion gas stream from furnace **A** are burned with additional air to completely oxidize the oxidizable components in the combustion gas stream; a steam boiler unit **C**, connected to furnace **B**, through which the hot combustion gas stream passes in contact with the boiler tubes; an air heater unit **D** connected to the boiler unit in which air is heated for introduction into the combustion chamber of furnace **A** and the oxidation chamber of furnace **B**; an aqueous solution scrubber unit **E** connected to the air heater in which residual water soluble components not previously removed from the combustion gas stream are removed; and a suction fan **F** connected to the scrubber unit for drawing the combustion gases through the apparatus and forcing the residual combustion gases through a stack (not shown) for discharging the combustion gases into the atmosphere where they are dissipated.

After the cooking process the black liquor may contain, as a typical example, about 16 to 17% solid matter and about 83 to 84% water and will test about 11°Bé at 60°F. Before burning, it is concentrated to about 60% solids and will contain liquid carbohydrates, rosin and fats which are combustible in addition to the nonburnable chemical soda compounds.

The waste black liquor from the digesters, after having been concentrated to the desired solids content is introduced through conduit **10**, through a burner **11**. Heated air is introduced through hot air conduit **12** through the burner **11** into the combustion chamber **13** of furnace **A**. The hot air is taken from air heater unit **D** through hot air conduit **14** which connects with air conduit **12**.

The concentrated black liquor which, in a typical example, will contain about 60% solids is burned in the combustion chamber **13** with insufficient air to completely oxidize the combustible parts of the liquor so that a reducing atmosphere is maintained in chamber **13** at a temperature in the neighborhood of 1950°F. In this first or initial burning step the ligneous matter and other combustible organic material is burned out of the black liquid and forms a combustion gas stream.

The unburnable chemicals are collected in the bottom of furnace **A**. At the temperature maintained there the unburned chemicals will form a molten product which may be drawn off through a suitable draw-off conduit **15**. If additional fuel is needed to maintain proper burning it may be introduced into the furnace chamber **13**.

For example, fuel oil may be passed through fuel conduit **16a**, through the burner **11**; such fuel being also burned for preheating the plant when starting up.

FIGURE 137: APPARATUS FOR RECOVERY OF CHEMICALS FROM THE INCIN-
ERATION OF WASTE BLACK LIQUOR AND
ELIMINATING NOXIOUS COMPOUNDS ENTRAINED IN THE
COMBUSTION GAS

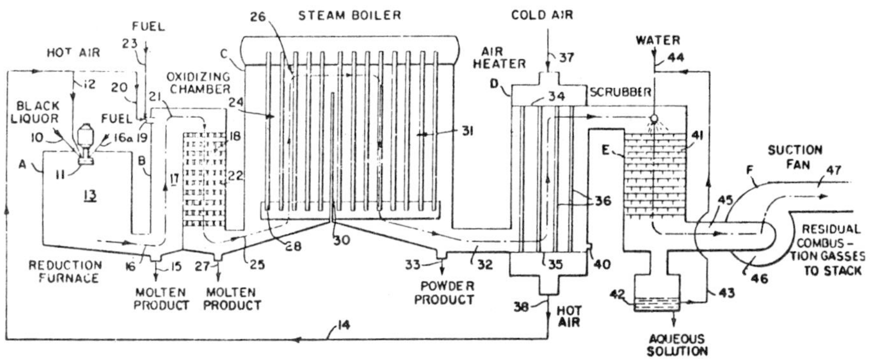

Source: U.S. Patent 3,163,495

Inasmuch as a reducing atmosphere is maintained in combustion chamber **13**, the molten
product, which flows to the bottom of the chamber is drawn off at **15**, will contain sodium
sulfide Na_2S produced by reducing sodium sulfate contained in the spent black liquor, sodium carbonate and other condensed sulfur compounds in small amounts, such as sodium
sulfite (Na_2SO_3), sodium thiosulfate ($Na_2S_2O_3$), sodium sulfate Na_2SO_4, sodium carbonate,
sodium hydroxide, possible minor amounts of other sulfur complexes and a small amount
of insoluble material.

In commercial operations, the combustion gases which pass from the combustion chamber
13, through flue **16** still contain chemical compounds either in solid, vapor or gaseous
forms which are carried along with the combustion gas stream. Some of these compounds
are valuable recoverable chemicals and also contained in the combustion gas stream are
malodorous compounds which, in accordance with the process may be eliminated by
further burning in an oxidizing atmosphere.

The combustion gases are then passed upwardly through flue **17** into the oxidizing chamber
18. At the upper end of combustion furnace unit **B** there is provided air introducing
ports **19** through which additional hot air may be introduced through hot air conduit **20**,
which is connected to hot air conduit **14**. Introduction of additional hot air into the mixing zone **21** of the oxidizing furnace **B** provides an excess of oxygen in the combustion
gas stream and the chemical compounds carried along in the combustion gas stream from
the reducing furnace **A** are burned and further oxidized as they pass downwardly through
the oxidizing combustion chamber **18**.

The oxidizing chamber is packed with refractory brick **22** laid up in checkered form so as
to expose as large a hot surface as possible to the stream of combustion gases to which
has been added an excess of oxygen by introducing hot air through ports **19**. Brick known
to the trade as chrome brick is preferred because it acts in the nature of a catalyst to
accelerate the desired oxidizing reaction, and, of course, the brick laid up in checker fashion
provides a large hot refractory surface which enhances the oxidizing reaction that is desired.

Any suitable means may be provided to provide a large hot surface area but it is preferred
to use a material which provides a catalytic effect to enhance the oxidation of the oxidizable

constituents that are carried in the combustion gases discharged from the reducing chamber **13**. The temperature of the combustion gas stream is raised in the oxidizing chamber **18** by reason of the oxidizing reaction which is exothermic and the temperature is considerably higher than the melting point of the chemicals so that if any chemical product drops out of the combustion gases as the gases pass through flue **25** into the first section **24** of boiler tube chamber **26** of the steam boiler unit **C**, the product is sufficiently hot to be molten and may be drawn off in molten condition through draw-off conduit **27** in the bottom of the oxidation chamber.

Inasmuch as the atmosphere in chamber **18** is oxidizing, H_2S carried into it is burned to H_2O and SO_2 or oxidized to the sulfate form and also any other oxidizable compounds are oxidized and this oxidation will convert entrained bad odor compounds to other compounds in which the bad odor does not exist, whereas otherwise the ill smelling sulfur compounds and ill smelling complexes would persist and be carried along with the combustion gas stream.

The hot combustion gas stream from the oxidation chamber **18** which may be in the neighborhood of 2200°F comes into contact with the boiler tubes in the steam boiler unit and gives up heat by heat interchange with the water in the boiler tubes. It will be observed that the boiler tube chamber **26** is provided with a vertical baffle **30** which terminates short of the top of the boiler tube chamber so the combustion gas stream passes through flue **25** then upwardly into the first section **24** of the boiler tube chamber then over the top of the baffle **30**; then downwardly through the second section **31** of the tube chamber and then through flue **32**.

In some instances it may be desirable to introduce additional fuel in addition to air into the oxidizing chamber **21**, if it is desired to produce hotter combustion gases to the boiler tube chamber **26**. This may be done by passing fuel, such as fuel oil, through pipe **23** through a suitable burner at port **19**.

The hot combustion gases which pass through the first section **24** of the boiler tube chamber are rapidly cooled because of their contact with the relatively cooler boiler tubes **28** and as the combustion gas stream passes over the baffle **30** and then downwardly in section **31** of the boiler tube chamber they give up further heat through the tubes to the water in them.

In this section **31** the gases are sufficiently cooled so that any chemical products that drop out of the combustion gas stream will be in powder form as distinguished from molten, and may be removed from time to time through a suitable removal port **33**. The chemical products which are recovered from oxidation chamber **18** and from the boiler tube chamber may be added to the black liquor which is introduced to the first burning step or to such other make-up solutions as best suited to the overall operation, but in any case they are returned to the cycle and hence ultimately recovered.

The combustion gas stream which passes from the steam boiler unit **C** through flue **32** still contains a considerable amount of heat. The heat then in the combustion gases is utilized to heat the air which is used to burn the black liquor in reducing chamber **13** and the air which is introduced into oxidation unit **18**. The air heater **D** comprises a heat exchange unit having headers **34, 35** into which air heater tubes **36** are secured. Atmospheric air is passed through a port **37** and then passes through the tubes in heat interchange relation with the hot combustion gases. The heated air then passes through the discharge end **38** of the air heater into conduit **14** which carries the hot air to the units **A** and **B**. If any chemical in powder form collects in the air heater it may be removed from time to time through a suitable port **40**.

The combustion gas stream which leaves the air heater is now cooled to a temperature approaching the temperature of the ambient atmosphere introduced into the air heater. The gas stream is then passed into the scrubber unit **E**, which is provided with a suitable packing **41** such as ceramic brick or tile, to increase contact surface. The combustion

gases are then scrubbed with water or an aqueous solution. The scrubber solution is sprayed
into the top of the scrubber unit and comes into contact with the combustion gases to dissolve or absorb any residual chemicals that may still be carried along with the gas stream.
Of course, any H_2S that was initially present in the combustion gas stream will have been
burned out in the oxidation chamber **22**.

If SO_2 or other soluble gases or entrained particles are present they are removed with the
scrubber solution. In practice it is desirable to circulate the aqueous solution through the
scrubber. The solution may be collected in a collector chamber **42** and returned to the
scrubber through suitable piping **43**. When the solution is sufficiently concentrated it may
be used as make-up for cooking solution or returned to the recovery cycle, as desired.
Make-up water for the scrubber solution may be introduced through line **44**. If desired,
sodium carbonate may be added to the circulating scrubber solution to absorb SO_2.

Residual combustion gases stripped of the recoverable chemicals and undesirable bad smelling or noxious fumes are then drawn through flue **45** and forced by fan **46** through its discharge conduit **47** into a stack (not shown) from which the residual gas stream is dissipated
into the atmosphere.

A process developed by V.P. Owens et al (4) is a chemical recovery unit in which black liquor
from a pulping process is burned including a direct contact evaporator for concentrating
the black liquor prior to its introduction into a furnace. Preheated air is used as the heating medium in the direct contact evaporator. The moisture laden air leaving the evaporator is used as secondary combustion air in the furnace. Figure 138 illustrates the application of this process.

Referring to the drawing, **10** denotes a recovery furnace having a hearth, or smelting zone
12. Air to support combustion within the furnace is introduced through primary air nozzles **15**, and secondary air nozzles **14**. These nozzles are supplied by ducts **16** and **17**,
respectively. Black liquor is sprayed into the furnace through nozzles **18**. The black liquor
comes from a direct contact cascade evaporator **20**, which is supplied by means of inlet
22 with incoming black liquor. The concentrated black liquor is transported to the furnace through line **26**.

After the black liquor is burned in the furnace, the hot combustion gases flow upwardly
through the furnace, first passing over superheater sections **28** and **30**, and then flowing
through the steam generating tubes **34**. Water is supplied to these tubes by drum **38**, and
steam is removed from upper drum **36** and flows to the superheaters. Steam is also generated in tubes **40**, which line the walls of the furnace. The combustion gases leave the
furnace through gas pass **42**, thereafter dividing into ducts **44** and **46**. These ducts carry
the combustion gases to air preheaters **48** and **50**, where the gases pass in heat exchange
relationship with air supplied by fan **56** through ducts **52** and **54**. The relatively cool
combustion gases leave the air preheaters by means of ducts **58** and **60**, and are exhausted
to atmosphere through the stack **62**.

Heated air leaves the air preheaters through ducts which merge into a plenum **64**. From
this plenum extends duct **17**, which supplies primary combustion air to the furnace. Also
extending from plenum **64** is duct **66**, which conveys hot air to the direct contact cascade
evaporator **20**. After this air passes through the evaporator, evaporating moisture from the
black liquor in the process, it enters duct **16**, which conveys it to the furnace to be used
as secondary combustion air. Bypass duct **68** containing damper **70** is provided for supplying air to the secondary combustion air nozzles **14** without the necessity of flowing through
the evaporator **20**, when desired.

Dampers **72** and **76** are located in air ducts **16** and **17** so that the amount of primary and
secondary air can be varied as desired. Likewise, dampers **78** and **80** are positioned in the
combustion gas ducts **58** and **60** so that the amount of combustion gases flowing through
the air preheaters **48** and **50** can be controlled.

FIGURE 138: INCINERATOR AND CHEMICAL RECOVERY UNIT FOR BLACK LIQUOR

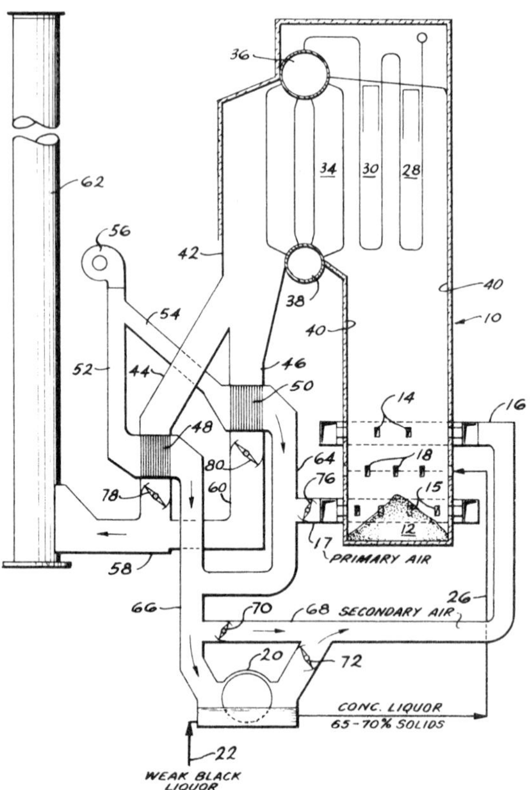

Source: U.S. Patent 3,703,919

The operation of the system will now be described. The moisture is evaporated from the black liquor supplied to direct contact evaporator **20**. The concentrated liquor, with a solids content of 65 to 70%, is sprayed into furnace **10** through nozzles **18**. The liquor in falling to the hearth or smelting zone **12** is dried and partially combusted, with some of the volatiles being driven off.

The combustion of the volatiles is supported by secondary combustion air supplied through nozzles **14**. The major portion of the combustion takes place in hearth **12**, with primary air to support combustion being introduced into the smelt zone through nozzles **15**.

The combustion gases after passing through the various heat exchangers associated with the furnace, divide and pass through the two air preheaters **48** and **50**. A portion of the air from the preheaters passes through the cascade evaporator **20**, evaporating moisture from the black liquor. The moisture laden air is introduced into the furnace through nozzles **14**, to act as secondary combustion air.

A second portion of the air from the preheaters is conveyed directly to the furnace, and introduced through nozzles **15** as primary combustion air. This dry, heated air stabilizes combustion of the black liquor, virtually eliminating the possibility of loss of flame. By-pass duct **68**, and the dampers located in the various ducts, can be used to vary the proportions of the primary and secondary air. The use of hot, dry air as primary air permits the reduction of temperature of the black liquor to the sprays **18**. This is an important factor because established data proves that low liquor spray temperatures are necessary to maintain minimum levels of malodorous compounds discharged from the furnace.

A process developed by D.R. Sheeley et al (5) is a process for treating concentrated spent liquor from a soda-based pulping process to recover sodium for reuse in the pulping process and to prevent the usual pollution of streams and air caused by the usual effluent from the pulping process. The process comprises the following steps. Concentrated spent liquor containing sodium is mixed with recycled finely-divided reactive alumina hydrate and formed into solid pellets by spraying the concentrated liquor-aluminum hydrate mixture onto a bed of recycled sodium aluminate furnace ash in an enclosed rotating container and tumbling the mixture and ash.

The solid pellets are combusted and reacted by feeding the pellets through a furnace operating at a temperature below the fusion temperature of sodium aluminate and at a sufficiently high temperature to combust the organic portion of the pellets and react the sodium content thereof with the alumina to form additional sodium aluminate as a particulate unfused ash.

The resulting furnace ash is pulverized and a portion thereof is recycled for subsequent reuse in the treating process. The remaining portion of the ash is dissolved in water to form a solution of sodium aluminate. The sodium aluminate solution is reacted with either sulfur dioxide or carbon dioxide or both to form sodium sulfite and/or sodium carbonate pulping chemical and alumina and alumina is separated for reuse in the treating process.

Ocean Disposal

The so-called black liquor produced by the sulfate process used in paper manufacture includes sodium carbonate, sodium sulfate, sodium hydroxide and other chemicals. Such wastes were discharged off the Texas coast in 1955 and contained about 45% solids, were water-soluble but quite viscous, and had a pH greater than 13 (6).

Laboratory studies using four species of phytoplankton showed the concentration of black liquor affecting photosynthesis (i.e., threshold toxicity) was about 0.05 gram per liter, and the lethal concentration was on the order of 1 gram per liter. Tests on zooplankton produced no acute toxicities in concentrations up to 0.4 gram per liter; greater concentrations were not tested because it was impossible to observe the zooplankton as a result of the opaque characteristics of the wastewater mixture.

Visual observations at sea confirmed the nontoxic character of the wastes, as zooplankton in samples taken immediately astern of the discharge barge were observed to be swimming about normally. Field evidence also showed that the absorbent component of the waste sank almost immediately, leaving only a slight discoloration at the surface that was visible for only 30 to 40 minutes after discharge.

A minimum dilution ratio of waste to water immediately behind the barge was calculated to be on the order of 1 to 300,000 or a concentration of 0.03 gram of waste per liter, which is well below the values given previously for threshold and acute toxicity on phytoplankton.

Bacterial action should result in the ultimate disintegration of the organic components of the black liquor wastes while the inorganic fraction, principally sodium sulfate, would rapidly lose its identity when mixed with seawater.

SULFITE LIQUOR

Coking

A process developed by W.F. Franz et al (7) is one in which wash water and digester liquor from a wood pulping operation are heated at a temperature in the range of 450° to 700°F under autogeneous pressures of about 1,000 to about 3,000 psig for a period of 0.5 minute to 6 hours to form coke and a purified effluent.

The coke is separated from the effluent and dewatered. Liquid from this dewatering step is combined with the effluent and depressured to flask vaporize and concentrate the combined stream to a soluble chemicals concentration suitable for reconstitution and recycle to the pulping operation.

Coke produced in the process may be burned as plant fuel and the ash recovered and worked for chemical values. The liquid stream is reconstituted for pulping purposes by the addition thereto of pulping chemicals derived, in part, by burning the coke produced in the process.

Incineration

A process developed by G.G. Copeland et al (8) involves treating waste sulfite liquors containing organic and inorganic materials to facilitate their disposal in a simple and economical manner, and to recover and reclaim the valuable residual chemicals in the waste liquors in a form readily amenable to treatment for subsequent reuse. Figure 139 shows the essential features of a fluidized-bed incinerator design which may be used to conduct such a process.

FIGURE 139: FLUIDIZED BED REACTOR FOR INCINERATION OF SULFITE LIQUOR

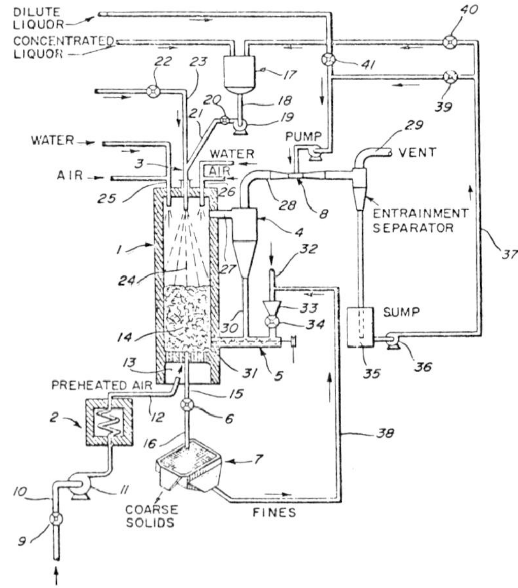

Source: U.S. Patent 3,309,262

There is shown a single-bed fluidized-bed roasting furnace **1** as the main reaction vessel which has, as associated equipment, a preheater **2** to heat incoming fluidizing air, waste liquor introduction facilities including a pneumatic atomizing or spray device **3**, a cyclone collector **4** to separate entrained particulate solids from the exhaust gases, a dust recycle system including a screw return **5** to return disentrained solids from collector **4** to the fluidized-bed roasting furnace **1**, a solids discharge mechanism including a valve **6** to discharge the granular inorganic agglomerates from the fluidized-bed roasting furnace, a dry solids classifying means **7**, such as screen classifier, for classifying the granular product removed from the roasting furnace, and a wet-scrubbing system **8** to remove the final traces of suspended material in the exhaust gases.

Within the roasting furance **1**, a bed of fluidized solids **14** is maintained by passing air through a control valve **9** and conduit **10** and then through a blower or compressing device **11** before introducing it to the preheater **2** to add the desired degree of sensible heat to the fluidizing air.

The preheated air is conveyed through conduit **12** to windbox **13** from which it passes through a plurality of orifices into the bottom of fluidized bed **14** of residual inorganic chemicals from waste liquor or a mixture of residual chemicals and inert materials. The air passes upward through the fluidized bed **14** maintaining it in continual agitation with violent action, commonly known as a state of fluidization.

The height of fluidized bed **14** is maintained at a desired level by the rate of discharge of the finished product through conduit **15**, valve **6** and conduit **16**. Waste pulping liquor, preconcentrated by conventional evaporation procedures, is stored in tank **17**, and by means of conduit **18**, pump **19**, valve **20** and conduit **21** it is introduced at a controlled rate to the pneumatic atomizing spray device **3** which receives a controlled supply of compressed air through valve **22** and conduit **23**.

Waste pulping liquor is introduced pneumatically into the freeboard **24** of the roasting furnace through the atomizing device **3** which directs the liquor downward toward the fluidized bed **14**. If desired, additional atomizing devices **25** and **26** can be employed to introduce water or weak waste liquor into the freeboard **24** of the reactor as a cooling medium. Exhaust gases containing the evaporated water, the products of combustion, and some entrained solids are conducted from the freeboard **24** of the reactor through conduit **27** to the cyclone collector **4**, and then through conduit **28** to the wet-scrubbing system **8** to remove particulate matter before exhausting the gases to the atmosphere through the conduit **29**.

Particulate matter removed from the exhaust gases by the cyclone collector **4** may be returned directly to the fluidized-bed roasting furnace **1** through conduit **30** and the screw return **5** at a point **31** below the top level of the fluidized bed **14**. Make-up inert material desired for the fluidized bed during processing can be introduced into the roasting furnace through conduit **32**, hopper **33**, a valve or feeder **34**, and the screw return **5**.

Particulate fines and fumes recovered in the wet-scrubbing system **8** can be recovered in sump **35** and recycled to the concentrated liquor storage tank **17**, or to the wet-scrubbing system **8** by pump **36** and conduit **37** as controlled by the valves **39**, **40** and **41**. Dry solid product is discharged from the fluidized-bed roasting furnace **1** through the conduit **15**, the solids valve **6** and the conduit **16** to the classifying means **7**. Coarse solids can be treated for chemical recovery or can be discarded; fine solids from the classifying means **7** are preferably recycled to the fluidized bed **14** through the conduit **38**, the hopper **33**, the valve or feeder **34**, and the screw return **5**.

Figure 140 is a flow sheet and schematic view of equipment suitable for treating magnesia-base waste sulfite liquors by this process. A fluidized-bed roasting furnace **80** forms the main reaction vessel and has as associated equipment a preheater **81** to heat incoming fluidizing air; waste liquor introduction facilities including one or more pneumatic atomizing devices **83** which can be a single device designed so that the proportion of finely atomized liquor feed to coarsely atomized liquor feed can be varied at will or a multiple device capable of producing both finely atomized and coarsely atomized liquor feed; a cyclone collector **44** to separate entrained particulate solids from the exhaust gases; a dust recycle system including a screw return **45** to return disentrained solids to the fluidized-bed roasting furnace; a solids discharge mechanism including a valve **46** to discharge the product from the fluidized-bed roaster; and a wet-scrubbing system **47** to remove the final traces of suspended material in the exhaust gases.

FIGURE 140: APPLICATION OF FLUIDIZED BED REACTOR OF FIGURE 139 TO MAGNESIA-BASE PULPING PROCESS

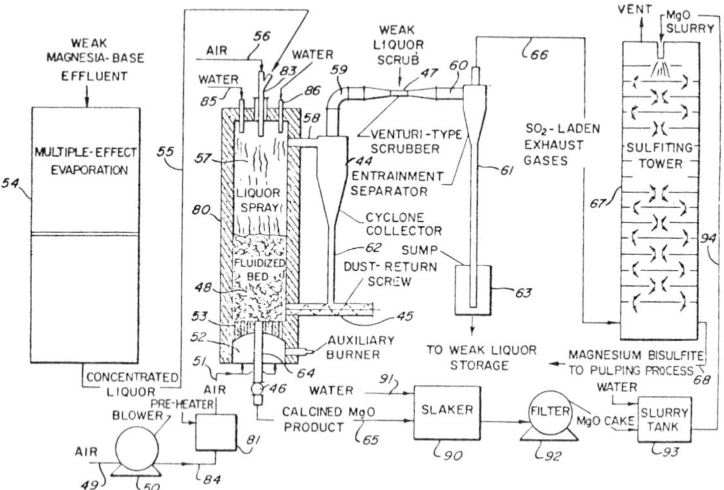

Source: U.S. Patent 3,309,262

Within the roaster **80**, a bed of magnesia **48** is maintained fluidized by passing air through conduit **49** and then through a blower or compressing device **50** before introducing it by conduit **84** to the preheater **81** to add the desired degree of sensible heat to the fluidizing air. The preheated air is conveyed through conduit **51** to windbox **52**, through constriction plate **53** and into the bottom of the fluidized bed **48** of magnesia.

Waste magnesia-base pulping liquor is preconcentrated to a 20 to 45% solids concentration in evaporator **54** and is introduced by conduit **55** at a controlled rate to the pneumatic atomizing device **83** which also receives a controlled supply of compressed air through conduit **56**. Waste pulping liquor is thus introduced pneumatically to the freeboard area **57** of the reactor **80** through the atomizing device **83** which directs the liquor downward toward the fluidized bed **48**.

If desired, water or weak waste liquor can be introduced into the freeboard **57** of the reactor as a cooling medium by conduits **85** and **86**. Exhaust gases containing the evaporated water, the products of combustion, sulfur dioxide, and some entrained solids are conducted from the freeboard area **57** of the reactor through conduit **58** to cyclone collector **44**, and then through conduit **59** to the wet-scrubbing system **47** to remove particulate matter before conducting the gases through conduit **60** to entrainment separator **61** and then by conduit **66** to absorbing tower **67**. Particulate matter removed from the exhaust gases by the cyclone collector **44** can be returned directly to the fluidized-bed roasting furnace **80** through conduit **62** and screw return **45** to a point below the level of the fluid-ized bed **48**. Particulate fines and fumes recovered in the wet-scrubbing system **47** can be recovered in sump **63** and recycled to a weak liquor storage tank.

Dry solid granular magnesia product is discharged from the fluidized-bed roasting furnace **80** through the conduit **64**, the solids valve **46** and conduit **65** to slaker **90** for mixing with water added by conduit **91**. The slurry from slaker **90** is then conveyed to filter **92** from which magnesia is removed and slurried with water in tank **93**.

By conduit **94** the slurry is introduced into absorbing tower **67**. The sulfur dioxide exhaust gases from reactor **80** are transported by conduit **66** to the bottom of the absorbing tower **67**. The sulfur dioxide flows upwardly in tower **67**, contacts the magnesia slurry and is thus converted to magnesium bisulfite which is drawn off through conduit **68** for use, if desired, as magnesia-base sulfite pulping cooking liquor.

A process developed by K.H. Wirths (9) is one in which sulfite liquor, for example, mag-nesium bisulfite liquor is burned in a combustion chamber which forms part of a steam generator. The burning of the liquor is accomplished at pressures of between 5 to 25 atmospheres excess pressure and at a temperature of about between 80° and 100°C.

The combustion chamber should preferably have constant height dimensions with a depth of at least 1.5 and at the most 5 meters. The combustion chamber is charged in quantities so as to correspond to a specific cross-sectional load of 3×10^6 to 5×10^6 kcal/m^2 hr.

A process developed by H.V. Hess et al (10) is a closed circuit process for treating waste liquors, such as spent sulfite liquors from wood digestion by coking the liquors in the liquid phase under pressure under turbulent flow conditions to form coke and a clear ef-fluent of low chemical oxygen demand. In the process, a heat exchange zone, a coking zone, a steam and flue gas producing zone, an absorbing zone, a wood digesting zone and a kilning zone are provided and interconnected by suitable conduits.

Pressurized spent liquor is sent to the heat exchange zone where it absorbs heat from the coking zone effluent before coke separation, thereby absorbing also heat content contained in the coke. The coke is discharged to the kilning zone where it is dried by hot flue gases produced by burning previously dried coke in the steam producing zone. Sulfur dioxide containing warm, flue gas from the kilning zone is contacted in an absorbing zone with cool coker effluent to produce warm ammonium bisulfite solution which is passed to a wood digestion zone as well as cool flue gas substantially free of all air pollutants.

Sulfur containing gases produced by coking and wood digestion are burned in the steam-producing zone to manufacture sulfur dioxide for recycling. Use of a closed system re-duces to a minimum odors in the digestion section of a pulp plant since all the odorous sulfur compounds are burned and the sulfur dioxide is recovered.

VAPORS FROM PULPING PROCESSES

Incineration

A process developed by C.S. Flynn (11) involves combusting waste fluids, particularly combustible vaporous and gaseous fluids as from kraft paper pulp processing, in a thorough and safe manner. Such fluids are passed at high velocity through manifold jets specially located adjacent special gaseous burners, into a hot gas flow zone in a stack for ignition and passage to the atmosphere in purified form. The processing kettle is preferably vented to the atmosphere through a kettle stack from which the fluids are drawn by suction to be forcefully blown through the manifold equipment and out a second stack.

A process developed by D.L. Brink et al (12) is a process involving the high temperature distillation and pyrolysis of an organic material such as kraft black liquor, formed during conventional chemical wood pulping, to break down the material to noncombustible solids and to a stable gaseous clean burning fuel. The process involves introduction of malodorous-containing gaseous emissions from the pulping operation as a source of oxygen for attaining the temperature necessary for the pyrolysis but insufficient to effect stoichiometric (complete) combustion of the organic material. These kraft mill malodorous emission gases are occasionally burned with inherent incomplete combustion, leaving significant amounts of obnoxious gases to pollute the atmosphere.

REFERENCES

(1) Booz-Allen Applied Research, Inc., *A Study of Hazardous Waste Materials, Hazardous Effects and Disposal Methods,* Vol. II, Report PB 221 466, Springfield, Va., Nat. Tech. Information Service (July 1973).
(2) G. Steinert; U.S. Patent 3,053,615; September 11, 1962; assigned to Kohlenscheidungs-GmbH.
(3) J.E. Greenawalt; U.S. Patent 3,163,495; December 29, 1964.
(4) V.P. Owens and G.J. Prohazka; U.S. Patent 3,703,919; November 28, 1972; assigned to Combustion Engineering, Inc.
(5) D.R. Sheeley, J.H. Rion, W.R. Cook and W.A. Biggs, Jr.; U.S. Patent 3,787,283; January 22, 1974; assigned to Sonoco Products Co.
(6) A.W. Reed, *Ocean Waste Disposal Practices,* Park Ridge, N.J., Noyes Data Corp. (1975).
(7) W.F. Franz, H.V. Hess and E.L. Cole; U.S. Patent 3,705,077; December 5, 1972; assigned to Texaco Inc.
(8) G.G. Copeland and J.E. Hanway, Jr.; U.S. Patent 3,309,262; March 14, 1967; assigned to Container Corp. of America.
(9) K.H. Wirths; U.S. Patent 3,618,571; Nov. 9, 1971; assigned to L & C Steinmuller GmbH.
(10) H.V. Hess and E.L. Cole; U.S. Patent 3,717,545; Feb. 20, 1973; assigned to Texaco, Inc.
(11) C.S. Flynn; U.S. Patent 3,497,308; February 24, 1970.
(12) D.L. Brink and J.F. Thomas; U.S. Patent 3,718,446; Feb. 27, 1973; assigned to The Regents of the University of California.

ROOFING INDUSTRY WASTES

ROOFING FACTORY FUMES AND SOLID WASTES

Incineration

A process developed by H.B. Johnson (1) is one in which the solid wastes of combustible and noncombustible roofing material are comminuted in an attrition mill, fed to a surge bin, and then used as fuel in a solid waste and fume incinerator which also receives and ecologically incinerates the exhaust fumes from a saturator. A fume incinerator also receives saturator exhaust fumes along with effluent gas from the oxidizer of the factory, and ecologically incinerates the same with a minimum consumption of conventional fuel. The fume ducts from the saturator exhausts to the incinerators include a cross duct and switching valves for selectively controlling the flow of the saturator exhaust fumes from one or both saturators to one or both incinerators for optimum operational efficiency of the system in accordance with current production of the roofing factory. Granules resulting from the solid scrap incineration are washed and recovered for reuse.

Figure 141 shows the sequence of operations in this process. Referring to the drawing, saturators **10** and **12**, and oxidizer **14** constitute components of a roofing factory which also includes a felt section **16**. A by-product of the factory is scrap, including tabs which may be termed solid waste that contains combustible material used in making the roofing products. The scrap from the factory is loaded on a platform **18** for movement by a conveyor **20** to an attrition mill **22**. The tabs **24** and **26** are also moved by conveyors **28** and **30** onto the conveyor **20** which, in turn, carries them to the mill **22**. The output of mill **22** goes to a surge or storage bin **32** on a conveyor **34**, and from bin **32** to a fuel input of incinerator **36** by a suitable conveyor **38**.

Fume exhaust blowers or fans **40** and **42** are associated with the saturators **10** and **12**, respectively, and are connected by ducts **44** and **46** to incinerators **48** and **36**, respectively. A cross duct **50** having a damper or valve **52**, is connected to the ducts **44** and **46** for selectively separating, or putting the latter ducts in communication with each other. The ducts **44** and **46** are also provided with switching valves **54** and **56**, and **58** and **60** at opposite sides of the cross duct **52** for selectively controlling the flow of exhaust fumes from either one or both saturators **10**, **12**, to either one or both of the incinerators **36**, **48**.

The effluent gas from the oxidizer **14** flows to the incinerator **46** through a duct **62**. The effluent of the incinerator **48** goes to a suitable stack or chimney by way of an air heater **64** from which the heated air flows by way of a duct **66** to the felt section **16** of the factory, providing auxiliary heat for the vapor absorption system of such felt section **16**.

FIGURE 141: ROOFING FACTORY FUME AND SOLID WASTE DISPOSAL SYSTEM

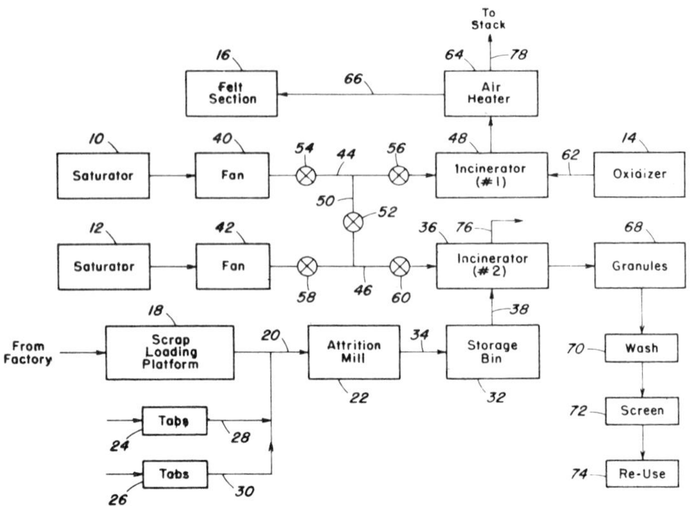

Source: U.S. Patent 3,662,695

Residual noncombustible granules produced as a by-product of the incinerator **36**, are collected at station **68**, then washed and screened at stations **70** and **72**, respectively, for reuse from a station **74** where they are finally collected for such purpose. Effluent from the incinerator **38** go directly to an individual stack from effluent outlet **76**. Such effluent as well as that of incinerator **48** going to its stack through outlet **78**, is ecologically clean due to complete combustion of the exhaust fumes, and the oxidizer effluent gas, as well as all of the combustible material contained in the solid wastes that are fed to the incinerators **36** and **48**. The residual granules of incinerator **36** are also ecologically clean even before washing by virtue of the operation of the complete incinerator **36**.

Briefly, the illustrated system of the process efficiently functions to dispose of the solid wastes in the form of various roofing scrap, and incinerates the saturator exhaust fumes, as well as the oxidizer effluent gas. The system takes full advantage of the fuel value of the roofing scrap, and also incorporates heat recovery to minimize the consumption of conventional fuel. The resulting stack effluent is ecologically acceptable, and the noncombustible solids are reclaimed.

Furthermore, the system is adjustable to handle any waste material that is available as a result of the particular roofing product being made at the time by the factory. Thus, such waste is taken care of efficiently and ecologically by the system regardless of such particular roofing product. Changeover to another type of roofing product requires merely the adjustment of the duct switching valves, as well as adjustment of conventional fuel and air (or oxygen) supplied to the incinerators to bring about complete combustion, and the resulting, clean stack effluent.

REFERENCE

(1) H.B. Johnson; U.S. Patent 3,662,695; May 16, 1972; assigned to GAF Corporation

RUBBER INDUSTRY WASTES

RUBBER PRODUCTION WASTES

Incineration

A simplified flowsheet for incineration of a flotation sludge containing rubber wastes is shown in Figure 142. In this example system, no information was obtainable for a typical quantity of wastes generated by plants in Ohio. Thus, capacity was arbitrarily selected at 10,000,000 pounds annually. A typical sludge, however, is known to contain 15.4% solids which have a heating value of about 15,000 Btu per pound of dry solids.

FIGURE 142: SIMPLIFIED FLOWSHEET OF INCINERATOR FOR RUBBER WASTES

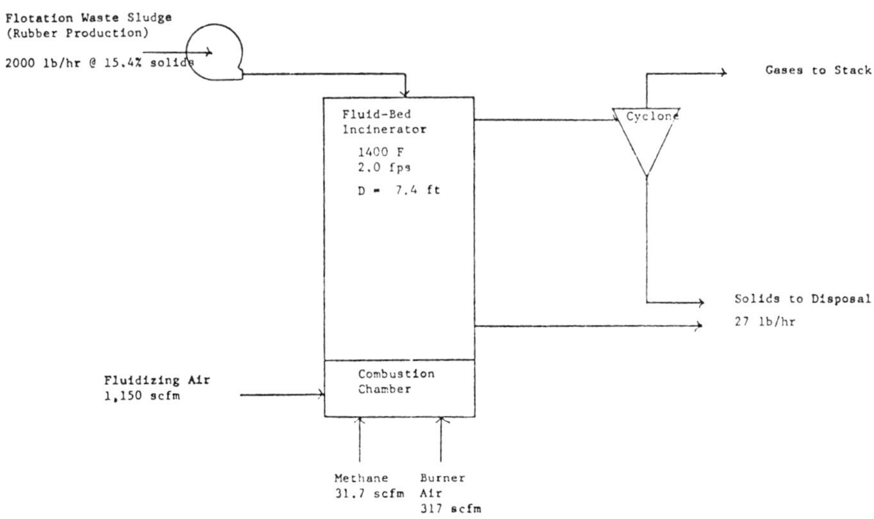

Source: Report 12/20 FYF 03/72

Operating conditions from experimental runs with this sludge would be bed temperature of 1400°F and superficial velocity of 2.0 fps. The results of design calculations indicate this incinerator would be about 7.5 feet in diameter with provision for burning about 32 scfm of methane in a combustion chamber.

USED RUBBER TIRES

Incineration

A small number of tires can be combined with large quantities of refuse in incineration as long as they do not comprise more than about 5% of the charge. Large percentages of tires result in damage to furnace walls and also require flue gas control. Tires represent only about 1.5% of the municipal waste stream but are often delivered to a disposal site in large batches rather than in an even day-to-day flow. Shredding of the tires would allow the percent of tires in the incinerator charge to be increased above 5% if the shredded rubber is thoroughly mixed with other waste materials.

Incineration of tires for steam generation in special furnaces has been examined by some of the major tire manufacturers and appears to be a promising disposal and recovery method. One such plant has been built. When in full operation, the plant will handle about 1 million tires per year. The incinerator is reported to be completely odorless and pollution free. The major drawback appears to be the high capital cost of the equipment.

A device developed by M.E.P. Hill (2) is a combustion apparatus particularly aimed to dispose of materials which are normally difficult to completely burn, such as used engine oil and used rubber tires. The apparatus has a refractory structure forming the combustion chamber into which the material to be burnt is fed. The material is subject to a current of forced air. Such a device may take the form shown in Figure 143.

FIGURE 143: FURNACE FOR INCINERATION OF USED TIRES

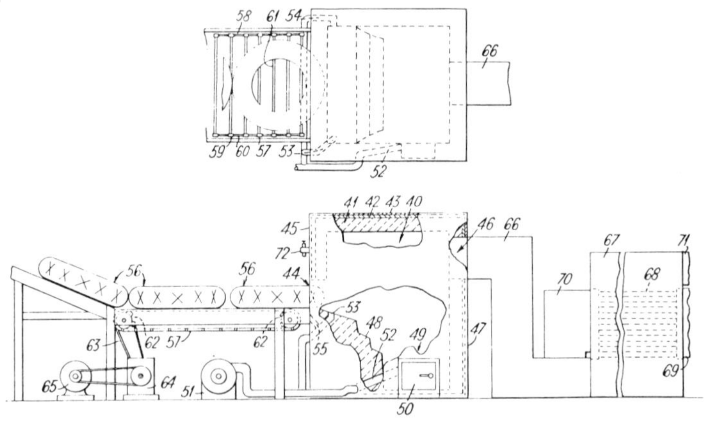

Source: U.S. Patent 3,565,021

The combustion chamber indicated at **40** is of a generally cylindrical shape and is formed of firebrick **41**, bearing externally a layer of insulation **42** consisting of vermiculite and cement fondu, and is housed in a metal casing **43**. The chamber is provided with an inlet **44** in the general form of a slot in an end wall **45** and with a flue exit **46** in the opposite end wall **47**. The floor of the chamber departs from the cylindrical form of the rest of the chamber and descends by steps **48** into a wall **49** constituting an ash pit, to which access is provided through a door **50** in the metal casing.

Forced air is provided by a compressor **51** through three inlet passages (shown in outline) in the combustion chamber, one passage **52** providing forced air in the ash pit, one passage **53** at the right hand side of the chamber as viewed from the inlet side of the installation, in the floor of the chamber directing forced air segmentally upwards therein, and one passage **54** in the left hand side of the chamber as viewed from the inlet side of the installation directing forced air horizontally across the chamber inlet so that immediately the leading part of a tire is received through the inlet slot **44** it is subjected to the current of air from passage **54**.

The inlet passage **53** is joined by a smaller passage **55** (shown in outline) through which water may be passed to be entrained in the forced air stream in inlet passage **53**. The tires, generally indicated at **56**, are fed, one after another, to and through the inlet slot **44** by an endless conveyor comprising two chains **57, 58** consisting of ordinary chain links **59** connected by metal straps **60** to which are secured bars or slats **61** to extend between the two chains; the chains being carried round sprockets **62** one of which is driven by a belt **63** from a reduction gear box **64** in turn driven by a belt drive from an electric motor **65**, to give a forward speed of the conveyor such that the tires are introduced at a rate equal to the rate at which they are burned.

The flue exit of the combustion chamber leads into a pipe **66** to conduct the effluent gases to a tubular heat exchanger **67** which receives material to be heated, by passage through the tubes **68** indicated in outline, through an inlet pipe **69**. The material passing through the tubes **68** is discharged through pipe **70** and the cooled effluent gases are discharged through pipe **71**. Water may be fed onto the tires before entry to the combustion chamber from the main supply through a tap **72** although if water is to be introduced with the tires any other means may be used for depositing water thereon.

In the operation of the above described tire-burning installation, combustion may readily be started by dousing the leading part of the first tire with an easily ignited fuel, e.g., petrol and igniting the fuel. The forced air supply to the combustion chamber not only provides a current of forced air for the tire as it enters the chamber but also gives rise to a turbulent or vortex movement of the gas stream in the combustion chamber thereby increasing the residence time of the gaseous materials and particles of the tire entrained therein.

As a result combustion is substantially complete by the time the combustion products are discharged from the chamber into the flue pipe **66**. In one experiment a combustion chamber was employed of dimensions of some 5 ft in axial length from the inner surface of one end wall to the like surface of the other end wall and some 1½ ft in internal radius giving a height of some 5 ft from the floor of the ash pit to the interior surface of the roof of the combustion chamber. Water was supplied at a rate of 12 gal/hr mostly by deposition on and within the tires and the compressor supplied air at approximately 3,200 ft³/min under a pressure of some 2¾ psi gauge pressure to provide approximately one-third this amount in each inlet passage. Tires were burned at about 720 lb/hr to give an effluent temperature of from 800° to 850°C of the combustion gases effluent from the combustion chamber and about 100°C of the combustion gases effluent from the heat exchanger.

A device developed by J. Brion (3) is an incinerator designed to burn whole tires representing 100% of the charge. A hearth which is smaller in width than the diameter of the tires to be burned has longitudinal vertical walls provided with lateral orifices for the admission of combustion air. A charging lock-chamber having internal dimensions slightly larger than the dimensions of a vertical tire has a sloping floor along which the tire rolls into the hearth and remains vertical. The combustion chamber of the incinerator is fitted with nozzles for the injection of air in order to induce high turbulence.

A device developed by K.W. Stookey (4) is a device for thermally reducing rubber or plastic materials in which the rubber or plastic material is comminuted to a predetermined size and it is then subjected to primary combustion in the lower portion of a first chamber with the gases and distilled combustible vapors in the upper portion of the first chamber being subjected to a secondary combustion and with the still only partially oxidized gases then being conducted to a second chamber and admixed with further air for tertiary combustion and the combustion thereof completed and with these last hot gases being used as the heat input for a boiler, such as a waste heat boiler. Automatic controls are provided to control the temperature in the upper portion of the first mentioned chamber to within predetermined limits and a further control is provided to insure substantially complete combustion in the second mentioned chamber, and with a bare minimum of excess air.

A process developed by G. Crane et al (5) is one in which solid waste vulcanized rubber, such as scrap tires, is heat treated to produce a fluid material which, in turn, is burned to produce heat energy and ash. Zinc and titanium are recovered from the ash.

Landfill Disposal

Tires are one of the most difficult consumer wastes to dispose of properly. In sanitary landfills, whole tires cannot be effectively compacted and tend to work their way up to the landfill surface. They also consume more landfill space per unit weight than other items and are not biodegradable (6). Tires on the surface of landfills or dumps or littered in urban or rural areas provide nesting places for rodents, flies and mosquitoes.

Shredding, splitting, or otherwise reducing the size of tires overcomes the problem of land filling to a large degree. Shredding is the most feasible method of physically altering the tires but is presently employed only in a few locations. Cost is the primary deterrent, and neither public nor private entities are likely to pay the price of processing in the absence of disposal regulations.

Pyrolysis

Destructive distillation, carbonization and hydrogenization are thermal conversion processes that may be used to recover the chemical constituents of tires. The first two processes are both forms of pyrolysis, i.e., thermal decomposition in a low oxygen atmosphere; hydrogenization is a process of chemical synthesis involving addition of hydrogen. The primary product of carbonization (high-temperature destructive distillation) is carbon black, a major tire raw material. Lower temperature destructive distillation yields a mixture of oils, gases and carbon residue with significant fuel value. Hydrogenization yields hydrocarbon products that can be used in tire manufacturing.

Pilot-plant-scale carbon black processes utilizing scrap rubber have been tested by some of the major rubber companies. It was reported that the processes are feasible but economically unattractive at present, with costs of about three to four times that of carbon black production by conventional means from petroleum. Destructive distillation of tires has been examined by the Bureau of Mines in conjunction with a major rubber company. The products consist primarily of heavy oils, light oils, gases and char.

It has been reported that the value of these products from 100 pounds of tires is approximately $1.50 (i.e., $30 per ton of tires processed). It has been further suggested that even at this value the process is uneconomical if shipping, handling, storage and preparation are included. Hydrogenation of the products of destructive distillation has been examined by Hydrocarbon Research Inc. at the pilot plant scale, and preliminary projections were made that such a plant could operate at a profit if tires could be obtained for $5/ton (6).

A process developed by C.E. Scott (7) is one in which waste vulcanized rubber is depolymerized in a liquid hydrocarbon medium by means of agitation, heat, free-radical initiators and molecular oxygen. The resulting solution of rubber-modified hydrocarbon can then be utilized in the manufacture of end-use products, thereby disposing of the waste rubber without pollution of the ecological environment. The rubber-modified hydrocarbon can, e.g., be vulcanized to produce a moisture barrier or insulation coating. It can also be mixed with asphalt to provide rubberized asphaltic compositions. In still another application, the rubber-modified hydrocarbon can be thermally decomposed to produce carbon black.

REFERENCES

(1) Tewksbury, A.K. Reed, H. Nack and G.R. Smithson, Jr., *Fluidized-Bed Incineration of Selected Carbonaceous Industrial Wastes,* Report 12/20 FYF 03/72, Wash., D.C., U.S. Environmental Protection Agency (March 1972).
(2) M.E.P. Hill; U.S. Patent 3,565,021; February 23, 1971.
(3) J. Brion; U.S. Patent 3,779,183; December 18, 1973; assigned to Commissariat a l'Energie Atomique, France
(4) K.W. Stookey; U.S. Patent 3,785,304; January 15, 1974.
(5) G. Crane, E.L. Kay and J.R. Laman; U.S. Patent 3,890,141; June 17, 1975; assigned to the Firestone Tire and Rubber Co.
(6) M. Sittig, *Pollution Control in the Plastics and Rubber Industry,* Park Ridge, N.J., Noyes Data Corp. (1975).
(7) C.E. Scott; U.S. Patent 3,700,615; October 24, 1972; assigned to Cities Service Co.

TEXTILE INDUSTRY WASTES

TEXTILE FINISHING EFFLUENTS

Incineration

A process developed by E.S. Monroe, Jr. (1) is one designed to dispose of the finish oils utilized in textile yarn manufacture, which are relatively high-boiling materials and also very stable and retentive, so that they survive even severe chemical treatments aimed at their destruction. Inevitably, finish oils find their way into waste collection sewers, heavily diluted with washdown water, so that they cannot be restored to the process and constitute a waste which has to be disposed of.

Attempts have been made to incinerate wastes such as dilute finish oils by introducing them into combustion flames in regulated amounts, but these have not always been successful, the chilling effect of the dilute aqueous waste frequently extinguishing the flame, or the organics being only partially burned to heavy soots or being incompletely vaporized, or burned, so that there is objectionable deposition on furnace walls, and other problems. Moreover, such combustion disposition processes have been exceedingly expensive, since approximately 5,000 Btu of fuel have hitherto been required for each pound of aqueous waste.

This present process provides a highly economical and exceedingly effective process and apparatus for disposing of heavily diluted organic wastes, which accomplishes substantially complete destruction of the organics to final stable inoffensive equilibrium products, such as CO_2 and water, without accompanying soot, deposition or stack plume formation. The process operates by an extremely efficient vaporization of the waste stream projected as a countercurrent spray droplet envelope predominantly enclosing the burner flame. However, a substantial portion of the liquid waste can be sprayed head on into the flame front while still obtaining the benefits of this process.

Under these circumstances, the dilution water is completely vaporized and a substantial portion of the organic waste is burned in situ, to thereby supply a considerable portion of the heat required for continued vaporization. The uncombusted remainder of the organics is vaporized and preheated to an optimum temperature enabling substantially complete catalytic oxidation to inoffensive equilibrium products as an immediately following step. Analysis of effluent products has shown that complete waste disposal is effected, and the heat consumption required has been as low as 1,300 Btu/lb of finish oil waste having a water dilution as great as 90%.

Figure 144 shows a suitable form of apparatus for the conduct of the present process. A wide variety of fuels can be employed, such as propane, natural gas or oil; however, the burner **10** is preferably one which produces an elongated flame and which can function with a range of 0 to 1,500% excess air, or more, in order to safeguard the catalyst against excessive temperatures. Actually, conventional burners operating on lower excess air-to-fuel ratios of 0 to 200% can be made to work, except that it is then usually necessary to provide bypasses around the catalyst beds to protect them from overheating when the apparatus is started up.

In a typical installation, using natural gas as the fuel, a North American Model 223 burner with pilot and flame scanner proved entirely suitable. The burner is preferably housed at one end of a horizontal cylindrical firebrick furnace chamber **11** and the water-diluted waste stream is introduced from the opposite end of the chamber as a countercurrent frustoconical spray envelope **12** preferably enclosing the flame cone **13** but not intermixing with it. Thus, the integrity of the flame is not disturbed, and quenching troubles are avoided. (A considerable amount of waste spray received head on into the outboard end of flame cone **13** can usually be tolerated without bad results, and therefore the design of waste spray nozzle is not particularly critical.)

FIGURE 144: INCINERATOR FOR FINISH OILS FROM TEXTILE YARN MANU-FACTURE

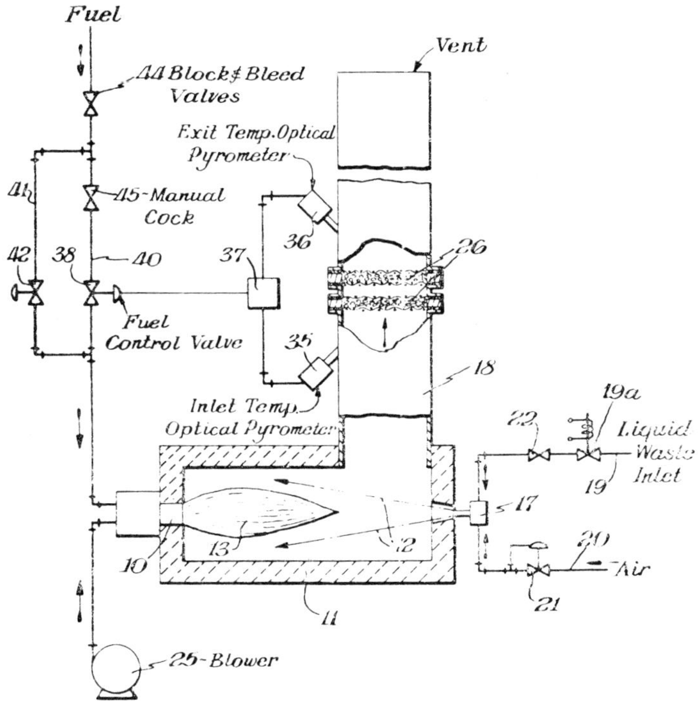

Source: U.S. Patent 3,611,954

The liquid waste is sprayed into chamber **11** through a conventional nozzle **17**, which can typically be a pneumatic atomizing type; however, mechanical or sonic designs are equally operable. In one series of tests, a pneumatic atomizer was employed, producing a round spray pattern in conical shape by external mixing of air and liquid waste. This atomizer had a small concentric annular air supply jet encircling a central liquid waste supply orifice, thereby producing a fine mist of droplets having diameters of 40 microns or less. This nozzle produced an envelope of quite consistent form and unit waste concentration, although, of course, some waste liquid droplets were incidentally deflected into the flame cone without, however, any detectable interference with operation.

It will be understood that rotary cup or pressure sprayers of a wide variety of designs can alternatively be utilized for the liquid waste introduction, and other atomizing media, such as steam, nitrogen or other gaseous phase substances can be substituted for air if the particular characteristics of the waste necessitate.

The spray cone encloses flame 13 as a near-tangent envelope 12, so that dilution water is speedily vaporized by the flame combustion products which exhaust, together with organic material surviving the combustion, out through vertical stack 18.

As shown, the organic waste to be disposed of is supplied under pressure to nozzle 17 through a line 19 and atomization air is supplied through a separate line 20 at a typical pressure of 23 psig, maintained by air pressure regulator 21.

Additional air required for the catalytic oxidation hereinafter described is furnished by blower 25, which, as hereinbefore described, supplies air to burner 10 in excess of that needed for combustion. If necessary, yet another air supply line can be fitted to furnace chamber 11.

The hot exhaust vapors and gases drawn off via stack 18 are subjected to catalytic oxidation to carbon dioxide and water. The catalytic oxidation is achieved by passing the exhaust through one or, preferably, several series-arranged, gas-permeable catalyst trays 26 at temperatures in the range of about 100° to 250°C. Trays 26, which may be about 1½ inches thick, each containing about 200 lb of oxidation catalyst, are loaded with platinum or palladium metal catalysts carried on a calcined support presenting a surface area in excess of about 100 m^2/g, such as alumina or silica granules, or a ceramic honeycomb. These catalysts are very effective in achieving substantially complete oxidation of organic material surviving exposure to the preceding combustion. Also, the catalysts are capable of treating gases at relatively high space velocities in the range of about 25,000 to 75,000 cubic feet of gas at standard condition per hour per cubic foot of the catalyst composition.

The oxidation catalyst in trays 26 has a relatively long service life, usually measured in months, depending, of course, upon the nature of the waste disposed of. However, after protracted operation, a progressive decline in organics elimination efficacy occurs, finally evincing itself in the appearance of fumes in the exhaust venting stack 18. At this point, either new trays 26 can be slipped in in substitution for the vitiated catalyst, or the apparatus may be simply taken out of waste disposal service for a long enough time, usually only several hours, to increase the flame and available air supply of burner 10 to burn off deposits collected on the catalyst.

Alternatively, the trays 26 can be removed and treated in a furnace wherein a two-step process of [1] steaming for 1 hour at about 475°C followed by [2] introducing air and conducting a deposits burnoff is effective in regenerating. Actually, one tray 26 has been found to have an 80 to 85% system effectiveness in waste elimination while the other tray is removed for regeneration, so that continuous waste disposal operation is entirely practicable with this method.

It is preferred to incorporate a temperature control safeguarding the oxidation catalyst in trays 26 against heat damage, and this is readily achieved by installing temperature sensors such as optical pyrometers 35 and 36 ahead of and immediately following the last tray 26, respectively. An optimum temperature for a Pt or Pd oxidation catalyst has been found to be in the range 300° to 500°F; therefore, pyrometer 35 is reserved to this

monitoring service by continuous signal output of a DC current of approximately 5 mv
to signal selector 37. Optical pyrometer 36 is adapted to measure the exit temperature of
the exhaust from the second catalyst tray 26 and to override the normal control imposed
by optical pyrometer 35 whenever this temperature reaches 800°F or higher. A conventional throttling valve 38 responsive to the output of signal selector 37 is mounted in the
main gas supply line 40, valve 38, under normal conditions, continuously increasing or decreasing gas flow to burner 10 to maintain the upside catalyst temperature within the
range of 300° to 500°F, but shutting off, as necessary, responsive to the override imposed
by optical pyrometer 36 whenever the upside stack temperature approaches 800°F.

There is an interlock control connection (not shown) to solenoid valve 19a in the liquid
waste input which cuts off waste supply whenever throttling valve 38 closes completely.
Since burner 10 is fitted with a pilot burner, there is automatic resumption of operation
after an emergency gas shutoff following normal temperature restoration. (Provision can
be readily made for automatic restoration of waste liquid supply after an outage, but,
ordinarily, manual restoration by an operator is preferred in order to verify the cause of
an unscheduled outage.)

It has been found advantageous to provide a low-fire startup auxiliary to facilitate startup
when the furnace is cold and before a steady-state combustion environment is attained.
This simply comprises a bypass line 41 around throttling valve 38 in series connection
with a pressure regulating valve 42. Manual valves 44 and 45 are provided in the main gas
supply line 40 ahead of pressure regulating valve 42 and throttling valve 38, respectively.
Thus, when valve 44 is opened and valve 45 is closed, gas is supplied at a sufficient rate
to maintain startup low-fire operation of burner 10.

During this period valve 22 in line 19 is closed so that waste liquid is not being introduced
to furnace chamber 11. Thereafter, when optical pyrometer 35 senses that the upside
catalyst tray 26 temperature is in the 300° to 500°F temperature range, it can be switched
into control by simply opening valve 45, after which waste liquid supply can be commenced
by opening valve 22.

A process developed by H. Fernandes et al (2) involves the removal of oxidizable fumes
from heated gases used for the treatment of temperature responsive yarns carrying an
oxidizable finish containing organic components.

In many textile processing operations, it is necessary to apply a finish to the fiber, particularly synthetic fibers, in order to improve the running characteristics during the processing in subsequent applications. Frequently, in the processing of synthetic fibers, it is
necessary to subject the fibers to a heat treatment in order to obtain or improve upon
the desirable properties of the fiber. Such heat treatments tend to volatize the finish,
which may be an organic material dispersed in water or may be a combination of volatile
oils and waxes, which escapes as a fume commonly referred to as "smoking." Such fumes
escaping into the atmosphere not only may be irritating to the workers in the area but
frequently condense on operating machinery causing problems in cleanliness, housekeep-
ing, potential flammable conditions and even degradation of the quality of the textile product being produced.

Many attempts have been made to eliminate the potential problems by the choice of formulation of the finish to avoid "smoking" components. However, it is not always possible
to arrive at a suitable compromise between adequate finish requirements and reduction of
"smoking" components. Other attempts to minimize the problems have included: extra-
ordinary measures to make process equipment air tight to prevent escape of fumes; more
sophisticated ventilation systems with increased capacity to exhaust fumes which may then
require scrubbing and taller stacks or other conventional means for minimizing the potential for air pollution. Frequently, increased ventilation reduces the heating efficiency of
the heat treatment process, thereby requiring excess amounts of energy in order to provide the required degree of heat treatment.

While catalytic oxidation of fumes and organic vapors has been known in other fields, it has been necessary to operate such systems at high temperatures in order for the catalytic action to be effective. Such temperatures are generally about 300°C or higher which are entirely unsuitable for synthetic textile heat treatments which generally require temperatures below about 250°C. The use of high temperature catalytic systems known in the prior art would require extensive cooling systems and an additional cooling step in a yarn heat treating process which is sensitive to temperature change and in which it is difficult to maintain the exact temperature conditions for proper heat treatment of the textile fibers.

Figure 145 is a simplified schematic diagram of the present process which overcomes many of the problems of earlier devices. It is seen that a thread line **10** passes through a heating chamber **2**. Hot gas heated by heat exchanger **3** is circulated by a blower **4**.

FIGURE 145: GENERAL OUTLINE OF PROCESS FOR INCINERATION OF TEXTILE FINISHING FUMES

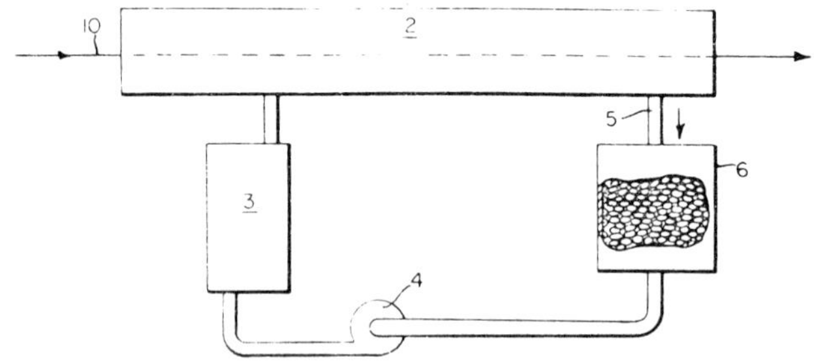

Source: U.S. Patent 3,725,532

The hot gas volatilizes a portion of the finish on thread line **1** and is exhausted through duct **5** connected to the chamber **6** containing suitable catalytic material and thence to the intake of blower **4** for recirculation. The catalytic material can be held in a wire-screen basket. A wire mesh canister in the shape of a truncated annular cone has been found to be quite efficient.

A more detailed design of apparatus which may be used to carry out the present process is shown in Figure 146. Two rotatably driven rolls **12, 14** are positioned inside housing **32**. Openings **34, 36** are provided for entry and exit of yarn **10**. A blower **40**, a heat exchanger unit **62** and a series of passages are arranged to form a substantially closed circuit which directs hot air from the heat exchanger **62** into the interior of the roll elements **12** and **14** through inner concentric tubular elements **47**, thence along the exterior of the roll elements and back to the inlet of blower **40**. A wire-screen basket **70** containing catalytic material is positioned in the inlet line **72** to blower **40**.

An access door **52** pivotally mounted at hinge **61** is provided for housing **32**. When access door **52** is opened, damper **50** which is linked to and actuated by door movement through link **56** moves toward its closed position and acts to prevent cold air from being pulled in by blower **40**. In addition, when door **42** is opened, air inlet **54** is opened by movement of the upper portion of door **52** above hinge **61**.

To prevent any outside cool air currents from entering through openings **34** or **36** or any other opening that may exist, the housing **32** is kept under a slight positive pressure. This is done by having a small air inlet opening **48** near the suction of blower **40**.

FIGURE 146: DETAIL OF FURNACE FOR INCINERATION OF TEXTILE FINISHING FUMES

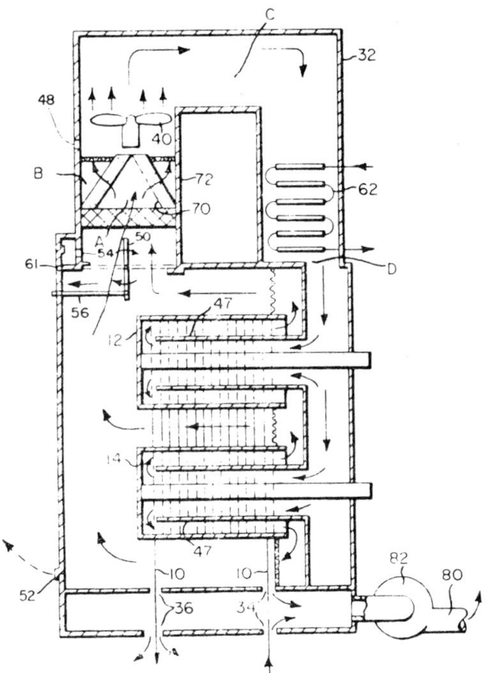

Source: U.S. Patent 3,725,532

Coil **62** is a condensing vapor heating means, wherein the temperature at which vapor in the coil condenses if held constant by holding coil pressure at a constant level. A variation in the heating demands on coil **62** (represented by a rise or fall of the input temperature of the air to be heated) merely affects the amount of vapor which condenses in the coil which varies the amount of heat (of condensation) given out by the heating means.

By the continuous catalytic oxidation of the finish fumes near the source of the fumes and at the temperature of operation of the heat treatment process the accumulation and decomposition of the volatilized finish is prevented. Without using the catalytic treatment, finish accumulates in the gaseous medium and the apparatus, necessitating frequent cleaning of the apparatus to remove the coating of finish on the apparatus parts. Without such cleaning sufficient finish may collect and drip on the textile product being produced. Such drips cause nonuniform spots on the product which often lead to streaks and other dyeing defects in articles produced from the textile product. Buildup of finish also leads to partial decomposition of the finish which forms a tacky coating on the apparatus and unsatisfactory operation of the apparatus.

In operation, the flow rate of the air is maintained at such a high level and the area of the coil **62** is such a large size that the desired temperature of heat transfer is maintained very accurately within small limits despite the exothermic reaction of the catalyst in basket **70**.

A process developed by F.S. Whitfield (3) is concerned with the disposal of liquid textile wastes. Broadly stated the method comprises: (a) heating at the temperature of its boiling point at atmospheric pressure a liquid waste containing combustible carbonaceous material and water, said liquid being substantially free of substances yielding solid residue; (b) simultaneously pressuring said liquid to maintain it in liquid state; (c) spraying into an open fire said heated liquid waste kept under pressure to flash out the liquid, thereby evaporating the water from said liquid and burning the combustible carbonaceous material.

There is also provided a relatively simple incinerator for carrying out this method comprising: (a) an open fire furnace provided with a spraying nozzle, said nozzle spaced in relationship to said open fire; (b) a means to heat and pumping means to feed under pressure into said spraying nozzle a heated liquid waste of carbonaceous material containing water and yielding substantially no solid residue, to evaporate said liquid and to burn said carbonaceous material.

Figure 147 shows a suitable form of apparatus for the conduct of such a process. The incinerator comprises a furnace partly shown at **10** provided with at least one burner having a nozzle **12** generating an open flame **13**. The burner is coupled with auxiliaries not shown. A spray jet **14** is spaced in relationship to the open flame **13**. The jet **14** sprays a liquid waste received from the line **16** via pump **18**. Generally the liquid waste is stored in tank **20**. Before pumping it into line **16** it is preheated to a temperature above its boiling point at atmospheric pressure, for instance, by means of a heat exchanger **19**. The pressure exerted by pump **18** must be such as to maintain the waste in liquid state until it reaches the spray jet **14**. Generally the pressure exerted on the waste may be in the order of 50 lbs/inch^2 to 60 lbs/inch^2.

FIGURE 147: APPARATUS FOR LIQUID TEXTILE WASTE INCINERATION AND
HEAT GENERATION

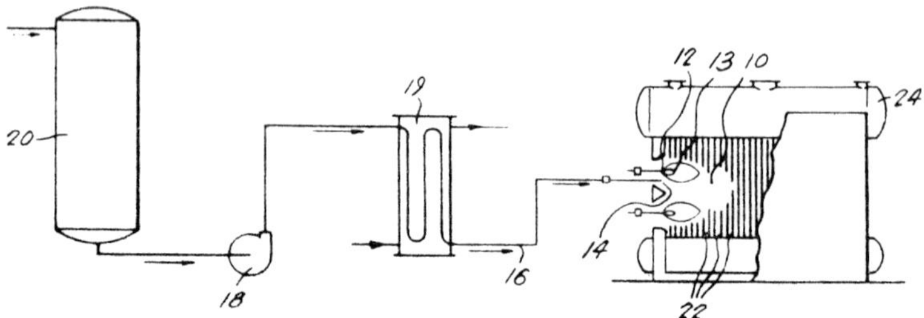

Source: U.S. Patent 3,734,035

On leaving the spray jet **14**, the liquid waste is sprayed into the fire or flame and the liquid is flashed out, thereby evaporating the water from said liquid waste while burning the combustible carbonaceous material. Thus the water content is converted into vapor at high temperature. This vapor together with the products of combustion circulate around a plurality of running pipes **22** and boiler **24** giving up most of their heat contents, then pass through the boiler stack to finally emerge to the atmosphere. These running pipes coupled with boilers are used to supply steam to the required units not shown. The furnace is generally provided with auxiliaries (not shown) such as manual or automatic controls for air intake for the furnace to compensate to suit.

In a typical example of the operation of such a process, a mixture of liquid carbonaceous waste was preheated to its boiling point at atmospheric pressure (approximately 212°F) and kept under 50 psi. The mixture of liquid carbonaceous waste was approximately: (a) about 7 parts alcohol still washing containing approximately 80% methanol, 10% glycol and 10% water, and (b) 22 parts of spin finish waste containing approximately 2% oil and 98% water.

The mixture was fed to a fine spray nozzle or jet located adjacent to the oil jets in the fire box of a boiler. The distances between the jet **14** and nozzle **12** were about one-half foot each respectively vertically center to center and horizontally face of spray to face of nozzle.

The pressure and delivery rates were arrived at by visual observation. Under these conditions, complete combustion was obtained and confirmed by analysis of the flue gas. The spray jet was provided with an automatic shut-off of the liquid in the event of a flame outage of the boiler. An approximate value of about 700 Btu/lb of the mixture was obtained. The boiler was rated at 30,000 lb/hr at 250 lb/inch^2. When incinerator equipment was in use at a rate of about 137 imperial gph at 50 lb/inch^2 delivery pressure, a minimum of approximately 15,000 lb/hr at 250 lb/inch^2 per boiler was obtained.

The incineration temperature was 2000° to 2200°F. The brick stack was 125'0" high above grade, 4'6" inside diameter. The flue gas temperature of the boiler breeching was approximately 500°F and at the top of the stack approximately 450°F. Flue gas by test indicated unburned combustible nil and approximately 3% excess oxygen.

REFERENCES

(1) E.S. Monroe, Jr.; U.S. Patent 3,611,954; October 12, 1971; assigned to E.I. du Pont de Nemours and Co.
(2) H. Fernandes and W.H. Walsh; U.S. Patent 3,725,532; April 3, 1973; assigned to E.I. du Pont de Nemours and Co.
(3) F.S. Whitfield; U.S. Patent 3,734,035; May 22, 1973; assigned to Millhaven Fibres, Ltd., Canada.

WOOD INDUSTRY WASTES

WOOD AND WOOD VENEER DRYER EFFLUENTS

Incineration

A process developed by A.J. Jones (1) is one in which combustion gases from a waste wood fired furnace are circulated through a wood veneer dryer or lumber dry kiln and returned to a mixing chamber for exposure to the hot gases leaving the furnace, causing incineration of volatile hydrocarbons and wood fiber picked up by the gas stream in its passage over the veneer or lumber. The only exhaust from the system is located between the outlet from the high temperature mixing chamber and the gas inlet to the dryer. This arrangement insures the incineration of all combustibles before they are vented to atmosphere or returned to the dryer so that no pollutants are discharged into the atmosphere and no potentially explosive substances are passed into the dryer.

A process developed by R.B. Burden, Jr. et al (2) is one in which heat generated within an incinerator is transferred to oil in a combustion gas-to-oil heat exchanger. The hot oil is circulated through a radiator in a material dryer to heat air used as the drying medium. The hot air passing in contact with the material to be dried picks up particulate solids and gases which heretofore have been exhausted to atmosphere. These gas-borne waste emissions are confined and conducted to the combustion chamber of the incinerator where they are burned, thus becoming a source of heat for the dryer.

A process developed by A.B. Baardson (3) is one in which waste wood, as well as the fumes emitted by veneer being dried, are burned in such a manner as to basically accomplish two things, namely, provide large amounts of heat energy that was formerly wasted and, secondly, substantially reduce the emission of pollutants into the atmosphere. The system finds particular application to the lumber mill industry.

Another process developed by A.B. Baardson (4) is one in which waste wood and the fumes emitted by drying wood can effectively be burned to not only provide additional heat energy but also to reduce atmospheric pollution. The burner basically comprises two chambers. The wood, fed in powdery form, is burned in the first chamber, the product of this combustion being passed to the second chamber where it is mixed with the fumes before being combusted a second time. A number of innovations makes this burner highly effective and useful in the lumber mill industry.

SAWDUST AND SCRAP WOOD

Incineration

Incineration of wood wastes and bark residues has generally been carried out by burning the wastes in conical or wigwam burners. The typical wigwam burner is a metal enclosure in the form of a truncated cone wherein wood wastes or bark residues are conveyed inside the burner and dropped onto a burning pile. Air is supplied to the combustion zone of the burner through a grate structure in the base of the burner. For the most part these wigwam burners have operated with little or no control and as a result the combustion efficiency of the burners has been very low. This has resulted in excessive smoke and particulate emissions neither of which is acceptable under many of the recently enacted state and federal air pollution regulations.

A recent study (5) describes the results of a two year study of means of modifying wigwam burners with particular attention to the prevention of air pollution. Modifications described by the study included a three-zone forced draft underfire system, a forced overfire air system with adjustable nozzles, natural draft vents, hot-gas receiving ducts, variable speed fuel conveyors and provision for adding auxiliary fuel. A temperature sensing system, such as described in U.S. Patent 3,472,184 for sensing the temperature of the exit gases and controlling the amount of air to the underfire air system was also evaluated.

The emission standards for particulate matter in combustion of wood wastes and bark residues in many of the states where wigwam burners are used specify that the burners operate with particulate emissions of 0.2 grains per cubic foot or less and a smoke capacity of Ringelmann No. 2 or less. One of the main problems in attaining these standards, particularly the standard relating to smoke capacity, has been the fact that most wigwam burners have operated at low temperatures with resultant low combustion efficiency and excessive smoke.

It has been determined according to the study mentioned previously that operation of a wigwam burner with exit-gas temperatures in the range of 700° to 900°F results in minimum emissions of particulates, smoke and air pollutants. It has remained a problem, however, to design a practical burner capable of efficiently combusting wet wood wastes and bark residues so that the exiting combustion gases exit the burner at a temperature of 700° to 900°F considering the variable fuel input generally necessary.

Such a burner has been developed by M.J. Leman (6) and is shown in Figure 148. The refuse burner comprises in general a conical or cylindrical burner **10** made up of a plurality of steel plates fabricated into the form of a cone, a grate system, a forced air underfire system **30** for directing primary combustion air under pressure into the interior of the burner through the grate structures of the grate system, a forced air overfire and recirculation system **70** for withdrawing hot gases from inside the burner, mixing the hot gases with ambient air and reinjecting the mixed gases into the interior of the burner, a temperature sensing means **50** operatively connected to a controller which is in turn connected to dampers of the overfire and recirculation air system, and a variable speed conveyor **60** for conveying the material to be burned into the interior of the burner.

The shell **10** of the burner is of conventional construction and is generally formed of heavy gauge sheet metal plates **11** formed with ribs **12**, the plates secured to a rigid frame structure. Exterior circular rings **13** assist in supporting the burner shell. Observation ports **16** providing visual access to the interior of the burner may be provided at spaced intervals around the shell so that an operator can visually determine the location and size of the burning pile of combustible wastes.

An access door to the interior of the burner is also provided for cleaning and maintenance. Because a lining of heat insulating material, preferably a refractory lining, is applied to the inner surface of the burner, the strength of the burner shell should be stronger than those conventionally used because of the added weight.

FIGURE 148: WIGWAM BURNER DESIGN FOR WOOD WASTE AND BARK RESIDUE DISPOSAL

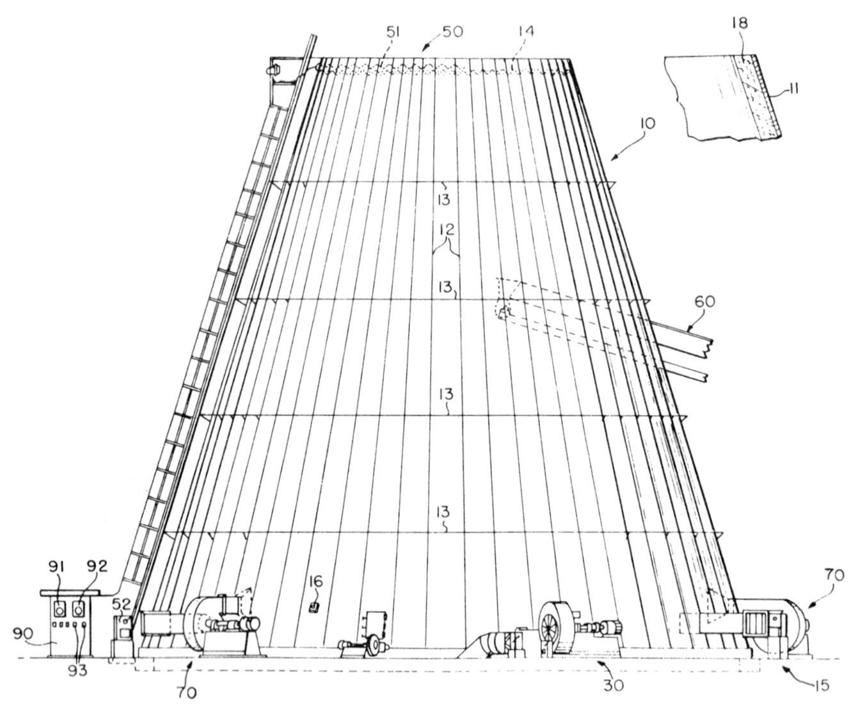

Source: U.S. Patent 3,669,039

A screen **14** covers the exhaust opening at the top of the burner to catch particulate matter which may become entrained in the exhaust gases. As shown in detail, a lining **18** of an insulating material is applied to the inner surface of the burner shell. The insulating lining may be applied to the inner shell walls of new or existing refuse burners. The lining serves to retard heat loss through the shell wall of the burner and to prevent leakage of outside air into the interior of the burner through openings between the plates, a problem encountered with many older burners.

The lining is applied to the overall inner surface of the shell wall or to the lower portion of the shell wall adjacent the burning zone. The lining is preferably a refractory lining conventionally applied after fabrication and erection of the shell wall or before. Gunite grade castable refractory materials are suitable. The refractory material integral with the shell wall on the lower portion thereof is subjected to higher temperatures than is the shell wall above the burning zone.

For this reason an insulating material of lesser quality can be applied to the upper portion of the shell wall, for example, calcium silicate or a thermo-asbestos type material. The lower portion of the lining, however, should be a high grade refractory material. An additional advantage obtained by providing an insulating lining on the inner surface of the shell wall is that of operating at higher temperatures than is normally possible. Most refuse burners of the conical type are operated so that the exit gas temperatures are in the range of 600° to 1000°F.

With the insulating lining this refuse burner can operate at exit gas temperatures of 1200°F or greater without affecting the structural integrity of the burner shell and without severe oxidation of the inner surface of the steel plates. Generally refractory linings are applied through and over a wire mesh held in spaced relation from the inner surface of the shell by studs at spaced intervals. The thickness of the lining can be as needed and may be varied from top to bottom, the thicker lining applied to the lower portion of the shell. The shell of the burner is rested on a base **15** of concrete or other suitable material. The waste material to be burned is conveyed into the interior of the burner by suitable conveying means **60** as shown. The wood wastes or bark residues are dropped in a pile on the floor of the interior of the burner, the floor incorporating a grate system through which air is supplied for combustion.

The figure also shows a control panel **90** having a temperature indicator-recorder **91** operatively connected to the temperature sensing means **51**, and a recording smoke density bolometer **92** for recording the density of smoke emission. Hand-operated controls **93** are also provided for operating the motor-operated damper controls. Conventional controls are also provided for starting and stopping the conveyor **60**, the underfire air fans and the overfire and recirculation air fans. The motor-operated damper controls, the temperature responsive control means and the electric circuitry needed for connecting them are within the skill of the art.

The control system **50** is connected so that the temperature responsive control means causes actuation of the motor-operated damper controls of each of the overfire and recirculation air systems to maintain the exit gas temperature, sensed by temperature sensing means **51**, within a predetermined range. If desired, the ratio of overfire to recirculated air may be controlled manually instead of automatically.

In operation, the fuel, consisting of wet or damp wood wastes and bark residue, feeds into the interior of the burner by conveyor **60** and drops in a conical pile on top of the grate system **20**. The fire is started manually and at the same time air is supplied to the underfire air system through the central grates of the grate structure by the blower. Air through the remaining grates is added as the fire builds up. As the temperature rises the temperature sensor **51** records it on recorder **91**.

An operator or automatic controller adjusts the ratio of hot recirculated gases to ambient air injected into the burner by the overfire and recirculation air system. The refractory lining or other heat insulating material applied to the inner surface of the shell of the burner acts to substantially retard heat loss through the walls of the burner and thereby aids in attaining an overall higher combustion temperature, this higher temperature resulting in greater combustion efficiency, less particulate emission and less smoke emission. Use of the insulating lining also allows greater latitude in sizing of the burner.

When the temperature reaches a predetermined set point, for example, 850°F, the controller responsive to the temperature sensing means **51**, a thermocouple, gas filled tube, etc. adjusts the dampers and varies the ratio of recirculated to ambient air through the overfire and recirculated air system to maintain the temperature at the desired set point. The net result of the combination of heat-insulating lining, overfire air system and underfire air system with an independently controlled air supply to the grate system results in a burner having greater combustion efficiency with less smoke and particulate emission. The burner is capable of meeting existing as well as forecast pollution regulations in the states where these burners are most utilized.

A system developed by R.E. Stone et al (7) is a control system for installation on existing waste burners at mill sites for the purpose of reducing emissions from the wood waste burned. A damper structure includes downwardly swingable doors positioned in response to the burner internal temperature. An underfire duct system supports combustion at the base of the waste pile while overfire blowers feed air about the perimeter of the pile in a cyclonic manner. Igniter units located about the burner shell retain burner temperature above a preset level by intermittent operation.

Control means operates the systems components at preset temperatures to insure highly efficient combustion within the burner to reduce smoke and particulate emissions to a level acceptable to regulating authorities.

It has been found that all tepee type incinerators emit tremendous and objectionable volumes of heavy and dense smoke during periods when the temperature of the steel skin of the tepee is cold, such as at the time the pile of waste material is first lit or fired, or at any time while the tepee is in operation and when the amount of waste material that is being fed into the tepee is not sufficient to maintain a degree of temperature sufficiently high to maintain the skin of the tepee uniformly hot.

It has been found that the temperature of the steel skin must be maintained at approximately 400°F around the lower three-fourths of the structure and at approximately 750°F at the top or crown of the structure in order to promote good combustion within the structure and to burn the smoke generated by the burning pile of waste material.

A cyclonic type burner design developed by K.V. Lutes et al (8) provides an excellent burner for preheating the steel skin of the tepee type incinerator; and by the use of regulating controls a method of maintaining the skin temperature at required levels throughout the entire operating period of the tepee incinerator is achieved.

A process developed by R.E.S. Thompson (9) is a process for the closed-circuit processing of all waste products from forest product plants, without air or water pollution. Solid wastes, typically bark, wood chips, sawdust or the like, are carbonized to a char. The char is at least in part treated in a second furnace to produce activated carbon, char not so treated being briquetted and sold. The activated carbon is used as a filter medium for plant effluent liquors, which can be recycled after solids removal. The carbon and entrained solids are returned to one of the furnaces. Off-gas from all furnaces is passed to a boiler for steam generation, the spent gas being passed through an absorber, if necessary, before release.

An apparatus developed by E.H. Davis (10) is a portable incinerator in which the burning of solid waste material such as slash and small growth, resulting from logging operations to acquire merchantable timber and wood products and/or to clear right of ways for vehicles or power lines, is accomplished in a manner whereby controlling the burning process in an incinerator eliminates completely or in varying degrees the amount of pollutants discharged into the atmosphere.

At the same time this method reduces the direct operating costs, possibilities of setting ancillary fires and the need for extended tending and/or dousing of several fires. This incinerator is completely transportable to the general area over public roads and then after arrival to specific multiple locations off public roads it is conveniently moved over unimproved land to places where the incinerator is needed.

An apparatus developed by R.M. Williams (11) is one in which the disposal of hygroscopic combustible material, such as scrap wood and bark, is carried on at a temperature of about 600°F or somewhat lower to prevent flashing off certain volatiles which at or above that temperature level cause the formation of a noxious blue smoke.

The system and apparatus is caused to move a large volume of air through a furnace to provide the needed heat for drying the material to a state where it can be consumed by serving as a fuel. The air volume so moved is dust laden, and the apparatus concentrates the dust and particulate material for burning, and effectively mixes the combustible products with cleansed air to produce a source of heat at a controlled temperature level.

An incinerator developed by N.K. Sowards (12) is one in which pieces of solid waste, such as fragments of wood, are conveyed to an influent vertical feed tube where an air jet pump injects them into an incinerating chamber or vessel and at the same time providing influent air for aiding combustion of volatile matter.

The falling waste particles are horizontally distributed by striking a cone shaped spreader and secondary air directed horizontally and radially from the spreader may or may not be used to aid in the particle distribution. The particles are predried as they pass through the high temperature vapor space before reaching a fluidized bed. A fluidized bed, situated immediately above an air delivery chamber at the bottom of the vessel, supports combustion of the solid wastes in the top layer of fine granular material. The top layer is supported by coarse stone and a perforated plate with cover caps in that order.

The air delivery system channels high temperature air into the fluidized bed until operating temperature is reached and so channels ambient air thereafter. Volatile matter given off by combustion of the solid waste in the fluidized bed is burned smokelessly in the vapor space immediately above the bed. The temperature of the vapor space and the bed are set by one or more controlled water spray nozzles. The vessel exhaust is processed through a cyclone separator to remove small particles. A fog nozzle spray system cools the exhaust before it reaches the cyclone.

REFERENCES

(1) A.J. Jones; U.S. Patent 3,675,600; July 11, 1972; assigned to Michel Lumber Company.

(2) R.B. Burden, Jr. and E.J. O'Gieblyn; U.S. Patent 3,749,030; July 31, 1973; assigned to Wasteco, Inc.

(3) A.B. Baardson; U.S. Patent 3,831,535; August 27, 1974; assigned to Mill Conversion Contractors, Inc.

(4) A.B. Baardson; U.S. Patent 3,837,303; September 24, 1974; assigned to Mill Conversion Contractors, Inc.

(5) S.E. Corder, et al, *Wood and Bark Residue Disposal in Wigwam Burners,* Bulletin No. 11, Forest Research Laboratory, School of Forestry, Oregon State University (March 1970).

(6) M.J. Leman; U.S. Patent 3,669,039; June 13, 1972; assigned to Simpson Timber Company.

(7) R.E. Stone and A.E. Wing; U.S. Patent 3,730,114; May 1, 1973; assigned to Industrial Construction Co., Ltd.

(8) K.V. Lutes, J.E. Lander, V.J. Shanahan and R.G. Mills; U.S. Patent 3,777,678; December 11, 1973; assigned to Macmillan Bloedel Ltd.

(9) R.E.S. Thompson; U.S. Patent 3,783,128; January 1, 1974.

(10) E.H. Davis; U.S. Patent 3,785,302; January 15, 1974.

(11) R.M. Williams; U.S. Patent 3,826,208; July 30, 1974; assigned to Williams Patent Crusher and Pulverizer Company.

(12) N.K. Sowards; U.S. Patent 3,834,326; September 10, 1974; assigned to Environmental Products, Inc.

APPENDIX

Off-site treatment and disposal of land-destined hazardous wastes from the inorganic chemicals industry are most often handled by private contractors and waste service organizations. The contractors may provide both hauling and disposal services or the hauling and disposal services may be provided by two different firms. In some cases the inorganic chemical companies supply their own hauling service. A list of private treatment and disposal contractors and waste service organizations is given in this Appendix. This list is not intended to be complete but does give a good cross-section of facilities, including most of the largest and most active, involved with treatment and disposal of inorganic chemical wastes.

PRIVATE WASTE CONTRACTORS AND SERVICE ORGANIZATIONS

Large Multisite Integrated Hazardous Waste Contractors

Rollins Environmental Services, Inc. P.O. Box 2349 Wilmington, DE 19899	Integrated multitreatment complexes and disposal services; 3 sites.
Browning-Ferris Industries, Inc. 300 Fannin Bank Bldg. Houston, TX 77002	Integrated national treatment and disposal facilities; about 100 sites, mostly landfill but including 5 chemical fixation stations.
Waste Management, Inc. 900 Jorie Boulevard Oak Brook, IL 60521	Integrated national treatment and disposal facilities; about 50 sites, mostly landfill.
SGA Services, Inc. 99 High Street Boston, MA 02109	Integrated national treatment and disposal facilities; about 50 sites, mostly landfill, but with waste treatment subsidiaries.

Hazardous Waste Treatment and Recovery Contractors

Chem-trol Pollution Services, Inc. P.O. Box 200 Model City, NY 14107	Liquid treatment and secured landfill disposal.

Hyon Waste Treatment Services
Chicago, IL 60607

Integrated waste treatment facility: biological, chemical incineration.

Chemical Control Corp.
Elizabeth, NJ 07207

Liquid and solid chemical processing and incineration.

Conservation Chemical Co.
P.O. Box 6066
Gary, IN 46406

Treatment and disposal of liquid and hazardous wastes.

Conservation Chemical Co.
P.O. Box 6304
Kansas City, MO 64126

Treatment and disposal of liquid and hazardous wastes.

Industrial Waste Disposal Co., Inc.
Dayton, OH 45401

Treatment and disposal plant.

Nelson Chemical Co.
12345 Schaefer Highway
Detroit, MI 48227

Wet chemical treatment of cyanides, chrome, plating solutions and similar wastes amenable to this approach.

Evor Phillips Leasing Co.
Old Bridge, NJ 08857

Industrial liquid transportation and treatment: ocean disposal.

Erieway Pollution Control, Inc.
33 Industry Drive
Cleveland, OH 44101

Industrial waste treatment and chemical fixation.

Industrial Tank, Inc.
210 Berrellesa Street
P.O. Box 831
Martinez, CA 94553

Liquid transporting, treatment, and disposal; Class I hazardous waste disposal sites, State of CA.

Bio-Ecology Systems, Inc.
4100 E. Jefferson Avenue
Grand Prairie, TX 75050

Liquid treatment of metal processing wastes (plating, etc.) chemicals, acids, incineration.

Systems Technology
Franklin, OH 45005

Hazardous waste treatments.

Environmental Waste Control, Inc.
26705 Michigan Avenue
Inkster, MI 48141

Liquid treatment of acids, alkalies, pickle liquor.

Western Processing Co.
Kent, WA 98031

Contractor for chemical reclamation.

Pollution Controls, Inc.
P.O. Box 238
Shacopee, MN 55379

Incineration and liquid waste treatment.

Pollution Abatement Wastes
Oswego, NY 13126

Industrial waste disposal; landfill and incineration.

Aztec Mercury
P.O. Box 1676
Alvin, TX 77511

Mercury recovery from industrial wastes.

Wood Ridge Chemical Corp.
Division of Troy Chemical Co.
Park Place East
Wood Ridge, NJ

Reprocessors of mercury wastes.

SGC Industries
Old Bridge, NJ 08857

Chemical fixation processes.

Renner, Inc. Box 3224 Port Arthur, TX 77640	Landfilling and chemical fixation.
Environmental Sciences, Inc. Chemfix Division 505 McNeilly Road Pittsburgh, PA 15226	Chemical fixation on-site and off-site mobile treatment units.

General Purpose Landfill Contractors and Services

Koski Construction Co. 5841 Woodman Avenue Ashtabula, OH 44004	Impervious base settling pond.
Knickerbocker Landfill Malvern, PA 19355	Lined landfill.
Pottstown Disposal Services Pottstown, PA 19464	Lined landfill.
Scientific Inc. 17 E. Second Street Scotch Plains, NJ 07076	Landfill operators accepting liquids.
Sanitas Waste Disposal of MI 15500 Schaeffer Road Detroit, MI	Contractor for disposal in commercial landfill.
Tork Construction Wisconsin Rapids, WS 54494	Contractor for disposal in private landfill.

General Purpose Secured Landfill Contractors and Services

Nuclear Engineering Co., Inc. Hurstbourne Park 9200 Shelby Road Louisville, KY 40222	Primarily nuclear waste disposal; hazardous containerized waste burial plus some liquid chemical treatment; 3 sites.
Wescon, Inc. P.O. Box 564 245 Third Avenue E. Twin Falls, IA 83301	Hazardous wastes secured disposal.
Industrial Hazardous Waste Land Disposal Site Route 2 Lindsay, OK 74052	Hazardous materials landfill.
Resource Recovery Corp. Pasco, WA 99301	Interrelated companies: liquid, sludge and solid hazardous chemical disposal by evaporation plus secured landfill.
Chemical Processors, Inc. 5501 Airport Way South Seattle, WA 98108	See above
County Sanitation Districts of Los Angeles County 2020 W. Beverly Boulevard Los Angeles, CA 90057	Municipal Class I disposal sites; State of CA.

Ventura Regional County Sanitation District 181 S. Ash Street Ventura, CA 93001	Municipal Class I disposal site, State of CA.
Wescon, Inc. P.O. Box 564 245 Third Ave., E. Twin Falls, IA 83301	Hazardous waste disposal; abandoned missile base.

Deep Welling Contractors

Sonics International P.O. Box 47088 Dallas, TX 75247	Deep welling of acids and other chemicals.
U.S. Pollution Control, Inc. 5024 S. Quaker Tulsa, OK 74101	Contractor for disposal of hazardous industrial wastes deep well and landfill.

Ocean Disposal Contractors

Modern Transportation Kearney, NJ 07032	Ocean disposal.
Safety Products and Engineering 3 Malden Street W. Quincy, MA 02169	Ocean disposal hazardous waste.

Solid Waste Consultants

Chemical Buyers Service, Inc. P.O. Box 2065 Berkeley, CA 97400	Chemical recycling contractors.
International Hydronics Princeton, NJ 08540	Consulting process engineers for waste treatment facilities.
Emcon Associates 326 Commercial Street San Jose, CA 95112	Environmental consultants–landfill geology and design.

Miscellaneous Services

Quality Septic Tank Service A & B Quality Drain Service Capital Heights, MD 20027	Contractors for waste sludge hauling.
Contee Sand & Gravel Co., Inc. Box 460 Laurel, MD 20810	Land disposal.